The Retinoids

VOLUME 2

Contributors

Kathleen O. Ashenfelter
William S. Blaner
C. D. B. Bridges
Frank Chytil
Gunther Dennert
Carl W. Ehmann
Charles A. Frolik
DeWitt S. Goodman
Loretta M. Itri
Jerome J. Kamm
Richard C. Moon
David E. Ong
Gary L. Peck
Anita B. Roberts
Michael B. Sporn

The Retinoids

VOLUME 2

Edited by

MICHAEL B. SPORN

ANITA B. ROBERTS

Laboratory of Chemoprevention
National Cancer Institute
National Institutes of Health
Bethesda, Maryland

DEWITT S. GOODMAN

Department of Medicine
College of Physicians & Surgeons
Columbia University
New York, New York

1984

ACADEMIC PRESS, INC.

(Harcourt Brace Jovanovich, Publishers)

Orlando San Diego San Francisco New York London
Toronto Montreal Sydney Tokyo São Paulo

ACADEMIC PRESS, INC.
Orlando, Florida 32887

United Kingdom Edition published by
ACADEMIC PRESS, INC. (LONDON) LTD.
24/28 Oval Road, London NW1 7DX

Library of Congress Cataloging in Publication Data
Main entry under title:

The Retinoids.

Includes index.
1. Vitamin A--Physiological effect. 2. Retinol--Physiological effect. I. Sporn, Michael B. II. Roberts, Anita B. III. Goodman, DeWitt S.
QP772.V5R47 1984 615'.7 83-15375
ISBN 0-12-658102-9 (v. 2 : alk. paper)

PRINTED IN THE UNITED STATES OF AMERICA

84 85 86 87 9 8 7 6 5 4 3 2 1

Contents

Contributors

Numbers in parentheses indicate the pages on which the authors' contributions begin.

Kathleen O. Ashenfelter (287), Department of Medical Research, Roche Research Center, Hoffmann–La Roche Inc., Nutley, New Jersey 07110

William S. Blaner (1), Department of Medicine, College of Physicians & Surgeons, Columbia University, New York, New York 10032

C. D. B. Bridges (125), Cullen Eye Institute and Program in Neuroscience, Department of Ophthalmology, Baylor College of Medicine, Houston, Texas 77030

Frank Chytil (89), Department of Biochemistry, Vanderbilt University School of Medicine, Nashville, Tennessee 37232

Gunther Dennert (373), The Salk Institute for Biological Studies, San Diego, California 92138

*Carl W. Ehmann** (287), Department of Medical Research, Roche Research Center, Hoffmann–La Roche Inc., Nutley, New Jersey 07110

Charles A. Frolik (177), Laboratory of Chemoprevention, National Cancer Institute, National Institutes of Health, Bethesda, Maryland 20205

DeWitt S. Goodman (1, 41), Department of Medicine, College of Physicians & Surgeons, Columbia University, New York, New York 10032

Loretta M. Itri (327), Department of Medical Oncology, Roche Research Center, Hoffmann–La Roche Inc., Nutley, New Jersey 07110

Jerome J. Kamm (287), Department of Toxicology and Pathology, Roche Research Center, Hoffmann–La Roche Inc., Nutley, New Jersey 07110

Richard C. Moon (327), Laboratory of Pathophysiology, Life Sciences Division, IIT Research Institute, Chicago, Illinois 60616

David E. Ong (89), Department of Biochemistry, Vanderbilt University School of Medicine, Nashville, Tennessee 37232

*Present address: Research and Development, Bristol-Myers Company, Buffalo, New York 14213.

Gary L. Peck (391), Dermatology Branch, National Cancer Institute, National Institutes of Health, Bethesda, Maryland 20205
Anita B. Roberts (209), Laboratory of Chemoprevention, National Cancer Institute, National Institutes of Health, Bethesda, Maryland 20205
Michael B. Sporn (209), Laboratory of Chemoprevention, National Cancer Institute, National Institutes of Health, Bethesda, Maryland 20205

Preface

This two-volume treatise, the collected effort of 19 authors, represents the first comprehensive survey of the chemistry and biology of the retinoids. Profound advances in organic chemistry, analytical methodology, cell biology, and biochemistry, as well as in clinical applications, have occurred since the topic of vitamin A was last reviewed in 1967 in a multi-authored volume (''The Vitamins,'' 2nd ed., W. H. Sebrell, Jr. and R. S. Harris, eds., Academic Press). There is now an entire domain of new substances, representing analogs of retinol, retinaldehyde, and retinoic acid, which were scarcely known in 1967; the entire set of molecules, including the naturally occurring parent compounds as well as their synthetic derivatives, have collectively been termed retinoids. The introductory chapter provides a brief overview of the impact that the retinoids have had on both basic and clinical investigations in the past 10 years. It is our intention that this treatise should be of interest to both basic scientists and clinicians.

The two volumes are organized to begin with basic chemistry and progress toward biological and then clinical considerations. The first chapters in Volume 1 deal with new advances in organic chemistry and tracer chemistry, followed by surveys of chemical and biological methods for analysis and assay of retinoids; the first volume concludes with a comprehensive survey of vitamin A in animal and human nutrition. Volume 2 begins with several chapters that review the biochemistry and metabolism of the retinoids and of specific retinoid-binding proteins that are found in plasma and in cells. These chapters deal with the biosynthesis and metabolism of retinol, with plasma retinol-binding protein, with intracellular retinoid-binding proteins, and with the metabolism and role of retinoids in the eye. Following a chapter on the metabolism of retinoic acid and synthetic retinoids, current knowledge of the cellular and molecular mechanisms of action of the retinoids is reviewed at length. The final four chapters deal with the applied biology of the retinoids and include reviews of toxicology and teratology, immunology, and applications of retinoids in the fields of cancer and dermatology. An Appendix showing the chemical structures which are referred to in the text appears in each volume and should be of major help to the reader. Since the chemistry and biology of carotenoids have been comprehensively reviewed elsewhere, the present volumes devote relatively little attention to these topics.

We would like to acknowledge the assistance of the many colleagues who have contributed to the reviews of the manuscripts of the various chapters. In particular, we wish to thank David Bridges, Gerald Chader, Frank Chytil, Marcia Dawson, Peter Elias, Fritz Frickel, Charles Frolik, William Lennarz, Reuben Lotan, Robert McIntire, James Olson, Gary Peck, and Barbara Underwood for their editorial help. We are indebted to Fritz Frickel and colleagues at BASF Aktiengesellschaft for assistance with the Appendix, and to the National Eye Institute and Hoffmann–La Roche Inc. for help with the color illustrations. We would like to express our sorrow at the untimely death in July 1983 of one of the authors, Hans Kaegi. Dr. Kaegi was one of the leading tracer chemists in his field, and at the time of his death was attempting to synthesize new retinoidal benzoic acid derivatives of extremely high specific activity. Finally, we would like to express our appreciation to the staff of Academic Press, for all of their efforts in bringing this treatise to publication.

Michael B. Sporn
Anita B. Roberts
DeWitt S. Goodman

Contents of Volume 1

7

Biosynthesis, Absorption, and Hepatic Metabolism of Retinol

DEWITT S. GOODMAN AND WILLIAM S. BLANER

Department of Medicine
College of Physicians & Surgeons
Columbia University
New York, New York

THE RETINOIDS, VOL. 2

ISBN 0-12-658102-9

I. INTRODUCTION: GENERAL SUMMARY

The major natural sources of vitamin A in the diet are certain plant carotenoid pigments, such as β-carotene, and the long-chain retinyl esters found in animal tissues. β-Carotene is converted to vitamin A primarily in the intestinal mucosa. The biosynthetic process involves two soluble enzymes: β-carotene 15,15′-dioxygenase and retinaldehyde reductase. The first enzyme catalyzes the cleavage of β-carotene at the central double bond by a dioxygenase mechanism to yield two molecules of retinaldehyde (C1); the aldehyde is then reduced to retinol (A1).

Dietary retinyl esters are hydrolyzed in the intestinal lumen, and the resulting retinol is absorbed into the mucosal cell. Retinol in the mucosal cell (newly absorbed or newly synthesized from carotene) is reesterified with long-chain, mainly saturated fatty acids; the retinyl esters, in association with chylomicrons, are then transported via the lymph into the general circulation. After entry into the vascular compartment, chylomicrons are metabolized in extrahepatic tissues by the lipolytic removal of much of the chylomicron triglyceride. The chylomicron remnant thus formed is a smaller and cholesterol-rich particle that contains essentially all the chylomicron retinyl esters and is removed from the circulation almost entirely by the liver.

After uptake of the chylomicron retinyl esters, hydrolysis and reesterification occur in the liver. The resulting retinyl esters (predominantly retinyl palmitate) are stored in the liver and can be mobilized as needed in a highly regulated process. Vitamin A mobilization from hepatic retinyl ester stores takes place as the free alcohol retinol bound to a specific plasma transport protein retinol-binding protein (RBP).

The aim of this chapter is to summarize our current knowledge of the biosynthesis, the intestinal absorption, and the hepatic metabolism of retinol.

II. BIOSYNTHESIS OF β-CAROTENE

Carotenoids are a widely distributed family of tetraterpenes that are synthesized by a large variety of photosynthetic microorganisms and members of the plant kingdom. Several excellent reviews are available that describe the biosynthesis of carotenoids in detail (Bauernfeind *et al.*, 1981; Goodwin, 1980; De Luca, 1978). In brief summary, the initial reactions leading to carotenoid biosynthesis are identical to those involved in the biosynthesis of sterols (Schroepfer, 1981; Cornforth, 1976; Bloch, 1965). These common reactions involve: (1) formation of 3-hydroxy-3-methylglutaryl coenzyme A from three molecules of acetyl coenzyme A; (2) conversion of 3-hydroxy-3-methylglutaryl coenzyme A to mevalonic acid; (3) phosphorylation and decarboxylation of mevalonic acid to yield isopentenyl pyrophosphate; and (4) condensation of two molecules of iso-

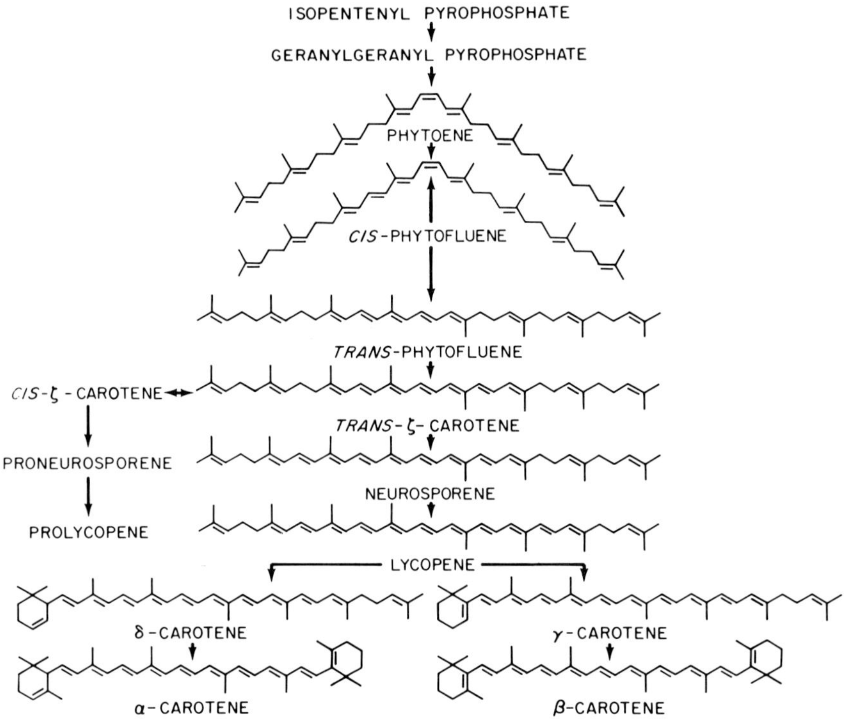

Fig. 1. The biosynthetic pathway from isopentenyl pyrophosphate to *cis*- and *trans*-carotene. (From Qureshi *et al.*, 1974.)

pentenyl pyrophosphate (after isomerization of one of them to dimethylallyl pyrophosphate) to yield the 10-carbon compound geranyl pyrophosphate. Subsequent addition of five-carbon isoprene units (via isopentenyl pyrophosphate) leads to the formation of the 15-carbon and 20-carbon intermediates, namely farnesyl pyrophosphate and geranylgeranyl pyrophosphate, respectively. In the pathway of sterol biosynthesis, two molecules of farnesyl pyrophosphate condense to form one molecule of squalene, which is subsequently cyclized to the sterol ring structure. In contrast, in carotenoid biosynthesis, two molecules of geranylgeranyl pyrophosphate condense to form the first common carotenoid precursor, the 40-carbon hydrocarbon phytoene.

Figure 1 summarizes the biosynthetic pathway between phytoene and the carotenes. Phytoene is converted to progressively more unsaturated compounds, leading to the formation of lycopene. Lycopene is then cyclized to the monocylic δ- and γ-carotenes, which in turn are further cyclized to yield α- and β-carotene, respectively.

β-Carotene has the highest provitamin A biological activity of any of the many known carotenoids. The relative provitamin A activities of different carotenoids in the foods commonly eaten by animals and humans is discussed in Chapter 6, Vol. 1.

III. BIOSYNTHESIS OF RETINOL FROM β-CAROTENE

It has been known for more than 50 years that β-carotene can replace the growth requirement of rats for vitamin A (von Euler *et al.*, 1928; Moore, 1930), suggesting a precursor role of β-carotene in vitamin A formation. The conversion of β-carotene to retinol takes place mainly in the intestinal mucosa during the absorption of dietary β-carotene. In addition, this conversion has been demonstrated to occur in liver (Zachman and Olson, 1963) and potentially in other tissues as well (see below).

The conversion of β-carotene to retinol (vitamin A) is an important biosynthetic process since it converts the plant product β-carotene to a product necessary for the growth and life of the animal organism. Throughout the course of evolution most, if not all, of the vitamin A utilized by animals was ultimately derived from this process. In recent years, however, retinol has been synthesized chemically by several procedures and chemical synthesis is now the main commercial source of vitamin A.

A. Cleavage of β-Carotene to Form Retinaldehyde

The enzymatic cleavage of β-carotene into retinaldehyde with soluble enzyme preparations from rat intestinal mucosa and liver was reported in 1965 (Goodman and Huang, 1965; Olson and Hayaishi, 1965). In subsequent studies, the characteristics of the reaction and of the enzyme that catalyzes it in rat intestinal mucosa were explored in some detail (Goodman *et al.*, 1966a, 1967). The mechanism of the reaction was shown to be that of a dioxygenase reaction, in which molecular oxygen reacts with the central two carbon atoms of β-carotene, followed by the cleavage of the central double bond of carotene to yield two molecules of retinaldehyde, as illustrated in Fig. 2. The evidence for this reaction mechanism includes the following: (1) the reaction shows an absolute requirement for molecular oxygen; (2) the stoichiometry of the results indicates the conversion of one molecule of β-carotene into two molecules of retinaldehyde; (3) studies with doubly labeled β-carotene demonstrated that the hydrogen atoms attached to the two central carbon atoms of β-carotene are completely retained during the conversion of β-carotene into retinaldehyde *in vitro* and during its conversion into retinol *in vivo*; and (4) the reaction is inhibited by the metal chelating agents α, α′-dipyridyl and *o*-phenanthroline. The enzyme has been designated as β-carotene 15,15′-dioxygenase and sometimes as the carotene cleavage enzyme.

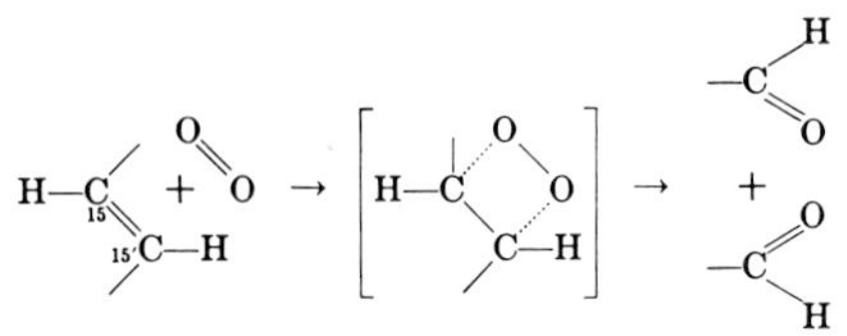

Fig. 2. Mechanism of the carotene 15,15′-dioxygenase reaction, in which molecular oxygen reacts with the central double bond of carotene to yield two molecules of retinaldehyde. (From Goodman *et al.*, (1966a).

Carotene 15,15′-dioxygenase has been partially purified from rat (Goodman *et al.*, 1967), hog (Fidge *et al.*, 1969), and rabbit (Lakshmanan *et al.*, 1972) intestines. However, because of lability during purification, the pure enzyme has not yet been isolated. The enzyme preparations from these three species all had generally similar properties. Thus, all three partially purified preparations were inhibited by sulfhydryl binding agents and by ferrous iron chelators, were stimulated by the addition of thiols, and had pH optima near pH 8; moreover, the preparations all displayed Michaelis kinetics and had V_{max} values for β-carotene in the range of 0.5–1 nmol β-carotene cleaved per milligram protein per hour at 37°C. Additionally, the rat and hog enzyme preparations required the addition of an appropriate detergent (or bile salt) or detergent–lipid mixture for *in vitro* activity in the systems employed. The molecular weights of the rat and hog enzymes were estimated to be in the 100,000–200,000 range.

The substrate specificity of the carotene dioxygenase enzyme from rabbit intestine was explored by Olson and colleagues (Lakshmanan *et al.*, 1968, 1972; Olson, 1969). Several β-apocarotenals were cleaved at more rapid rates than β-carotene, although this observation could not be confirmed by Singh and Cama (1974). α-Carotene was reported to be cleaved somewhat more slowly than β-carotene. The diepoxy derivative of β-carotene, which is inert biologically, was not cleaved by the enzyme preparation as used by Olson and colleagues (Olson, 1969). The monoepoxy derivative also was not cleaved in these *in vitro* studies, although this compound has been reported to show 21% of the biological activity of β-carotene (Subbarayan *et al.*, 1966) and possesses one unsubstituted β-ionone ring. All compounds that were cleaved contained one unsubstituted β-ionone moiety.

The carotenoid cleavage enzyme activity is found widely distributed throughout the vertebrates. In a study of a variety of species, enzyme activity was observed in the intestinal mucosa of several mammals, the chicken, the tortoise, and a freshwater fish but was absent from cat intestinal tissue (Lakshmanan *et al.*, 1972). The herbivores examined (the guinea pig and the rabbit) had the highest activities, the omnivores (the chicken, tortoise, and fish) had intermediate activities, and the only carnivore studied, the cat, had none at all. It was pointed out by these authors that it is known that cats become vitamin A deficient on a vitamin A-free diet supplemented with β-carotene. It would be interesting to determine whether other carnivores also lack the ability to convert carotenoids into retinaldehyde.

The possibility that the same dioxygenase enzyme or a different enzyme can convert small amounts of biologically active carotenoids into retinaldehyde by a mechanism other than central cleavage at the 15,15′ bond has not been entirely excluded. Sharma *et al.* (1977) reported the presence of small amounts of β-apo-carotenals in the intestine of chickens fed β-carotene and suggested (Sharma *et al.*, 1976, 1977) that the dioxygenase enzyme may be able to attack any one of the double bonds of β-carotene (the primary attack being at the central double bond), giving the corresponding β-apocarotenal or retinaldehyde. This suggestion must, however, be viewed with some skepticism without more direct evidence being available. It has also been pointed out (Lakshmanan *et al.*, 1972) that the epoxycarotenoids, some of which have biological activity apparently without being cleaved by β-carotene 15,15′-dioxygenase, may possibly be metabolized by other pathways. More information is needed concerning this overall question. In any event, it is clear that the carotenoid 15,15′-dioxygenase reaction represents the process responsible for the formation of most, if not all, of the retinaldehyde formed from biologically active carotenoids.

The intestinal mucosa is not the only tissue that can enzymatically cleave β-carotene into retinaldehyde. Enzymatic activities with similar properties were observed in homogenate preparations of both rat liver and intestine by Olson and Hayaishi (1965). The conversion of β-carotene to retinaldehyde and retinol by slices of bovine corpus luteum has been reported (Gawienowski *et al.*, 1974; Stacewicz-Sapuncakis *et al.*, 1975). The potential physiological significance of this reaction in the corpus luteum remains to be determined.

Cleavage of β-carotene, once thought unique to the animal kingdom, has been observed in the bacterium *Halobacterium halobium* and in related halobacteria (Kushwaha *et al.*, 1974; Oesterhelt, 1976). These bacteria subsequently utilize the retinaldehyde, coupled with an opsin-like protein, to form bacteriorhodopsin, an energy-generating light-dependent proton pump (Stoeckenius *et al.*, 1979; see also Chapter 10).

B. Reduction of Retinaldehyde to Retinol

Retinaldehyde produced by β-carotene cleavage in the intestinal mucosa is reduced to retinol, then transferred into the lymph, predominantly in the form of retinyl esters (Huang and Goodman, 1965; Goodman *et al.*, 1966b). Labeled retinaldehyde, introduced into the rat duodenum in small amounts in the form of bile–lipid mixtures resembling normal intestinal contents, was found in lymph predominantly as labeled retinyl esters, indicating that retinaldehyde absorption involved its reduction to retinol (and subsequent esterification) (Fidge *et al.*, 1968).

The reduction of retinaldehyde to retinol was studied with an approximately 13-fold purified soluble enzyme preparation from rat intestinal mucosa (Fidge and Goodman, 1968). The enzyme was relatively heat stable and had a molecular weight approximately in the range of 60,000–80,000. The partly purified reductase was unable to oxidize ethanol in the presence of NAD^+. Retinaldehyde reduction required NADH or NADPH as cofactor; both reduced nucleotides were effective. The reaction was stimulated by glutathione and inhibited by thiol inhibitors. There was a sharp pH optimum near 6.3. Retinaldehyde reduction displayed typical Michaelis kinetics, with a V_{max} of 2 μmol of retinol formed per hour per milligram of protein and with an apparent K_m of 20 μM. The enzyme appeared to be a relatively nonspecific aldehyde reductase, exhibiting greatest activity with aliphatic aldehydes of C_4–C_8 length.

IV. INTESTINAL ABSORPTION OF RETINOL

A. Lipid Digestion and Absorption: An Overview

Retinyl esters, retinol, and β-carotene are water-insoluble lipids. Like other exogenous and endogenous lipids that enter the gastrointestinal tract, these compounds must be processed by a series of physical and often chemical changes within the bowel lumen before absorption can take place. These changes involve a complex but coordinated series of gastric, intestinal, biliary, and pancreatic events. In particular, dietary lipid is emulsified, triglycerides undergo lipolysis to form monoglycerides and fatty acids, and these products of lipolysis are solubilized with bile salts to form a micellar solution. The resulting mixed micelles serve to help transport lipid digestive products across the aqueous diffusion barrier (the unstirred water layer) to the mucosal epithelial cell. Only a brief summary of these processes is provided, without references, in this chapter. Excellent detailed reviews of lipid digestion and absorption are available for more information and for appropriate references (Patton, 1981; Thomson and Dietschy, 1981; Friedman and Nylund, 1980; Davidson and Glickman, 1983). A review of the intestinal absorption of the fat-soluble vitamins (vitamins A, E, D, and K) has been provided by Hollander (1981).

Hydrolysis of esterified lipids is essential in order for lipid absorption to occur. Anatomically, lipolysis begins in the stomach where a limited amount of partial hydrolysis of triglycerides provides more polar lipids (fatty acids and partial glycerides) to assist in emulsification. A lingual lipase is thought to be involved in this process. The net result of this gastric phase of fat digestion is the initiation of lipolysis and the preparation of a stable emulsion on which the major pancreatic lipolytic enzymes can act in the milieu of the small intestine.

Several enzymes involved in fat digestion act in the small intestine. These include the pancreatic enzymes lipase (and colipase), phospholipase A_2, and cholesterol esterase. Human breast milk also contains a lipase that is active in the small intestine. Pancreatic lipase has specificity for hydrolysis of fatty acid ester bonds at positions 1 and 3 of the triglyceride molecule. Phospholipase A_2 contributes to the efficiency of lipolysis by pancreatic lipase by digesting the phospholipid envelope that surrounds the triglyceride emulsion.

Another pancreatic enzyme, cholesterol esterase, is responsible for intraluminal cholesteryl ester hydrolysis. Complete hydrolysis of dietary cholesteryl esters occurs in the lumen before absorption of cholesterol can occur (Shiratori and Goodman, 1965). This enzyme is discussed further below.

A micellar phase is formed in the intestinal lumen when the bile salt concentration exceeds the critical micellar concentration (approximately 3–4 m*M*). This concentration of bile salts is usually exceeded during normal digestion. Mixed micelles contain bile salts, fatty acids, monoglycerides, cholesterol, and other lipid-soluble molecules (including fat-soluble vitamins) and are considered to be the major route of delivery of the products of fat digestion to the absorptive mucosal cell. Other nonmicellar phases may coexist in the intestinal lumen with the micellar phase; these include an oil phase and a viscous isotropic phase.

Micellar solubilization of lipids within the intestinal lumen permits the lipids to diffuse to the surface of the intestinal epithelium and make contact with the microvillus membrane. This is important because there is a major physiological diffusion barrier called the unstirred water layer, that covers the surface of the epithelial cell membranes. The unstirred layer of water is a feature common to all biological membranes. Micelles penetrate this relatively immobile aqueous layer more readily, thus increasing the efficiency of lipid uptake into the intestinal mucosal cell.

Following the uptake of lipids across the microvillus membrane, thought to occur mainly by passive diffusion, the process of absorption becomes dependent on a series of intracellular events within the mucosal cell. These events include the resynthesis of esterified lipids (triglycerides, phospholipids, cholesteryl esters), the synthesis of chylomicron apolipoproteins (particularly apolipoproteins B, A–I, and A–IV), and the assembly of lipids and apoproteins into chylomicrons. These particles represent the principal lipoprotein vehicles for transporting absorbed lipid out of the intestine and into the general circulation.

The intracellular events that occur during lipid absorption and during the production of chylomicrons and other intestinal lipoproteins have been reviewed in detail (Zilversmit, 1978; Green and Glickman, 1981; Friedman and Nylund, 1980; Thomson and Dietschy, 1981; Davidson and Glickman, 1983). As described by Davidson and Glickman (1983), the chylomicron is a relatively large droplet composed of a coat of protein and phospholipid surrounding a core of nonpolar lipid. Triglycerides, resynthesized from fatty acids and monoglyceride in the smooth endoplasmic reticulum of the mucosal cell, constitute most of the nonpolar core. Newly synthesized phospholipid and apoprotein are added to the developing triglyceride droplet, resulting in formation of a particle within the interconnecting profiles of the endoplasmic reticulum. The particles then move to the Golgi apparatus; after fusion has occurred between the Golgi membranes and the basolateral plasma membrane, they are secreted by reverse pinocytosis. The chylomicrons then pass through intercellular spaces to mesenteric lymphatics and ultimately enter the blood and gain access to the circulation via the thoracic duct.

B. Retinol Absorption: Intraluminal Events

Dietary retinyl esters, retinol, and provitamin A carotenoids are dispersed and emulsified in the stomach during the gastric phase of lipid digestion. They then enter the lumen of the duodenum where extensive hydrolysis of retinyl esters takes place. In infants, the bile salt-stimulated lipase in human milk may play a role in retinyl ester hydrolysis in the duodenum (Fredrikzon *et al.*, 1978). The major retinyl ester hydrolase activity that acts in the intestines is, however, derived from the pancreas.

It has been known for many years that extensive hydrolysis of ingested retinyl esters occurs in the intestinal lumen, followed by reesterification during absorption (Gray *et al.*, 1940; Eden and Sellers, 1950). In a study with doubly labeled retinyl palmitate in rats, data were obtained that indicated that repeated hydrolysis and reesterification of retinyl ester occurred during its digestion and assimilation (Lawrence *et al.*, 1966). Both hydrolytic and synthetic activities for retinyl esters were found by Mahadevan *et al.* (1961) in luminal contents and in preparations of intestinal mucosa and of pancreas. Further *in vivo* and *in vitro* studies in rats suggested that prior hydrolysis of retinyl esters is essential for absorption, and that only the alcohol retinol crosses the mucosal cell membrane to be reesterified inside the cell (Mahadevan *et al.*, 1963a,b). Analogous to studies with cholesteryl esters (Shiratori and Goodman, 1965), it is likely that complete hydrolysis of long-chain retinyl esters occurs before retinol is absorbed into the mucosal cell.

The pancreatic enzyme mainly responsible for retinyl ester hydrolysis appears to be the same enzyme that catalyzes intraluminal cholesteryl ester hydrolysis (Erlanson and Borgström, 1968; Lombardo and Guy, 1980). This enzyme has been purified from rat (Calame *et al.*, 1975) and from porcine (Momsen and Brockman, 1977) pancreas, and from human pancreatic juice (Lombardo *et al.*, 1979). The enzyme appears to be a relatively nonspecific carboxylic ester hydrolase that can act on a wide variety of esters as substrates. Thus the purified human enzyme hydrolyzes triacetin, tributyrin, *p*-nitrophenylacetate, and lysophosphatidylcholine, as well as esters of cholesterol and of vitamins A, D_3, and E and glycerides solubilized by bile salts. Its molecular weight (approximately 100,000) is greater than that of the rat or pig enzyme, and it can hydrolyze cholesteryl esters in the presence of appropriate bile salts. This enzyme has been given a variety of names in the literature, the most common being pancreatic nonspecific lipase and cholesterol esterase.

Retinol formed by retinyl ester hydrolysis (or originating as such in the diet) and dietary β-carotene are solubilized in mixed micelles as discussed above, thus enabling these molecules to reach the microvillus membrane. In studies with everted rat gut sacs *in vitro,* El-Gorab *et al.* (1975) reported that micellar solutions significantly enhance uptake of both retinol and β-carotene over emulsions. Maximal uptake occurred at the critical micellar concentration of the bile salt mixture. At higher detergent concentrations, β-carotene uptake declined whereas retinol absorption remained high.

The intestinal absorption of retinol and of β-carotene and the effects of luminal and other factors on these processes were examined in a series of studies by Hollander and his colleagues (Hollander and Muralidhara, 1977; Hollander and Ruble, 1978; Hollander and Morgan, 1979; Hollander, 1980, 1981). These studies were conducted with a variety of preparations, including *in vivo* studies with lymph cannulated and other rats, studies with isolated perfused intestinal segments *in situ* in intact rats, and studies with everted gut sacs *in vitro.* The effects of luminal physical and chemical characteristics, including variations in pH, bile salt concentrations, fatty acid content, and in the effective thickness of the unstirred water layer were explored. Apparent saturation kinetics were found with physiological concentrations of retinol, whereas a linear relationship was found at high (pharmacological) concentrations of retinol in the intestinal perfusate (Hollander and Muralidhara, 1977). In physiological concentrations, retinol uptake *in vitro* by everted gut sacs was unaffected by anoxia or metabolic inhibitors and uncouplers. It was concluded that *in vivo* absorption of retinol in physiological concentrations is mediated by a saturable, carrier-mediated passive absorption mechanism that is modified by the presence of fatty acids of varying chain length. At higher concentrations of retinol, transport appeared to occur by simple passive diffusion. Hollander (1981) suggested that penetration of the microvillus cell membrane by retinol and other lipid micronutrients probably occurs by passive diffusion but that after penetration of the membrane, these compounds may be transferred through the cytosol by a carrier protein present on the cytosol side of the absorptive cell membrane. Cellular retinol-binding protein (CRBP, see Chapter 9) is a potential candidate to serve this latter function in the mucosal cell, although no direct information concerning this is available.

The intestinal absorption of β-carotene was investigated in unanesthetized rats with isolated perfused intestinal loops (Hollander and Ruble, 1978). A linear relationship was found between the concentration of β-carotene and its rate of absorption at perfusate levels of 0.5–11 μM. Increases in perfusate acidity, additions of fatty acids, and an increase in the perfusate flow rate caused higher rates of β-carotene absorption. It was concluded that β-carotene absorption takes place by passive diffusion, and that this process can be modulated by intraluminal factors that change the physical characteristics of the perfusate and possibly by factors that stimulate the intracellular cleavage of carotene to retinaldehyde.

C. Retinol Absorption: Intracellular Events

After the uptake of retinol and β-carotene into the mucosal cell, a number of metabolic events occur. For β-carotene these events include the dioxygenase cleavage of carotene to form retinaldehyde, which is then reduced to retinol. These reactions have been discussed in detail earlier in this chapter. Data have also been reported that suggest that a small portion of newly biosynthesized (or dietary) retinaldehyde may be oxidized in the mucosal cell to retinoic acid (D1), which is then transported to the liver via the portal vein (see below) (Fidge *et al.*, 1968).

Retinol within the mucosal cell is largely reesterified with long-chain fatty acids. The retinyl esters are then incorporated, together with other lipids and with apolipoproteins, into chylomicron particles that are secreted from the cell into the lymph.

In a detailed study in rats, it was found that retinyl esters were the predominant labeled compounds in lymph and contained approximately 90% of the recovered radioactivity, after feeding either labeled retinol or β-carotene to rats with lymph fistulas (Huang and Goodman, 1965). Small amounts of labeled retinol were also recovered in lymph. Similar findings were obtained in humans, in studies in which labeled retinol and β-carotene were fed to patients containing cannulae in their thoracic ducts in the neck (Goodman *et al.*, 1966b; Blomstrand and Werner, 1967). Thus, in these human studies, labeled retinyl esters predominated in lymph and represented 80–90% of the absorbed radioactivity after feeding labeled retinol and 60–70% after β-carotene (Goodman *et al.*, 1966b). Man differed from the rat in that the human intestine was able to absorb a small amount of unchanged dietary β-carotene into the lymph. The quantitative difference in the ability of the two species to absorb intact β-carotene was, however, not very great.

In both man (Goodman *et al.*, 1966b) and the rat (Huang and Goodman, 1965), the composition of lymph retinyl esters was remarkably constant, regardless of the fatty acid composition of the test meal, and regardless of whether the retinyl esters were derived from retinol or from β-carotene. Retinyl palmitate predominated, and saturated retinyl esters (retinyl palmitate and stearate, in a ratio of about 2:1) comprised approximately three-fourths of the labeled esters. Small amounts of retinyl oleate and linoleate were also found.

The reaction responsible for retinol esterification within the enterocyte has not been fully defined. One possibility is that retinol esterification is catalyzed by the reversible action of pancreatic cholesterol esterase, acting within the mucosal cell. It has long been known that pancreatic cholesterol esterase can catalyze both the hydrolysis of cholesteryl esters and the esterification of free cholesterol (see Goodman, 1965, for review and references). The mechanism appears to involve the direct reaction of free cholesterol and free fatty acid. A similar esterification reaction between retinol and fatty acids independent of coenzyme A (CoA) has been demonstrated with acetone powders of rat pancreas (Pollard and Bieri, 1960), with preparations of rat small intestine (Mahadevan *et al.*, 1961), and with purified human pancreatic carboxyl ester hydrolase (Lombardo *et al.*, 1980). Immunocytochemical studies with antisera to rat pancreas cholesterol esterase demonstrated the localization of specific immunoreactivity within the intestinal absorptive cell (Gallo *et al.*, 1980). These observations suggest that intestinal cholesterol esterase is of pancreatic origin, and that cholesteryl ester hydrolysis in the lumen and cholesteryl ester synthesis in the rat small intestine may be mediated by the same enzyme. Similar considerations may apply to retinyl ester hydrolysis and formation. On the other hand, the physiological significance of this reaction mechanism has been questioned (Huang and Goodman, 1965; Lombardo *et al.*, 1980), particularly since *in vitro* esterification studies (e.g., Murthy *et al.*, 1961) have shown a lack of fatty acid specificity for retinol esterification, whereas the composition of lymph retinyl esters is relatively constant regardless of the diet fed (see above).

An alternative reaction mechanism for retinol esterification has been proposed (Helgerud *et al.*, 1982, 1983). This mechanism involves the reaction of retinol with a fatty acyl-CoA ester, catalyzed by an acyl-CoA:retinol acyltransferase (ARAT) enzyme. Similarly, acyl-CoA:cholesterol acyltransferase (ACAT) activity may be involved in cholesterol esterification within the mucosal cell (Norum *et al.*, 1979; Bennett Clark, 1979). ARAT activity was found in the microsomal fraction of rat mucosal homogenates (Helgerud *et al.*, 1982) and in microsomal preparations from human intestine (Helgerud *et al.*, 1983). With both rat and human preparations, ARAT activity had a pH optimum close to 7.4 and was inhibited by taurocholate and by a thiol-blocking reagent. In both species the activity was of sufficient magnitude to be able to account for all retinyl esters absorbed into the lymph. In addition, the rat activity showed a relative specificity for palmitoyl-CoA as compared to oleyl-CoA. These properties suggest that the enzyme is physiologically important. A number of similarities were noted between the properties of intestinal ARAT and ACAT activities (Helgerud *et al.*, 1982). These findings raised the possibility that both activities might be due to the same enzyme. In human small intestine, the activity of ARAT was an order of magnitude higher than that of ACAT (Helgerud *et al.*, 1983). More information is clearly needed in order to explore the relationship between ARAT and ACAT and to define better the reaction(s) and enzyme(s) involved in retinol esterification in the mucosal cell.

Retinyl esters are secreted from absorptive cells mainly in the hydrophobic core of chylomicron particles. In species that are able to absorb some β-carotene into the body beyond the intestine, the carotene also is absorbed mainly in association with lymph chylomicrons. As indicated above, different mammalian species differ in their intestinal metabolism of β-carotene. Some species, such as man, absorb unchanged carotenoids to some extent. In contrast, many other species, such as the rat and the pig, absorb virtually no unchanged carotene beyond the intestinal mucosa. The basis for this difference is not understood.

D. Chylomicron Metabolism

Chylomicrons secreted by the enterocytes into the intercellular space traverse the adjacent lamina propria and enter the intestinal lacteals (see Friedman and Nylund, 1980, for discussion and references). They are then transported through the intestinal lymphatics ultimately to enter the plasma compartment, mainly via the thoracic duct.

Chylomicron catabolism is normally a rapid process that proceeds in at least two phases. Detailed reviews of the events that occur during this process, together with full appropriate references, are available in reviews of lipoprotein metabolism (Jackson *et al.*, 1976; Smith *et al.*, 1978; Green and Glickman, 1981). On entering the plasma, chylomicrons acquire the C and E apolipoproteins (not intially present in the newly formed particle) from plasma high-density lipoproteins (HDL). The acquisition of apoC-II serves to activate the enzyme lipoprotein lipase, leading to the hydrolysis of chylomicron core triglycerides. This phase of chylomicron metabolism takes place at the surface of capillary endothelium of extrahepatic tissues. Triglyceride hydrolysis (lipolysis) results in shrinkage of the chylomicron core and transfer of redundant surface components (apoA-I, A-II, and some phospholipid) into HDL. ApoA-IV is lost into the lipoprotein-free density >1.21 g/ml fraction of plasma (Tall *et al.*, 1979). Fatty acids produced by lipolysis either circulate as free fatty acids bound to serum albumin or are taken up by adjacent tissues (e.g., muscle, adipose tissue) for oxidation or reesterification.

These processes lead to the formation of a smaller spherical lipoprotein particle called a chylomicron remnant (Redgrave, 1970). The remnant is depleted in triglyceride and more enriched in cholesteryl ester, phospholipid, and protein, as percent of total mass, than chylomicrons. The major apoproteins of the remnant are apoB (specifically, the lower molecular weight or B48 form) and apoE. Newly absorbed retinyl esters found in the lymph chylomicron are almost entirely retained in the particle during its processing to a chylomicron remnant, so that remnants are relatively enriched in vitamin A as compared to chylomicrons (Hazzard and Bierman, 1976). Chylomicron remnants are almost entirely removed from the circulation by the liver. Hepatic uptake of the remnant is a rapid, high-affinity, receptor-mediated process in which the major determinant appears to be apoE on the remnant. This uptake process is discussed in more detail below.

Retinyl and cholesteryl esters can undergo exchange reactions between lipoproteins (including chylomicrons) in rabbit or human plasma (Zilversmit *et al.*, 1982; Goodman *et al.*, 1983). This exchange reaction appears to be catalyzed by the cholesteryl ester transfer (or exchange) protein that is found in the density >1.21 fraction of rabbit and human plasma but not in plasma of several other mammalian species (Zilversmit *et al.*, 1975; Chajek and Fielding, 1978; Barter and Lally, 1979). In one study (Zilversmith *et al.*, 1982), normal or hypercholesterolemic rabbit plasma was found to stimulate the transfer of retinyl ester as well as cholesteryl ester from rabbit lymph chylomicrons, chylomicron remnants, or from cholesteryl ester-rich very low-density lipoprotein (VLDL) of plasma to lipoprotein fractions with density >1.019. Partially purified transfer protein from rabbit or from human plasma also accelerated the transfer of these esters. The exchange of cholesteryl esters between lipoproteins is probably of physiological importance in the overall dynamics of lipoprotein metabolism. On the other hand, exchange of retinyl esters between chylomicrons and other lipoproteins is probably normally not a physiologically significant process in view of the normal rapid clearing of chylomicrons and their remnants from the circulation, as compared to the much slower kinetics of the transfer process (Zilversmit *et al.*, 1982).

E. Alternate Pathways of Retinol Absorption

Although it has been clearly demonstrated that under normal physiological conditions retinol absorption occurs mainly via the lymphatic route (see above; Fidge *et al.*, 1968), the existence of an alternative route for retinol absorption has been suggested both by experimental studies in the rat and by features of the genetic disease abetalipoproteinemia in man. Information available suggests that under abnormal conditions retinol may be able to be absorbed via a nonlymphatic pathway and in a biochemical form other than the chylomicron in amounts sufficient to meet nutritional requirements for vitamin A.

In a study in rats, Murray and Grice (1961) observed that ligation of the thoracic duct did not prevent the absorption of vitamin A [fed as retinyl acetate (A3) or palmitate (A4)]. Using vitamin A-deficient rats, these authors found that thoracic duct ligation did not significantly change the amount of vitamin A stored in the liver 48 h after an oral dose. Moreover, after an oral dose of esterified retinol, its concentration in portal blood was greater than in aortic blood of rats with ligated thoracic ducts. These observations led these authors to suggest that the portal system may be an important alternative route for the absorption of vitamin A when the normal pathway via the lymph is blocked. Subsequent studies by Lawrence *et al.* (1966) (who fed doubly labeled retinyl palmitate to rats with thoracic duct cannulae) and by Yeung and Veen-Baigent (1972) (who fed [11,12-^{3}H]retinyl acetate to rats with ligated thoracic ducts) provided further evidence for the potential role of the portal system as an alternative route for retinol absorption.

Abetalipoproteinemia is a rare genetic disease characterized clinically by malabsorption of fat and also, in reports in the literature, by acanthocytosis, retinitis pigmentosa, and ataxic neuropathic disease (Kayden, 1972; Herbert *et al.*, 1983). Apolipoprotein B appears to be completely absent in these patients, and the most likely genetic defect is considered to be one deleting the synthesis of apoB. As a consequence, plasma of these patients has no apoB-containing lipoproteins, i.e., no low-density lipoprotein (LDL), no VLDL, and no chylomicrons. Severe hypolipidemia is present (total plasma cholesterol approximately 20–50 ml/dl and triglyceride 1–10 mg/dl), and plasma contains HDL as the sole lipoprotein class.

The problem of vitamin A status and absorption in these patients has received considerable attention, particularly in view of their risk of developing retinal degeneration. Most measurements of plasma vitamin A levels have shown low values in untreated patients; with oral supplementation, however, plasma levels in the normal range are observed (Kayden, 1972, 1980; Herbert *et al.*, 1983; D. S. Goodman, unpublished observations). Since these patients do not produce chylomicrons, retinol must be absorbed by a pathway other than that involving lymph chylomicrons. One possibility is that retinol absorption in these patients occurs via intestinally produced HDL (Green and Glickman, 1981), and perhaps via the portal venous route. Direct uptake of retinol from the mucosal cell by RBP should also be considered as a possibility. Direct evidence is needed concerning such possible pathways before the mechanism of retinol absorption in abetalipoproteinemia can be understood.

F. Absorption and Enterohepatic Circulation of Retinoic Acid

Retinoic acid demonstrates selective vitamin A biological activity. Animals maintained on retinoic acid as the only source of vitamin A activity will grow at a normal rate and be in general good health but will be blind and will show impaired reproductive functions. Thus, retinoic acid is neither able to replace retinol as a precursor for the chromophore of the visual pigment nor is it able to support reproduction [see DeLuca (1979) for key references]. It is, however, highly effective in supporting growth and epithelial differentiation.

It is now well established that retinoic acid is present in low concentrations as an endogenous retinoid in a number of tissues (Ito *et al.*, 1974; DeLuca, 1979; Saari *et al.*, 1982). Retinoic acid has been reported to be present as an endogenous compound in human plasma at concentrations of approximately 3–4 ng/ml (10^{-8} *M*) (De Leenheer *et al.*, 1982).

Studies on the biosynthesis of retinoic acid in rat intestinal mucosa were reported by Crain *et al.* (1967). These studies employed ^{14}C-labeled β-carotene and retinaldehyde, in both *in vivo* and *in vitro* experiments. It was suggested that some of the retinaldehyde that is produced in the intestinal mucosa by cleavage of β-carotene may be oxidized to retinoic acid, which is then transported via the portal circulation to the liver.

The chemical and anatomical pathways of absorption of dietary retinoic acid (as compared to dietary retinaldehyde and retinol) were examined in detail by Fidge *et al.* (1968). It should be noted that there is probably very little retinoic acid in the foods that form the usual diets of animals and man. Some retinoic acid may normally be absorbed via the enterohepatic cycle (see below). The extent to which this may occur under physiological conditions is not known. The studies of Fidge *et al.* (1968) employed rats containing lymph, bile, and duodenal cannulae and were designed to maintain physiological conditions to the greatest extent possible. Labeled retinoids were introduced into the duodenum in very small amounts in the form of a bile–lipid mixture resembling normal luminal contents. Most (70–80%) of the radioactivity recovered after the feeding of labeled retinol or retinaldehyde was found in the lymph, predominantly in saturated retinyl esters. In contrast, 92–95% of the radioactivity recovered after feeding labeled retinoic acid was found in the bile in a mixture of polar metabolites, most of them more polar than free retinoic acid. Two-thirds of the small amount of radioactivity found in lymph after feeding labeled retinoic acid was found as free retinoic acid. It is likely that the radioactivity found in bile derived from retinoic acid absorbed via the portal vein and transported via this route to the liver. These findings thus indicated that absorption of retinoic acid occurs via the portal venous route. The authors speculated that retinoic acid may have been absorbed as the free acid, perhaps in the form of its carboxylate anion bound to serum albumin. Alternatively, it was considered that retinoic acid may have been converted to its glucuronide conjugate in the intestinal mucosa and then transported in portal blood in this form. Retinoyl glucuronide has been identified as a major metabolite of retinoic acid in the rat intestine (Zile *et al.*, 1982b). More information is needed in order to define fully the chemical form(s) in which retinoic acid is absorbed via the portal vein.

The findings of Fidge *et al.* (1968) were consistent with earlier observations (Zachman and Olson, 1965; Zachman *et al.*, 1966) that retinoic acid, injected into the portal vein, was rapidly metabolized and excreted in the bile. The labeled compounds found in bile consisted of a mixture of polar metabolites, most of which were more polar than free retinoic acid. One of the metabolites of retinoic acid in bile has been identified as retinoyl β-glucuronide (Dunagin *et al.*, 1965; Zile *et al.*, 1980, 1982a; Swanson *et al.*, 1981). Other polar metabolites account for most of the radioactivity in bile after administration of labeled retinoic acid to rats (Zile *et al.*, 1980, 1982a); one of these has been identified as a taurine conjugate of a retinoic acid metabolite (Skare *et al.*, 1982).

Retinoic acid has been shown to undergo a variety of conversions *in vivo* to new metabolites (see Chapter 11 for an extensive discussion). Oxidative decarboxylation of retinoic acid was demonstrated by Roberts and DeLuca (1967). More recently, a number of polar metabolites have been isolated and identified from various biological sources. These include 5,6-epoxyretinoic acid (D6) (McCormick *et al.*, 1978), 4-hydroxy- (D3) and 4-aporetinoic acid (D4) (Roberts and Frolik, 1979). The production of 5,6-epoxyretinoic acid as an endogenous metabolite of labeled retinol has also been reported (McCormick and Napoli, 1982), as has the formation of 5,6-epoxyretinoyl β-glucuronide as a major metabolite in the small intestinal mucosa of vitamin A-deficient rats dosed intrajugularly with labeled 5,6-epoxyretinoic acid (Napoli *et al.*, 1982). The extent to which these and other metabolites might participate in the enterohepatic circulation of retinoic acid and its metabolites is not known. Other reactions of retinoic acid include esterification (Fidge *et al.*, 1968) and isomerization to 13-*cis*-retinoic acid. The formation, to a small extent, and enterohepatic circulation of water-soluble metabolites of retinol in the rat has also been reported (Zachman and Olson, 1964, 1965; Zachman *et al.*, 1966).

V. HEPATIC UPTAKE, STORAGE, AND METABOLISM OF RETINOL

A. General Summary

Newly absorbed retinyl esters are largely taken up from the circulation by the liver, mainly in association with the uptake of chylomicron remnants. Chylomicron remnant uptake appears to occur mainly by receptor-mediated endocytosis, leading to lysosomal degradation of the remnant constituents. After uptake of the retinyl esters, hydrolysis and reesterficiation occur in the liver. Hepatic vitamin A (95% as long-chain retinyl esters, mostly retinyl palmitate) normally represents over 90% of the total body reserves of vitamin A. Hepatic retinyl esters are stored both in parenchymal liver cells and in nonparenchymal fat-storing cells; the relative roles of these different cell types have not been fully defined. Retinyl esters may be stored in several forms within liver cells, including lipid droplets or granules, and a high-molecular-weight lipid–protein aggregate that may be present in the cytosol. From these liver stores, vitamin A is mobilized as the free alcohol (retinol) bound to a specific plasma transport protein, retinol-binding protein (RBP). Within the hepatocyte, these various processes require enzymatic activities able to synthesize retinyl esters and to hydrolyze them, specific binding proteins for intracellular transport of retinol and for its mobilization from the cell, and possibly specialized subcellular structures to coordinate and facilitate these events.

B. Chylomicron Remnant Uptake by the Liver

On reaching the liver, intact chylomicron remnants are selectively and rapidly taken up by a finite number of transport sites on the sinusoidal membranes (Sherrill and Dietschy, 1978). Studies have demonstrated that remnant uptake involves high-affinity cell-surface lipoprotein receptors, and that apolipoprotein E plays an important role in this uptake process. Thus, Sherrill *et al.* (1980) observed that when canine lipoproteins were perfused through a rat liver, only lipoproteins containing apoE displayed high-affinity receptor-mediated uptake and saturation kinetics. Competition experiments between apoE-containing HDL and rat chylomicron remnants indicated that the hepatic uptake mechanism for the two lipoproteins was identical. Other studies with perfused rat livers in several laboratories have provided considerable further evidence that apoE plays a critical role in the recognition and binding, and subsequent uptake, of chylomicron remnants by a receptor-mediated process (Shelburne *et al.*, 1980; Windler *et al.*, 1980; Havel *et al.*, 1980). The presence of C apoproteins (particularly apoC-III) appears to interfere with uptake, so that the fastest uptake rates are obtained with particles rich in apoE and depleted of apoC (Windler *et al.*, 1980; Shelburne *et al.*, 1980).

More general information about the potential role of apoE in cellular lipoprotein uptake has come from studies with cultured cells (e.g., fibroblasts). Such cells manifest surface receptors for LDL that bind apoB, the protein component of LDL. This is followed by receptor-mediated endocytosis, fusion of the endocytic vesicles with lysosomes, and LDL degradation within the lysosomes (see Goldstein and Brown, 1979; Brown *et al.*, 1981, for reviews and references). Cholesteryl esters taken into cells in this manner are hydrolyzed by a lysosomal acid lipase. The liberated cholesterol then leaves the lysosome and is used in the cell for membrane synthesis and as a regulator of intracellular cholesterol homeostasis.

There is considerable evidence available that apoE can interact with the same cell-surface receptor as does LDL, thereby delivering lipoprotein cholesterol to cells [see Brown *et al.* (1981) for review and references]. In fact, the affinity of the fibroblast receptor for apoE-containing particles appears to be an order of magnitude greater than its affinity for apoB-containing particles (Pitas *et al.*, 1979). Studies of apoE-containing lipoproteins in patients with abetalipoproteinemia support these conclusions (Blum *et al.*, 1982).

Chylomicron remnant uptake by liver cells presumably involves a similar receptor-mediated process. Livers of experimental animal species that have been investigated (including the rat, the rabbit, and the dog) have been found to possess a lipoprotein receptor that resembles the LDL receptor [see Brown *et al.* (1981) for review]. Studies by Hui *et al.* (1981) have suggested that there are two kinds of high-affinity receptors on canine liver membranes. One of these appeared to specifically bind apoE-containing lipoproteins (an apoE-receptor), whereas the other bound both apoB- and apoE-containing lipoproteins (the apoB,E-receptor, identical to the classic LDL receptor). Both receptors were found in young puppies, whereas in the normal adult dog only the apoE-receptor was observed. Treatment with cholestyramine resin induced the expression of the apoB,E-receptor in adult dog liver. The relative roles of these two kinds of receptors in chylomicron remnant uptake in other species needs further study.

C. Chylomicron Cholesteryl Ester Metabolism

Much information is available about the metabolism of chylomicron cholesteryl esters taken up by the liver in association with the chylomicron remnant. This information may be relevant to the issue of chylomicron retinyl ester metabolism in the liver, about which much less direct information is on hand. Hepatic uptake of chylomicron cholesteryl esters occurs without hydrolysis of the cholesteryl esters (Goodman, 1965; Quarfordt and Goodman, 1967; Stein *et al.*, 1969). In studies with chylomicrons containing doubly labeled cholesteryl esters injected intravenously into rats, Quarfordt and Goodman (1967) observed that 80–90% of the chylomicron cholesteryl esters were removed by the liver without hydrolysis. In the liver, the newly absorbed cholesteryl esters underwent slow but extensive hydrolysis, to the extent of about 60% after 1 h and about 85–90% after 3.5 h. Subsequent to hydrolysis, most of the labeled free cholesterol slowly left the liver and was extensively redistributed in the body. Thus, 24 h later, only 20–28% of the labeled cholesterol found in the entire animal body was present in the liver. Since newly absorbed retinol, which is retained in the liver, is only mobilized slowly (see below), it is clear that following ester hydrolysis the hepatic metabolism of chylomicron cholesterol and retinol diverge in a major way.

Further information was obtained in a combined radioautographic–electron microscopic and biochemical study by Stein *et al.* (1969). Chylomicrons containing labeled cholesterol, mainly (70%) present as cholesteryl ester, were injected intravenously into rats. After 30 min, most of the cholesteryl ester had been taken up by the liver without ester hydrolysis. At this time, the radioautographic reaction was localized mainly over the region of the cell boundary of the hepatic parenchymal cells, with fewer grains being present over intracellular organelles. At later time intervals when most of the cholesteryl ester had been hydrolyzed, the labeled cholesterol was found to be distributed widely and generally within the cell. Information was also obtained about the role of the hepatic Kupffer cells in the uptake and metabolism of chylomicron cholesterol. Under physiological conditions, when the load of chylomicron lipid presented to the liver was small, only a very small amount of chylomicron cholesterol seemed to be taken up by the Kupffer cells. Phagocytosed labeled lipid was, however, seen in Kupffer cells after larger lipid loads.

The anatomic sites (subcellularly) and the details of the enzymatic processes involved in the hydrolysis of chylomicron cholesteryl esters newly taken up by the liver have not been fully defined. It is clear that one of the major processes consists of receptor-mediated endocytosis of chylomicron remnants, followed by hydrolysis of cholesteryl esters and other remnant components within lysosomes. In rare genetic diseases characterized by lysosomal acid lipase deficiency (Wolman's disease and cholesteryl ester storage disease), cholesteryl esters accumulate in liver cells and in other tissues as well [see Assmann and Frederickson (1983) for review and references]. An acid cholesteryl ester hydrolase from rat liver lysosomes has been partially purified and characterized (Brown and Sgoutas, 1980; Van Berkel *et al.*, 1980). Enzymatic activity was found in preparations of both parenchymal and nonparenchymal liver cells (Van Berkel *et al.*, 1980). Hydrolysis of chylomicron cholesteryl esters taken up by isolated rat hepatocytes was inhibited by chloroquine (Florin and Nilsson, 1977), an agent which inhibits the action of acid hydrolases in lysosomes. Finally, there is also evidence that the rate of cholesteryl ester hydrolysis may be limited by the rate at which internalized remnant particles are moved to the presumably lysosomal site of hydrolysis (Nilsson, 1977; Florin and Nilsson, 1977; Cooper and Yu, 1978).

On the other hand, there are other processes that may participate as well, to some extent, in the initial metabolism and hydrolysis of chylomicron cholesteryl esters in the liver. Liver homogenates and homogenate fractions display cholesteryl ester hydrolase activity at neutral pH, and the enzyme(s) responsible for such activity have been partially purified and characterized (Deykin and Goodman, 1962; Stein *et al.*, 1969; Tuhackova *et al.*, 1980). It is possible that some uptake of cholesteryl esters can occur without uptake of the entire remnant particle [see, e.g., Chajek-Shaul *et al.* (1981a,b) for such evidence in other tissues]. It is also possible that dissociation of the constituents of the remnant can occur to some extent, permitting cholesteryl ester hydrolysis to take place before remnants are delivered to lysosomes. The extent to which these alternative processes might occur in normal physiology is not known.

D. Retinyl Ester Uptake and Initial Metabolism

The liver plays the major role in the body in the uptake and metabolism of newly absorbed retinol (retinyl esters). Thus, retinyl esters appear to remain almost completely with the hydrophobic core of the chylomicron during its extrahepatic conversion to a remnant particle, and then to be taken up by the liver in association with the chylomicron remnant. A quantitative and detailed study of these processes was reported by Goodman *et al.* (1965). Chylomicrons containing newly absorbed labeled retinol (93% as retinyl esters) were injected intravenously into normal intact rats, and the tissue distribution and metabolism of the labeled retinol was observed for several days. At 17 min after injection of the labeled chylomicrons, 68% of the recovered lipid-soluble radioactivity was found in the liver. The total recovery of lipid-soluble radioactivity in the entire animal ranged from 92% at 17 min to 56% after 6 days. At all time intervals, approximately two-thirds of the recovered radioactivity was found in the liver. After 17 min the fatty acid composition of the labeled hepatic retinyl esters was almost identical to that of the injected material. However, during the first 3 h, the composition of the labeled retinyl esters in the liver changed from one reflecting that of the injected chylomicrons to one with a much greater preponderance of saturated esters (85–90% of the labeled esters compared to 66% in the chylomicrons), and in particular of retinyl palmitate (70–75% of the labeled esters). The composition of the labeled retinyl esters in the liver did not change after 3 h. The rapid change in composition of the labeled retinyl esters seen in the first 3 h suggested that on entering the liver, the retinyl esters had undergone hydrolysis followed by reesterification of the labeled retinol. The constant composition beyond 3 h suggested that the labeled esters had equilibrated with the entire liver pool of retinyl esters, and that the final composition of the labeled esters reflected the total composition of retinyl esters in liver. This suggestion was supported by the similarity in fatty acid composition of the labeled retinyl esters after 3 h to that reported by Futterman and Andrews (1964) for naturally occurring retinyl esters in rat liver.

Additional support for these conclusions is available from the earlier studies of Lawrence *et al.* (1966). These workers fed doubly labeled retinyl-15-^{14}C palmitate-9,10-^{3}H to rats and determined the $^{14}C/^{3}H$ ratio in the total lipids and retinyl esters isolated from intestine, lymph, and liver at various time intervals after dosing. They observed a progressive increase in the $^{14}C/^{3}H$ ratio for the retinyl esters in the intestinal contents, intestinal mucosa, lymph, and liver at both 3 and 24 h after administration of the isotopes. The ratio of $^{14}C/^{3}H$ of liver retinyl esters was approximately five times greater than that of retinyl esters in lymph. These data further indicate that hydrolysis of retinyl esters and reesterification of retinol occur during hepatic uptake and storage of newly absorbed retinol.

Further information about chylomicron retinyl ester uptake was obtained in a study using ether analogs of both retinyl and cholesteryl esters (Goodman *et al.*, 1983). Rat lymph chylomicrons were doubly labeled *in vitro* with ether analogs of retinyl palmitate and cholesteryl linoleate. The ether analogs represent stable, nonhydrolzable lipids that can be used to explore some aspects of movement of their respective esters, without having to deal with the problem of ester hydrolysis. Three hours after injection of the labeled chylomicrons intravenously into intact rats, 85–90% of the cholesteryl ether and 80–85% of the retinyl ether were recovered in the liver. Thereafter, during the subsequent 4 weeks, the cholesteryl and retinyl ethers were lost progressively from the liver at very slow but differing rates.

Although only very limited direct data are available, the evidence on hand strongly suggests that the hepatic uptake and initial metabolism of chylomicron retinyl esters occurs in a manner that is analogous to that of chylomicron cholesteryl esters. Thus, it is likely that intact retinyl esters are taken up by the liver cell, followed mainly by retinyl ester hydrolysis along with the degradation of other remnant constituents within lysosomes. No direct data concerning retinyl ester hydrolysis by lysosomal acid hydrolase(s) are available. The extent to which extralysosomal hydrolysis of newly taken up retinyl esters, at a more neutral pH, might occur is not known. In any event, under normal conditions after ester hydrolysis the liberated retinol is reesterified and stored, as discussed below. The influence of abnormal nutritional status, particularly retinol depletion, on these processes is discussed later in this chapter.

E. Retinol Esterification

It has long been known that vitamin A is stored mainly in liver as retinyl esters, and that the liver is capable of storing large quantities of retinol in ester form. A study of the composition of retinyl esters isolated from the livers of members of a number of vertebrate species (calf, sheep, rabbit, rat, human, frog, trout, and cat) was reported by Futterman and Andrews (1964). In all species examined, except the cat, livers contained substantial stores of retinol, of which 95% or more was present in esterified form. Cat liver contained very little vitamin A, with approximately equal amounts present as free retinol and as retinyl ester. The palmitate ester was the predominant retinyl ester in all species, comprising 66% of the retinyl esters in the human and 69% in the rat. Retinyl stearate and oleate were the next most common esters; together with retinyl palmitate these three esters accounted for approximately 90% of all retinyl esters in the species examined.

In one study, Tomassi and Olson (1983) examined the effects of dietary fatty acid composition on hepatic retinyl ester composition and on vitamin A utilization in the rat. The fatty acid composition of retinyl esters in the liver was not perturbed by the presence of a large amount of polyunsaturated fat in the diet over a 10-day feeding period. The predominant retinyl ester in the liver was retinyl palmitate, regardless of whether rats were fed 10% corn oil, coconut oil, or linseed oil. In addition, the ingestion of polyunsaturated fat did not significantly influence the mobilization rate of vitamin A from the liver.

The ability of the liver to synthesize retinyl esters has been well established in studies with the isolated perfused rat liver (Worker, 1959; Zachman and Olson, 1963, 1965). Only limited information is, however, available about the enzymatic process involved. In 1964, Futterman and Andrews reported studies on the *in vitro* esterification of retinol with homogenate fractions of cat liver (Futterman and Andrews, 1964) and of calf retina (Andrews and Futterman, 1964). Enzymatic activity was localized mainly in the microsomal fractions from both tissues. It was reported that retinol esterification by microsomes did not require the addition of supplements or cofactors, such as ATP or CoA, and was not increased by the addition of preformed palmitoyl-CoA. Two pH optima, one close to pH 4.5 and the other close to pH 8.2, were observed. However, only at the higher pH did the fatty acid composition of the reaction product resemble that of naturally occurring retinyl esters in the liver of the cat. The authors concluded that the esterification of retinol was not dependent on energy or activated fatty acid, and therefore that it occurred via a different enzyme and mechanism than that involved in the esterification of cholesterol.

Cholesterol esterification in liver (Goodman *et al.*, 1964; Erickson and Cooper, 1980) as well as in many other types of tissues (see Spector *et al.*, 1979; Erickson and Cooper, 1980, for references) is catalyzed by the enzyme acyl-CoA:cholesterol acyltransferase (EC 2.3.1.26; ACAT), which reacts free cholesterol with a fatty acyl-CoA to form cholesteryl ester. This membrane-bound enzyme is found mainly in liver microsomes and has been partially characterized.

As discussed earlier in this chapter, studies by Helgerud *et al.* (1982, 1983) have demonstrated the formation of retinyl esters in rat intestinal mucosa by way of an analogous reaction, involving the reaction of retinol with a fatty acyl-CoA ester, catalyzed by an acyl CoA:retinol acyltransferase (ARAT) enzyme. Similarly, Ross (1982a) demonstrated the formation of retinyl esters by mammary gland microsomes from the lactating rat in a reaction that was greatly stimulated by the formation or addition of fatty acyl-CoA thioesters, and thus had the characteristics of an ARAT-catalyzed reaction.

It is likely that retinol esterification in liver proceeds by way of a comparable reaction. Direct evidence for retinol esterification catalyzed by a microsomal ARAT enzyme was reported by Ross (1982b). Using microsomal membrane preparations from rat or cat liver incubated with [^{3}H]retinol dispersed in dimethyl sulfoxide, some [^{3}H]retinol esterification was consistently observed. However, the rate of esterification could be increased 6- to 11-fold by addition of either palmitoyl-CoA or a fatty acyl-CoA-generating system. Studies with hydroxylamine suggested that the fatty acid used to esterify retinol under basal conditions was derived from an endogenous pool of fatty acyl-CoA associated with the microsomal preparations. Maximum esterification was observed between pH 7 and 8. When microsomes were incubated with both [^{3}H]retinol and [^{14}C]palmitoyl-CoA of known specific radioactivities, the ratio of ^{14}C/^{3}H in newly synthesized retinyl palmitate was essentially equal to that of its assumed substrates, indicating that [^{14}C]palmitate did not undergo significant isotope dilution prior to acylation of [^{3}H]retinol. The possible relationships between the liver acyltransferases that catalyze the esterification of cholesterol and of retinol remain to be explored.

F. Retinyl Ester Hydrolysis

Hydrolysis of retinyl esters occurs in the liver both during the hepatic uptake of dietary vitamin A and during the mobilization of retinol from its stores in the liver. The hydrolysis of chylomicron retinyl esters that occurs during hepatic uptake has been discussed above. In addition, retinyl ester hydrolysis must precede the mobilization of retinol from hepatic stores of retinyl ester since retinol is mobilized in the form of the unesterified alcohol (retinol) bound to RBP. Accordingly, it is clear that the enzymatic hydrolysis of retinyl esters in liver represents an important process in the overall metabolism of retinol in the body.

Nevertheless, in early studies from several laboratories, a number of investigators were unable to demonstrate the *in vitro* hydrolysis of retinyl palmitate with liver homogenate preparations (Yeung and Veen-Baigent, 1971; see also Olson, 1964; Mahadevan *et al.*, 1966, for earlier references and discussion). Compounding this apparent inability of the liver to hydrolyze the naturally occurring, long-chain retinyl ester was the observation (Seshadri Sastry and Ganguly, 1961) that the nonnaturally occurring ester retinyl acetate was actively hydrolyzed by liver microsomes.

In 1966, Mahadevan *et al.* (1966) reported that retinyl palmitate hydrolyzing activity was found in the "nuclear" and "mitochondria–lysosome-rich" fractions of rat liver homogenates. The activity was assayed using a reaction mixture containing 0.6 m*M* retinyl palmitate, 1% sodium taurocholate, and 0.2% Triton X-100. Enzyme activity required the addition of a bile salt and was partially characterized with extracts of acetone powders of rat liver. The pH optimum was 8.6. Enzyme preparations hydrolyzed a variety of long-chain retinyl esters, with the greatest relative activity being seen with retinyl palmitate.

Hydrolysis of retinyl acetate by highly purified carboxylesterase from liver of several species (pig, ox, man) was reported by Bertram and Krisch (1969). The purified enzyme did not hydrolyze long-chain fatty acid esters of retinol. These findings suggested that previously observed hydrolysis of retinyl acetate by crude liver preparations was due to nonspecific esterases of little or no physiological relevance to hepatic retinyl ester metabolism.

Studies from the authors' laboratory have provided more detailed information about the enzymatic hydrolysis of retinyl esters in rat liver. In the initial work (Harrison *et al.*, 1979), the hydrolysis of retinyl palmitate was explored with liver homogenate fractions from retinol-depleted, retinoic acid-fed rats. Retinol-depleted animals were used in order to avoid the problem posed by large and variable amounts of endogenous substrate (retinyl esters) present in the liver of normal animals. Retinyl palmitate hydrolase (RPH) activity was maximal near pH 8 and required a bile salt for stimulation. The enzymatic activity showed an unusual subcellular distribution, with about 40% of total RPH activity recovered in the washed "nuclear" fraction and about 30–35% in the 105,000 *g* supernatant. This unusual distribution was not observed for marker constituents for plasma membranes, nuclei, mitochondria, lysosomes, Golgi apparatus, or endoplasmic reticulum. Despite its enrichment in the "nuclear" fraction, RPH activity was not enriched in purified preparations of nuclei or plasma membranes. Thus, RPH activity was not localized in any single characterized subcellular structure.

Another striking feature of the hepatic RPH activity was an extreme variability from rat to rat as assayed *in vitro*. Thus, when individual rat liver homogenates were assayed, RPH activity varied over a more than 40-fold range, yet the activity of any given homogenate remained remarkably constant with time. Of ten lipid and nonlipid esters tested as substrates for hepatic hydrolase activities, only activities against cholesteryl oleate and phytyl oleate correlated with RPH activity and resembled RPH activity with respect to subcellular distribution and interanimal variability.

Subsequent work by Prystowsky *et al.* (1981) extended these observations to the liver of the normal rat and added new information about the characteristics of the RPH activity. These studies employed a highly sensitive and accurate assay that could be carried out effectively with microgram amounts of liver homogenate protein and nanogram amounts of substrate. This sensitive assay permitted the use of small enough quantities of liver homogenate so that the endogenous retinyl esters present in normal liver did not interfere with the quantitative accuracy of the assay. The unusual subcellular distribution of RPH activity previously observed in retinol-depleted rats was found to exist as well in the normal rat, indicating that the earlier observations were not a function of the retinol-depleted status of the animals. In addition, great variability in RPH activity from rat to rat was seen in the normal animals, as it had been seen in the retinol-depleted animals. Thus, hydrolase activity varied over 50-fold in individual rat livers. Hydrolytic activities against cholesteryl oleate and triolein showed a similar variation and correlated strongly with RPH activity, whereas hydrolytic activities against retinyl acetate and *p*-nitrophenyl acetate did not. RPH activity was partially purified by hydrophobic interaction chromatography. Copurification, with comparable enrichment, of the hydrolytic activities against cholesteryl oleate and triolein were observed. Moreover, chromatographic and other data indicated that the RPH activity had strongly hydrophobic physical properties.

Preliminary experiments were conducted to see if cyclic AMP-mediated protein phosphorylation mechanisms might be involved in the variation (and regulation) of enzyme activity, as had been reported for lipase and cholesteryl ester hydrolase in other tissues (Khoo *et al.*, 1976; Pittman and Steinberg, 1977). These experiments did not provide evidence for the involvement of such mechanisms in the observed interanimal variation of RPH activity. This variation, additionally, was not related to the age of the animals, time of day of death, order of animal kill, or strain of rat. At present, an explanation for the observed variability in RPH activity between individual animals is not available, nor is it known whether and how RPH activity might be regulated in a given animal. The possibility has, however, been raised that the observed variation in activity might have, in part, a genetic basis (Harrison *et al.*, 1979; Prystowsky *et al.*, 1981). No data are available on this issue.

In addition, the *in situ* localization of RPH activity within the liver cell still remains to be defined. It was suggested (Prystowsky *et al.*, 1981) that the unusual subcellular distribution observed might have resulted in part from the hydrophobic physical properties of the enzyme, which led it to become absorbed to particulate material (i.e., membranous components) that sediments in the nuclear (and mitochondrial–lysosomal) fraction. Within the cell, these hydrophobic properties may well direct the enzyme to be associated with retinyl ester-containing lipid droplets or lipid–protein aggregates (discussed further below). Thus, the hydrophobic properties of the enzyme may be physiologically involved in enabling it to perform its role of catalyzing lipid ester hydrolysis.

The nature of the relationship between the three correlated lipid ester hydrolase activities (against retinyl palmitate, cholesteryl oleate, and triolein) is of considerable interest. In studies in this laboratory (Blaner *et al.*, 1981; W. S. Blaner and D. S. Goodman, unpublished observations), more extensive purification of RPH activity was achieved without detergents from an acetone powder prepared from livers prescreened for high RPH activity levels. A more than 200-fold purification was obtained by sequential Phenyl-Sepharose, DEAE-Sepharose, and Heparin-Sepharose chromatography. Copurifying with the RPH activity throughout these steps were hydrolytic activities against cholesteryl oleate and triolein. These three copurifying activities showed similarities in their animal-to-animal variability, requirement for trihydroxy bile salt for *in vitro* activity, pH optimum (approximately 8), and inhibition by high ionic strength ($>$ 85% by 0.2 *M* NaCl). However, it was possible to dissociate the three activities by the use of selected solubility conditions and inhibitors. Thus, differential solubility studies showed that extraction of the liver acetone powder preparation with acetate buffer at pH 5.0 resulted in good recovery of the hydrolytic activities against retinyl palmitate and triolein with very little recovery of cholesteryl oleate hydrolytic activity. The latter activity could then be recovered in good yield by extraction of the powder with Tris buffer at neutral pH. Similarly, differential inhibition studies with the serine esterase inhibitors diisopropyl fluorophosphate or phenylmethylsulfonyl fluoride showed a much more marked inhibition of the hydrolytic activities against retinyl palmitate and cholesteryl oleate than of that against triolein. These results suggest that three closely related but different enzymes are involved in the three hydrolase activities.

G. The Role of Cellular Retinol-Binding Protein (CRBP)

Liver contains fairly high concentrations of the specific intracellular binding protein for retinol, cellular (or cytosol) retinol-binding protein (CRBP). Information available about CRBP (and about the intracellular binding protein for retinoic acid, CRABP) has been summarized in several reviews (Chytil and Ong, 1978, 1979; Chytil, 1982) and the subject of cellular retinoid-binding proteins is discussed in full in Chapter 9. CRBP from several tissue sources is a single polypeptide chain with a molecular weight of approximately 14,600. CRBP differs from plasma RBP with regard to a number of characteristics (Ross and Goodman, 1979). Although CRBP and CRABP are similar with regard to many physical and chemical characteristics, these two intracellular proteins have different retinoid binding specificities and are immunochemically distinct from each other (see Chapter 9).

CRBP has been isolated from livers of several species, including the rat (Ong and Chytil, 1978), dog (Liou *et al.*, 1981), and human (Ong, 1982; Fex and Johannesson, 1982). The mean concentration of CRBP in rat liver in radioimmunoassay studies was reported as 49.6 μg/g by Adachi *et al.* (1981) and as 16.8 μg/g by Ong *et al.* (1982). The basis for these different reported values is not understood. Most (90%) of the liver CRBP was found in the 105,000 × *g* supernatant after differential centrifugation of liver homogenates (Adachi *et al.*, 1981).

There are several potential roles that CRBP may play within the cell. It has been suggested that the intracellular binding protein may play a direct role in the biological expression of vitamin A activity (e.g., analogous to steroid hormone receptors (Ong and Chytil, 1978; see Chapter 9). Evidence has been reported that suggests that CRBP may be involved in facilitating the specific interaction of retinol with binding sites for retinol in the cell nucleus (Takase *et al.*, 1979; Liau *et al.*, 1981). Another possibility is that CRBP in the cell serves as an intracellular transport protein, acting to transport retinol from one locus to another between metabolic events in the liver. This transport may, for example, be from the lysosome to the site of retinol esterification, or from the site of retinyl palmitate hydrolysis prior to mobilization to a newly synthesized molecule of apo-RBP for secretion as the retinol–RBP complex. Thus, retinol bound to CRBP may represent a metabolically very active pool of intracellular retinol. CRBP may also serve to present retinol in an appropriate or "correct" manner to a specific site, such as the active center of the ARAT enzyme or the binding site of apo-RBP, and in this way further facilitate the metabolic process involved. Future research is needed in order to explore these and other possibilities.

H. The Role of Different Cell Types

The liver comprises several different types of cells, including the hepatic parenchymal cells, the sinusoidal cells, and cells that line the bile ductules and ducts. The sinusoidal (and perisinusoidal) cells include the Kupffer cells, the endothelial cells, and the fat-storing cells. Most of the liver, both in terms of total cell number, cell volume, and mass (total protein) is composed of parenchymal cells (hepatocytes). Sinusoidal cells account for about 30% of the total number of cells in the liver. Because sinusoidal cells are much smaller than hepatocytes, however, only 2–10% of hepatic protein is in sinusoidal cells. Of the nonparenchymal (sinusoidal) cells, endothelial cells represent about half and Kupffer cells about one-third. It has been reported that in rat liver the fat-storing cells constitute 13% of the nonparenchymal cells (Widmann *et al.*, 1972), and that in human liver there is approximately 1 fat-storing cell for every 20 parenchymal cells (Bronfenmajer *et al.*, 1966). For general reviews and references concerning the sinusoidal cells see Fahimi (1982) and Jones and Summerfield (1982).

A number of investigators have explored the potential roles of the nonparenchymal cells in the storage and metabolism of vitamin A in the liver. It has been shown convincingly that the Kupffer cells do not play an important role in the storage of vitamin A in the liver under normal conditions. Thus, when Kupffer cells were isolated from livers of rats on normal or on excess vitamin A-containing diets, the content of vitamin A in the Kupffer cells was found to be substantially less than that in hepatocytes (expressed on a per cell or per milligram DNA basis) (Linder *et al.*, 1971; Hori and Kutamura, 1972; Peterson *et al.*, 1973).

In contrast, evidence now available suggests that the fat-storing cells do play a significant role in vitamin A storage under both normal and hypervitaminotic conditions. Fat-storing cells have been referred to in the literature by a variety of names including: perisinusoidal stellate cells (Wake, 1980), lipocytes (Bronfenmajer *et al.*, 1966), vitamin A-storing cells (Yamada and Hirosawa, 1976), and Ito cells (Ito, 1973); an extensive review of the literature on the fat-storing cell has been published (Wake, 1980). Similar cells are also found in a number of extrahepatic organs.

Fat-storing cells are located in the perisinusoidal space and contain numerous lipid droplets. The size and number of lipid droplets vary with the nutritional status (especially of vitamin A, see below); there are also species differences. Humans and rats have numerous small (generally 0.5–3 μm in diameter) droplets. When large doses of vitamin A are given to experimental animals, the number and size of these lipid droplets increase considerably, and the Ito cells become larger and more prominent histologically (Wake, 1974, 1980). Similar changes have been seen in human liver fat-storing cells in hypervitaminosis A (Russell *et al.*, 1974; Hruban *et al.*, 1974). Under these conditions, the lipid droplets of the fat-storing cells display a prominent green fluorescence under fluorescence microscopy that has been considered characteristic of vitamin A (Wake, 1974, 1980; Russell *et al.*, 1974). Localization of vitamin A in the lipid droplets of the fat-storing cells in mice injected with large doses of [^{3}H]retinyl acetate was also demonstrated by electron microscope radioautography by Hirosawa and Yamada (1973). These observations have led to the conclusion that the fat-storing cells may play a significant role in vitamin A storage in the liver (Wake, 1980), and that they may act as a "sump" for excess vitamin A (McLaren, 1981).

Fat-storing cells may also be involved in fibrogenesis within the liver [see Wake (1980) for review and references]. It has been suggested that upon hepatic injury the lipocytes undergo transition to the fibroblasts responsible for collagen formation and intralobular fibrosis (Kent *et al.*, 1976). Russell *et al.* (1974) reported two patients with chronic hypervitaminosis A who presented with ascites and clinical signs of cirrhosis. Liver biopsy revealed many Ito cells with prominent lipid droplets associated with perisinusoidal fibrosis and central vein sclerosis. They suggested that chronic ingestion of large amounts of vitamin A may stimulate fibrogenesis and result in a disorder of hepatic function resembling cirrhosis. Thus, the fat-storing cell may be involved in hepatic pathology resulting from hypervitaminosis A.

Evidence has been obtained that the fat-storing cells may play an important role in vitamin A storage in the liver under normal as well as under hypervitaminotic conditions. In one study, Knook *et al.* (1982) isolated fat-storing cells from rat liver by enzyme digestion, followed by Metrizamide density centrifugation and centrifugal elutriation. The isolated cells showed a well-preserved ultrastructure and contained lipid droplets with a fluorescence characteristic of vitamin A. Preliminary analyses by HPLC demonstrated the presence of large quantities of retinyl palmitate and retinol (in a ratio of approximately 25:1) in the isolated fat-storing cells. In a related study, nonparenchymal cells (including fat-storing cells) were isolated by a similar method (Yumoto, 1982). Lipid granules similar in size to those seen in the cytoplasm of fat-storing cells were obtained after cell homogenization. The granules demonstrated marked fluorescence characteristic of vitamin A and were highly enriched in retinyl palmitate as determined by HPLC analysis.

Information about the role of various liver cell types has also been obtained by the studies of Olson and Gunning (1983). Different cells and cellular organelles were isolated by differential and density gradient centrifugation after treatment of the liver with collagenase or with pronase and DNase. By density gradient centrifugation, hepatocytes were divided into six fractions that contained fewer lipid globules per cell and less vitamin A per cell as the cell density increased. Fat-storing cells could not be isolated by the procedures used. Of the total liver vitamin A [predominantly (>95%) present as retinyl ester], 40–80% was found in vitamin A-containing lipid globules. The authors concluded that fat-storing cells and hepatocytes that contain a significant number of lipid globules are the major storage cells for vitamin A in the liver. They estimated that in a well-nourished rat the total lipid globules in liver might arise in roughly equal numbers from fat-storing cells and from hepatocytes.

Evidence for retinol transfer, after its hepatic uptake, from parenchymal to nonparenchymal cells has been reported by Blomhoff *et al.* (1982). Lymph chylomicrons containing newly absorbed, ^{3}H-labeled retinyl ester were injected intravenously into intact normal or retinol-deficient rats. At various times after injection, livers were analyzed for total radioactivity, and for the distribution of ^{3}H (retinol) between isolated hepatocytes and nonparenchymal cells (which included fat-storing cells). As expected, retinyl ester was taken up rapidly by the liver, which, after 30 min, contained 80–90% of the radioactivity injected. Initially, most of the radioactivity was found in hepatocytes. After 30 min, however, in normal rats the amount of ^{3}H label in hepatocytes gradually decreased, whereas that in the nonparenchymal cells increased. After 2 h, the nonparenchymal cells contained more radioactivity than the hepatocytes. In contrast, when labeled lymph was injected into retinol-deficient rats, the nonparenchymal cells did not accumulate radioactivity. Instead, in these animals, [^{3}H]retinol gradually left the entire liver and was transported to extrahepatic tissues, after its initial uptake. As a control normal rats were injected intravenously with lymph containing newly absorbed [^{14}C]cholesterol (mainly presented as cholesteryl ester). In this instance, radioactivity taken up by and found in the liver was recovered mostly in hepatocytes throughout the 4-h study. Unlike the case with retinol, the nonparenchymal cells accumulated only a small amount of radioactive cholesterol. From these data, Blomhoff *et al.* (1982) suggested that retinol (in chylomicron remnants) was taken up initially by hepatocytes and then, in normal animals, was transferred from these cells in substantial amount to nonparenchymal liver cells. In preliminary observations using centrifugal elutriation they observed that the labeled retinol was associated with a fraction of nonparenchymal cells that was neither Kupffer nor endothelial cells. They speculated that the labeled cells might be the fat-storing cells.

Taken together, these various studies as well as observations by Redgrave and Vakakis (1976) suggest that chylomicron retinol is first taken up by parenchymal cells and then may be transferred to fat-storing cells for storage (as retinyl ester in its lipid droplets). Under normal nutritional conditions, a significant fraction of total liver retinol may be stored in the fat-storing cells. In the retinol-depleted state, incoming retinol is mobilized from the hepatocyte for transport to extrahepatic tissues without being transferred to the fat-storing cells. With dietary retinol excess, large quantities of retinyl ester are stored in the fat-storing cells.

Much more information is needed about the metabolic characteristics of the fat-storing cells and about the metabolic interrelationships between these cells and parenchymal cells. The mechanism whereby retinol is transferred between these two types of cells requires exploration. This mechanism may involve one of the known retinoid-binding proteins or a different protein present in the intercellular space. Studies in the authors' laboratory (M. Kato and D. S. Goodman, unpublished observations) using immunohistochemical methods and antibodies specific for CRBP have shown that CRBP is concentrated in the fat-storing cells. Information is needed about the potential enzymatic capability of the fat-storing cells for retinyl ester synthesis or hydrolysis. Thus, many questions remain to be addressed by future investigation.

The hepatic parenchymal cell clearly plays the major role in most aspects of retinol metabolism within the liver. Several studies have indicated that chylomicron remnants or their constitutent lipids (particularly chylomicron cholesterol) are mainly taken up by parenchymal cells in the liver (Stein *et al.*, 1969; Redgrave and Vakakis, 1976; Floren and Nilsson, 1977; Blomhoff *et al.*, 1982). These studies and the direct studies of chylomicron retinol uptake discussed above (Blomhoff *et al.*, 1982) clearly demonstrate that the parenchymal cell is the major cell type involved in the uptake and initial metabolism of newly absorbed retinol by the liver. It is highly likely that the parenchymal cells (particularly those with significant numbers of lipid globules) also play an important role in the storage of retinol (retinyl ester) in the liver. The relative roles of the parenchymal and the fat-storing cells in retinol storage under different conditions need to be delineated. Finally, it has been clearly established that plasma RBP is synthesized and secreted by hepatic parenchymal cells. The evidence includes immunofluorescent microscopy studies with rat liver (Poole *et al.*, 1975) and studies with isolated liver cells in culture (Smith *et al.*, 1978; Borek *et al.*, 1981; D. R. Soprano and D. S. Goodman, unpublished observations with primary rat hepatocytes in culture). Since retinol is mobilized and RBP is secreted as the retinol-RBP complex, these studies demonstrate that the parenchymal cell is the cell type responsible for mobilization of retinol from the liver.

I. Subcellular Localization and Forms

Under normal conditions, approximately 95% of the retinol found in the liver is present as retinyl ester. Most of the stored retinol (retinyl ester) is found in association with lipid droplets within the liver cells. These lipid droplets are found within both the fat-storing cells and within parenchymal cells (Olson and Gunning, 1983). The characteristics of the lipid droplets that are found in the fat-storing cells are discussed above. Less information is available about the lipid droplets within hepatocytes with which retinyl esters and retinol are associated. The studies of Olson and Gunning (1983) discussed above suggest, however, that the vitamin A-containing lipid globules in hepatocytes may have generally similar characteristics as do those in the fat-storing cells. Thus, isolated vitamin A-containing globules, thought to be derived from both fat-storing cells and from "light" hepatocytes, had a median diameter of 1.7 μm (range 0.4–4.6 μm), showed intense yellow-green fluorescence under UV light, and contained more than 95% of their vitamin A as retinyl ester. It was emphasized that the vitamin A-containing lipid droplet or granule appears to be the major subcellular entity involved in the storage of vitamin A in the liver.

Other kinds of observations are consistent with this conclusion. When rat liver is homogenized and subjected to subcellular fractionation by differential centrifugation, more than half of the total retinyl ester plus retinol can be recovered in the floating lipid layer from the cytosol (Berman *et al.*, 1979). (In unpublished observations from the authors' laboratory over many years, the recovery of retinyl ester plus retinol in the floating lipid layer has usually been found to be substantially more than half the total liver content of these compounds.)

Only limited information is available concerning the extent to which retinol and its esters may be present within the liver cell in association with other specific subcellular organelles or membrane fractions. In one study, significant amounts of both retinyl ester and retinol were found in each of several crude membrane fractions isolated by differential centrifugation of liver homogenates (Berman *et al.*, 1979). Lesser, but still significant, amounts of retinyl ester and retinol were also found by Nyquist *et al.* (1971) in the large particle fraction sedimenting at 2000 × *g* from the liver homogenate of a fed male rat. These investigators reported, moreover, that both retinyl esters and retinol were concentrated in the purified Golgi apparatus, whereas no retinyl compounds were detectable (with the spectrophotometric method used) in extracts of purified mitochondria or endoplasmic reticulum. More information is needed, using currently available methodology, concerning the content and extent of localization of retinol and its esters in different defined subcellular structures.

A fluorescent, high-molecular-weight, lipid–protein aggregate was partially isolated from the cytosol fraction of rat liver and found to contain a significant proportion of the total liver content of retinyl esters, of retinyl palmitate hydrolase activity, and of CRBP (Sklan *et al.*, 1982). The partial isolation and characterization of a soluble, lipid-protein aggregate containing retinyl esters from rat liver cytosol was first reported by Heller (1979). This entity, called the "cytosol retinyl ester lipoprotein complex" by Heller (1979), contained about 10% of the total liver retinyl compounds and also displayed retinyl ester hydrolase activity (Heller, 1979; Chen and Heller, 1979). It was suggested that this entity might represent a specific vitamin A-metabolizing lipoprotein particle, and that it might play a physiological role in the liver cell in the storage and hydrolysis of retinyl esters and in retinol mobilization. In further studies, Chen *et al.* (1981) incubated *in vitro* the cytosol retinyl ester–lipoprotein complex together with serum apo-RBP and serum albumin. After 24 h at 30°C, 68% of the retinyl ester was hydrolyzed and up to 30% of the unesterified retinol was transferred from the lipid–protein aggregate to serum apo-RBP (to form holo-RBP). From these data the authors suggested that in addition to serving as a storage form for retinyl ester and retinol and as an enzyme for hydrolyzing its associated retinyl ester, the "lipoprotein complex" might also serve as a medium for transfer of retinol to RBP. While this suggestion is an interesting one, it should be noted that serum apo-RBP in solution is quite different from membrane-bound apo-RBP in the liver cell. It is well established that retinol added (in many different ways) to a solution of apo-RBP will bind readily to the apo-RBP (see Chapter 8). However, attempts to add retinol to apo-RBP associated with liver microsomes, in order to form holo-RBP and release the RBP from its membrane-bound form, have so far not been successful (J. E. Smith and D. S. Goodman, unpublished observations). More studies concerning this question are needed.

Further characterization of this lipid–protein aggregate by Sklan *et al.* (1982) indicated that it was composed of approximately equal parts of protein and of lipid (mainly triglycerides) and contained approximately 19% of the total liver vitamin A (predominantly as retinyl esters). Much of the liver CRBP and of the RPH activity present in the cytosol was also found associated with the fluorescent lipid–protein aggregate. The lipid–protein aggregate and its several retinol-related components displayed an apparent hydrated density between 1.052 and 1.090 in the ultracentrifuge. Three other lipid hydrolytic activities were also found in association with the lipid–protein aggregate, namely, triolein, cholesteryl oleate, and phosphatidylcholine hydrolase activities. These several hydrolytic activities were all found to be stimulated optimally by the addition of either sodium cholate or bovine serum albumin.

The existence of a vitamin A-containing high-molecular-weight lipid-protein aggregate in liver cytosol has also been observed in studies with homogenates of equine liver (Sklan and Donoghue, 1982a) and of chicken liver (Sklan and Donoghue, 1982b). In general, the characteristics of these lipid–protein aggregates resembled those of the aggregate observed in rat liver homogenates.

With the information availabe, it is not clear whether the lipid–protein aggregate represents a specific lipoprotein that exists *in vivo* in the liver cell, or whether it represents an entity found only in a liver homogenate. For example, because retinyl palmitate hydrolase has hydrophobic physical properties (discussed above), this enzyme might become associated nonspecifically with the lipid–protein complex during the process of liver cell homogenization. Thus, it is possible that the high-molecular-weight lipid–protein complex might represent an artifact formed during homogenization. The hypothesis that the lipid–protein aggregate represents a specific vitamin A-metabolizing lipoprotein particle is conceptually appealing, but much more information will be needed before the significance and functional role of this entity can be evaluated effectively.

J. Retinol Glycolipids and Glycoprotein Metabolism

Another set of reactions involving retinol that has been studied in liver concerns the formation of retinyl phosphate and of retinyl phosphomannose conjugates. These studies have been directed toward the hypothesis that retinoid-containing glycolipids serve as intermediates for specific glycosylation reactions. It is well established that the biosynthesis of some glycoproteins is markedly decreased in a number of tissues in retinoid deficiency, and that glycoprotein synthesis is stimulated upon administration of retinoids. In order to explain these observations, it has been suggested that retinol or a derivative of retinol may serve as the lipid portion of a glycolipid intermediate involved in certain glycosylation reactions. Thus, in this hypothesis, retinol is thought to function in a manner analogous to the well-documented role of dolichol, which is converted to dolichyl phosphoglycosides that are used for glycoprotein formation. For reviews of work dealing with this subject see De Luca (1977, 1978, 1982), Wolf (1977), De Luca *et al.* (1979), Wolf *et al.* (1979), and Shidoji *et al.* (1981). This topic and the published literature dealing with this area are reviewed and discussed in detail in another chapter of this book (Chapter 12).

Retinyl phosphate has been shown to be formed in mammalian cells both *in vitro* and *in vivo* and has been identified as a trace constitutent of certain mammalian membranes. Its glycosylated derivative mannosyl retinyl phosphate has also been shown to be formed both *in vivo* and *in vitro*. The enzyme system that forms mannosyl retinyl phosphate is located primarily in the rough endoplasmic reticulum of rat liver cells. Under appropriate conditions, glycosyl transfer can be demonstrated from the retinyl glycolipid to membrane glycoproteins. It is not known, however, whether this occurs *in vivo* under normal conditions.

The hypothesis that retinoid-containing glycolipids are obligatory intermediates in the synthesis of specific glycoproteins is an intriguing one that has led to much interesting biochemical research. As indicated previously (Goodman, 1980), with the information available it seems prudent to defer judgment as to the possible validity of this hypothesis.

K. Mobilization of Retinol from the Liver

Retinol is mobilized from the liver, and delivered to peripheral target tissues, as the retinol–RBP complex. Retinol mobilization and delivery are highly regulated processes that are particularly controlled by processes that regulate the rates of synthesis and secretion of RBP by the liver. These processes are discussed in full in Chapter 8, this volume.

ACKNOWLEDGMENT

The studies from the author's laboratory discussed in this chapter were supported by NIH grants HL 21006 (SCOR) and AM 05968.

REFERENCES

Adachi, N., Smith, J. E., Sklan, D., and Goodman, D. S. (1981). *J. Biol. Chem.* **256,** 9471–9476.

Andrews, J. S., and Futterman, S. (1964). *J. Biol. Chem.* **239,** 4073–4076.

Assmann, G., and Fredrickson, D. S. (1983). *In* "The Metabolic Basis of Inherited Disease" (J. B. Stanbury, J. B. Wyngaarden, D. S. Fredrickson, J. L. Goldstein, and M. S. Brown, eds.), pp. 803–819. McGraw-Hill, New York.

Barter, P. J., and Lally, J. I, (1979). *Metab. Clin. Exp.* **28,** 230–236.

Bauernfeind, J. C., Adams, C. R., and Marusich, W. L. (1981). *In* "Carotenoids as Colorants and Vitamin A Precursors: Technological and Nutritional Applications" (C. Bauernfeind, ed.), pp. 563–743. Academic Press, New York.

Bennett Clark, S. (1979). *J. Biol. Chem.* **254,** 1534–1536.

Berman, E. R., Segal, N., and Feeney, L. (1979). *Biochim. Biophys. Acta* **572,** 167–177.

Bertram, J., and Krisch, K. (1969). *Eur. J. Biochem.* **11,** 122–126.

Blaner, W. S., Smith, J. E., and Goodman, D. S. (1981). *Fed. Proc., Fed. Am. Soc. Exp. Biol.* **40,** 1584. (Abstr.)

Bloch, K. (1965). *Science* **150,** 19–28.

Blomhoff, R., Helgerud, P., Rasmussen, M., Berg, T., and Norum, K. R. (1982). *Proc. Natl. Acad. Sci. U.S.A.* **79,** 7326–7330.

Blomstrand, R., and Werner, B. (1967). *Scand. J. Clin. Lab. Invest.* **19,** 339–345.

Blum, C. B., Deckelbaum, R. J., Witte, L. D., Tall, A. R., and Cornicelli, J. (1982). *J. Clin. Invest.* **70,** 1157–1169.

Borek, C., Smith, J. E., Soprano, D. R., and Goodman, D. S. (1981). *Endocrinology (Baltimore)* **109,** 386–391.

Bronfenmajer, S., Schaffner, F., and Popper, H. (1966). *Arch. Pathol.* **82,** 447–453.

Brown, M. S., Kovanen, P. T., and Goldstein, J. L. (1981). *Science* **212,** 628–635.

Brown, W. J., and Sgoutas, D. S. (1980). *Biochim. Biophys. Acta* **617,** 305–317.

Calame, K. B., Gallo, L., Cheriathundam, E., Vahouny, G. V., and Treadwell, C. R. (1975). *Arch. Biochem. Biophys.* **168,** 57–65.

Chajek, T., and Fielding, C. J. (1978). *Proc. Natl. Acad. Sci. U.S.A.* **75,** 3445–3449.

Chajek-Shaul, T., Friedman, G., Halperin, G., Stein, O., and Stein, Y. (1981a). *Biochim. Biophys. Acta* **666,** 147–155.

Chajek-Shaul, T., Friedman, G., Halperin, G., Stein, O., and Stein, Y. (1981b). *Biochim. Biophys. Acta* **666,** 216–222.

Chen, C.-C., and Heller, J. (1979). *Arch. Biochem. Biophys.* **198,** 572–579.

Chen, C.-C., Heller, J., Ding, L. L., and Horwitz, J. (1981). *Arch. Biochem. Biophys* **207,** 392–398.

Chytil, F. (1982). *Hepatology* **2,** 282–287.

Chytil, F., and Ong, D. E. (1978). *Vitam. Horm. (N. Y.)* **36,** 1–32.

Chytil, F., and Ong, D. E. (1979). *Fed. Proc., Fed. Am. Soc. Exp. Biol.* **38,** 2510–2514.

Cooper, A. D., and Yu, P. Y. S. (1978). *J. Lipid Res.* **19,** 635–643.

Cornforth, J. W. (1976). *Science* **193,** 121–125.

Crain, F. D., Lotspech, F. J., and Krause, R. F. (1967). *J. Lipid Res.* **8,** 249–254.

Davidson, N. O., and Glickman, R. M. (1983). *In* "Progress in Gastroenterology" (P. Sherlock, and G. J. Glass, ed.), Vol. IV, pp. 57–75. Grune & Stratton, New York.

DeLuca, H. F. (1979). *Fed. Proc., Fed. Am. Soc. Exp. Biol.* **38,** 2519–2523.

De Luca, L. M. (1977). *Vitam. Horm. (N. Y.)* **34,** 1–57.

De Luca, L. M. (1978). *In* "Handbook of Lipid Research" (H. F. DeLuca, ed.), Vol. 2, pp. 1–67. Plenum, New York.

De Luca, L. M. (1982). *J. Am. Acad. Dermatol.* **6,** 611–618.

De Luca, L. M., Bhat, P. V., Sasak, W., and Adamo, S. (1979). *Fed. Proc., Fed. Am. Soc. Exp. Biol.* **38,** 2535–2539.

De Leenheer, A. P., Lambert, W. E., and Claeys, I. (1982). *J. Lipid Res.* **23,** 1362–1367.

Deykin, D., and Goodman, D. S. (1962). *J. Biol. Chem.* **237,** 3649–3656.

Dunagin, P. E., Jr., Meadows, E. H., Jr., and Olson, J. A. (1965). *Science* **148,** 86–87.

Eden, E., and Sellers, K. C. (1950). *Biochem J.* **46,** 261–266.

El-Gorab, M. I., Underwood, B. A., and Loerch, J. D. (1975). *Biochim. Biophys. Acta* **401,** 265–277.

Erickson, S. K., and Cooper, A. D. (1980). *Metab. Clin. Exp.* **29,** 991–996.

Erlanson, C., and Borgström, B. (1968). *Biochim. Biophys. Acta* **167,** 629–631.

Fahimi, H. D. (1982). *In* "The Liver, Biology and Pathobiology" (I. Arias, H. Popper, D. Schachter, and D. A. Shafritz, eds.), pp. 495–506. Raven, New York.

Fex, G., and Johannesson, G. (1982). *Biochim. Biophys. Acta* **714,** 536–542.

Fidge, N. H., and Goodman, D. S. (1968). *J. Biol. Chem.* **243,** 4372–4379.

Fidge, N. H., Shiratori, T., Ganguly, J., and Goodman, D. S. (1968). *J. Lipid Res.* **9,** 103–109.

Fidge, N. H., Smith, F. R., and Goodman, D. S. (1969). *Biochem. J.* **114,** 689–694.

Floren, C. H., and Nilsson, A. (1977). *Biochem. J.* **168,** 483–494.

Fredrikzon, B., Hernell, O., Bläckberg, L., and Olivecrona, T. (1978). *Pediatr. Res.* **12,** 1048–1052.

Friedman, H. I., and Nylund, B. (1980). *Am. J. Clin. Nutr.* **33,** 1108–1139.

Futterman, S., and Andrews, J. S. (1964). *J. Biol. Chem.* **239,** 4077–4080.

Gallo, L., Chiang, Y., Vahouny, G. V., and Treadwell, C. R. (1980). *J. Lipid Res.* **21,** 537–545.

Gawienowski, A. M., Stacewicz-Sapuncakis, M., and Longley, R. (1974). *J. Lipid Res.* **15,** 375–379.

Goldstein, J. L., and Brown, M. S. (1979). *Annu. Rev. Genet.* **13,** 259–289.

Goodman, D. S. (1965). *Physiol. Rev.* **45,** 747–839.

Goodman, D. S. (1980). *Fed. Proc., Fed. Am. Soc. Exp. Biol.* **39,** 2716–2722.

Goodman, D. S., and Huang, H. S. (1965). *Science* **149,** 879–880.

Goodman, D. S., Deykin, D., and Shiratori, T. (1964). *J. Biol. Chem.* **239,** 1335–1345.

Goodman, D. S., Huang, H. S., and Shiratori, T. (1965). *J. Lipid Res.* **6,** 390–396.

Goodman, D. S., Huang, H. S., and Shiratori, T. (1966a). *J. Biol. Chem.* **241,** 1929–1932.

Goodman, D. S., Blomstrand, R., Werner, B., Huang, H. S., and Shiratori, T. (1966b). *J. Clin. Invest.* **45,** 1615–1623.

Goodman, D. S., Huang, H. S., Kanai, M., and Shiratori, T. (1967). *J. Biol. Chem.* **242,** 3543–3554.

Goodman, D. S., Stein, O., Halperin, G., and Stein, Y. (1983). *Biochim. Biophys Acta* **750,** 223–230.

Goodwin, T. W. (1980). *In* "The Biochemistry of the Carotenoids" (T. W. Goodwin, ed.), 2nd ed., Vol. I, pp. 33–76. Chapman and Hall, London.

Gray, E., Morgareidge, K., and Cawley, J. D. (1940). *J. Nutr.* **20,** 67–74.

Green, P. H. R., and Glickman, R. M. (1981). *J. Lipid Res.* **22,** 1153–1173.

Harrison, E. H., Smith, J. E., and Goodman, D. S. (1979). *J. Lipid Res.* **20,** 760–771.

Havel, R. J., Chao, Y., Windler, E. E., Kotite, L., and Guo, L. S. S. (1980). *Proc. Natl. Acad. Sci. U.S.A.* **77,** 4349–4353.

Hazzard, W. R., and Bierman, E. L. (1976). *Metab. Clin. Exp.* **25,** 777–801.

Helgerud, P., Petersen, L. B., and Norum, K. R. (1982). *J. Lipid Res.* **23,** 609–618.

Helgerud, P., Petersen, L. B., and Norum, K. R. (1983). *J. Clin. Invest.* **71,** 747–753.

Heller, J. (1979). *Arch. Biochem. Biophys.* **198,** 562–571.

Herbert, P. N., Assmann, G., Gotto, A. M., Jr., and Fredrickson, D. S. (1983). *In* "The Metabolic Basis of Inherited Disease" (J. B. Stanbury, J. B. Wyngaarden, D. S. Fredrickson, J. L. Goldstein, and M. S. Brown, eds.), pp. 589–621. McGraw-Hill, New York.

Hirosawa, K., and Yamada, E. (1973). *J. Electron Micros.* **22,** 337–346.

Hollander, D. (1981). *J. Lab. Clin. Med.* **97,** 449–462.

Hollander, D. (1980). *Am. J. Physiol.* **239,** G210–G214.

Hollander, D., and Morgan, D. (1979). *Exp. Gerontol.* **14,** 301–305.

Hollander, D., and Muraldihara, K. S. (1977). *Am. J. Physiol.* **232,** E471–E477.

Hollander, D., and Ruble, P. E., Jr. (1978). *Am. J. Physiol.* **235,** E686–E691.

Hori, S. H., and Kitamura, T. (1972). *J. Histochem. Cytochem.* **20,** 811–816.

Hruban, Z., Russell, R. M., Boyer, J. L., Glagov, S., and Bagheri, S. A. (1974). *Am. J. Pathol.* **76,** 451–461.

Huang, H. S., and Goodman, D. S. (1965). *J. Biol. Chem.* **240,** 2839–2844.

Hui, D. Y., Innerarity, T. L., and Mahley, R. W. (1981). *J. Biol. Chem.* **256,** 5646–5655.

Ito, T. (1973). *Gunma Rep. Med. Sci.* **6,** 119–163.

Ito, Y. L., Zile, M., Ahrens, H., and DeLuca H. F. (1974). *J. Lipid Res.* **15,** 517–524.

Jackson, R. L., Morrisett, J. D., and Gotto, A. M., Jr. (1976). *Physiol. Rev.* **56,** 259–316.

Jones, E. A., and Summerfield, J. A. (1982). *In* "The Liver, Biology and Pathobiology" (I. Arias, H. Popper, D. Schachter, and D. A. Shafritz, eds.), pp. 507–523. Raven, New York.

Kayden, H. J. (1972). *Annu. Rev. Med.* **23,** 285–296.

Kayden, H. J. (1980). *Nutr. Rev.* **38,** 244–246.

Kent, G., Gay, S., Inouye, T., Bahu, R., Minick, O. T., and Popper, H. (1976). *Proc. Natl. Acad. Sci. U.S.A.* **73,** 3719–3722.

Khoo, J. C., Steinberg, D., Huang, J. J., and Vagelos, P. R. (1976). *J. Biol. Chem.* **251,** 2882–2890.

Knook, D. L., Seffelaar, A. M., and Leeuw, A. M. (1982). *Exp. Cell Res.* **139,** 468–471.

Kushwaha, S. C., Gochnauer, M. B., Kushner, D. J., and Kates, M. (1974). *Can. J. Microbiol.* **20,** 241–245.

Lakshmanan, M. R., Pope, J. L., and Olson, J. A. (1968). *Biochem. Biophys. Res. Commun.* **33,** 347–352.

Lakshmanan, M. R., Chansang, H., and Olson, J. A. (1972). *J. Lipid Res.* **13,** 477–482.

Lawrence, C. W., Crain, F. D., Lotspeich, F. J., and Krause, R. F. (1966). *J. Lipid Res.* **7,** 226–229.

Liau, G., Ong, D. E., and Chytil, F. (1981). *J. Cell Biol.* **91,** 63–68.

Linder, M. C., Anderson, G. H., and Ascarelli, M. (1971). *J. Biol. Chem.* **246,** 5538–5540.

Liou, G. I., Fong, L. S., and Bridges, C. D. B. (1981). *J. Biol. Chem.* **256,** 3153–3155.

Lombardo, D., and Guy, O. (1980). *Biochim. Biophys. Acta* **611,** 147–155.

Lombardo, D., Fauvel, J., and Guy, O. (1979). *Biochim. Biophys. Acta* **611,** 136–146.

Lombardo, D., Deprez, P., and Guy, O. (1980). *Biochimie* **62,** 427–432.

McCormick, A. M., and Napoli, J. L. (1982). *J. Biol. Chem.* **257,** 1730–1735.

McCormick, A. M., Napoli, J. L., Schnoes, H. K., and DeLuca, H. F. (1978). *Biochemistry* **17,** 4085–4090.

McLaren, D. S. (1981). *Am. J. Clin. Nutr.* **34,** 1611–1616.

Mahadevan, S., Murthy, S. K., Krishnamurthy, S., and Ganguly, J. (1961). *Biochem. J.* **79,** 416–424.

Mahadevan, S., Seshardri Sastry, P., and Ganguly, J. (1963a). *Biochem. J.* **88,** 531–534.

Mahadevan, S., Seshardri Sastry, P., and Ganguly, J. (1963b). *Biochem. J.* **88,** 534–539.

Mahadevan, S., Ayyoub, N. I., and Roels, O. A. (1966). *J. Biol. Chem.* **241,** 57–64.

Momsen, W. E., and Brockman, H. L. (1977). *Biochim. Biophys. Acta* **486,** 103–113.

Moore, T. (1930). *Biochem. J.* **24,** 692–702.

Murray, T. K., and Grice, H. C. (1961). *Can. J. Biochem. Physiol.* **39,** 1103–1106.

Murthy, S. K., Mahadevan, S., Seshardri Sastry, P., and Ganguly, J. (1961). *Nature (London)* **189,** 482.

Napoli, J. L., Khalil, H., and McCormick, A. M. (1982). *Biochemistry* **21,** 1942–1949.

Nilsson, A. (1977). *Biochem. J.* **162,** 367–377.

Norum, K. R., Lilljeqvist, A. C., Helgerud, P., Normann, E. R., Mo, A., and Selbekk, B. (1979). *Eur. J. Clin. Invest.* **9,** 55–62.

Nyquist, S. E., Cranes, F. L., and Morre, D. J. (1971). *Science* **173,** 939–941.
Oesterhelt, D. (1976). *Prog. Mol. Subcell. Biol.* **4,** 133–166.
Olson, J. A. (1964). *J. Lipid Res.* **5,** 281–299.
Olson, J. A. (1969). *Am. J. Clin. Nutr.* **22,** 953–962.
Olson, J. A., and Gunning, D. (1983). *J. Nutr.* **113,** 2184–2191.
Olson, J. A., and Hayaishi, O. (1965). *Proc. Natl. Acad. Sci. U.S.A.* **54,** 1364–1370.
Ong, D. E. (1982). *Cancer Res.* **42,** 1033–1037.
Ong, D. E., and Chytil, F. (1978). *J. Biol. Chem.* **253,** 828–832.
Ong, D. E., Crow, J. A., and Chytil, F. (1982). *J. Biol. Chem.* **257,** 13385–13389.
Patton, J. S. (1981). *In* "Physiology of the Gastrointestinal Tract" (L. R. Johnson, ed.), pp. 1123–1146. Raven, New York.
Peterson, P. A., Rask, L., Ostberg, L., Anderson, L., Kamwendo, F., and Pertoft, H. (1973). *J. Biol. Chem.* **248,** 4009–4022.
Pitas, R. E., Innerarity, T. L., Arnold, K. S., and Mahley, R. W. (1979). *Proc. Natl. Acad. Sci. U.S.A.* **76,** 2311–2315.
Pittman, R. C., and Steinberg, D. (1977). *Biochim. Biophys. Acta* **487,** 431–444.
Pollard, C. J., and Bieri, J. G. (1960). *Arch. Biochem. Biophys.* **87,** 9–12.
Poole, A. R., Dingle, J. T., Mallia, A. K., and Goodman, D. S. (1975). *J. Cell Sci.* **19,** 379–394.
Prystowsky, J. H., Smith, J. E., and Goodman, D. S. (1981). *J. Biol. Chem.* **256,** 4498–4503.
Quarfordt, S. H., and Goodman, D. S. (1967). *J. Lipid Res.* **8,** 264–273.
Qureshi, A. A., Kim, M., Qureshi, N., and Porter, J. W. (1974). *Biochim. Biophys. Acta* **162,** 108–116.
Redgrave, T. G. (1970). *J. Clin. Invest.* **49,** 465–471.
Redgrave, T. G., and Vakakis, N. (1976). *Aust. J. Exp. Biol. Med. Sci.* **54,** 519–525.
Roberts, A. B., and DeLuca, H. F. (1967). *Biochem. J.* **102,** 600–605.
Roberts, A. B., and Frolik, C. A. (1979). *Fed. Proc., Fed. Am. Soc. Exp. Biol.* **38,** 2524–2527.
Ross, A. C. (1982a). *J. Lipid Res.* **23,** 133–144.
Ross, A. C. (1982b). *J. Biol. Chem.* **257,** 2453–2459.
Ross, A. C., and Goodman, D. S. (1979). *Fed. Proc., Fed. Am. Soc. Exp. Biol.* **38,** 2515–2518.
Russell, R. M., Boyer, J. L., Bagheri, S. A., and Hruban, Z. (1974). *N. Engl. J. Med.* **291,** 435–440.
Saari, J. C., Bredberg, L., and Garwin, G. G. (1982). *J. Biol. Chem.* **257,** 13329–13333.
Schroepfer, G. J. (1981). *Annu. Rev. Biochem.* **50,** 585–621.
Seshardri Sastry, P., and Ganguly, J. (1961). *Biochem. J.* **80,** 397–406.
Sharma, R. V., Mathur, S. N., and Ganguly, J. (1976). *Biochem. J.* **158,** 377–383.
Sharma, R. V., Mathur, S. N., Dmitrovskii, A. A., Das, R. C., and Ganguly, J. (1977). *Biochim. Biophys. Acta* **486,** 183–194.
Shelburne, F., Hanks, J., Meyers, W., and Quarfordt, S. (1980). *J. Clin. Invest.* **65,** 652–658.
Sherrill, B. C., and Dietschy, J. M. (1978). *J. Biol. Chem.* **253,** 1859–1867.
Sherrill, B. C., Innerarity, T. L., and Mahley, R. W. (1980). *J. Biol. Chem.* **255,** 1804–1807.
Shidoji, Y., Sasak, W., Silverman-Jones, C. S., and De Luca, L. M. (1981). *Ann. N. Y. Acad. Sci.* **359,** 345–357.
Shiratori, T., and Goodman, D. S. (1965). *Biochim. Biophys. Acta* **106,** 625–627.
Singh, H., and Cama, H. R. (1974). *Biochim. Biophys. Acta* **370,** 49–61.
Skare, K. L., Schnoes, H. K., and DeLuca, H. F. (1982). *Biochemistry* **21,** 3308–3317.
Sklan, D., and Donoghue, S. (1982a). *Biochim. Biophys. Acta* **711,** 532–538.
Sklan, D., and Donoghue, S. (1982b). *Br. J. Nutr.* **47,** 273–280.
Sklan, D., Blaner, W. S., Adachi, N., Smith, J. E., and Goodman, D. S. (1982). *Arch. Biochem. Biophys.* **214,** 35–44.
Smith, J. E., Borek, C., and Goodman, D. S. (1978). *Cell* **15,** 865–873.

Smith, L. C., Pownall, J. H., and Gotto, A. M., Jr. (1978). *Annu. Rev. Biochem.* **47,** 751–777.
Stacewicz-Sapuncakis, M., Wang, H.-H., C., and Gawienowski, A. M. (1975). *Biochim. Biophys. Acta* **380,** 264–269.
Stein, O., Stein, Y., Goodman, D. S., and Fidge, N. H. (1969). *J. Cell Biol.* **43,** 410–431.
Stoeckenius, W., Lozier, R. H., and Bogomolni, R. A. (1979). *Biochim. Biophys. Acta* **505,** 215–278.
Subbarayan, C., Lakshmanan, M. R., and Cama, H. R. (1966). *Biochem. J.* **99,** 308–311.
Swanson, B. N., Frolik, C. A., Zaharevitz, D. W., Roller, P. P., and Sporn, M. B. (1981). *Biochem. Pharmacol.* **30,** 107–113.
Takase, S., Ong, D. E., and Chytil, F. (1979). *Proc. Natl. Acad. Sci. U.S.A.* **76,** 2204–2208.
Tall, A. R., Green, P. H. R., Glickman, R. M., and Riley, J. W. (1979). *J. Clin. Invest.* **64,** 977–989.
Thomson, A. B. R., and Dietschy, J. M. (1981). *In* ''Physiology of the Gastrointestinal Tract'' (L. R. Johnson, ed.), pp. 1147–1220. Raven, New York.
Tomassi, G., and Olson, J. A. (1983). *J. Nutr.* **113,** 697–703.
Tuhackova, Z., Kriz, W., and Hradec, J. (1980). *Biochim. Biophys. Acta* **617,** 439–445.
Van Berkel, T. J. C., Vaandrager, J., Kruijt, J. K., and Koster, J. F. (1980). *Biochim. Biophys. Acta* **617,** 446–457.
von Euler, H., von Euler, B., and Hellstrom, H. (1928). *Biochem. Z.* **203,** 370–384.
Wake, K. (1974). *J. Cell Biol.* **63,** 683–691.
Wake, K. (1980). *Int. Rev. Cytol.* **66,** 303–353.
Widmann, J. J., Cotran, R. S., and Fahimi, H. D. (1972). *J. Cell Biol.* **52,** 159–170.
Windler, E., Chao, Y., and Havel, R. J. (1980). *J. Biol. Chem.* **255,** 5475–5480.
Wolf, G. (1977). *Nutr. Rev.* **35,** 97–99.
Wolf, G., Kiorpes, T. C., Masushige, S., Schreiber, J. B., Smith M. J., and Anderson, R. S. (1979). *Fed. Proc., Fed. Am. Soc. Exp. Biol.* **38,** 2540–2543.
Worker, N. A. (1959). *Br. J. Nutr.* **13,** 400–418.
Yamada, E., and Hirosawa, K. (1976). *Cell Struct. Func.* **1,** 201–204.
Yeung, D. L., and Veen-Baigent, M. L. (1971). *Am. J. Clin. Nutr.* **24,** 172–173.
Yeung, D. L., and Veen-Baigent, M. J. (1972). *Can. J. Physiol. Pharmacol.* **50,** 753–760.
Yumoto, S. (1982). *Proc. Jpn. Acad.* **58,** 173–176.
Zachman, R. D., and Olson, J. A. (1963). *J. Biol. Chem.* **238,** 541–546.
Zachman, R. D., and Olson, J. A. (1964). *Nature (London)* **201,** 1222–1223.
Zachman, R. D., and Olson, J. A. (1965). *J. Lipid Res.* **6,** 27–32.
Zachman, R. D., Dunagin, P. E., Jr., and Olson, J. A. (1966). *J. Lipid Res.* **7,** 3–9.
Zile, M. H., Schnoes, H. K., and DeLuca, H. F. (1980). *Proc. Natl. Acad. Sci. U.S.A.* **77,** 3230–3233.
Zile, M. H., Inhorn, R. C., and DeLuca, H. F. (1982a). *J. Biol. Chem.* **257,** 3537–3543.
Zile, M. H., Inhorn, R. C., and DeLuca, H. F. (1982b). *J. Biol. Chem.* **257,** 3544–3550.
Zilversmit, D. B. (1978). *In* ''Disturbances in Lipid and Lipoprotein Metabolism'' (J. M. Dietschy, A. M. Gotto, Jr., J. A. Ontko, eds.), pp. 69–81. Am. Physiol. Soc., Bethesda, Maryland.
Zilversmith, D. B., Hughes, L. B., and Balmer, J. (1975). *Biochim. Biophys. Acta* **409,** 393–398.
Zilversmit, D. B., Morton, R. E., Hughes, L. B., and Thompson, K. M. (1982). *Biochim. Biophys. Acta* **712,** 88–93.

8

Plasma Retinol-Binding Protein

DEWITT S. GOODMAN

Department of Medicine
College of Physicians & Surgeons
Columbia University
New York, New York

THE RETINOIDS, VOL. 2

ISBN 0-12-658102-9

I. INTRODUCTION

Vitamin A is transported normally in postabsorptive plasma as the lipid alcohol retinol bound to a specific transport protein plasma retinol-binding protein (RBP). The isolation and partial characterization of RBP was first reported in 1968 (Kanai *et al.*, 1968). Since then, extensive studies in many laboratories have provided a considerable amount of information about the structure, metabolism, and biological roles of this protein. Human RBP is a single polypeptide chain with a molecular weight close to 21,000, α_1-mobility on electrophoresis, and a single binding site for one molecule of retinol. In plasma, most of RBP normally circulates as the retinol–RBP complex (holo-RBP). RBP preparations resembling human RBP have been isolated from the serum of many mammalian species other than man and from the chicken. In contrast, a rather different species of RBP with differing properties has been isolated from fish.

RBP interacts strongly with another protein, plasma transthyretin (TTR, also more commonly referred to as plasma prealbumin) and normally circulates as a 1:1 molar RBP–TTR complex. [The name transthyretin has been suggested (Nomenclature Committee of IUB *et al.*, 1981) for the protein usually called prealbumin and will be used instead of the term prealbumin in this and other chapters.] The usual level of RBP in plasma is about 40–50 μg/ml and that of TTR is about 200–300 μg/ml. In addition to its role in vitamin A transport, TTR plays a role in the binding and plasma transport of thyroid hormones. The formation of the RBP–TTR complex serves to reduce the glomerular filtration and renal catabolism of RBP.

Vitamin A is mobilized from the liver and delivered to peripheral target tissues as the retinol–RBP complex. Retinol mobilization and delivery are highly regulated processes that are particularly controlled by processes that regulate the rates of synthesis and secretion of RBP by the liver. Some information is available about the cellular and molecular mechanisms that mediate these phenomena. In addition, there is evidence that delivery of retinol to peripheral target tissues may involve specific cell-surface receptors that recognize RBP.

The retinol transport system provides an interesting model for the study of protein–protein and protein–retinoid interactions and of the characteristics and metabolic regulation of a specific binding and transport system. The aim of this chapter is to summarize the information available about this transport system, including information about the structure and chemistry, biochemistry, and metabolism of RBP, and about related clinical phenomena. Brief comments are also made about serum albumin as a retinoid-binding protein for retinoic acid and other acidic retinoids.

II. ISOLATION OF RBP

Most investigators have isolated RBP from serum by first isolating the RBP–TTR complex and then dissociating the protein–protein complex into its two components. The human RBP–TTR complex can be isolated readily by a combination of ion-exchange chromatography (e.g., on DEAE–Sephadex) and gel filtration (Kanai *et al.*, 1968; Raz and Goodman, 1969; Peterson, 1971a). The complex can then be dissociated, enabling the separated RBP (and TTR) to be isolated by procedures such as electrophoresis (Kanai *et al.*, 1968) or by gel filtration in the presence of low ionic strength (Peterson, 1971a) or 6 *M* urea (Haupt and Heide, 1972). A detailed description of the isolation of human RBP using a sequence of procedures including DEAE–Sephadex chromatography, gel filtration on Sephadex G-200, preparative polyacrylamide gel electrophoresis, and gel filtration on Sephadex G-75, has been published (Muto *et al.*, 1982). The reported yield from 500 ml of serum was about 25%.

Isolation of rat RBP has been more difficult than that of human RBP because the rat RBP–TTR complex does not separate readily from rat serum albumin on ion-exchange chromatography (Muto and Goodman, 1972; Peterson *et al.*, 1973). A relatively simple, three-step procedure for isolation of rat serum RBP, using DEAE–cellulose chromatography at pH 6.0, Sephadex G-75 gel filtration in the presence of 3 *M* urea (to separate the RBP-TTR complex), and finally DEAE–cellulose chromatography at pH 8.3, has been reported (McGuire and Chytil, 1980).

Both human (Vahlquist *et al.*, 1971) and rat (Poole *et al.*, 1975) RBP have been isolated by affinity chromatography on human TTR coupled to Sepharose. Affinity chromatography with retinoic acid linked to Sepharose has also been used (Fex and Hansson, 1978) to isolate delipidated apo-RBP from serum and urine. Finally, RBP has been isolated by a number of investigators from the urine of patients with tubular proteinuria, using procedures similar to those employed for the isolation of RBP from serum (Peterson and Berggård, 1971; Vahlquist *et al.*, 1971; Poulik *et al.*, 1975; Fex and Hansson, 1979).

III. THE PHYSICOCHEMICAL PROPERTIES OF RBP

Human RBP has a molecular weight of 21,200 (Peterson, 1971a) to 21,300 (Raz *et al.*, 1970), as determined by sedimentation equilibrium analysis in the analytical ultracentrifuge. The molecular weight calculated from the reported amino acid sequence (Rask *et al.*, 1979) is 21,230. RBP preparations isolated from serum of other mammalian and avian species have been found to have similar molecular weights, generally close to 20,000–21,000 (see below). Human RBP has a sedimentation coefficient ($s_{20,w}$) of 2.13 (Raz *et al.*, 1970) to 2.3 (Peterson, 1971a) and an estimated Stokes' molecular radius of 20.5 Å.

RBP from mammalian serum migrates with α-mobility on electrophoresis. Purified RBP displays charge microheterogeneity on polyacrylamide gel electrophoresis (Raz *et al.*, 1970; Peterson, 1971a) or on isoelectric focusing (Peterson and Berggård, 1971). By these procedures, three or four different bands of immunologically identical RBP have been observed. Two of the bands are usually fluorescent under ultraviolet light, representing different forms of holo-RBP. Apo-RBP has a slightly faster electrophoretic mobility than does holo-RBP (Raz *et al.*, 1970; Kopelman *et al.*, 1976a; Fex and Hansson, 1979), and it has been suggested that the four bands represent two bands of holo-RBP and two of apo-RBP (Kopelman *et al.*, 1976a). There is no evidence for genetic polymorphism of RBP, and it has been observed that the microheterogeneity is more pronounced in RBP isolated from less-fresh samples of plasma or after extensive handling during purification (Raz *et al.*, 1970). It has been suggested that the microheterogeneity is due to loss of amide groups (and of some of the ligand retinol) during serum storage and/or fractionation (Raz *et al.*, 1970).

The ultraviolet absorption spectrum of holo-RBP in solution exhibits two peaks, with maxima at approximately 280 nm (due to the protein moiety) and at 330 nm (due to the protein-bound retinol). These two peaks of absorbance are of approximately the same magnitude (A_{330}: A_{280} ratio of about 1.0) for the retinol–RBP complex (holo-RBP). Holo-RBP has relatively high absorbance at 280 nm, reflecting the relatively high content of tryptophan and tyrosine residues in the protein. The $E^{1\%}_{1\text{cm}}$ for RBP (mainly holo-RBP) was estimated as 19.4 (Kanai *et al.*, 1968) and as 18.6 (Peterson, 1971a) (calculated from the reported molar extinction coefficient at 280 nm of 39,100). The molar extinction and the absorption peak position of retinol when bound to RBP and in aqueous solution are almost the same as for retinol itself in solution in benzene or hexane (Goodman and Raz, 1972). A molecule of RBP possesses one binding site for one molecule of retinol.

Solutions of the retinol–RBP complex are highly fluorescent. The fluorescence properties of human RBP have been studied in detail (Peterson and Rask, 1971; Goodman and Leslie, 1972), and information is also available about the fluorescence spectral properties of RBP isolated from several other species. Retinol bound to RBP has uncorrected peak excitation and emission maxima of 334 and 463 nm, respectively. The relative fluorescence intensity of retinol bound to RBP (in aqueous solution) is an order of magnitude greater than that of retinol in solution in any one of a variety of organic solvents (Goodman and Leslie, 1972; Futterman *et al.*, 1975). In addition, for retinol bound to RBP, the emission peak due to retinol fluorescence shows a blue shift of about 15–20 nm compared with retinol itself in solution in organic solvents. These data are consistent with the interpretation that retinol is tightly bound to RBP in a fixed position with restricted mobility. Excitation of holo-RBP at 280 nm yields a fluorescence emission spectrum with two peaks, one at 335 nm caused by the fluorescence of the protein, and one at approximately 465 nm representing the retinol bound to RBP. This emission spectrum indicates that there is efficient transfer of energy within the holoprotein, probably mainly from excited tryptophanyl residues to the bound retinol. The efficiency of transfer of excitation energy was estimated at 60% (Peterson and Rask, 1971) for human holo-RBP. In another study, the efficiency of energy transfer was estimated as 42% for rabbit RBP, 41% for the chicken RBP, and 24% for piscine RBP (Shidoji and Muto, 1977). A direct study quantitatively comparing energy-transfer efficiency of human RBP with RBP from other species would be of value.

Studies employing circular dichroism and optical rotary dispersion have shown that RBP appears to have a relatively high content of unordered conformation, a significant but small complement of β conformation, and little or no α-helix (Rask *et al.*, 1972; Gotto *et al.*, 1972). The crystallization of RBP as long needles has been reported (Haupt and Heide, 1972).

IV. THE CHEMICAL STRUCTURE OF RBP

The RBP molecule is a single polypeptide chain of about 180–185 amino acid residues containing three intramolecular disulfide bonds (as cystine residues) (Rask *et al.*, 1981). No fatty acid or fatty acyl chains were detected in purified RBP (Kanai *et al.*, 1968). Analyses for carbohydrate showed that RBP was free of neutral sugar and hexosamine (Peterson, 1971) and had a maximum of 0.9 mol of hexose per mol of protein (Raz *et al.*, 1970). It is likely that RBP contains no bound lipid (other than retinol) and no carbohydrate.

The complete primary structure of human RBP has been reported (Rask *et al.*, 1979, 1981). RBP is cleaved by cyanogen bromide into five fragments; of these, the carboxy-terminal fragment represents slightly more than half the molecule. In a study of the partial amino acid sequence of human RBP, the five cyanogen bromide fragments were isolated and aligned, the amino acid sequences of four of the fragments were determined, and the sequence of almost two-thirds (the amino-terminal portion) of the RBP molecule was reported (Kanda and Goodman, 1979). Slight discrepancies exist between this partial and the reported complete sequence proposals, although in general the agreement between them is quite good. The reported complete sequence, shown in Fig. 1, consists of 182 amino acids, with the carboxy-terminal sequence: -Gly-Arg-Ser-Glu-Arg-Asn-Leu-COOH. This proposed carboxy-terminal sequence differs from those pre-

A. Human RBP

```
    1                                  10                                                20
H2N-Glu-Arg-Asp-Cys-Arg-Val-Ser-Ser-Phe-Arg-Val-Lys-Glu-Asn-Phe-Asp-Lys-Ala-Arg-Phe-Ser-Gly-Thr-Trp-Tyr-

                30                                      40                                      50
   -Ala-Met-Ala-Lys-Lys-Asp-Pro-Glu-Gly-Leu-Phe-Leu-Gln-Asp-Asn-Ile-Val-Ala-Glu-Phe-Ser-Val-Asp-Glu-Thr-

                                        60                                      70
   -Gly-Gln-Met-Ser-Ala-Thr-Ala-Lys-Gly-Arg-Val-Arg-Leu-Leu-Asn-Asn-Trp-Asp-Val-Cys-Ala-Asp-Met-Val-Gly-

                    80                                      90                                      100
   -Thr-Phe-Thr-Asp-Thr-Glu-Asp-Pro-Ala-Lys-Phe-Lys-Met-Lys-Tyr-Trp-Gly-Val-Ala-Ser-Phe-Leu-Gln-Lys-Gly-

                                        110                                     120
   -Asn-Asp-Asp-His-Trp-Ile-Val-Asp-Thr-Asp-Tyr-Asp-Thr-Tyr-Ala-Val-Gln-Tyr-Ser-Cys-Arg-Leu-Leu-Asn-Leu-

                    130                                     140                                     150
   -Asp-Gly-Thr-Cys-Ala-Asp-Ser-Tyr-Ser-Phe-Val-Phe-Ser-Arg-Asp-Pro-Asn-Gly-Leu-Pro-Pro-Gln-Ala-Gln-Lys-

                                        160                                     170
   -Ile-Val-Arg-Gln-Arg-Gln-Glu-Glu-Leu-Cys-Leu-Ala-Arg-Gln-Tyr-Arg-Leu-Ile-Val-His-Asn-Gly-Tyr-Cys-Asp-

                    180
   -Gly-Arg-Ser-Glu-Arg-Asn-Leu-COOH
```

B. Human TTR Subunit

```
    1                                  10                                                20
NH2-Gly-Pro-Thr-Gly-Thr-Gly-Glu-Ser-Lys-Cys-Pro-Leu-Met-Val-Lys-Val-Leu-Asp-Ala-Val-Arg-Gly-Ser-Pro-Ala-

            30                                              40                                      50
   -Ile-Asn-Val-Ala-Val-His-Val-Phe-Arg-Lys-Ala-Ala-Asp-Asp-Thr-Trp-Glu-Pro-Phe-Ala-Ser-Gly-Lys-Thr-Ser-

                                        60                                      70
   -Glu-Ser-Gly-Glu-Leu-His-Gly-Leu-Thr-Thr-Glx-Glx-Gln-Phe-Val-Glu-Gly-Ile-Tyr-Lys-Val-Glu-Ile-Asp-Thr-

                    80                                      90                                      100
   -Lys-Ser-Tyr-Trp-Lys-Ala-Leu-Gly-Ile-Ser-Pro-Phe-His-Glu-His-Ala-Glu-Val-Val-Phe-Thr-Ala-Asn-Asp-Ser-

                                        110                                     120
   -Gly-Pro-Arg-Arg-Tyr-Thr-Ile-Ala-Ala-Leu-Leu-Ser-Pro-Tyr-Ser-Tyr-Ser-Thr-Thr-Ala-Val-Val-Thr-Asn-Pro-

   -Lys-Glu-COOH
```

Fig. 1. (A) Primary structure of human RBP (Rask *et al.*, 1979) and (B) of the human TTR subunit. (Kanda *et al.*, 1974.)

viously reported from the same (Rask *et al.*, 1971) and other (White *et al.*, 1972; Fex and Hansson, 1979) laboratories.

Confirmation of the proposed carboxy-terminal sequence was very recently obtained from the nucleotide sequence of a complementary DNA (cDNA) segment coding for the carboxy-terminal portion of RBP (Costanzo *et al.*, 1983). A human liver cDNA library was constructed and a large number of individual clones were screened directly by cDNA sequence analysis. One of the 236 sequences screened coded for the region of human RBP from amino acid residue 159 to 182.

Unlike the case for TTR (see below), detailed information is not available about the three-dimensional structure of RBP. Very recently, the crystallization and preliminary X-ray data of human plasma RBP were reported (Ottonello *et al.*, 1983). The crystals diffracted to a resolution of 2.0 Å.

Information has been sought concerning the possible existence of sequence homologies between the RBP amino acid sequence and the amino acid sequences of other proteins of known primary structure. The initial computer searches carried out concerning this question failed to reveal homologies in sequence between RBP and any other known sequenced protein (Kanda and Goodman, 1979; Rask *et al.*, 1981). Significant, albeit limited, sequence homologies have been found between human RBP, human α_1-microglobulin, bovine β-lactoglobulin, and rat α_{2u}-globulin (P. Feigelson, personal communication). The significance of this observation needs further evaluation, although it is possible that these four proteins may be evolutionarily related as part of the same protein super-family. It is of interest that β-lactoglobulin is known to bind and to form a structured complex with retinol (Hemley *et al.*, 1979). The other two proteins (other than RBP) do not appear to bind retinol. Sequence similarities between limited portions of the RBP and TTR structures have also been reported (Jörnvall *et al.*, 1981).

It has been reported that different portions of the RBP sequence display statistically significant homology with each other (Rask *et al.*, 1981; Jörnvall *et al.*, 1981), suggesting that mammalian RBP may have arisen by internal duplication of a primordial gene (for a smaller protein). It was also reported by Rask *et al.* (1981) that the amino acid sequence of RBP is not homologous with that of the intracellular retinol-binding protein (CRBP, M_r 14,600, isolated from rat liver). However, Crabb and Saari (1981) reported that there does appear to be some structural similarity between the amino-terminal sequences of human RBP and those of CRBP and CRABP (cellular retinoic acid-binding protein) isolated from bovine retina. More information is needed about the primary structures of the intracellular retinoid-binding proteins and about the possibility of some limited relationship between these structures and those of serum RBP. It has also been pointed out (Rask *et al.*, 1981) that since piscine serum RBP has a molecular weight of about 16,000 (Shidoji and Muto, 1977), the possibility should be considered that the gene for serum RBP underwent a partial duplication after the divergence of fish and terrestrial vertebrates. Potential interspecies and evolutionary relationships are discussed further below.

V. THE STRUCTURE OF TRANSTHYRETIN (PREALBUMIN)

Transthyretin (TTR) is one of the most completely characterized human proteins. The TTR molecule is a stable and symmetrical tetramer composed of four identical subunits with a molecular weight of 54,980 (Kanda *et al.*, 1974). The complete amino acid sequence of the human TTR subunit is shown in Fig. 1. The full three-dimensional structure of the human TTR molecule has been elucidated by X-ray crystallographic studies at 6-Å (Blake *et al.*, 1971), at 2.5-Å (Blake *et al.*, 1974), and at 1.8-Å (Blake *et al.*, 1978) resolution. These studies have shown that the subunits have extensive β-sheet structure, are arranged tetrahedrally, and are linked into stable dimers, each comprising two of the four subunits. A channel runs through the center of the TTR molecule in which are located two symmetry-related binding sites for iodothyronine molecules. The high-affinity binding of only one molecule of thyroxine to a molecule of TTR can, however, be observed. Thus, the apparent association constants for the two binding sites on TTR for thyroxine have been estimated as $k_{a1} = 1 \times 10^8$ and $k_{a2} = 9.5 \times 10^5$ (Ferguson *et al.*, 1975). Binding of thyroxine to TTR apparently represents an example of negative cooperativity that is due to ligand–ligand interactions (Ferguson *et al.*, 1975; Blake *et al.*, 1974).

In the human, TTR is one of three plasma proteins involved in the transport of the thyroid hormones in blood (Ingbar, 1963; Oppenheimer, 1968). The principal transport protein for thyroid hormones in human plasma is thyroxine-binding globulin, and TTR plays only a lesser role in the normal transport and metabolism of thyroid hormones in man (Woeber and Ingbar, 1968). In the rat, however, TTR appears to be the major thyroid hormone transport protein (Davis *et al.*, 1970; Sutherland and Brandon, 1976).

TTR isolated from serum of other mammalian and avian species greatly resembles human TTR (see below). Rhesus monkey TTR displays genetic polymorphism, which is not present in the human. The partial amino-terminal sequences of the TTR subunit from the rhesus monkey (van Jaarsveld *et al.*, 1973b) and from the rat (Navab *et al.*, 1977a) are extremely similar to that of human TTR.

High-resolution X-ray crystallographic studies have shown that the TTR molecule contains two surface sites with structural complementarity to double-helical DNA (Blake and Oatley, 1977). Although the binding of TTR to DNA has not been reported, it has been suggested that the TTR molecule may serve as a model for the kind of structure that may be involved in hormone receptors with nuclear effects on DNA transcription (Blake and Oatley, 1977; Blake, 1981).

VI. INTERACTION OF RBP WITH TTR

The interaction of RBP with TTR is very sensitive to ionic strength, with dissociation of the protein–protein complex occurring at low ionic strength (Peterson, 1971b; van Jaarsveld *et al.*, 1973a). At ionic strength of 0.002 *M* or below, there is virtually no complex formation between the two proteins. The interaction of RBP with TTR is also strongly pH-dependent. The maximum binding occurs near physiological pH and falls gradually at lower and higher pH values (van Jaarsveld *et al.*, 1973a). It was observed that the protein–protein affinity fell rapidly above pH 9 and below pH 5, so that there was no binding present at pH 11.4 and only an extremely small amount at pH 3.5.

Studies employing polarization of fluorescence (van Jaarsveld *et al.*, 1973a) and equilibrium dialysis (Raz and Goodman, 1969) have shown that there is no

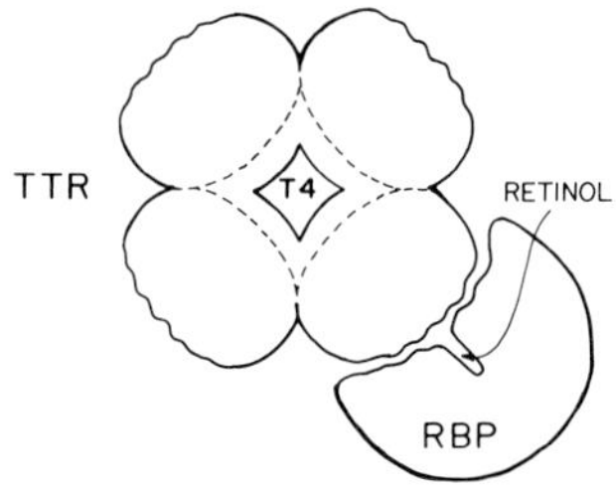

Fig. 2. Cartoon model of the RBP–TTR complex. TTR is shown as a tetramer containing a central channel that is the site of thyroxine (T4) binding. The locations of the different binding sites suggest the known relations between the several interactions: RBP and T4 bind at independent sites on TTR, whereas the binding of retinol to RBP is stabilized by formation of the protein–protein complex. (Reproduced with permission from Smith and Goodman, 1979.)

interdependence of the binding of thyroxine and of RBP to TTR. In contrast, the interaction of retinol with RBP appears to be stabilized by the formation of the RBP–TTR complex (Peterson, 1971b; Goodman and Raz, 1972; Goodman and Leslie, 1972). Figure 2 shows our diagram for the RBP–TTR complex.

Circular dichroic (CD) spectral studies have been conducted to examine the effects of the RBP–TTR interaction on the secondary structures of the two proteins. The CD spectra of mixtures of RBP and TTR in the 200- and 240-nm region were additive, suggesting that formation of the RBP–TTR complex results in very little if any alteration in the secondary structures of the two proteins (Rask *et al.*, 1972; Gotto *et al.*, 1972). In another study (Heller and Horwitz, 1973), it was reported that in the 240- to 300-nm region the CD spectra are not additive upon binding of holo-RBP to TTR, suggesting that some conformational changes in one or both proteins may occur on formation of the protein–protein complex. Further work is needed fully to resolve this question.

The effects of a number of other perturbations on the interaction of RBP with TTR (and on the interactions of these proteins with their ligands retinol and thyroxine, respectively) have been explored. The other manipulations that have been examined include effects of changes in temperature and urea concentration and the effects of reduction and alkylation of disulfide bonds and of iodination (Raz *et al.*, 1970); the effects of modifications of tyrosine, lysine, and tryptophan residues were also examined (Heller and Horwitz, 1975; Horwitz and Heller, 1974b). Addition of 6 *M* urea completely disrupted the RBP–TTR complex, markedly reduced the affinity of TTR for thyroxine, but did not interfere with the association of RBP with retinol. Iodination of isolated RBP decreased its affinity for TTR (Raz *et al.*, 1970; Vahlquist, 1972; Vahlquist *et al.*, 1973). However, it was found (Vahlquist, 1972; Vahlquist *et al.*, 1973) that iodination of the RBP–TTR complex, followed by the dissociation of the complex and separate isolation of the two proteins, yielded iodinated RBP with full affinity for

TTR. These observations suggest that one or more tyrosine residues of RBP may be present in the protein–protein binding domain.

A number of studies have investigated the stoichiometry of the interaction of RBP with TTR and have attempted to define quantitatively the affinity of TTR for RBP. Although RBP normally circulates in plasma as a 1:1 molar complex with TTR, studies employing polarization of retinol fluorescence and velocity ultracentrifugation suggested that human TTR may contain four binding sites for human RBP (van Jaarsveld *et al.*, 1973a). However, a variety of studies in other laboratories have found human TTR to possess only 1–1.35 (Heller and Horwitz, 1974a), 1.6–2.0 (Kopelman *et al.*, 1976b), or approximately 2 (Trägårdh *et al.*, 1980) binding sites for human RBP. Thus, it is clear that TTR possesses multiple binding sites for RBP, although the number of binding sites that can be demonstrated has not been the same in different laboratories and with different techniques.

This apparent discrepancy has been largely resolved by fluorescence polarization studies of the interaction between RBPs and TTRs of human and chicken (Kopelman *et al.*, 1976b). The binding affinity between chicken plasma RBP and chicken TTR was essentially the same as that between the respective human proteins. The RBPs and TTRs of human and chicken were found to cross-interact, displaying an affinity similar to that displayed by the proteins of the same species. Data analysis suggested that human TTR displayed approximately two binding sites for RBP and that chicken TTR displayed approximately four such sites. It was suggested that TTR possesses four identical binding sites for RBP, but that the binding of RBP to these sites is of a negative cooperative nature; this negative cooperativity is much more pronounced with the human than with the chicken proteins. A similar conclusion (multiple binding sites with negative cooperativity) was reached by Trägårdh *et al.* (1980). Moreover, it has been reported that the isolated TTR subunit, covalently attached to Sepharose, retained some affinity for RBP (Nilsson *et al.*, 1975).

Taken together, these various studies strongly suggest that TTR contains four binding sites for human RBP. It seems reasonable to assume that each TTR subunit might contain one binding site for RBP, although additional direct evidence is needed to support this conclusion. In general, more information is needed about the structures and the characteristics of the binding sites on RBP and TTR that are involved in the interaction between these two proteins.

Estimates of the apparent association constant of human TTR for human holo-RBP have varied from $1.2 \times 10^6\ M^{-1}$ (van Jaarsveld *et al.*, 1973a) to approximately $8 \times 10^6\ M^{-1}$ (Nilsson *et al.*, 1975; Kopelman *et al.*, 1976b; Trägårdh *et al.*, 1980) to 1.3×10^7 (Fex *et al.*, 1979). A number of studies using different methodologies have demonstrated that retinol-free apo-RBP also binds well to TTR (Raz *et al.*, 1970; Peterson and Rask, 1971; van Jaarsveld *et al.*, 1973a; Fex *et al.*, 1979), although one study (Heller and Horowitz, 1973) reported a lack of affinity between apo-RBP and TTR. Despite this one report, it seems clear that the presence of retinol is not essential for the interaction of RBP and TTR. Thus, in addition to the reports cited above, it was shown on gel filtration of serum from retinol-deficient rats that most of the apo-RBP in such serum is present as the RBP–TTR complex (Muto *et al.*, 1972). There is some uncertainty regarding the relative affinity of apo-RBP, as compared to holo-RBP, for TTR. Some reports have suggested that there is no important quantitative difference between the affinity of apo- and holo-RBP for TTR (Peterson and Rask, 1971; van Jaarsveld *et al.*, 1973a). The possibility that the affinity of apo-RBP for TTR is reduced as compared to that of holo-RBP was, however, suggested by Raz *et al.* (1970). Affinity chromatography studies in the author's laboratory using TTR coupled to Sepharose have also indicated that the affinity of apo-RBP for TTR is less than that of holo-RBP (Goodman, 1980). In one quantitative study, it was estimated that the apparent association constant (for binding to TTR) for holo-RBP was 4.4 times that for apo-RBP (Fex *et al.*, 1979). More precise information is needed on this point since (as discussed later) the putative difference in affinity for TTR between holo- and apo-RBP is likely to be of physiological significance with regard to the overall metabolism *in vivo* of RBP.

VII. INTERACTION OF RBP WITH RETINOL AND OTHER RETINOIDS

Methods are available for the removal of retinol from RBP in order to form the retinol-free apoprotein and for the recombination of apo-RBP with retinol to form again the holoprotein. Using these methods, a large number of studies have explored the binding of a variety of retinoids and related compounds to apo-RBP (Goodman, and Raz, 1972; Heller and Horwitz, 1973, 1974b; Horwitz and Heller, 1973, 1974a; Cogan *et al.*, 1976; Hase *et al.*, 1976). These studies have shown that the structural features required for the binding of all-*trans*-retinol to RBP appear to be moderately, but far from absolutely, specific. It has also been shown in circular dichroic studies that compared the CD spectra of apo- and holo-RBP (Rask *et al.*, 1972; Gotto *et al.*, 1972), that the removal of retinol from RBP does not lead to any major alteration of the secondary structure of the protein.

Information is available about the general importance, for binding to RBP, of different portions of the retinol molecule, including the functional end group, the polyene chain, and the cyclohexene ring. Changing the functional end group of retinol from a hydroxyl group to a carboxylic acid does not significantly affect the binding of the retinoid to RBP (Goodman and Raz, 1972; Cogan *et al.*, 1976). Thus, the apparent dissociation constant for the interaction of human RBP with retinol (A1) and with retinoic acid (D1) was estimated, respectively, as 1.9

$\times 10^{-7}$ *M* and as 2.1 $\times 10^{-7}$ *M* (Cogan *et al.*, 1976). Retinaldehyde (C1) and retinyl acetate (A3) also bind to RBP, although some investigators have reported less (Goodman and Raz, 1972; Hase *et al.*, 1976) whereas others have reported similar binding (Horwitz and Heller, 1973; Cogan *et al.*, 1976) of these compounds to RBP, as compared to the binding of retinol. The discrepancy between these results may be due to the various experimental methods employed and the different criteria used to assay for binding activity. The long-chain retinyl ester retinyl palmitate (A4) showed virtually no affinity for RBP (Hase *et al.*, 1976; Cogan *et al.*, 1976); retinoic acid esters also do not appear to form a complex with RBP (Hase *et al.*, 1976). From these studies it can be concluded that the functional groups affecting the binding of these retinoids can be listed tentatively in order of decreasing ligand affinity: retinol and retinoic acid have the greatest affinity for RBP, followed by retinaldehyde and retinyl acetate, with long-chain retinyl esters and retinoic acid esters being the least likely to bind.

The 13-cis and 11,13-di-cis isomers of retinol bind to RBP somewhat less well than did the all-trans isomer (Goodman and Raz, 1972). In studying the binding of retinaldehyde to RBP, it was observed that the 13-cis, 11-cis, and 9-cis isomers bound virtually as well as did the all-trans isomer (Horwitz and Heller, 1973). It hence appears that the retinol binding site is not very sensitive to configurational changes in the side chain of the retinyl ligand. Both 15-methyl- and 15-dimethylretinol (A10) have been reported to bind to RBP *in vivo* in rats (Tosukhowong and Olson, 1978). In addition, retinyl derivatives with shortened side chains, such as β-ionone and β-ionylideneacetic acid (Hase *et al.*, 1976), have also been shown to bind well to RBP.

Attempts to form an RBP complex with compounds such as phytol and a number of other terpenoids without the cyclohexene ring structure (Goodman and Raz, 1972; Hase *et al.*, 1976) and with cholesterol were not successful. These results suggest that a structural similarity to retinol in the area of the cyclohexene ring may be necessary for binding to RBP.

These studies with retinoids have provided some information about the structural requirements of the retinol binding site on RBP. No information is, however, available about the amino acid residues in RBP that are involved in the binding site. Acetylation of lysine residues of RBP did not affect its binding of retinol (Heller and Horwitz, 1975). Modification of one of eight tyrosine residues and two of four tryptophan residues of RBP also had no effect on the retinol–RBP interaction (Heller and Horwitz, 1975; Horwitz and Heller, 1974b). The binding site was, however, disrupted by reduction and alkylation of disulfide bonds (Raz *et al.*, 1970).

The synthesis of a retinoid affinity label for the retinol binding site of RBP has been reported (Gawinowicz and Goodman, 1982). Three radioactive retinoid bromoacetates were synthesized as potential retinoid affinity labels. The ^{3}H-labeled compounds synthesized were the bromoacetate derivatives of retinol (retinyl bromoacetate) and of two retinol analogs with shorter side chains [β-ionylideneethyl bromoacetate (or IEBA) (A15) and β-ionyl bromoacetate]. When excess ligand was incubated with RBP, both of the shorter-chain retinoids formed nearly 1:1 molar complexes with RBP, whereas retinyl bromoacetate bound only approximately one-third as well. Further studies showed that of the three compounds, only IEBA formed a covalent link with the protein. Covalent bonding appeared to occur slowly, over the course of several hours, following the initial formation of a reversible (noncovalent) IEBA–RBP complex. It was pointed out that the retinoid affinity label could be used in future studies to explore the structure of the retinol-binding site of RBP.

VIII. COMPARATIVE BIOCHEMISTRY OF RBP; PHYLOGENETIC CONSIDERATIONS

RBP has been isolated and partly characterized from serum (or urine) of seven mammalian species in addition to man: the cynomologus monkey (Vahlquist and Peterson, 1972), the rat (Muto and Goodman, 1972; Peterson *et al.*, 1973), the cow (Heller, 1975b; Fex and Lindgren, 1977), the ox (Glover, 1973), the pig (Rask, 1974), the rabbit (Muto *et al.*, 1976), and the dog (Poulik *et al.*, 1975). The characteristics of RBP in all these species and of RBP isolated from chicken plasma (Mokady and Tal, 1974; Abe *et al.*, 1975) are, in general, similar to those of human RBP. The RBP from all these species has been found to be a small protein of approximately 20,000 daltons that binds one molecule of retinol per molecule of RBP. In spite of occasional reports to the contrary (Heller, 1975b), the RBP of all these species appears to circulate in plasma as a protein–protein complex together with a larger protein. In the cases where the larger protein has been isolated, it has been found to have a tetrameric structure and generally to resemble human TTR. In most instances, this second protein (like human TTR) has electrophoretic mobility somewhat greater than that of serum albumin. This is not always the case, however, as illustrated by the bovine counterpart to human TTR which has α_1-mobility on agarose gel electrophoresis at pH 8.6 (Fex and Lindgren, 1977).

The amino acid compositions reported for RBP preparations purified from these different species are generally quite similar to each other (see Table I). More than 90% of the amino acid sequence of rabbit RBP has been reported (Rask *et al.*, 1981) in preliminary form. Of the 170 positions available for comparison with the sequence of human RBP, 160 were found to be identical. A comparison of the amino-terminal sequences of human and dog RBP, isolated from urine, has also been reported (Poulik *et al.*, 1975). The two sequences were demonstrably different at only 5 of the first 49 amino-terminal positions. The amino-terminal sequence of 29 residues of bovine serum RBP has been reported

TABLE I

The Amino Acid Composition of Purified RBP as Reported for Various Species[a]

	Human	Monkey	Rat	Cow	Ox	Pig	Rabbit	Dog	Chicken	Fish
Lys	10	10	12	11	10–11	13	10	11	15	8
His	2	2	5	3	3–4	5	4	3	3	4
Arg	14	14	12	12	10	11	14	12	10	11
Asp	27	26	22	27	24	28	30	27	31	21–22
Thr	9	9	10	9	9	7	9	9	9	9
Ser	11	11	12	11	12	15	10	11	12	8
Glu	18	19	20	19	20	18	15	18	19	15
Pro	5	6	7	4	8	10	7	7	4	5
Gly	11	11	10	12	14	9	10	12	14	10–11
Ala	13	14	14	14	15	11	12	14	13	10–11
Half-Cys	6	5	5	4	5	5	6	7	3	4
Val	13	12	9	12	10	13	13	12	14	6
Met	4	3	4	4	3	3	4	4	3	4
Ile	4	5	6	5	6	5	3	4	3	5
Leu	13	13	11	13	12–13	14	12	13	10	13
Tyr	8	8	6	7	8	5	7	8	7	7
Phe	10	10	8	12	11	6	12	10	10	4
Trp	4	5	4	4	3	ND[b]	4	6	6	4
Total	182	183	177	183	183–186	—	182	188	186	148–151

[a] The values listed are those reported by Rask *et al*. (1979) for human; Vahlquist and Peterson (1972) for monkey; Peterson *et al*. (1973) for rat; Heller (1975b) for cow; Glover (1973) for ox; Rask (1974) for pig; Muto *et al*. (1976) for rabbit; Poulik *et al*. (1975) for dog; Abe *et al*. (1975) for chicken; and Shidoji and Muto (1977) for fish RBP.

[b] ND = Not determined.

to be identical to that of rabbit RBP and, except for an Ala/Ser interchange at position 21, also to that of human RBP (Crabb and Saari, 1981).

There is evidence that the structures that characterize the binding sites on RBP and TTR are very similar across a range of mammalian and avian species. As discussed above, RBPs and TTRs of human and chicken cross-interact and show interspecies affinities similar to those displayed by the proteins of the same species (Kopelman *et al.*, 1976b). In addition, the affinities of human and of chicken RBP were found to be similar for both retinol and retinoic acid (Cogan *et al.*, 1976). Rat RBP has been isolated by affinity chromatography on human TTR coupled to agarose (Poole *et al.*, 1975), and conversely, rat TTR has been isolated by affinity chromatography on human RBP coupled to agarose (Navab *et al.*, 1977a). These studies indicate that the binding sites on TTR for RBP, and on RBP for TTR, are structurally similar in the human, rat, and chicken proteins. It appears that the structural features that characterize the binding sites and that are essential for the biological roles of RBP and TTR, were maintained during much of vertebrate evolution.

A study was conducted by Shidoji and Muto (1977) to explore vitamin A transport and RBP in the nonmammalian vertebrates. Retinol-containing proteins in fresh plasma obtained from bullfrogs and a turtle exhibited similar properties to those found in mammalian and chicken plasma: molecular weight of about 60,000–80,000 (i.e., for a protein–protein complex) and binding affinity to TTR as shown by affinity chromatography on human TTR coupled to agarose. In contrast, retinol-containing proteins in plasma from frog larvae (tadpoles) and from three fish (carp, blue sharks, and young yellowtails) appeared to be present as monomeric RBP without any affinity to human TTR. It was also noted (Shidoji *et al.*, 1979) that exogenous thyroxine appeared to induce the production of RBP with affinity for TTR in tadpoles. Piscine RBP isolated from pooled plasma of young yellowtails (Shidoji and Muto, 1977) was smaller than mammalian RBP (M_r ~16,000, see Table I) and lacked binding affinity for human TTR. It was suggested that piscine RBP represents a prototype of the specific plasma vitamin A-transporting protein, which became modified later in evolution during phylogenetic development of the vertebrates, to acquire a binding site for TTR.

IX. RBP METABOLISM

A. General Summary

RBP is synthesized and secreted by the liver and circulates in plasma mainly as a protein–protein complex with TTR. Normally, RBP is secreted almost entirely as the holoprotein, containing a molecule of bound retinol. TTR is also produced

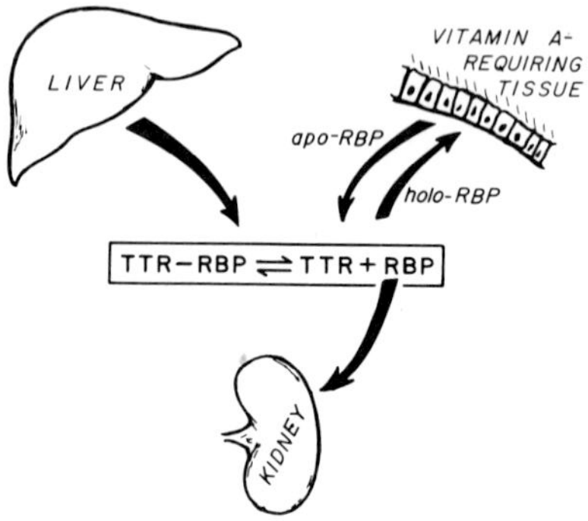

Fig. 3. Schematic summary of the metabolism of RBP. See text for description and discussion.

by the liver. Studies in the rat suggest that RBP and TTR are secreted independently by the liver and combine in serum to form the RBP–TTR complex. The plasma levels of both proteins are decreased in patients with parenchymal liver disease, reflecting the fact that both proteins are produced in the liver.

Retinol is delivered to target tissues by holo-RBP, possibly by the very small amount of holo-RBP normally present in the free uncomplexed state. Delivery of retinol to extrahepatic tissues may involve specific cell-surface receptors for RBP. The apo-RBP that results after delivery of retinol probably has a reduced affinity for TTR and is thus selectively enriched in the free RBP fraction.

Free uncomplexed RBP is small enough to be filtered by the renal glomeruli, whereas TTR and the RBP–TTR complex are not. Although very little RBP is normally present in the free state, the rates of glomerular filtration and renal metabolism of RBP are sufficiently large enough to constitute the major pathway of RBP catabolism. Patients with severe chronic renal disease show a reduced metabolic clearance rate and an elevated plasma concentration of RBP, with normal levels of TTR.

Figure 3 provides a schematic summary of the metabolism of RBP.

B. Assay; Normal Levels

Effective immunoassays have been used to measure RBP levels in plasma in clinical human studies and in plasma and tissue homogenates in studies in experimental animals. The most widely used methods have been the radioimmunoassay (Smith *et al.*, 1970; Muto *et al.*, 1972) and the radial immunodiffusion assay (e.g., Peterson, 1971c; Vahlquist *et al.*, 1978a). The plasma levels of TTR have been measured by similar immunoassays (e.g., Smith and Goodman, 1971; Navab *et al.*, 1977b). Other methods that have been used to assay RBP levels include a latex immunoassay (Bernard *et al.*, 1982) and fluorimetric assays that measure the level of protein-bound retinol in the RBP region after disc gel electrophoresis (Glover *et al.*, 1974; Glover, 1980) or in diluted serum (Futterman *et al.*, 1975), and hence provide an estimate for the level of holo-RBP.

The normal mean plasma concentrations of RBP and TTR in well-nourished human adults are approximately 45 μg/ml and 250 μg/ml, respectively. The mean ± SEM values reported for 109 normal adults in New York were 46.2 ± 1.0 μg/ml for RBP and 250 ± 5 μg/ml for TTR (Smith and Goodman, 1971). Similar values for RBP and similar or slightly higher values for TTR have been observed in Sweden (Peterson, 1971c; Vahlquist *et al.*, 1978a), in Japan (Kanai *et al.*, 1972a), and in other countries as well.

The total plasma concentrations of vitamin A and RBP are highly correlated with each other (Smith *et al.*, 1970; Smith and Goodman, 1971; Vahlquist *et al.*, 1978a). The mean molar ratio of RBP to vitamin A for normal subjects was reported as 1.22 by Smith and Goodman (1971) (for 109 subjects) and as 1.1 by Peterson (1971c) (for 50 subjects). Mean molar ratios (RBP:vitamin A) of 1.14 were found in 71 patients with diseases of the liver, and of 1.16 in 21 patients with thyroid disease (Smith and Goodman, 1971). The results indicate that under normal (and under many abnormal) circumstances, RBP circulates almost exclusively as the holoprotein, with only approximately 10–15% of total plasma RBP being present in the form of the retinol-free apoprotein. Significant correlations between the plasma levels of TTR and of both RBP and vitamin A have also been found (Smith and Goodman, 1971; Vahlquist *et al.*, 1978a), although these correlations are weaker than those observed between RBP and vitamin A.

Plasma RBP and vitamin A levels are significantly lower in young children than in adults (Smith *et al.*, 1972; Vahlquist *et al.*, 1975). In children between the ages of 2 and 10 years, the plasma values of RBP were found to be approximately 60% of the adult levels (Vahlquist *et al.*, 1975). Plasma RBP levels rise during puberty and adolescence to reach adult values. Significantly lower plasma RBP levels were observed in premature infants, 36 weeks or less in gestation, than in term neonates (Shenai *et al.*, 1981).

An estimate can be made of the concentration of free uncomplexed RBP normally present in plasma in equilibrium with the circulating RBP–TTR complex. Using values for the total plasma concentration of RBP of 45 μg/ml (2.14 μ*M*) and of TTR of 250 μg/ml (4.56 μ*M*) and an estimate of 10^7 for the association constant for the protein–protein complex, it can be calculated that the free RBP concentration is approximately 0.082 μ*M* or 3.8% of the total plasma RBP. If, as discussed above, the affinity of apo-RBP for TTR is less than that of holo-RBP, free RBP will comprise somewhat more than 3.8% of total plasma RBP and will be relatively enriched in apo-RBP as compared to the total RBP present in plasma.

C. RBP Biosynthesis: Role of the Liver

RBP is biosynthesized in and secreted (into the plasma) by the liver. In patients with chronic liver disease, plasma RBP levels are decreased, and the extent of decrease from normal is correlated with the extent of hepatic parenchymal dysfunction (Smith and Goodman, 1971). In the rat, fairly high immunoreactive RBP levels can be demonstrated in liver homogenates and homogenate fractions; when RBP secretion from the liver is blocked or is stimulated acutely, reciprocal changes occur in the RBP levels in rat liver and serum (see discussion below). Immunoreactive RBP has been localized in rat liver by immunofluorescence microscopy (Poole *et al.*, 1975); with the techniques used, specific localization of RBP was observed within liver parenchymal cells. More extensive immunohistochemical studies with both normal and retinol-deficient rat liver have confirmed these observations (M. Kato and D. S. Goodman, unpublished observations).

The synthesis and secretion of RBP has been observed and studied with isolated liver cells in culture *in vitro*. Detailed studies have been carried out with two differentiated rat hepatoma cell lines (Smith *et al.*, 1978). The production of RBP by primary rat hepatocytes in culture *in vitro* has also been observed (unpublished studies from the author's laboratory). Finally, the specific *in vitro* synthesis of rat RBP using isolated messenger RNA from rat liver and the rabbit reticulocyte *in vitro* protein-synthesizing system has been demonstrated (Soprano *et al.*, 1981). Taken together, these various studies clearly establish the liver, and specifically the hepatic parenchymal cell, as the locus of RBP synthesis. The molecular and cellular mechanisms that are involved in the processes of RBP synthesis and secretion by the liver and the factors that regulate these processes are discussed in detail below. While it is clear that the liver is the major site of RBP synthesis, there is no evidence as to whether or not the liver is the only site of RBP synthesis in the body.

D. Tissue Distribution

The tissue distribution and levels of RBP in normal and in retinol-deficient rats were measured in order to explore the role of different tissues in the metabolism of RBP (J. E. Smith *et al.*, 1975). The tissues examined included liver, kidney, fat, muscle, brain, eye, salivary gland, thymus, lung, heart, intestine, spleen, adrenal, testes, thyroid, and red blood cells. The RBP levels were low or very low in tissues other than liver, kidney, and serum and varied from 12 μg/g of tissue for normal spleen to an undectable level in red blood cells. Much of the RBP in the tissues with low levels was most likely due to residual serum in the samples. In general, except for liver, RBP levels were lower in tissues from retinol-deficient rats than in those from normal rats. In normal rats, the liver, kidney, and serum levels were 30 ± 4 (mean ± SEM), 151 ± 22, and 44 ± 3 μg/g, respectively. In retinol-deficient rats, the liver RBP level was about three times the normal level whereas the kidney and serum levels were about one-fifth the normal values. It was suggested that the levels of RBP in normal as compared to deficient liver, serum, and kidney appear to reflect the relative rates of RBP secretion and turnover (see later discussion).

The tissue distribution and levels of TTR have also been determined in normal and in retinol-deficient rats (Navab *et al.*, 1977b). Except for serum, thyroid glands from both control and deficient rats had the highest concentrations (approximately 82 μg/g wet wt) of TTR. Lungs, liver, kidneys, and heart contained intermediate levels of TTR, and other tissues had lower levels, with red blood cells devoid of any detectable TTR. In contrast to the case with RBP, no significant differences in TTR levels were observed between the tissues from the control and the deficient rats, except for the intestine and testes. In view of the very low levels of RBP in the thyroid glands (< 3 μg/g), and the evidence that TTR is the primary thyroxine-binding plasma protein in the rat (Davis *et al.*, 1970; Sutherland and Brandon, 1976), it was suggested that the high levels of TTR in the thyroid may be related to the role of TTR in the plasma transport of thyroid hormones in this animal.

E. RBP Catabolism: Role of the Kidneys

The kidneys play an important role in the normal catabolism of RBP. In patients with chronic renal disease, the levels of both RBP and vitamin A were found to be greatly elevated, although the TTR levels remained normal (Smith and Goodman, 1971). These findings demonstrate the substantial role of the kidneys in RBP metabolism. In the initial studies with RBP (Kanai *et al.*, 1968; Raz *et al.*, 1970), it was pointed out that in view of the relatively small size of RBP, free uncomplexed RBP would be filtered fairly readily by the renal glomeruli. It was suggested that the formation of the RBP–TTR complex served to protect RBP by preventing its glomerular filtration. Although the proportion of RBP in plasma present in the free state, not as a complex with TTR, is normally very small, it is apparently sufficient to permit a significant amount of RBP to be filtered by the glomeruli and metabolized by the kidneys each day. Patients with severe chronic renal disease have a reduced overall glomerular filtration rate, and hence a reduced metabolic clearance rate and an elevated plasma concentration of RBP.

Considerable information is available about the role of the kidney in the metabolism of low-molecular-weight serum proteins (see, e.g., Mogielnicki *et al.*, 1971). In normal states, small proteins that are filtered with relative ease through the glomeruli are taken up and catabolized in the tubular cells. In tubular proteinuria when the glomerular filtration rate is normal, protein filtration continues unchanged, but tubular uptake and catabolism decrease; more protein hence appears in the urine, but there is no change in the overall protein metabolic rate. As expected from these considerations, patients with impaired tubular function and tubular proteinuria excrete considerable amounts of low-molecular-weight plasma proteins (Peterson *et al.*, 1969). In the presence of renal disease with greatly decreased glomerular filtration rate, less protein passes into the tubular lumen, with a resultant decline in the rate of catabolism (and excretion) of the protein. These latter abnormalities can explain the high serum levels and prolonged survival of certain small proteins in uremia.

These considerations effectively describe the role of the kidney in the metabolism of free uncomplexed RBP. However, since neither TTR nor the RBP–TTR complex would normally be filtered appreciably (because of their larger size), the quantitative importance of the kidneys in the metabolism of RBP would depend to a major extent on the proportion of RBP present in plasma in the free uncomplexed state. If, as estimated above, free RBP normally represents approximately 4% of circulating RBP, and if the glomerular membrane were fully permeable to uncomplexed RBP, then glomerular filtration and renal catabolism would account for the metabolic clearance of 7–8 liters of plasma per day in a 70-kg human. This represents a large proportion of the normal daily turnover of RBP (see discussion below) and confirms the major quantitative role of the kidney in the overall metabolism of RBP.

RBP has been found in relatively large amounts in the urine from patients with tubular proteinuria (Peterson and Berggård, 1971). Much is known about urinary protein excretion of Japanese patients with chronic cadmium poisoning (Kanai *et al.*, 1972a,b). These patients manifest tubular proteinuria and the excretion of considerable amounts of low-molecular-weight proteins, including RBP. The urinary excretion of RBP has been induced in a rabbit (Muto *et al.*, 1976) and in a rhesus monkey (Nomiyama *et al.*, 1981) by chronic poisoning with cadmium. Studies of the role of the kidney in RBP metabolism have also been carried out in rats with various kinds of experimentally induced renal lesions (Peterson *et al.*, 1974). These various reported observations are all consistent with the above postulated role of the kidney in RBP metabolism.

Further information has been provided by direct tissue studies of the kidney in RBP (and vitamin A) metabolism. Kidneys were found to contain high levels of RBP by radioimmunoassay (J. E. Smith, 1975); most (76%) of the RBP in kidney homogenates was recovered in the soluble supernatant fraction. These data support the conclusion that kidney RBP represents RBP that had undergone glomerular filtration and tubular reabsorption. Moreover, it was found that kidney lymph (mainly the product of filtration and reabsorption) contained a significant level (12 μg/ml) of RBP (J. E. Smith *et al.*, 1975). Cell-fractionation studies of the distribution of vitamin A in rat kidneys showed that most of the vitamin A resided in cortex tubular cells (Peterson *et al.*, 1973). Finally, immunohistochemical studies with human kidney tissue have shown that RBP is localized mainly in proximal convoluted tubular cells (Katoh *et al.*, 1982). Similar studies with similar findings have also been carried out in the rat (M. Kato and D. S. Goodman, unpublished observations).

No information is available about the possible role of tissues other than the kidney in the catabolism of RBP. The tissue sites and mechanisms involved in the catabolism of TTR are not known.

F. RBP Turnover

Kinetic studies of the turnover and metabolism of RBP *in vivo* have been carried out in humans (Smith and Goodman, 1970; Vahlquist *et al.*, 1973), in cynomolgus monkeys (Vahlquist, 1972), and in rats (Peterson *et al.*, 1974). Valquist *et al.* (1973) employed RBP labeled with radioactive iodine that retained its full affinity for TTR; this labeled RBP was prepared by iodination of the RBP–TTR complex. Kinetic studies were conducted by injecting labeled RBP (and TTR) intravenously into normal persons and into three subjects with renal failure on maintenance hemodialysis. Biological half-lives for RBP of 11.1 and 11.7 h with synthetic rates of 190 mg/m^2/day body surface area were reported for two normal young men. Turnover studies were also carried out by Vahlquist *et al.* (1973) using a species of RBP that had been isolated from human urine and that lacked affinity for TTR. These studies provided information about the metabolism of plasma free RBP, not complexed to TTR. The biological half-life of this "free" RBP was very short in normal persons (about 4 h). In contrast, in patients with renal failure this parameter was increased 10- to 15-fold. These data are consistent with the conclusion that most (if not all) free RBP in plasma is catabolized by the kidney. For TTR, similar estimates of the biological half-life (2–3 days) and the total body turnover rate (synthetic rate) (approximately 250–300 mg/m^2/day were obtained in normal subjects and in patients with renal failure.

For cynomolgus monkeys (2.8–4.0 kg body weight), the biological half-times observed for RBP and TTR (using RBP with normal affinity for TTR) were 6.6 h and 22–23 h, respectively. Free RBP, without affinity for TTR, showed a biological half-time of only 1.9 h.

These data indicate that the turnover of RBP *in vivo* is quite rapid, with a fairly high body synthetic (production) rate for a protein of such low plasma concentration. This rapid turnover rate underlies the potential usefulness of RBP measurements in helping to assess the functional state of the liver in patients with hepatic disease, or the nutritional status of patients with borderline or actual malnutrition (see discussion below under clinical studies). It is of interest to compare the turnover of RBP with that of other plasma proteins. It has been pointed out (F. R. Smith *et al.*, 1975) that the respective half-lives (in days) and synthetic rates (in milligrams per kilogram per day) in normal adult human subjects are approximately 0.5 and 5 for RBP; 2–3 and 8–9 for TTR (Vahlquist *et al.*, 1973; Socolow *et al.*, 1965); and 14 and 200 for albumin (Beeken *et al.*, 1962).

The estimated total body turnover rate of RBP can be compared with available estimates for total body vitamin A utilization and turnover in normal well-nourished adults. A body turnover rate for RBP of about 5 mg kg^{-1} day^{-1} is equivalent to 16–17 μmol turnover per day for a 70-kg man. A comparable turnover for retinol would represent the daily turnover of 4.6–4.8 mg, or approximately 15–16 × 10^3 IU of vitamin A per day. This value is at least double the estimated average intake of vitamin A in well-nourished adults in Western countries (U. S. Dept. of Agriculture, 1980). These calculations suggest that the daily body turnover of RBP (in molar terms) may exceed that of vitamin A by a factor of as much as 2. Accordingly, it is extremely unlikely that RBP molecules are reutilized in the body, i.e., that a molecule of RBP that has delivered its retinol to a target cell returns to the liver to collect (and transport) another molecule of retinol. Instead, the data are consistent with the hypothesis (see also above) that an RBP molecule that has delivered its retinol subsequently undergoes catabolism (presumably in the kidney). Moreover, if we assume that well-nourished adults are in a steady state with regard to vitamin A nutriture, these estimates suggest that a significant portion of the retinol that is transported by RBP may be reutilized in the body, accounting for its lesser body turnover rate as compared to that of RBP. The possibility that quantitatively significant recycling and reutilization of retinol may occur in the animal body has also been suggested by others (De Luca *et al.*, 1979; Lewis *et al.*, 1981) on the basis of other kinds of data and considerations.

X. RBP SYNTHESIS AND SECRETION BY THE LIVER AND ITS REGULATION

A. General Comments; Biochemical Processes Involved

Vitamin A (retinol) mobilization from the liver is highly regulated by factors that control the rates of RBP synthesis and secretion by the liver. Considerable information is now available about the molecular and cellular events involved in the synthesis and secretion of RBP by the liver cell and about the mechanisms that regulate these events. The studies directed toward defining these phenomena have employed the rat as an animal model and a sensitive and specific radioimmunoassay for rat RBP (Muto *et al.*, 1972). Studies have been conducted both with intact animals and with isolated liver cells in culture.

A number of biochemical processes are involved in the phenomena of RBP synthesis and secretion. These include: (1) biosynthesis of RBP; (2) processing of the newly synthesized RBP molecule; (3) translocation (directed movement) of the newly synthesized RBP molecule in the cell; (4) complex formation of apo-RBP with retinol; and (5) secretion of the retinol–RBP complex (holo-RBP) from the cell. The information currently available about these processes is reviewed below.

One factor that exerts control over RBP secretion from the liver is the nutritional vitamin A (retinol) status of the animal. It is now well established that retinol deficiency specifically blocks the secretion of RBP from the liver so that plasma RBP levels fall and liver RBP levels rise. Repletion of retinol-deficient rats with retinol stimulates the rapid secretion of RBP from the expanded liver pool into the plasma. Thus, the secretion of RBP by the liver of the deficient rat appears to be strongly controlled by the availability of retinol at the liver cell for the formation of the retinol–RBP complex.

B. RBP Biosynthesis: Molecular Events

Studies have been conducted to investigate the molecular events that occur during the biosynthesis of RBP in the liver and to examine the effects of regulatory perturbations on these events. A study was performed to identify the first translated product of messenger RNA for RBP (Soprano *et al.*, 1981). Poly(A)$^+$RNA was isolated from rat liver and translated in the rabbit reticulocyte lysate *in vitro* protein-synthesizing system. The translation system used a mixture of unlabeled amino acids plus [^{35}S]methionine. RBP was identified and separated from other translated products by immunoprecipitation with specific rabbit anti-rat RBP antiserum. SDS–polyacrylamide gel electrophoresis and fluorography of the immunoprecipitate consistently revealed one major product that migrated more slowly than purified rat serum RBP. This protein product (preRBP) had an approximate molecular weight of 24,000. When dog pancreas microsomal membranes were cotranslationally present, the newly synthesized preRBP was processed to a protein that migrated coincidentally with purified rat serum RBP, approximately 20,500 daltons. In contrast, posttranslational addition of the pancreas microsomal membranes did not result in the formation of a small protein (of the size of serum RBP). These results indicate that RBP is initially synthesized as a larger molecular weight precursor (preRBP) that is rapidly processed by the removal of a signal peptide of about 3500 daltons to the size of the final RBP molecule that circulates in plasma. Thus, RBP resembles a number of other secretory proteins (Jackson and Blobel, 1980) by being synthesized initially as a larger molecular weight preprotein, which is processed cotranslationally by removal of the amino-terminal signal or leader peptide during translocation of the protein across the endoplasmic reticulum bilayer.

Several secretory proteins (e.g., albumin; Strauss *et al.*, 1977) are synthesized initially as preproproteins which are rapidly processed cotranslationally to more long-lived proproteins. These represent intermediate biosynthetic precursors of the final secreted protein. Most of the immunoreactive albumin present in liver microsomes consists of the larger protein proalbumin (Quinn *et al.*, 1975; Russell and Geller, 1975). Studies conducted with immunoreactive RBP solubilized and isolated from rat liver microsomes have suggested that RBP is not present in the microsomes as a larger proRBP precursor (Smith *et al.*, 1981). The results of *in vitro* and *in vivo* biosynthetic experiments with labeled amino acids are consistent with this conclusion. First, as indicated above, cotranslational processing of preRBP resulted in the production of a protein that appeared identical in size with purified RBP from serum (Soprano *et al.*, 1981). In addition, labeled RBP immunoprecipitated from liver homogenates of rats injected with labeled amino acids *in vivo* also appeared identical in size with purified RBP from serum (Soprano *et al.*, 1982a). These several lines of evidence all suggest that a pro-RBP intermediate does not exist during the biosynthesis of RBP in liver. Definitive evidence for this conclusion would, however, require the obtaining of amino acid sequence information about the immediate product of the processing of pre-RBP, as compared to RBP isolated from serum. The physiological significance, if any, of the apparent absence of a proRBP intermediate during the biosynthesis of RBP is not known.

The TTR subunit has also been shown to be initially biosynthesized in liver as a larger molecular size preTTR percursor, approximately 16,000 daltons in weight (Soprano *et al.*, 1982b). The preTTR subunit, like preRBP, is processed cotranslationally to a smaller polypeptide, of the size of the final, secreted TTR subunit. The mechanism and site of formation of the TTR tetramer (the final protein that circulates in plasma) remain to be defined.

C. Role of Retinol in RBP Secretion

Extensive studies conducted during the past decade have demonstrated that the availability of retinol to the liver cell plays a critical role in the control of RBP secretion from the cell. The effects of retinol depletion and of retinol repletion are illustrated in Fig. 4. In the retinol-deficient state, the secretion of RBP from the liver is blocked, resulting in the accumulation of an enlarged pool of apo-RBP in the liver, and a concomitant decline in serum RBP level (Muto *et al.*, 1972; J. E. Smith *et al.*, 1973a). Conversely, repletion of retinol-deficient rats with retinol stimulates the rapid secretion of RBP from the expanded liver pool into the plasma.

A detailed study of the repletion of retinol-deficient rats with lymph chylomicrons containing graded amounts of vitamin A (mainly as retinyl esters) was reported by J. E. Smith *et al.* (1973a). After the intravenous injection of chylomicrons containing vitamin A, a rapid increase in the serum levels of RBP and vitamin A occurred, with maximal levels seen at 2–4 h. The magnitude of the response was directly related to the amount of retinol given, in the dose range 0–17 μg. When a group of injected rats was killed after 2 h, the dose–response relationship of the increase in serum RBP was mirrored by a complementary dose-related decrease in the level of RBP in the liver (see also Fig. 4, right-hand side). Release of RBP from liver into serum, which was a function of the amount of retinol given, apparently occurred. Rats pretreated with inhibitors of protein synthesis, either puromycin or cycloheximide, also showed a rapid and substantial rise in serum RBP and vitamin A levels after the injection of vitamin A.

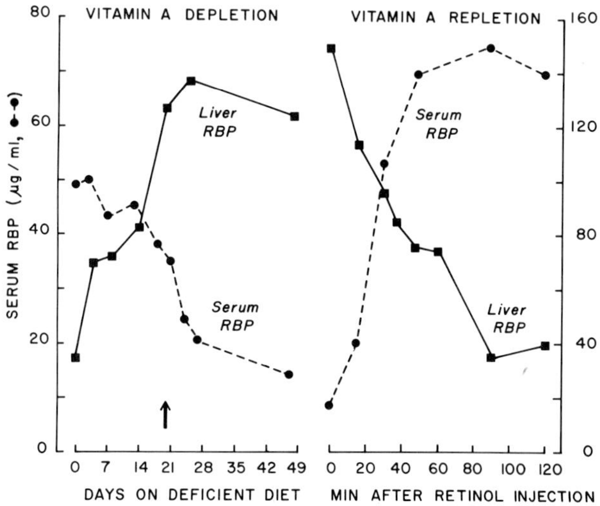

Fig. 4. The effects of vitamin A (retinol) depletion and repletion on the plasma and liver levels of RBP in the rat. Left panel: Vitamin A depletion (adopted from Muto *et al.*, 1972). The vitamin A stores of the liver were depleted on day 20, as indicated by the arrow. Right panel: Vitamin A repletion of vitamin A-deficient rats. The rats were each given an intravenous injection of 150 μg of retinol dispersed in 1 ml of an aqueous solution of 20% Tween 40. Composite data from several experiments are shown. (Reproduced with permission from Smith and Goodman, 1979.)

These data indicated that the increased level of RBP in serum after retinol injection mainly represented the release of RBP from an existing pool in the liver, rather than newly synthesized protein. Similarly, it was observed (Peterson *et al.*, 1973) that actinomycin D did not block the retinol-stimulated mobilization of RBP from the livers of vitamin A-deficient rats.

Repletion of retinol-deficient rats can also be effectively achieved by the intravenous injection of retinol dispersed in a 20% Tween 40 solution (Smith *et al.*, 1980; Fig. 4). Such an injection produces a rapid, dose-related increase in the serum concentration of RBP. The changes in serum RBP levels seen after the injection of retinol in a 20% Tween 40 solution closely resembled those previously seen after the injection of vitamin A (retinyl esters) in association with lymph chylomicrons. However, the amount of retinol required to stimulate the secretion of a given amount of RBP from the liver was about two to three times that required when retinol (retinyl esters) was injected in chylomicrons. As discussed by Smith *et al.* (1980), this quantitative difference is probably due to differences in the tissue distribution pattern of retinol when injected in the Tween 40 solution, compared to its administration in the form of chylomicrons.

The block in RBP secretion seen after retinol depletion is highly specific for RBP. Thus, neither retinol depletion and deficiency nor retinol repletion of deficient rats significantly altered plasma levels of TTR (Navab *et al.*, 1977b). The secretion of RBP and TTR appear to be independently regulated processes, with formation of the RBP–TTR complex occurring in plasma after secretion of the two proteins from the liver cell.

A study was conducted to explore the role of retinol in the control of the rate of synthesis of RBP in the liver (Soprano *et al.*, 1982a). In this study, normal rats were compared with retinol-depleted rats and with retinol-depleted rats that had been acutely repleted intravenously with retinol. Both the *in vivo* relative rate of RBP synthesis and the *in vitro* translatable level of RBP–specific messenger RNA (mRNA) were measured. Both the relative rate of RBP synthesis (approximately 0.26%) and the translatable level of RBP–specific mRNA (approximately 0.14%) were found to be constant regardless of the retinol status of the rats. These results indicate that retinol, the molecule that RBP specifically binds and transports, does not appear to control the rate of synthesis of RBP or the translatable level of RBP–specific mRNA in the liver of the rat. RBP thus appears to be unusual since the synthesis of other binding proteins for which information is available is, in general, controlled by their specific ligands (see Soprano *et al.*, 1982a, for references). This study further demonstrates that control of the rate of RBP secretion and of plasma RBP levels by retinol must be exercised fully at a locus beyond that of RBP synthesis.

Since RBP synthesis rate remains normal in retinol-deficient rats, whereas RBP secretion from the liver is blocked in these animals, this study suggests that the RBP degradation rate in the liver must be increased sufficiently in the deficient rats to maintain the elevated steady-state levels of RBP seen in their livers. No information is available about the mechanisms that may be involved in RBP catabolism in the liver or about their possible regulation.

D. Role of Subcellular Organelles and Structures

The roles of various subcellular organelles and structures in the hepatic metabolism and secretion of RBP have been studied with both normal and retinol-deficient rats. These studies have employed assays for a variety of marker enzymes and other constituents. After differential centrifugation of liver homogenates, 79 ± 1% of the RBP was found associated with the liver microsomes (Harrison *et al.*, 1979). Similar proportions of total liver RBP were found in the microsomal fractions of livers from both normal and retinol-deficient rats (J. E. Smith *et al.*, 1975). Further subfractionation of the microsomal fraction showed that RBP was particularly enriched in the rough microsomal fraction (3.8 ± 0.5-fold over the homogenate), which contained 49 ± 4% of the liver microsomal RBP (Smith and Goodman, 1979). RBP was also enriched in the smooth microsomal fraction (3.2 ± 0.2-fold over the homogenate).

The role of the Golgi apparatus was explored in a study that measured the levels of RBP, TTR, and several marker enzymes in homogenates, crude subcellular fractions, and isolated Golgi-rich fractions prepared from livers of retinol-deficient and control rats (Harrison *et al.*, 1980). Vitamin A (retinol) deficiency did not appear to alter the relative amount, the enzymatic composition, or the general sedimentation properties of any of the major subcellular organelles (including the Golgi apparatus) of liver. RBP was highly enriched (9.1 ± 2.0-fold) in the Golgi-rich fractions from control rats and was less enriched (3.7 ± 0.5-fold) in the comparable fractions from retinol-deficient rats. The Golgi apparatus contained ~ 23% of the liver RBP pool in normal rats and a maximum of < 10% of the total liver RBP in the deficient rats.

The effects of colchicine on RBP secretion and metabolism by the liver were explored by Smith *et al.* (1980). Colchicine treatment of retinol-deficient rats markedly inhibited the retinol-stimulated secretion of RBP from the liver into the serum. The inhibition of RBP secretion was quantitatively quite similar to the inhibition of very low-density lipoprotein secretion by colchicine seen in parallel experiments. In contrast, colchicine did not affect the overall rate of protein synthesis within the liver. The inhibition of RBP secretion by colchicine suggests that the microtubules play a role in RBP secretion.

When retinol-deficient rats were first treated with colchicine and then injected with retinol to stimulate RBP secretion, the RBP content of a Golgi-rich fraction from liver increased markedly, to a maximum of 34% of the total liver RBP. The level of TTR in the Golgi was not influenced by retinol injection.

Taken together, these studies are consistent with the conclusion that the Golgi apparatus, Golgi-derived secretory vesicles, and microtubules are involved in the normal pathway of RBP secretion in the liver cell. Thus these studies suggest that the RBP secretory process involves the same subcellular organelles and pathways previously shown to be involved in the secretion of other serum proteins such as albumin (Redman *et al.*, 1975). These data also suggest that the block in RBP secretion found in retinol deficiency occurs at a site before the RBP molecule reaches the major portion of the Golgi apparatus.

Further support for this conclusion was reported very recently by Rask *et al.* (1983) and Ronne *et al.* (1983) from studies of the subcellular localization of RBP and other marker components in normal and vitamin A-deficient rat liver, and from studies of the effects of retinol on RBP synthesis and secretion by primary rat hepatocytes in culture. From these studies, it was concluded that the transport of RBP from the endoplasmic reticulum to the Golgi complex is regulated by retinol. It is also possible that retinol is necessary not for transport of RBP from the endoplasmic reticulum to the Golgi, but for the appropriate directed movement of RBP through the Golgi apparatus. Thus, it has been reported that formation of the retinol–RBP complex may be necessary for the normal passage of RBP from the cis to the trans components of the Golgi apparatus, and hence for subsequent transport to Golgi-derived vesicles (Handler and Smith, 1983) for secretion from the cell. These recent reports need further development in order to define precisely the site of blockage of RBP secretion observed in retinol deficiency, and the exact anatomic locus where retinol normally complexes with RBP in the liver cell.

E. Complex Formation of Apo-RBP with Retinol

As discussed above, in retinol deficiency newly synthesized apo-RBP accumulates in the hepatic endoplasmic reticulum (microsomal fraction), whereas the secretion of serum albumin, TTR, and other plasma proteins appears to continue at a normal rate. The mechanisms responsible for the selective retention of RBP in the endoplasmic reticulum and for the specific stimulation of RBP secretion when retinol is made available, are not known.

In addition to retinol deficiency, a number of factors may affect the extent to which retinol is available, at the appropriate anatomic locus within the liver cell, for complex formation with apo-RBP. In view of the critical role played by retinol in influencing the rate of RBP secretion, these factors can be considered as potentially significant with regard to the regulation of RBP secretion. Thus, these factors warrant exploration and delineation.

Retinyl ester stored in the liver must be hydrolyzed before the retinol can be mobilized from the liver as the retinol–RBP complex. Information is needed as to whether or not the processes of retinyl ester hydrolysis and of RBP production and secretion might, under some circumstances, be coordinated in some way. The question to be asked is whether or not the hydrolysis of retinyl esters might at times serve as one of the regulatory steps in the overall process of retinol mobilization from the liver. Information available about the enzymatic hydrolysis of retinyl esters in liver is reviewed in Chapter 7.

Other factors that may affect the availability of retinol for complex formation with apo-RBP are: (1) the manner in which retinol is transported within the cell from the site(s) of retinyl ester hydrolysis to a molecule of apo-RBP; and (2) the manner in which the retinol molecule is presented to the membrane-bound molecule of apo-RBP. In this regard, the possible roles of cellular retinol-binding protein (CRBP) and of the so-called "cytosol retinyl ester lipoprotein complex" in delivering retinol to apo-RBP need exploration (Chen *et al.*, 1981; Sklan *et al.*, 1982). More information is also needed about the movement of retinol between hepatocytes and fat-storing cells (lipocytes) under normal and abnormal conditions. These various topics are also reviewed in Chapter 7.

Figure 5 is a diagram that summarizes information available about retinol and RBP metabolism in the hepatocyte.

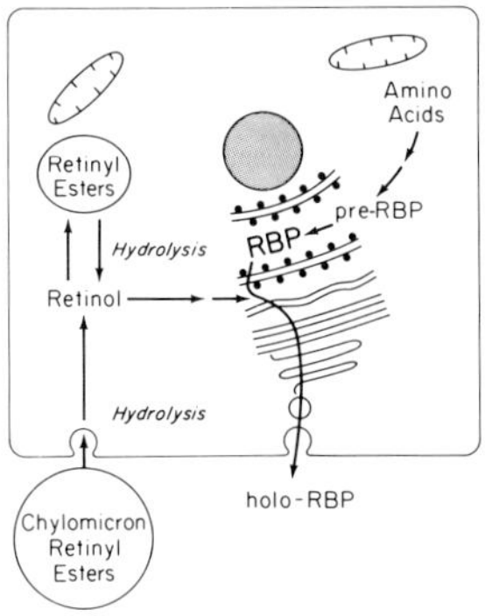

Fig. 5. Diagram summarizing the metabolism of vitamin A and RBP in the liver cell. The various procedures and events summarized in this drawing are discussed in the text.

F. Studies with Isolated Liver Cells in Culture

While much has been learned from whole-animal studies about the regulation of RBP metabolism, the interpretation of these results is sometimes clouded owing to the interplay between organs. To avoid this problem, studies have been conducted *in vitro* with isolated liver cells in culture.

Two lines of differentiated rat hepatoma cells were found that synthesized RBP during culture *in vitro* (Smith *et al.*, 1978). When the cells were incubated in a vitamin A-free serumless medium, a relatively large proportion of the synthesized RBP was retained within the cells. Addition of retinol to the medium (at levels of 0.1 or 1 μg/ml) stimulated the release of RBP. In contrast, retinol had no effect on either the synthesis or secretion of rat serum albumin or TTR by these cells.

Several reports in the literature have suggested that glucocorticoid hormones accelerate the rate of vitamin A mobilization from the liver. Accordingly, studies were conducted to explore the effects of glucocorticoid hormones on RBP metabolism by one of the lines of hepatoma cells in culture (Borek *et al.*, 1981). Cortisol, corticosterone, and the synthetic analog dexamethasone all induced a two- to three-fold increase in the net synthesis of RBP. Half-maximal stimulation occurred at concentrations of dexamethasone in the range of 1–5 n*M*. Dexamethasone, like retinol, had no effect on the net synthesis or the cell-to-medium distribution of rat serum albumin or TTR. The effects of retinol and dexamethasone studied over a wide range of concentrations were approximately additive on the net synthesis of RBP. Moreover, dexamethasone did not affect the retinol-induced secretion of RBP. Thus, retinol and dexamethasone appear to function via different and independent mechanisms to regulate the metabolism of RBP by the liver cell.

Because these cell lines respond to retinol depletion and repletion in a manner similar to the intact rat liver cell *in vivo,* they should provide a good model to study other hormonal and metabolic factors that may be involved in the regulation of RBP synthesis and secretion. In addition, preliminary experiments have been carried out with primary rat hepatocytes, which have been found to synthesize RBP for several days in culture *in vitro* (unpublished studies from the author's laboratory). The primary hepatocyte may be useful for studying regulatory factors that cannot be studied with the hepatoma cells. An important general question which these systems may help to address is: What are the molecular signals from peripheral tissues that normally stimulate or depress the synthesis and secretion of RBP?

XI. RETINOL DELIVERY; RBP RECEPTORS

RBP is responsible for the delivery of retinol from the liver to the extrahepatic sites of action of the vitamin. Evidence is available that this delivery process may involve cell-surface receptors for RBP. Thus, studies have been reported that suggest that there are specific cell-surface receptors for RBP on monkey small intestine mucosal cells, on bovine pigment epithelial cells from the eye, and on chicken testicular cell membranes.

In the studies with intestinal mucosal cells, Rask and Peterson (1976) explored the *in vitro* uptake of ^{3}H-labeled retinol, from the [^{3}H]retinol–RBP complex, into isolated small intestinal mucosal epithelial cells from the cynomolgus monkey. [^{3}H]Retinol accumulation in the cells was linear for about 45 min, was dependent on temperature, and was not inhibited by several metabolic inhibitors. RBP linked to Sepharose beads was also able to deliver retinol to the epithelial cells, indicating that the delivery of [^{3}H]retinol occurred without a concomitant cellular uptake of the RBP. Fab′ fragments of antibodies against RBP inhibited the cellular uptake of retinol. In other studies (Sege and Peterson, 1978), anti-idiotypic antibodies against antibodies to RBP also impeded the RBP-mediated [^{3}H]retinol uptake into intestinal mucosal cells. Although direct binding of RBP to the cells could not be documented (Rask and Peterson, 1976), it was suggested that there is a receptor for retinol on the cell surface that recognizes the protein part of the protein–ligand complex. The apparent transitory interaction of RBP with the cell-surface receptor was explained by suggesting that on binding to the receptor RBP gives up retinol and becomes structurally modified so that it cannot interact any longer with the receptor.

Direct binding of ^{125}I-labeled RBP to isolated pigment epithelial cells of bovine retinas was demonstrated by Heller (1975a) using a much lower concentration of labeled RBP [than that used by Rask and Peterson (1976)] of high specific radioactivity. Nonspecific binding was corrected for by carrying out each incubation with ^{125}I-labeled holo-RBP alone and also in the presence of a 600-fold molar excess of unlabeled holo-RBP. Apparently specific binding was observed that was very rapid (essentially complete after 1 min) and that was temperature-dependent (optimal at 22° C). The rate of dissociation of specifically bound holo-RBP from the cell surface was estimated to have a half-life of 24 min and a rate constant of 5×10^{-4} s^{-1}. Apo-RBP was less effective in displacing bound ^{125}I-labeled holo-RBP than was holo-RBP, suggesting that after delivering its retinol to the cell, apo-RBP is displaced from the receptor by another holo-RBP molecule.

In a subsequent study, Chen and Heller (1977) showed that bovine pigment epithelial cells were capable of taking up retinol and retinoic acid when these retinoids were presented to the cells as complexes with serum RBP. Neither free retinol nor retinol or retinoic acid presented as serum albumin complexes were taken up by the pigment epithelial cells. The uptake of the retinoids was accompanied by dissociation of the retinoid–RBP complex, with the retinoid being taken up by the cells while the protein remained outside the cell.

Autoradiographic studies with ^{125}I-labeled holo-RBP have provided data that have been interpreted as showing the presence of binding sites for RBP on the surface of pigment epithelial cells. These studies were carried out both *in vivo* (injecting rats with ^{125}I-labeled holo-RBP; Bok and Heller, 1976) and *in vitro* (incubating isolated cells with ^{125}I-labeled holo-RBP; Heller and Bok, 1976). Autoradiographic studies of the localization of RBP in rat testis have also been carried out (McGuire *et al.*, 1981). Finally, the apparently specific binding of RBP to a not well-characterized membrane fraction from chicken testes has also been reported (Bhat and Cama, 1979).

In summary, although the data available are limited, these several studies are consistent in suggesting that RBP delivers retinol to target tissues at specific cell-surface sites that "recognize" RBP, and that retinol is released from RBP at these locations. After its delivery, the retinol enters the cell for subsequent metabolism and action. The apo-RBP does not appear to enter the cell but rather dissociates from the receptor and returns to the circulation, where it shows a reduced affinity for TTR and is selectively filtered by the renal glomeruli. Much more information is needed about the putative RBP receptors on target cells, in order to understand more fully the phenomena involved in the retinol delivery process and their regulation.

XII. CLINICAL STUDIES

A considerable literature now exists describing the effects of a variety of diseases on the plasma levels of RBP and TTR. In general, many of these studies have tried to evaluate both the effects of a given disease on the plasma vitamin A transport system, and the possible role that disordered vitamin A transport (if present) might play in the pathophysiological manifestations of the disease. Although RBP has been studied in many diseases, to date no disease has been found where RBP was totally absent or had abnormal immunological (or physicochemical) properties.

A. Diseases of the Liver

A number of studies have examined RBP levels (and usually TTR and vitamin A levels as well) in patients with various kinds of acute and chronic diseases of the liver (Smith and Goodman, 1971; Kindler, 1972; Prellwitz *et al.*, 1974; Skrede *et al.*, 1975; Brissot *et al.*, 1978; Vahlquist *et al.*, 1978a; Russell *et al.*, 1978; McClain *et al.*, 1979). In patients with clinically significant hepatic parenchymal disease, the plasma levels of vitamin A, RBP, and TTR have usually been found to be substantially depressed. The low levels of RBP and TTR presumably reflect a reduced rate of production of the proteins by the diseased liver.

In 63 patients with liver disease, the levels of vitamin A, RBP, and TTR were all markedly decreased and were highly significantly correlated with each other over a wide range of concentrations (Smith and Goodman, 1971). Nineteen patients with acute hepatitis were studied with serial samples; as the disease improved, the plasma levels of vitamin A, RBP, and TTR all increased. In these patients, the RBP concentrations correlated negatively with the values of standard tests of liver function (plasma bilirubin, glutamic-oxaloacetic transaminase, and alkaline phosphatase). It has been reported that the level of RBP is of value clinically in assessing the course of acute infectious hepatitis and, to a limited extent, in the differentiation of various forms of jaundice (Kindler, 1972). It seems clear that measurements of plasma RBP levels in patients with liver disease could, if available, be used as an index of hepatic parenchymal functional status and hence could serve as a useful clinical liver function test.

Several studies have explored relationships between vitamin A transport and visual dark adaptation in patients with chronic liver disease. In one study (Vahlquist *et al.*, 1978a), patients with liver disease and low plasma RBP levels (below 20 μg/ml) were found to have impaired dark adaptation, suggesting that these patients had peripheral vitamin A deficiency symptoms secondary to their inability to mobilize vitamin A from the liver. In these patients, vitamin A therapy did not affect either the reduced dark adaptation ability or the low plasma RBP levels. In another study (McClain *et al.*, 1979) many patients with alcoholic cirrhosis, with significantly depressed serum levels of zinc, vitamin A, RBP, and TTR, manifested impaired dark adaptation. Some of these patients did not correct their abnormal dark adaptation with vitamin A supplementation but did with zinc therapy. In a third study (Russell *et al.*, 1978) vitamin A therapy of patients with alcohol-associated cirrhosis and marginal vitamin A status appeared to stimulate RBP release from the liver and led to improvements in dark adaptation. Differences observed between these various studies probably reflect differences in the patient populations studied, with regard to the availability of vitamin A within the liver and the extent of hepatic functional derangement. In general, however, these various studies illustrate the key role played by RBP in controlling the delivery of vitamin A to extrahepatic tissues, and the possibility that pathophysiological manifestations of vitamin A deficiency might result from impaired production of RBP.

B. Diseases of the Kidneys

Since the kidneys are the major site of RBP catabolism (as discussed earlier in this chapter), patients with functionally significant renal disease show disordered RBP and vitamin A metabolism. The effects of renal disease on the vitamin A transport system have been explored in some detail in a number of laboratories and clinics throughout the world.

In a study of 26 patients with chronic renal disease of varying etiologies (Smith and Goodman, 1971), plasma RBP and vitamin A levels were found to be markedly elevated [116 ± 9 (SEM) μg/ml and 98 ± 9 μg/dl, respectively], while TTR levels remained normal. Both the molar ratio of RBP:TTR (1.06 ± 0.10), and that of RBP:vitamin A (1.92 ± 0.22) were markedly elevated as compared to normals (or to patients with liver disease). In many patients, RBP was present in molar excess as compared with TTR. The presence of a relatively large proportion of free RBP, not complexed to TTR, in some patients was confirmed by gel filtration. On average, circulating RBP was less saturated with retinol than normal.

As discussed earlier, patients with impaired tubular function and tubular proteinuria excrete increased amounts of RBP and other low-molecular-weight plasma proteins in the urine. The assay of RBP in urine has been reported to be a sensitive and reliable test for renal tubular dysfunction (Bernard and Lauwerys, 1981). Extensive studies have been conducted on the serum and urinary levels of RBP in Japanese patients with chronic cadmium poisoning and "Itai-Itai" disease (Kanai *et al.*, 1971, 1972a,b). The differentiation of patterns of proteinuria was investigated by measuring urinary excretion of total protein, albumin, and RBP (Kanai *et al.*, 1972b). Four types of proteinuria (normal, glomerular, tubular, and mixed glomerular–tubular) were differentiated by the double logarithmic plot of albumin and RBP excretion in urine. In "pure" tubular proteinuria, plasma RBP levels and turnover remain normal. In contrast, plasma RBP levels rise and turnover is slowed in patients with azotemic renal disease. In a study of 91 patients with different kinds of kidney diseases and varying levels of renal function, a highly significant linear correlation ($r = 0.78$) was observed between the logarithms of serum RBP and creatinine concentrations (Scarpioni *et al.*, 1976).

Several studies have examined RBP and vitamin A metabolism in patients with severe renal disease undergoing chronic hemodialysis treatment (Schultz *et al.*, 1977; De Bevere *et al.*, 1981; Stewart and Fleming, 1982; Vahlquist *et al.*, 1982a). Elevated plasma levels of vitamin A and RBP have been observed in all these studies. A question repeatedly addressed in these and other studies has been the question of whether the elevated vitamin A levels found in patients with chronic renal disease is associated with any pathophysiological consequences, i.e., with manifestations of hypervitaminosis A. It has been pointed out that multivitamin supplements containing vitamin A should be avoided in patients on maintenance hemodialysis (Stewart and Fleming, 1982). The elevated plasma vitamin A in renal disease appears to be present entirely as retinol bound to RBP. Three studies failed to find elevated levels of retinyl esters in renal failure patients (De Bevere *et al.*, 1981; Stewart and Fleming, 1982; Vahlquist *et al.*, 1982a). As discussed later in this chapter, these biochemical conditions are not those that would be expected to lead to generalized manifestations of vitamin A toxicity. On the other hand, in another study (Vahlquist *et al.*, 1982a), increased amounts of vitamin A were found in the skin of uremic patients. In 10 patients (11 samples) a fairly strong linear correlation was observed between serum and epidermal concentrations of vitamin A. It has been pointed out that uremic patients frequently show skin symptoms that are also seen in patients with hypervitaminosis A (i.e., pruritis, xerosis, and hair loss). These symptoms may be due in part to increased skin levels of vitamin A. The possibility was also raised (Vahlquist *et al.*, 1982a) that uremic patients might develop increased vitamin A levels in other tissues (as well as in skin) with receptors for RBP. It was suggested that it might be meaningful to distinguish two types of hypervitaminosis A—a generalized form (i.e., classical hypervitaminosis A) and a localized form. The localized form would only be manifest in certain target tissues for vitamin A as the result of high serum RBP levels as in uremia. This interesting suggestion warrants investigation.

C. Malnutrition

The effects of protein–energy malnutrition (PEM), and its treatment, on the plasma retinol transport system have been investigated in a large number of studies during the past decade. Patients with PEM have decreased plasma concentrations of RBP, TTR, and vitamin A. Two major factors can contribute to these low plasma concentrations. First, patients with PEM manifest a defective hepatic production of RBP because of a lack of substrate (calories, amino acids from dietary protein) needed for RBP synthesis. Thus, PEM per se is associated with impaired production of RBP and TTR and defective vitamin A mobilization from the liver. Second, however, PEM is often accompanied by inadequate dietary intake of vitamin A. The extent of nutritional vitamin A deficiency seen in children with PEM varies greatly in different geographic locations and under different circumstances. If physiologically significant vitamin A deficiency coexists with PEM, however, it will exercise an important influence on the plasma levels of RBP and vitamin A. In a given patient, the levels of plasma RBP and vitamin A and the response of these levels to a given therapy will depend on the severity and form of the PEM and the extent of accompanying vitamin A deficiency in that particular patient. In addition, other factors (e.g., nutritional zinc status, presence or absence of infections; see discussion below and Chapter 6, Vol. 1) will also influence plasma retinol transport in these patients.

The critical role of a deficient supply of substrate for protein synthesis in PEM was shown in a study of 21 Egyptian children with classical kwashiorkor and without any clinical (conjunctival or corneal) evidence of vitamin A deficiency (F. R. Smith *et al.*, 1973a). Earlier studies by Arroyave *et al.*, (1961) had suggested that an impairment in vitamin A transport because of decreased plasma transport proteins might contribute significantly to low plasma vitamin A levels in children with protein malnutrition. After treatment with calories and proteins but without supplemental vitamin A, 11 of the 21 Egyptian patients were judged clinically to be cured. These patients showed highly significant increases in the concentrations of vitamin A, RBP, and TTR in plasma by the end of the second week, with further progressive increases occurring by the fourth week. Over the wide concentration range seen in kwashiorkor during treatment, the serum vitamin A, RBP, and TTR concentrations were highly significantly correlated with each other. These findings suggested that the low serum vitamin A levels in these patients with kwashiorkor largely reflected a functional impairment in the hepatic release of vitamin A, rather than vitamin A deficiency per se. Hepatic release was impaired because of defective hepatic production of plasma proteins, including the transport proteins for retinol, because of a limiting supply of substrate. When substrate was provided by dietary protein and calories, hepatic production of plasma proteins increased, plasma RBP and TTR levels rose, and hence plasma vitamin A increased.

A very high correlation between plasma retinol and RBP levels during nutritional therapy for severe malnutrition was also observed in 39 young (18–30 months) Senegalese children (Ingenbleek *et al.*, 1975a). It was suggested that RBP was acting as the limiting factor for retinol transport in these children. Further evidence of the need for adequate calories and for dietary protein as a source of amino acids for the synthesis of RBP before normal retinol transport can be reestablished comes from a variety of studies including: (1) studies of 34 xerophthalmic malnourished children treated at a nutrition center in Madurai, India (Venkataswamy *et al.*, 1977); (2) studies of children with PEM in Thailand (F. R. Smith *et al.*, 1975; Large *et al.*, 1980); (3) other studies of malnourished Senegalese children (Ingenbleek *et al.*, 1975b).

Most studies of the treatment of malnourished children have included therapy with supplemental vitamin A as well as with graded regimens of calories and protein (e.g., F. R. Smith *et al.*, 1973b, 1975; Reddy *et al.*, 1979; Large *et al.*, 1980). In two studies, intramuscular administration of a large dose of vitamin A (30 mg as water-miscible retinyl palmitate) to children with both evident vitamin A deficiency and PEM led to a rapid rise in the plasma levels of both vitamin A and RBP. In one study (Reddy *et al.*, 1979) a significant rise in RBP levels was seen at 4 h, with a further rise noted at 24 h after vitamin A injection. In the other study (Large *et al.*, 1980) a rapid rise in RBP levels was seen at 3 h, with a subsequent decline toward initial values at 24 h. The rapid rise in RBP level was not accompanied by a rise in TTR (Large *et al.*, 1980), similar to results seen in retinol-deficient rats (Navab *et al.*, 1977b). These two studies suggest that in the subjects studied, hepatic retinol deficiency was present and was accompanied by some degree of expansion in the hepatic pool of apo-RBP. This pool was presumably mobilized acutely on administration of vitamin A. The differences seen at 24 h may have reflected a more severe protein malnutrition in the patients studied by Large *et al.*, 1980). Because it is often very difficult to assess the potential degree of vitamin A deficiency that may coexist with PEM, it is recommended that treatment of patients with PEM routinely include supplemental vitamin A therapy.

Since both RBP and TTR are rapidly turning over plasma proteins, it can be anticipated that their plasma levels might provide a sensitive index for mild degrees of PEM (as compared to plasma levels of more slowly turning over proteins such as albumin). Conversely, it has been shown that plasma RBP and TTR levels rise more rapidly than do levels of albumin (or other more slowly turning over proteins) during therapy for PEM. Several groups of investigators have suggested that the plasma levels of TTR or RBP might serve as useful indexes for assessing nutritional protein status and mild degrees of PEM (Ingenbleek *et al.*, 1975b; Large *et al.*, 1980; Ogunshina and Hussain, 1980). Although there are limitations in the extent to which such measurements can be used as indexes of protein status (Golden, 1982), in general their use appears to have considerable potential value.

Zinc represents another nutrient whose nutritional status may influence plasma vitamin A and RBP levels in malnourished children. Interrelationships between zinc and vitamin A have been reviewed in detail by Solomons and Russell (1980) and by Smith (1980), and are discussed and summarized in Chapter 6, Vol. 1. Studies in experimental animals have shown that zinc deficiency is associated with low plasma vitamin A and RBP levels. It has been suggested that this may be due in part to impaired hepatic synthesis of RBP (Smith *et al.*, 1974). This effect on RBP production has not been thought to be in any way specific for RBP; rather hepatic synthesis of plasma proteins in general appears to be impaired in zinc deficiency, with RBP being one of the hepatic proteins most sensitive to this deficiency. The main effect of zinc deficiency on vitamin A and RBP metabolism appears, however, to be secondary to the depressed food intake and growth retardation that accompanies the zinc deficiency (Carney *et al.*, 1976; Smith *et al.*, 1976).

Relationships between plasma vitamin A and zinc were explored in 45 children with clinical vitamin A deficiency and 20 children with PEM (Shingwekar *et al.*, 1979). Mean levels of plasma vitamin A, RBP, and zinc were low in these malnourished children compared to controls. Supplementation with zinc resulted in a small but significant increase in plasma vitamin A and RBP levels in children with PEM but not in the vitamin A-deficient group. There was no correlation between plasma levels of vitamin A and zinc. It was suggested that in children with PEM, apart from deficiencies of protein and vitamin A, zinc deficiency may also contribute in part to low plasma vitamin A levels. Other clinical studies on the interaction between zinc and vitamin A are discussed in the above-mentioned reviews. Zinc deficiency is an uncommon but potential contributor to abnormal vitamin A transport in a number of gastrointestinal and hepatic diseases, as well as in PEM.

D. Cystic Fibrosis; Other GI Disorders

Plasma vitamin A and RBP levels have been investigated in patients with cystic fibrosis of the pancreas (Smith *et al.*, 1972; Knöpfle *et al.*, 1975; Palin *et al.*, 1979). Plasma vitamin A and RBP levels have been found to be lower than normal in patients with cystic fibrosis, despite the administration of oral vitamin A supplements adequate to maintain normal hepatic stores. In one study (Smith *et al.*, 1972), the plasma vitamin A transport system was studied in 43 patients with cystic fibrosis receiving oral supplements of vitamin A and in 95 normal control subjects of a similar age range. The mean plasma concentrations of vitamin A and RBP were significantly lower in the patients than in the controls. Moreover, in cystic fibrosis patients each component of the transport system failed to show the normal age-related rise. It is not known whether these abnormalities of the retinol transport system are primary or secondary features of cystic fibrosis; the abnormalities may, however, play a role in the pathophysiology of the disease.

Serum RBP and vitamin A levels have also been studied in patients with intestinal disorders and fat malabsorption. Although some patients with such disorders show low plasma retinol and RBP levels (Vahlquist *et al.*, 1978a), in other studies normal RBP levels were found in patients with malabsorption and steatorrhea (Smith and Lindenbaum, 1974; Prellwitz *et al.*, 1974). It can be anticipated that the effects of a given disorder would depend on the nature, severity, and duration of the disease.

E. Other Disorders (Eye, Skin, Thyroid, etc.)

A number of investigators have examined plasma RBP concentrations and characteristics in patients with different varieties of retinitis pigmentosa. Normal or nearly normal RBP levels have been found in all the studies (Futterman *et al.*, 1974; Rahi, 1972; Maraini *et al.*, 1975; F. R. Smith and D. S. Goodman, unpublished observations). Plasma RBP from patients with retinitis pigmentosa displays normal fluorescence and other characteristics (Futterman *et al.*, 1974; Maraini, 1974). It has also been reported that the interaction of RBP, isolated from plasma of patients with the recessive form of retinitis pigmentosa, with bovine pigment epithelium is similar to that of normal RBP (Maraini and Gozzoli, 1976).

Plasma RBP levels have been measured in patients with a number of different dermatologic disorders. Normal levels of RBP were observed in patients with Darier's disease (Vahlquist *et al.*, 1982b) and psoriasis (Benoldi *et al.*, 1981). Mild to moderate reductions of RBP levels have been reported in patients with severe acne (Michaëlsson *et al.*, 1977; Vahlquist *et al.*, 1978b) and in eczema and porphyria cutanea tarda (Benoldi *et al.*, 1981). These findings may, for the most part, represent effects of chronic inflammation and infection seen in many of these patients. There is no evidence that abnormal vitamin A transport participates in the pathophysiology of any of these disorders. A low plasma RBP level in patients with pityriasis rubra pilaris has been reported (Finzi *et al.*, 1981); it was also reported that treatment of this condition with stanozolol was associated with a rise in RBP level to normal (Bergamaschini *et al.*, 1982). The validity of the conclusion that low RBP may be a characteristic of this disorder has been questioned (Vahlquist, 1982), however, and normal serum levels of RBP have been reported in patients with pityriasis rubra pilaris (Stoll *et al.*, 1983). More observations are needed in such patients.

Levels of RBP, TTR, and vitamin A were measured in 14 patients with hyperthyroidism and in 7 with hypothyroidism (Smith and Goodman, 1971). Both RBP and TTR levels were significantly lower than normal in hyperthyroid patients. In hypothyroid patients, plasma vitamin A was increased significantly; RBP levels were somewhat higher than normal, but the difference was not statistically significant. In both hyper- and hypothyroidism, approximately normal molar ratios of RBP:TTR and of RBP:vitamin A were observed. The significance of these observations in thyroid disorders is not clear.

In other endocrinological studies, women taking oral contraceptives have been found to have elevated plasma levels of both RBP and vitamin A (Vahlquist *et al.*, 1979; Nonavinakere *et al.*, 1981). The estrogen component seems to be largely responsible for these increased levels. The RBP level was also found to show a bicyclic variation during the menstrual cycle (Vahlquist *et al.*, 1979).

Levels of RBP have been measured in patients with infections of various kinds (Arroyave and Calcaño, 1979), with burn injury (of between 2 and 37% of the body surface area) (Moody, 1982), and with "stress" arising from multiple traumas, cranial–cerebral lesions, or surgery (Kasper *et al.*, 1975). Low RBP levels were found in all of these conditions. The RBP levels returned toward or to normal following recovery from infection or from the burn injury.

Finally, relatively normal retinol and RBP levels were reported to be present in 91 children with hearing impairments of different etiologies (Biesalski *et al.*, 1981).

XIII. RBP AND VITAMIN A TOXICITY

Excessive retinol intake by animals and humans produces distinct toxic manifestations, as discussed in other chapters (Chapter 6, Vol. 1; Chapter 13, Vol. 2). Considerable information is available about the mechanism of the toxic effects of vitamin A on tissues. Studies both *in vivo* and *in vitro* have shown that excess retinol results in increased lability of biological membranes (Roels *et al.*, 1969; Fell, 1970). This effect of retinol is believed to be due to its surface-active "membranolytic" properties (Bangham *et al.*, 1964). In particular, excess retinol has been shown to lead to increased synthesis and release of lysosomal enzymes, and these hydrolases have been shown to be critically involved in the effects of vitamin A on cartilage and limb-bone rudiments (Fell, 1970; Fell and Dingle, 1963; Dingle *et al.*, 1971).

Retinol bound to RBP does not appear to manifest its surface-active effects on biological membranes. Thus, retinol bound to RBP was found to produce no discernible effects on chick limb-bone rudiments grown in organ culture (Dingle *et al.*, 1972). These observations suggested that the mode of transport of retinol (and other retinoids) may be an important determinant in the development of the manifestations of hypervitaminosis A.

An extensive study was carried out in rats given excessive vitamin A in doses sufficient to cause both growth disturbances and fatty livers (Mallia *et al.*, 1975). The hypervitaminotic rats showed large increases in serum vitamin A levels, which were mainly due to elevations in the circulating levels of retinyl esters, and decreased levels of serum RBP. In these toxic rats, most of the serum vitamin A and virtually all of the retinyl esters were found in association with the serum lipoproteins of hydrated density < 1.21. It was concluded that the serum lipoproteins play an important role in the transport of the vitamin A that accumulates in the serum in hypervitaminosis A. It was also suggested that serum lipoproteins may nonspecifically deliver vitamin A to biological membranes and, hence, lead to vitamin A toxicity.

Data consistent with this interpretation have been obtained in studies on vitamin A transport in human vitamin A toxicity (Smith and Goodman, 1976). In three patients with chronic hypervitaminosis A, the toxic state in each was associated with increased plasma levels of total vitamin A and particularly of retinyl esters. In contrast, plasma RBP levels were normal, and there was a molar excess of total vitamin A in relation to RBP. The data suggest that vitamin A toxicity appears to occur *in vivo* only when the level of vitamin A in the body is such that retinol begins to circulate in plasma and is presented to membranes in a form other than bound to RBP. The nonspecific and unregulated delivery of vitamin A to tissues in this manner may lead to vitamin A toxicity.

These studies provide considerable insight into metabolic phenomena that may be involved in the development of vitamin A toxicity caused by the "natural" form of the vitamin retinol. It may be asked, however, whether similar considerations apply to the development of vitamin A toxicity as a result of the administration of retinoic acid or of other retinoids.

Retinoic acid and other acidic retinoids are highly active in promoting the lysis of the extracellular matrix of cartilage grown in organ culture *in vitro* (Goodman *et al.*, 1974). This phenomenon is presumably caused by release of lysosomal enzymes as a result of the membrane-labilizing effects of the compounds. Thus, retinoic acid and its analogs resemble retinol in being able to cause generally similar (though not identical) toxic manifestations; in fact, retinoic acid was quantitatively more active than was retinol in producing lysis of the cartilage matrix.

In contrast to retinol, retinoic acid is absorbed through the portal system and is transported in plasma bound to serum albumin (see below). It does not accumulate in the liver or other tissues in any appreciable amounts. Retinoic acid appears to be quite rapidly metabolized, mainly to more polar compounds, which are then largely excreted in the urine and bile.

Since retinoic acid (and other acidic retinoids) circulates bound to albumin rather than to RBP, it is likely that retinoic acid is not delivered to tissues via a specific receptor-regulated process. Accordingly, one would anticipate that there might be a fairly close and consistent relationship between the dose of retinoic acid administered on the one hand and the rate of tissue delivery and risk of toxicity on the other. If this is true, then the plasma level of retinoic acid (or its analogs) should correlate well with the risk and development of toxicity and should provide a means of assessing and monitoring toxicity with these compounds in a clinical setting.

XIV. PHYSIOLOGICAL ROLES OF RBP

RBP plays a number of important physiological roles. First, RBP serves to solubilize the water-insoluble retinol molecule and to provide a vehicle to transport retinol from the liver to peripheral tissues. Second, RBP also serves to protect the reactive retinol molecule from oxidative damage while it is transported in plasma. Thus, free retinol is unstable in aqueous dispersion, whereas the retinol in the retinol–RBP complex is quite stable in stored plasma for weeks to months. Retinol bound to RBP is unavailable for oxidation by liver alcohol dehydrogenase, in contrast to retinol more weakly (and less specifically) complexed with either bovine serum albumin or β-lactoglobulin (Futterman and Heller, 1972).

Third, as discussed earlier in this chapter, RBP plays an important role in the regulation of the mobilization of retinol from the liver stores and hence of its delivery to peripheral tissues. Thus, the rate of release of retinol from the liver appears to be controlled mainly by factors that regulate the rates of RBP synthesis and secretion by the liver.

Fourth, as discussed above, RBP also appears to direct the delivery of retinol to specific sites at the surface of cells requiring vitamin A. This directed delivery may be necessary for the proper utilization of retinol within the cell and also may serve to prevent the toxic effects (see above) that can occur from the unregulated delivery of retinol to biological membranes.

Finally, RBP may play an important role in fetal development, by serving to deliver retinol from the maternal circulation to the developing fetus. Transfer of vitamin A from the dam to the fetus is known to be a highly regulated process. Thus, in a study in rats, the amount of vitamin A transferred to the fetus was found to be relatively constant over a wide range of maternal vitamin A intakes (Takahashi *et al.*, 1975). In the rat, the fetal requirement for vitamin A becomes particularly critical during the period of rapid tissue differentiation after day 11 of gestation. In the absence of retinol (even when retinoic acid was provided in the maternal diet) extensive fetal resorption and death were observed after day 14 of gestation (Takahashi *et al.*, 1975). There is evidence that, in the rat, retinol is probably transported from dam to fetus, from and after day 11 of gestation mainly by transplacental transport of the maternal retinol–RBP complex (Takahashi *et al.*, 1977). By day 16 of gestation, the rat fetal liver begins to synthesize its own RBP.

Evidence is also available that the developing chicken embryo obtains vitamin A needed for fetal development from maternal retinol–RBP complex that had been incorporated into the yolk of the egg (Heller, 1976; Abe *et al.*, 1976; Sreekrishna and Cama, 1978). It has been found in humans that vitamin A in cord blood is present as the retinol–RBP complex (Vahlquist *et al.*, 1975). RBP in human fetal serum, collected at the time of birth, appeared to be identical with RBP present in serum of adults (Ismadi and Olson, 1975). The extent to which cord blood RBP might represent RBP transferred from maternal to fetal circulation, compared to RBP synthesized by the term fetus, is not known. Retinol present in holo-RBP in cord blood certainly must have had a maternal origin. Information is needed about the mechanisms involved in the transport of RBP across the placenta in the developing mammalian fetus at different stages of development.

XV. SERUM ALBUMIN AS A RETINOID-BINDING AND TRANSPORT PROTEIN

In contrast to retinol, retinoic acid and other exogenous acidic retinoids appear to be transported in plasma bound to serum albumin, rather than as a complex with RBP. Retinoic acid has been reported to be present as an endogenous compound in human plasma, at concentrations of approximately 3–4 ng/ml (De Leenheer *et al.*, 1982). The plasma transport of retinoic acid and related retinoids has, however, been studied after exogenous administration of the particular retinoid of interest.

Plasma retinoic acid transport was studied with vitamin A-deficient rats maintained on small doses of radioactively labeled retinoic acid (J. E. Smith *et al.*, 1973b). On polyacrylamide gel electrophoresis almost all of the recovered radioactivity migrated with serum albumin. Similar results were obtained with serum from a normal control rat given a single oral dose of [^{14}C]retinoic acid. These findings indicated that retinoic acid is transported in rat plasma bound to albumin, and not by RBP.

It is well established that serum albumin has a small number of high-affinity binding sites for free fatty acids (Goodman, 1958; Spector, 1975). It is likely that retinoic acid is transported in plasma in a manner similar to the free fatty acids, presumably in the form of the carboxylate anion, and complexed with the same binding sites on the albumin molecule. Although retinoic acid can bind effectively to apo-RBP (discussed earlier in this chapter), in view of the much higher concentration of albumin than of RBP in plasma, there are many more binding sites in plasma available for retinoic acid on albumin molecules than on RBP. Moreover, RBP is normally secreted from the liver with its binding site occupied as the retinol–RBP complex. These considerations provide an explanation for the transport of retinoic acid and other acidic retinoids in plasma in association with serum albumin.

ACKNOWLEDGMENT

The studies from the author's laboratory discussed in this chapter were supported by NIH grants HL 21006 (SCOR) and AM 05968.

REFERENCES

Abe, T., Muto, Y., and Hosoya, N. (1975). *J. Lipid Res.* **16,** 200–210.

Abe, T., Muto, Y., and Hosoya, N. (1976). *Life Sci.* **17,** 1579–1587.

Arroyave, G., and Calcaño, M. (1979). *Arch. Latinoamer. Nutr.* **29,** 233–260.

Arroyave, G., Wilson, D., Mendez, J., Behar, M., and Scrimshaw, N. S. (1961). *Am. J. Clin. Nutr.* **9,** 180–185.

Bangham, A. D., Dingle, J. T., and Lucy, J. A. (1964). *Biochem. J.* **90,** 133–140.

Beeken, W. L., Volwiler, W., Goldsworthy, P. D., Garby, L. E., Reynolds, W. E., Stogsdill, R., and Stemler, R. S. (1962). *J. Clin. Invest.* **41,** 1312–1333.

Benoldi, D., Manfredi, G., Pezzarossa, E., and Allegra, F. (1981). *Br. J. Dermatol.* **105,** 659–664.

Bergamaschini, L., Tucci, A., Colombo, A., Agostini, A., Finzi, A. G., Altomare, G. F., and Pigatto, P. D. (1982). *N. Engl. J. Med.* **306,** 546–547.

Bernard, A. M., and Lauwerys, R. R. (1981). *Clin. Chem.* **27,** 1781–1782.
Bernard, A. M., Moreau, D., and Lauwerys, R. R. (1982). *Clin. Chem.* **28,** 1167–1171.
Bhat, A. K., and Cama, H. R. (1979). *Biochim. Biophys. Acta* **587,** 273–281.
Biesalski, H. K., Gross, M., Ehrenthal, W., and Harth, O. (1981). *Laryngol. Rhinol. Otol.* **60,** 631–635.
Blake, C. C. F. (1981). *Proc. R. Soc. London, Ser. B* **211,** 413–431.
Blake, C. C. F., and Oatley, S. J. (1977). *Nature (London)* **268,** 115–120.
Blake, C. C. F., Swan, I. D. A., Rerat, C., Berthou, J., Laurent, A., and Rerat, B. (1971). *J. Mol. Biol.* **61,** 217–224.
Blake, C. C. F., Geisow, M. J., Swan, I. D. A., Rérat, C., and Rérat, B. (1974). *J. Mol. Biol.* **88,** 1–12.
Blake, C. C. F., Geisow, M. J., Oatley, S. J., Rérat, B., and Rérat, C. (1978). *J. Mol. Biol.* **121,** 339–356.
Bok, D., and Heller, J. (1976). *Exp. Eye Res.* **22,** 395–402.
Borek, C., Smith, J. E., Soprano, D. R., and Goodman, D. S. (1981). *Endocrinology (Baltimore)* **109,** 386–391.
Brissot, P., Le Treut, A., Dien, G., Cottencin, M., Simon, M., and Bourel, M. (1978). *Digestion* **17,** 469–478.
Carney, S. M., Underwood, B. A., and Loerch, J. D. (1976). *J. Nutr.* **106,** 1773–1781.
Chen, C.-C., and Heller, J. (1977). *J. Biol. Chem.* **252,** 5216–5221.
Chen, C.-C., Heller, J., Ding, L.-L., and Horwitz, J. (1981). *Arch. Biochem. Biophys.* **207,** 392–398.
Cogan, U., Kopelman, M., Mokady, S., and Shinitzky, M. (1976). *Eur. J. Biochem.* **65,** 71–78.
Costanzo, F., Castagnoli, L., Dente, L., Arcari, P., Smith, M., Costanzo, P., Raugei, G., Izzo, P., Pietropaolo, T. C., Bougueleret, L., Cimino, F., Salvatore, F., and Cortese, R. (1983). *EMBO J.* **2,** 57–61.
Crabb, J. W., and Saari, J. C. (1981). *FEBS Lett.* **130,** 15–18.
Davis, P. J., Spaulding, S. W., and Gregerman, R. I. (1970). *Endocrinology (Baltimore)* **87,** 978–986.
De Bevere, V. O. R. C., De Paepe, M., De Leenheer, A. P., Nelis, H. J. C. F., Lambert, W. E. E. S., Claeys, A. G., and Ringoir, S. (1981). *Clin. Chim. Acta* **114,** 249–256.
De Leenheer, A. P., Lambert, W. E., and Claeys, I. (1982). *J. Lipid Res.* **23,** 1362–1367.
De Luca, L. M., Glover, J., Heller, J., Olson, J. A., and Underwood, B. (1979). Report of the International Vitamin A Consultative Group (IVACG). VI. The Nutrition Foundation, Inc., New York, pp. 30–32.
Dingle, J. T., Barrett, A. J., and Weston, P. D. (1971). *Biochem. J.* **123,** 1–13.
Dingle, J. T., Fell, H. B., and Goodman, D. S. (1972). *J. Cell Sci.* **11,** 393–402.
Fell, H. B. (1970). *In* "The Fat Soluble Vitamins" (H. F. DeLuca and J. W. Suttie, eds.), pp. 187–202. Univ. of Wisconsin Press, Madison.
Fell, H. B., and Dingle, J. T. (1963). *Biochem. J.* **87,** 403–408.
Ferguson, R. N., Edelhoch, H., Saroff, H. A., and Robbins, J. (1975). *Biochemistry* **14,** 282–289.
Fex, G., and Hansson, B. (1978). *Biochim. Biophys. Acta* **537,** 358–365.
Fex, G., and Hansson B. (1979). *Eur. J. Biochem.* **94,** 307–313.
Fex, G., and Lindgren, R. (1977). *Biochim. Biophys. Acta* **493,** 410–417.
Fex, G., Albertsson, P.-Å., and Hansson, B. (1979). *Eur. J. Biochem.* **99,** 353–360.
Finzi, A. F., Altomare, G., Bergamaschini, L., and Tucci, A. (1981). *Br. J. Dermatol.* **104,** 253–256.
Futterman, S., and Heller, J. (1972). *J. Biol. Chem.* **247,** 5168–5172.
Futterman, S., Swanson, D., and Kalina, R. E. (1974). *Invest. Ophthalmol.* **13,** 798–801.
Futterman, S., Swanson, D., and Kalina, R. E. (1975). *Invest. Ophthalmol.* **14,** 125–130.
Gawinowicz, M. A., and Goodman, D. S. (1982). *Biochemistry* **21,** 1899–1905.

Glover, J. (1973). *Vitam. Horm.* **31,** 1–42.
Glover, J., (1980). *In* "Methods in Enzymology" (D. B. McCormick and L. D. Wright, eds.), Vol. 67, pp. 282–287. Academic Press, New York.
Glover, J., Moxley, L., Muhilal, H., and Weston, S. (1974). *Clin. Chim. Acta* **50,** 371–380.
Golden, M. H. N., (1982). *Am. J. Clin. Nutr.* **35,** 1159–1165.
Goodman, D. S. (1958). *J. Am. Chem. Soc.* **80,** 3892–3898.
Goodman, D. S. (1980). *Ann. N. Y. Acad. Sci.* **348,** 378–390.
Goodman, D. S., and Leslie, R. B. (1972). *Biochim. Biophys. Acta* **260,** 670–678.
Goodman, D. S., and Raz, A. (1972). *J. Lipid Res.* **13,** 338–347.
Goodman, D. S., Smith, J. E., Hembry, R. M., and Dingle, J. T. (1974). *J. Lipid Res.* **13,** 338–347.
Gotto, A. M., Lux, S. E., and Goodman, D. S. (1972). *Biochim. Biophys. Acta* **271,**429–435.
Handler, C. E., and Smith, J. E. (1983). *Fed. Proc., Fed. Am. Soc. Exp. Biol.* **42,** 395.
Harrison, E. H., Smith, J. E., and Goodman, D. S. (1979). *J. Lipid Res.* **20,** 760–771.
Harrison, E. H., Smith, J. E., and Goodman, D. S. (1980). *Biochim. Biophys. Acta* **628,** 489–497.
Hase, J., Kobashi, K., Nakai, H., and Onaska, S. (1976). *J. Biochem. (Tokyo)* **79,** 373–380.
Haupt, H., and Heide, K. (1972). *Blut* **24,** 94–101.
Heller, J. (1975a). *J. Biol. Chem.* **250,** 3613–3619.
Heller, J. (1975b). *J. Biol. Chem.* **250,** 6549–6554.
Heller, J. (1976). *Dev. Biol.* **51,** 1–9.
Heller, J., and Bok, D. (1976). *Am. J. Ophthalmol.* **81,** 93–97.
Heller, J., and Horwitz, J. (1973). *J. Biol. Chem.* **248,** 6308–6316.
Heller, J., and Horwitz, J. (1974a). *J. Biol. Chem.* **249,** 5933–5938.
Heller, J., and Horwitz, J. (1974b). *Exp. Eye Res.* **18,** 41–49.
Heller, J., and Horwitz, J. (1975). *J. Biol. Chem.* **250,** 3019–3023.
Hemley, R., Kohler, B. E., and Siviski, P. (1979). *Biophys. J.* **28,** 447–455.
Horwitz, J., and Heller, J. (1973). *J. Biol. Chem.* **248,** 6317–6324.
Horwitz, J., and Heller, J. (1974a). *J. Biol. Chem.* **249,** 4712–4719.
Horwitz, J., and Heller, J. (1974b). *J. Biol. Chem.* **249,** 7181–7185.
Ingbar, S. H. (1963). *J. Clin. Invest.* **42,** 143–160.
Ingenbleek, Y., Van Den Schriek, H.-G., De Nayer, P., and De Visscher, M. (1975a). *Metabolism* **24,** 633–641.
Ingenbleek, Y., Van Den Schriek, H.-G., De Nayer, P., and De Visscher, M. (1975b). *Clin. Chim. Acta* **63,** 61–67.
Ismadi, S. D., and Olson, J. A. (1975). *Am. J. Clin. Nutr.* **28,** 167–172.
Jackson, R. C., and Blobel, G. (1980). *Ann. N. Y. Acad. Sci.* **343,** 391–404.
Jörnvall, H., Carlström, A., Pettersson, T., Jacobsson, B., Persson, M., and Mutt, V. (1981). *Nature (London)* **291,** 261–263.
Kanai, M., Raz, A., and Goodman, D. S. (1968). *J. Clin. Invest.* **47,** 2025–2044.
Kanai, M., Nomoto, S., Sasaoka, S., and Naiki, M. (1971). *Rinsho Kagaku Shimpojuma* **11,** 194–199.
Kanai, M., Nomoto, S., Sasaoka, S., and Muto, Y. (1972a). *Rinsho Kagaku Shimpojuma* **12,** 319–324.
Kanai, M., Sasaoka, S., and Naiki, M. (1972b). *Rinsho Kagaku Shimpojuma* **12,** 325–330.
Kanda, Y., and Goodman, D. S. (1979). *J. Lipid Res.* **20,** 865–878.
Kanda, Y., Goodman, D. S., Canfield, R. E., and Morgan, F. J. (1974). *J. Biol. Chem.* **249,** 6796–6805.
Kasper, H., Brodersen, M., and Schedel, R. (1975). *Acta Hepato-Gastroenterol.* **22,** 403–408.
Katoh, M., Kanai, M., Kameko, M., Ohno, S., Fujii, Y., and Nagata, T. (1982). *Acta Histochem. Cytochem.* **15,** 68–75.

Kindler, U. (1972). *Dtsch. Med. Wochenschr.* **97,** 1821–1823.
Knöpfle, G., Rotthauwe, H. W., and Odenthal, A. (1975). *Z. Kinderheilkd.* **119,** 279–291.
Kopelman, M., Mokady, S., and Cogan, U. (1976a). *Biochim. Biophys. Acta* **439,** 442–448.
Kopelman, M., Cogan, U., Mokady, S., and Shinitzky, M. (1976b). *Biochim. Biophys. Acta* **439,** 449–460.
Large, S., Neal, G., Glover, J., Thanangkul, O., and Olson, R. E. (1980). *Br. J. Nutr.* **43,** 393–402.
Lewis, K. C., Green, M. H., and Underwood, B. A. (1981). *J. Nutr.* **111,** 1135–1144.
McClain, C. J., Van Thiel, D. H., Parker, S., Badzin, L. K., and Gilbert, H. (1979). *Alcoholism: Clin. Exp. Res.* **3,** 135–141.
McGuire, B. W., and Chytil, F. (1980). *Biochim. Biophys. Acta* **621,** 324–331.
McGuire, B. W., Orbegin-Crist, M.-C., and Chytil, F. (1981). *Endocrinology (Baltimore)* **108,** 658–667.
Mallia, A. K., Smith, J. E., and Goodman, D. S. (1975). *J. Lipid Res.* **16,** 180–188.
Maraini, G. (1974). *Invest. Ophthalmol.* **13,** 288–290.
Maraini, G., and Gozzoli, F. (1976). *Invest. Ophthalmol.* **15,** 1020–1022.
Maraini, G., Fadda, G., and Gozzoli, F. (1975). *Invest. Ophthalmol.* **14,** 236–237.
Michaëlsson, G., Vahlquist, A., and Jublin, L. (1977). *Br. J. Dermatol.* **96,** 283–286.
Mogielnicki, R. P., Waldmann, T. A., and Strober, W. (1971). *J. Clin. Invest.* **50,** 901–909.
Mokady, S., and Tal, M. (1974). *Biochim. Biophys. Acta* **336,** 361–366.
Moody, B. J. (1982). *Clin. Chim. Acta* **118,** 87–92.
Muto, Y., and Goodman, D. S. (1972). *J. Biol. Chem.* **247,** 2533–2541.
Muto, Y., Smith, J. E., Milch, P. O., and Goodman, D. S. (1972). *J. Biol. Chem.* **247,** 2542–2550.
Muto, Y., Nakanishi, M., and Shidoji, Y. (1976). *J. Biochem. (Tokyo)* **79,** 775–785.
Muto, Y., Shidoji, Y., and Kanda, Y. (1982). *In* "Methods in Enzymology" (L. Parker, ed.), Vol. 81, pp. 840–852. Academic Press, New York.
Navab, M., Mallia, A. K., Kanda, Y., and Goodman, D. S. (1977a). *J. Biol. Chem.* **252,** 5100–5106.
Navab, M., Smith, J. E., and Goodman, D. S. (1977b). *J. Biol. Chem.* **252,** 5107–5114.
Nilsson, S. F., Rask, L., and Peterson, P. A. (1975). *J. Biol. Chem.* **250,** 8554–8563.
Nomenclature Committee of the IUB and IUPAC-IUB Joint Commission on Biochemical Nomenclature (1981). *J. Biol. Chem.* **256,** 12–14.
Nomiyama, K., Yotoriyama, M., and Nomiyama, H. (1981). *Arch. Environ. Contam. Toxicol.* **10,** 297–304.
Nonavinakere, V. K., Man, Y. M., and Lei, K. Y. (1981). *Nutr. Rep. Int.* **23,** 697–704.
Ogunshina, S. O., and Hussain, M. A. (1980). *Am. J. Clin. Nutr.* **33,** 794–800.
Oppenheimer, J. H. (1968). *N. Engl. J. Med.* **278,** 1153–1162.
Ottonello, S., Maraini, G., Mammi, M., Monaco, H. L., Spadon, P., and Zanotti, G. (1983). *J. Mol. Biol.* **163,** 679–681.
Palin, D., Underwood, B. A., and Denning, C. R. (1979). *Am. J. Clin. Nutr.* **32,** 1253–1259.
Peterson, P. A. (1971a). *J. Biol. Chem.* **246,** 34–43.
Peterson, P. A. (1971b). *J. Biol. Chem.* **246,** 44–49.
Peterson, P. A. (1971c). *Eur. J. Clin. Invest.* **1,** 437–444.
Peterson, P. A., and Berggård, I. (1971). *J. Biol. Chem.* **246,** 25–33.
Peterson, P. A., and Rask, L. (1971). *J. Biol. Chem.* **246,** 7544–7550.
Peterson, P. A., Evrin, P.-E., and Berggård, I. (1969). *J. Clin. Invest.* **48,** 1189–1198.
Peterson, P. A., Rask, L., Östberg, L., Andersson, L. Kamwendo, F., and Pertoft, H. (1973). *J. Biol. Chem.* **248,** 4009–4022.
Peterson, P. A., Nilsson, S. F., Östberg, L., Rask, L., and Vahlquist, A. (1974). *Vitam. Horm.* **32,** 181–214.

Poole, A. R., Dingle, J. T., Mallia, A. K., and Goodman, D. S. (1975). *J. Cell Sci.* **19,** 379–394.
Poulik, M. D., Farrah, D., Malek, G. H., Shinnick, C. J., and Smithies, O. (1975). *Biochim. Biophys. Acta* **412,** 326–334.
Prellwitz, W., Olbermann, M., Kapp, S., Müller, D., and Groth, U. (1974). *Klin. Wochenschr.* **52,** 921–925.
Quinn, P. S., Gamble, M., and Judah, J. D. (1975). *Biochem. J.* **146,** 389–393.
Rahi, A. H. S. (1972). *Br. J. Ophthalmol.* **56,** 647–651.
Rask, L. (1974). *Eur. J. Biochem.* **44,** 1–5.
Rask, L., and Peterson, P. A. (1976). *J. Biol. Chem.* **251,** 6360–6366.
Rask, L., Vahlquist, A., and Peterson, P. A. (1971). *J. Biol. Chem.* **246,** 6638–6646.
Rask, L., Peterson, P. A., and Björk, I. (1972). *Biochemistry* **11,** 264–268.
Rask, L., Anundi, H., and Peterson, P. A. (1979). *FEBS Lett.* **104,** 55–58.
Rask, L., Anundi, H., Böhme, H., Eriksson, U., Ronne, H., Sege, K., and Peterson, P. A. (1981). *Ann. N. Y. Acad. Sci.* **359,** 79–90.
Rask, L., Valtersson, C., Anundi, H., Kvist, S., Eriksson, U., Dallner, G., and Peterson, P. A. (1983). *Exp. Cell Res.* **143,** 91–102.
Raz, A., and Goodman, D. S. (1969). *J. Biol. Chem.* **244,** 3230–3237.
Raz, A., Shiratori, T., and Goodman, D. S. (1970). *J. Biol. Chem.* **245,** 1903–1912.
Reddy, V., Mohanram, M., and Raghuramulu, N. (1979). *Acta Paediatr. Scand.* **68,** 65–69.
Redman, C. M., Banerjee, D., Howell, K., and Palade, G. E. (1975). *J. Cell Biol.* **66,** 42–59.
Roels, O. A., Anderson, O. R., Lui, N. S. T., Shah, D. O., and Trout, M. E. (1969). *Am. J. Clin. Nutr.* **22,** 1020–1O32.
Ronne, H., Ocklind, C., Wiman, K., Rask, L., Öbrink, B., and Peterson, P. A. (1983). *J. Cell Biol.* **96,** 907–910.
Russell, J. H., and Geller, D. M. (1975). *J. Biol. Chem.* **250,** 3409–3413.
Russell, R. M., Morrison, S. A., Smith, F. R., Oaks, E. V., and Carney, E. A. (1978). *Ann. Intern Med.* **88,** 622–626.
Scarpioni, L., Dall'Aglio, P. P., Poisetti, P. G., and Buzio, C. (1976). *Clin. Chim. Acta* **68,** 107–113.
Schultz, W., Delling, G., Buchel, C. G., and Kasper, H. (1977). *Inn. Med.* **4,** *Prax. Klin. (2. Aufl.)* 349–360.
Sege, K., and Peterson, P. A. (1978). *Proc. Natl. Acad. Sci. U.S.A.* **75,** 2443–2447.
Shenai, J. P., Chytil, F., Jhaveri, A., and Stahlman, M. T. (1981). *J. Pediatr. (St. Louis)* **99,** 302–305.
Shidoji, Y., and Muto, Y. (1977). *J. Lipid Res.* **18,** 679–691.
Shidoji, Y., Watanabe, M., and Muto, Y. (1979). *J. Nutr. Sci. Vitaminol.* **25,** 53–57.
Shingwekar, A. G., Mohanaram, M., and Reddy, V. (1979). *Clin. Chim. Acta* **93,** 97–100.
Sklan, D., Blaner, W. S., Adachi, N., Smith, J. E., and Goodman, D. S. (1982). *Arch. Biochem. Biophys.* **214,** 35–44.
Skrede, S., Blomhoff, J. P., Elgjo, K., and Gjone, E. (1975). *Scand. J. Clin. Lab. Invest.* **35,** 399–406.
Smith, F. R., and Goodman, D. S. (1970). *J. Clin. Invest.* **49,** 90a.
Smith, F. R., and Goodman, D. S. (1971). *J. Clin. Invest.* **50,** 2426–2436.
Smith, F. R., and Goodman, D. S. (1976). *N. Engl. J. Med.* **294,** 805–808.
Smith, F. R., and Lindenbaum, J. (1974). *Am. J. Clin. Nutr.* **27,** 700–705.
Smith, F. R., Raz, A., and Goodman, D. S. (1970). *J. Clin. Invest.* **49,** 1754–1761.
Smith, F. R., Underwood, B. A., Denning, C. R., Varma, A., and Goodman, D. S. (1972). *J. Lab. Clin. Med.* **80,** 423–433.
Smith, F. R., Goodman, D. S., Zaklama, M. S., Gabr, M. K., El Maraghy, S., and Patwardhan, V. N. (1973a). *Am. J. Clin. Nutr.* **26,** 973–981.

Smith, F. R., Goodman, D. S., Arroyave, G., and Viteri, F. (1973b). *Am. J. Clin. Nutr.* **26,** 982–987.

Smith, F. R. Suskind, R., Thanangkul, O., Leitzmann, C., Goodman, D. S., and Olson, R. E. (1975). *Am. J. Clin. Nutr.* **28,** 732–738.

Smith, J. C., Jr. (1980). *Ann. N. Y. Acad. Sci.* **355,** 62–75.

Smith, J. C., Jr., Brown, E. D., McDaniel, E. G., and Chan, W. (1976). *J. Nutr.* **106,** 569–574.

Smith, J. E., and Goodman, D. S. (1979). *Fed. Proc., Fed. Am. Soc. Exp. Biol.* **38,** 2504–2509.

Smith, J. E., Muto, Y., Milch, P. O., and Goodman, D. S. (1973a). *J. Biol. Chem.* **248,** 1544–1549.

Smith, J. E., Milch, P. O., Muto, Y., and Goodman, D. S. (1973b). *Biochem. J.* **132,** 821–827.

Smith, J. E., Brown, E. D., and Smith, J. C., Jr. (1974). *J. Lab. Clin. Med.* **84,** 692–697.

Smith, J. E., Muto, Y., and Goodman, D. S. (1975). *J. Lipid Res.* **16,** 318–323.

Smith, J. E., Borek, C., and Goodman, D. S. (1978). *Cell* **15,** 865–873.

Smith, J. E., Deen, D. D., Jr., Sklan, D., and Goodman, D. S. (1980). *J. Lipid Res.* **21,** 229–237.

Smith, J. E., Borek, C., and Goodman, D. S. (1981). *Ann. N. Y. Acad. Sci.* **359,** 171–180.

Socolow, E. L., Woeber, K. A., Purdy, R. H., Holloway, M. T., and Ingbar, S. H. (1965). *J. Clin. Invest.* **44,** 1600–1609.

Solomons, N. W., and Russell, R. M. (1980). *Am. J. Clin. Nutr.* **33,** 2031–2040.

Soprano, D. R., Pickett, C. B., Smith, J. E., and Goodman, D. S. (1981). *J. Biol. Chem.* **256,** 8256–8258.

Soprano, D. R., Smith, J. E., and Goodman, D. S. (1982a). *J. Biol. Chem.* **257,** 7693–7697.

Soprano, D. R., Kato, M., and Goodman, D. S. (1982b). *Fed. Proc., Fed. Am. Soc. Exp. Biol.* **41,** 518.

Spector, A. A. (1975). *J. Lipid Res.* **16,** 165–179.

Sreekrishna, K., and Cama, H. R. (1978). *Indian J. Biochem. Biophys.* **15,** 255–259.

Stewart, W. K., and Fleming, L. W. (1982). *Nephron* **30,** 15–21.

Stoll, D. M., King, L. E., Jr., and Chytil, F. (1983). *Br. J. Dermatol,* **108,** 375–378.

Strauss, A. W., Bennett, C. D., Donohue, A. M., Rodkey, J. A., and Alberts, A. W. (1977). *J. Biol. Chem.* **252,** 6846–6855.

Sutherland, R. L., and Brandon, M. R. (1976). *Endocrinology (Baltimore)* **98,** 91–98.

Takahashi, Y. I., Smith, J. E., Winick, M., and Goodman, D. S. (1975). *J. Nutr.* **105,** 1299–1310.

Takahashi, Y. I., Smith, J. E., and Goodman, D. S. (1977). *Am. J. Physiol.* **233,** E263–E272.

Tosukhowong, P., and Olson, J. A. (1978). *Biochim. Biophys. Acta* **529,** 438–453.

Trägårdh, L., Anundi, H., Rask, L., Sege, K., and Peterson, P. A. (1980). *J. Biol. Chem.* **255,** 9243–9248.

USDA (1980). Nationwide Food Consumption Survey, 1977–1978; Preliminary Report No. 2: Food and Nutrient Intakes of Individuals in One Day in the United States, Spring, 1977.

Vahlquist, A. (1972). *Scand. J. Clin. Lab. Invest.* **30,** 349–360.

Vahlquist, A. (1982). *Br. J. Dermatol.* **107,** 125–126.

Vahlquist, A., and Peterson, P. A. (1972). *Biochemistry* **11,** 4526–4532.

Vahlquist, A., Nilsson, S. F., and Peterson, P. A. (1971). *Eur. J. Biochem.* **20,** 160–168.

Vahlquist, A., Peterson, P. A., and Wibell, L. (1973). *Eur. J. Clin. Invest.* **3,** 352–362.

Vahlquist, A., Rask, L., Peterson, P. A., and Berg, T. (1975). *Scand. J. Clin. Lab. Invest.* **35,** 569–575.

Vahlquist, A., Sjölund, K., Nordén, Å, Peterson, P. A., Stigmar, G., and Johansson, B. (1978a). *Scand. J. Clin. Lab. Invest.* **38,** 301–308.

Vahlquist, A., Michaëlsson, G., and Juhlin, L. (1978b). *Acta Derm. Venereol.* **58,** 437–442.

Vahlquist, A., Johnsson, A., and Nygren, K.-G. (1979). *Am. J. Clin. Nutr.* **32,** 1433–1438.

Vahlquist, A., Berne, B., and Berne, C. (1982a). *Eur. J. Clin. Invest.* **12,** 63–67.

Vahlquist, A., Lee, J. B., and Michaëlsson, G. (1982b). *Acta Dermatol.* **118,** 389–392.

van Jaarsveld, P. P., Edelhoch, H., Goodman, D. S., and Robbins, J. (1973a). *J. Biol. Chem.* **248,** 4698–4705.

van Jaarsveld, P., Branch, W. T., Robbins, J., Morgan, F. J., Kanda, K., and Canfield, R. E. (1973b). *J. Biol. Chem.* **248,** 7898–7903.

Venkataswamy, G., Glover, J., Cobby, M., and Pirie, A. (1977). *Am. J. Clin. Nutr.* **30,** 1968–1973.

White, G. H., Weston, S. M., and Glover, J. (1972). *FEBS Lett.* **27,** 107–110.

Woeber, K. A., and Ingbar, S. H. (1968). *J. Clin. Invest.* **47,** 1710–1721.

9

Cellular Retinoid-Binding Proteins

FRANK CHYTIL AND DAVID E. ONG

Department of Biochemistry
Vanderbilt University School of Medicine
Nashville, Tennessee

THE RETINOIDS, VOL. 2

ISBN 0-12-658102-9

I. INTRODUCTION

Many efforts designed to determine the molecular mechanisms of retinoid action in directing differentiation of epithelia as well as to understand their effects on malignant growth have centered on intracellular retinoid-binding proteins. The best characterized are cellular retinol-binding protein (CRBP) and cellular retinoic acid-binding protein (CRABP). While this nomenclature is widely accepted, other synonyms used for CRBP are intracellular retinol-binding protein, cytosol retinol-binding protein, or cytosol retinol receptor. CRABP is sometimes called retinoic acid-binding protein or retinoic acid receptor. A less well-characterized protein of this family is CRBP(F), reported originally in fish eye and now also in malignant, fetal, and regenerating rat liver specimens (Shidoji and Muto, 1978; Sato *et al.*, 1979; Muto *et al.*, 1979, 1981; Muto and Omori, 1981; Omori *et al.*, 1981). This protein is not discussed further in this chapter.

These proteins have been detected in a variety of tissues. The following discussion emphasizes their properties and possible function in nonvisual tissues. The possible functions of CRBP and CRABP and several other cellular retinoid-binding proteins in visual tissue are reviewed in Chapter 10 of this volume. Review articles about CRBP and CRABP in nonocular tissue are also available (Chytil and Ong, 1978a,b, 1979, 1982, 1983; Ross and Goodman, 1979; Lotan, 1980; Chytil, 1982; Goodman, 1981, 1982).

II. DETECTION AND QUANTITATION OF CELLULAR RETINOL- AND RETINOIC ACID-BINDING PROTEINS

Retinoid-binding proteins have been found in cytosols of many tissues from a number of species. CRBP was first detected by sucrose gradient centrifugation (Bashor *et al.*, 1973). Subsequently, this method was used for the successful detection of CRABP (Sani and Hill, 1974; Ong and Chytil, 1975a). Gel filtration on Sephadex G-100 (Bashor *et al.*, 1973), Sephadex G-75 (Futterman *et al.*, 1976), and Sepharose 4B has also been used (Wiggert *et al.*, 1977a). Agarose gel electrophoresis (Huber *et al.*, 1978; Küng *et al.*, 1980) as well as a batch-assay by removal of the free ligand by charcoal-coated dextran have been described (Trown *et al.*, 1980).

These methods require preincubation of the tissue extracts or cytosols with radioactive ligands. Retinol (A1) tritiated in the C-15 position and retinoic acid (D1) tritiated in the C-11 and C-12 positions have been used most frequently. The free ligand is then separated from that bound to the respective protein by one of the methods mentioned above. The amount of specifically bound retinol or retinoic acid is determined by the difference between the amount of radioactivity recovered compared to a sample which was incubated with the radioactive ligand and an excess (usually 100- to 200-fold) of nonradioactive retinoid. Thus the sensitivity of the assay depends primarily on the specific activity of the respective ligand and on the effectiveness of the separation of the free from the bound retinoid. The main component contributing to nonspecific binding in tissue extracts has been identified as serum albumin (Bashor and Chytil, 1975; Sani and Hill, 1976). Consequently, some methods incorporate the removal of albumin from the samples (Sani *et al.*, 1978; Trown *et al.*, 1980).

Specific antibodies to CRBP have permitted the development of sensitive radioimmunoassays (Ong and Chytil, 1979, 1981; Adachi *et al.*, 1981; Ong *et al.*, 1982b). A radioimmunoassay for CRABP is now also available (Ong *et al.*, 1982b). These radioimmunoassays appear to be more sensitive than the nonimmunochemical methods.

III. CHARACTERISTICS OF CELLULAR RETINOL-BINDING PROTEIN

A. Physicochemical Properties

CRBP is available as a homogeneous preparation from a number of tissues and species. The protein has been purified from rat liver (Ong and Chytil, 1978a, 1980; Liou *et al.*, 1981; Rask *et al.*, 1981), human liver (Ong, 1982; Fex and Johannesson, 1982), dog liver (Liou *et al.*, 1981), rat testes (Ong and Chytil, 1978b, 1980, Ross *et al.*, 1978), and bovine retina (Saari *et al.*, 1978a).

Other preparations of CRBP from transplantable mouse colon (Sani *et al.*, 1980a), oviduct magnum of laying hens (Das *et al.*, 1978; Rao *et al.*, 1979), rat testes cytosol, nucleosol, and chromatin (Shinde *et al.*, 1980) are less well characterized.

CRBP has a sedimentation coefficient of 2 S (Bashor *et al.*, 1973). The molecular weight of purified preparations from rat liver as well as testes has been determined independently by two groups to be 14,600 in a single polypeptide chain (Ong and Chytil, 1978a; Ross *et al.*, 1978). Human liver CRBP has a molecular weight of 14,600 (Ong, 1982) or 15,000 (Fex and Johannesson, 1982). The protein from bovine retina showed a molecular weight of 16,600 (Saari *et al.*, 1978a), and that from hen oviduct 13,180 (Das *et al.*, 1978). Thus the molecular weight in all cases is similar if not the same (see Liou *et al.*, 1981) and smaller than that of the blood transport protein for retinol, RBP (see Chapter 8). Also, CRBP does not bind to transthyretin (Ross *et al.*, 1978). Isoelectric focusing of the CRBP purified from rat testis separated that preparation into two bands with p*I* 4.8 and 4.9 (Ross *et al.*, 1978). Preparations from various sources have different mobilities when electrophoresed in acrylamide gels (Huber *et al.*,

1978; Liou *et al.*, 1981). A preliminary report has announced the crystallization of CRBP from rat liver and presented early X-ray data on these crystals (Newcomer *et al.*, 1981). The crystals were grown from solutions of polyethylene glycol-6,000 and $CdCl_2$. The crystals are orthorhombic, space group $P_{2,2,2}$, a = 40.09, b = 47.39, and c = 69.34. An assymetric unit contains one molecule. Diffraction has been observed to Bragg spacings of 1.7-Å resolution.

B. Spectral Properties

The absorption spectrum of purified human CRBP complexed with retinol is shown in Fig. 1. The spectrum is dominated by the absorbance of bound retinol with λ_{max} at 350 nm. This spectrum of retinol is considerably altered from its spectrum in organic solvents, having λ_{max} red shifted 25 nm and with the introduction of fine structure as shown by the shoulder at 330 nm and the second peak with λ_{max} at 367 nm. The molar extinction coefficient for bound retinol is increased by a factor of 1.14 compared to the value for retinol in ethanol. The protein absorbance peaks at 284 nm and the ratio of A_{350}/A_{284} for the complex is 1.8.

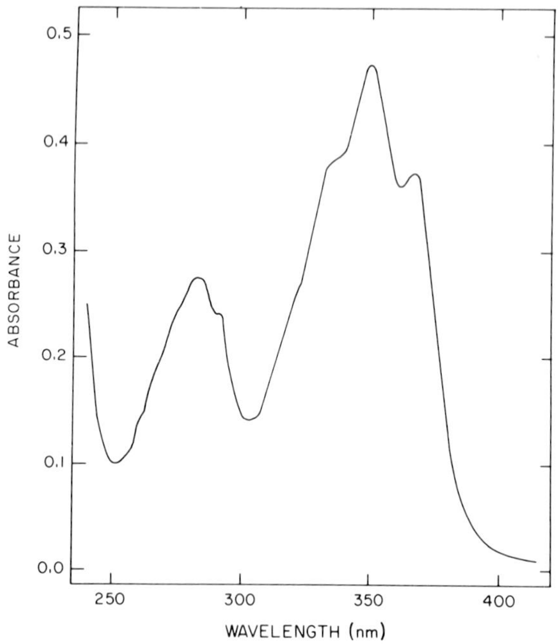

Fig. 1. Absorption spectrum of human liver CRBP with bound retinol, 8.0×10^{-6} *M*, in 0.05 *M* imidazole acetate, pH 6.4. (Reproduced with permission from Ong, 1982.)

It has been suggested that the spectrum of all-*trans*-retinol in organic solution is atypical for a conjugated pentaene, owing to the steric hindrance in the molecule that prevents coplanarity of the ring and side chain and, consequently, full conjugation of the double bond of the ring with the four double bonds of the side chain (Oroshnik *et al.*, 1952). Compared to what would be predicted for full conjugation, the spectrum of all-*trans*-retinol exhibits loss of fine structure, a displacement of λ_{max} to the blue of more than 20 nm, and a decrease in extinction coefficient (Oroshnik *et al.*, 1952; Braude *et al.*, 1949). Consequently, the presence of fine structure, the 25-nm shift in λ_{max} to the red, and the higher extinction coefficient observed in the spectrum of all-*trans*-retinol bound to CRBP suggests the protein binds retinol in a conformation that allows fuller conjugation of its double-bond system, i.e., a conformation more nearly planar than the average orientation of retinol in solutions in organic solvents.

As was observed previously for some other proteins that bind retinol, such as β-lactoglobulin and RBP (see Chapter 8; Futterman and Heller, 1972), there is considerable enhancement of the fluorescence of retinol when it binds to CRBP compared to its fluorescence in a solvent such as ethanol. This increase in intensity is about 7- to 8-fold. The fluorescence excitation and emission spectra of the bound retinol are shown in Fig. 2. Both spectra and uncorrected. The excitation spectrum has a major peak at 350 nm, a second peak at 368 nm, and a slight shoulder at about 330 nm. This compared well with the absorption spectrum, as would be expected. The small peak at 290 nm suggests there is energy

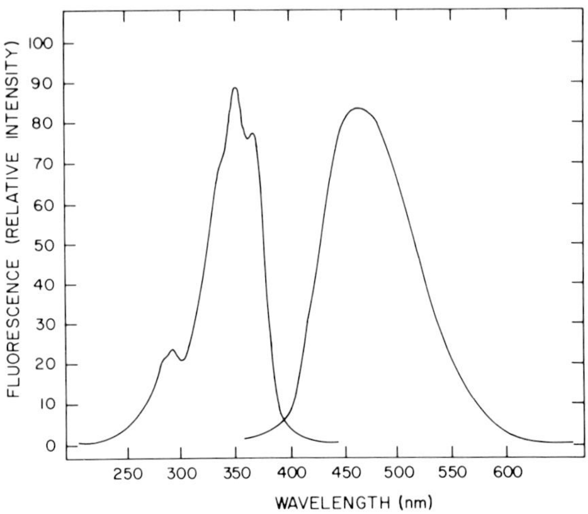

Fig. 2. Uncorrected fluorescence spectra of retinol bound to human liver CRBP, $2.8 \times 10^{-6} M$, in 0.05 *M* imidazole acetate, pH 6.4. Excitation spectrum (left) was determined by monitoring emission at 470 nm. Emission spectrum (right) was determined with excitation at 350 nm. (Reproduced with permission from Ong, 1982.)

transfer from tryptophan residues to the retinol. The emission spectrum peaks at 470 nm. Under the same conditions of buffer and pH, the spectra of rat CRBP are identical to those shown here for human CRBP.

C. Immunochemical Properties

An early study showed that liver CRBP is a protein different from plasma RBP when antibodies directed against plasma RBP failed to react with a crude preparation of CRBP (Bashor and Chytil, 1975). This finding has been corroborated using CRBP purified from rat testis (Ross *et al.*, 1978). Specific antisera against rat CRBP, obtained in rabbits or turkeys, were found not to react with plasma RBP (Ong and Chytil, 1979, 1981; Adachi *et al.*, 1981).

Immunochemical studies have also shown that CRBP does not appear to be tissue or sex specific (Ong and Chytil, 1979; Adachi *et al.*, 1981; Ong *et al.*, 1982b), but species differences have been detected using antisera against rat liver CRBP (Ong and Chytil, 1981). In contrast, other investigators did not observe any immunochemical differences in CRBP prepared from various species (Liou *et al.*, 1981). These findings can be explained by differences in the specificity of the antisera employed in the various studies. It should be stressed that CRBP has proved to be a poor immunogen (Ong and Chytil, 1979; Adachi *et al.*, 1981).

D. Amino Acid Sequences

The amino terminal amino acid sequences of CRBP isolated from rat liver (Rask *et al.*, 1981), human liver (Fex and Johannesson, 1982), and bovine retina (Crabb and Saari, 1981) have been reported. As shown in Table I, all three partial sequences show a striking degree of homology. This sequence homology may contribute to the poor immunogenicity of CRBP and could explain the failure in one case to raise species-specific antibodies (Liou *et al.*, 1981). Rask *et al.* (1981) observed that the sequence of rat liver CRBP also has some homology with myelin protein P_2. Rat liver Z-protein has similar homology (Takahashi *et al.*, 1982). The significance of this homology remains to be established. Nevertheless, it suggests some phylogenetic relation of the genes coding for these proteins (as well as CRABP, see Section IV,D).

The sequence data demonstrated again that CRBP is different from plasma RBP. Some similarity does exist, however, suggesting the possibility of a common ancestry for the genes for CRBP and RBP (Crabb and Saari, 1981).

E. Binding Specificity and Ligand Affinity

Earlier experiments have shown that CRBP has a striking specificity for binding retinol (Bashor *et al.*, 1973). It was established that preparations of CRBP

TABLE I

Amino Terminal Amino Acid Sequences of Cellular Retinol-Binding (CRBP) Protein from Various Species[a]

Source	Sequence (5, 10, 15, 20, 25, 30)
Rat[b] liver	Pro-Val-Asp-Phe-Asn-Gly-Tyr-Trp-Lys-Met-Leu-Ser-Asn-Glu-Asn-Phe-Glu-Glu-Tyr-Leu-Arg-Ala-Leu-Asp-Val-Asn-Val-Ala-Leu-Arg-
Bovine[c] retina	Pro-Val-Asp-Phe-Thr-Gly-Tyr-Trp-Lys-Met-Leu-Ala-Asp-Glu-Asn-Phe-Glu-Glu-Tyr-Leu-Arg-Gly-Leu-Asp-Val-Asn-Val-Ala-Leu-Arg-
Human[d] liver	Pro-Val-Asp-Phe-Thr-Gly-Tyr-X-Lys-Met-Leu-Val-Asn-Glu-Asn-Phe-

[a] Identical sequences are boxed.
[b] Rask *et al.* (1981).
[c] Crabb and Saari (1981).
[d] Fex and Johannesson (1982).

bound retinol, but not retinaldehyde (C1) or retinoic acid (Bashor *et al.*, 1973; Ong and Chytil, 1975a) in contrast to RBP. An apparent dissociation constant for retinol was determined by fluorometric titration of an homogeneous preparation of liver CRBP with all-*trans*-retinol and found to be 1.6×10^{-8} *M* (Ong and Chytil, 1978a).

Subsequently, the binding specificity of CRBP has been more extensively studied. The ability of various retinol derivatives to compete for the binding of all-*trans*-[^{3}H]retinol to partially purified CRBP was determined (Ong and Chytil, 1975b). For the cis isomers of retinol tested, 13-*cis*-retinol was the most effective competitor, although less effective than all-*trans*-retinol. Both 9-*cis*- and 9,13-di-*cis*-retinol had significantly less but approximately equal abilities to compete, compared with all-*trans* retinol. It was of interest that the ability of CRBP to bind cis derivatives of retinol reflected the growth-promoting activity of these compounds *in vivo* (Ong and Chytil, 1975b). This led to the suggestion that CRBP may mediate the action of retinol *in vivo*. A consequence of that suggestion is that the ability to interact with CRBP would then be one of the factors that determines whether a compound possesses vitamin A biological activity (Ong and Chytil, 1975b).

An apparent contradiction to this point was the fact that both α-retinol (A5) and 3,4-didehydraretinol (A9) competed as well as all-*trans*-retinol in the binding assay, whereas it was reported that their growth-promoting activities *in vivo* were 2% (Ames *et al.*, 1955a) and 40% (Shantz and Brinkman, 1950) of the activity of all-*trans*-retinol, respectively. This apparent paradox may be explained by the following observations. When administered to the rat, α-retinol is stored in the liver with high efficiency (Ames *et al.*, 1955b) and high doses fully support growth and development (Goodman *et al.*, 1974). Goodman *et al.* (1974) observed that α-retinol could not enter into the normal serum transport system, so delivery to target tissue is probably inefficient. Apparently the change in ring structure prevents the release of α-retinol from the liver to the blood via the RBP system, leading to a low physiological activity *in vivo*. However, this does not necessarily reflect the compound's potency in target tissues or conflict with its demonstrated ability to bind to CRBP. For example, Clamon *et al.* (1974) have shown that when α-retinyl acetate is added directly to the serum-free medium of organ cultures of hamster treacheas with keratinized squamous lesions, induced by retinoid deficiency, it is virtually as effective as all-*trans*-retinyl acetate (A3) in reversing the metaplasia. The sensitivity of the transport system to ring conformation may also explain why 3-dehydroretinol exhibits less activity than all-*trans*-retinol *in vivo*.

A requirement for binding to CRBP appears to be a free alcohol in the C-15 position (Ong and Chytil, 1975b). Retinyl acetate or retinyl palmitate (A4) do not compete for the binding of retinol to partially purified CRBP. In unfractionated cytosols, however, an excess of retinyl acetate or palmitate competed effectively for retinol binding (Bashor and Chytil, 1975). It is most likely that the competition observed was due to free retinol generated by tissue ester hydrolases.

The demonstrated binding specificity was consistent with but not proof that all-*trans*-retinol was the physiological ligand that interacts with the binding protein. Consequently, the partial purification of CRBP with the endogeneous ligand present allowed the development of evidence that the ligand was indeed all-*trans*-retinol (Ong and Chytil, 1974, 1975a; Ong *et al.*, 1976; Ross *et al.*, 1978; Saari, 1982; Saari *et al.*, 1982). It was also shown that testicular CRBP is totally depleted of retinol during vitamin A deficiency (Ong *et al.*, 1976) compared to 40–100% saturation with retinol in pair-fed or normally nourished animals (Ong *et al.*, 1976; Ross *et al.*, 1978; Saari, 1982; Saari *et al.*, 1982).

F. Species and Tissue Distribution

Since its discovery, CRBP has been detected in many species (see Table II). Although CRBP appears to be different in these various species as assessed by immunochemical methods, the partial amino acid sequences have shown significant homology when preparations from various species were compared, as discussed above. The presence of CRBP in all species thus far examined strongly suggests that it is important in the physiological function of retinol in animals.

Originally only a limited number of adult rat tissues were tested for the presence of CRBP, revealed by its ability to bind [^{3}H]retinol (Bashor *et al.*,

TABLE II

Species in Which Cellular Retinol-Binding (CRBP) and Retinoic Acid-Binding (CRABP) Proteins Have Been Detected[a]

Species	CRBP	Reference	CRABP	Reference
Chicken	+	Sani and Titus (1977)	+	Sani and Hill (1974)
Cow	+	Saari and Futterman (1976)	+	Saari and Futterman (1976)
Dog	+	Liou *et al.* (1981)	NT[b]	
Fish	+	Shidoji and Muto (1978)	+	Shidoji and Muto (1978)
Guinea pig	+	Ong and Chytil (1981)	NT	
Hamster	+	Ong and Chytil (1981)	+	Chytil and Ong (1976)
Human	+	Bashor *et al.* (1973)	+	Chytil *et al.* (1975)
Mouse	+	Ong and Chytil (1981)	+	Ong and Chytil (1975b)
Rabbit	+	Ong and Chytil (1981)	+	Rattanapanone *et al.* (1981)
Rat	+	Bashor *et al.* (1973)	+	Ong and Chytil (1975a)
Sheep	+	D. E. Ong (unpublished)	+	D. E. Ong (unpublished)
Swine	+	Ong and Chytil (1981)	NT	

[a] Detection by binding ability.
[b] NT, not tested.

1973). Subsequently, the widespread occurrence of CRBP was established by the more-sensitive radioimmunoassay procedure. Table III summarizes the distribution of this protein in adult rat tissues. Organs particularily rich in CRBP include epididymis, eye, kidney, liver, lung, and testis. CRBP has also been detected in all rat fetal tissues examined (Ong and Chytil, 1976a). Human tissues found positive for CRBP include breast, colon, kidney, liver, lung, testis, and uterus (see Chytil and Ong, 1983, for references). Some information on human tissue is also given in Table V.

Conflicting data have appeared concerning the actual levels of CRBP in rat tissues when determined by radioimmunoassay in two different laboratories. As determined by Ong *et al.* (1982b), the levels are comparable or somewhat higher than those determined by the sucrose gradient method. Considerably higher levels were observed by Adachi *et al.* (1981), however, which the authors attributed to an association of CRBP with a high-molecular-weight aggregate of lipid and protein. This association appears to occur under certain conditions in the extracts of some tissues (Sklan *et al.*, 1982). CRBP in that aggregate was not immunoreactive. If the association was prevented by detergent, higher levels were observed (Adachi *et al.*, 1981). Further work is needed to clarify the differences reported.

When analyzing the distribution of CRBP in various tissues, one point should be stressed. While in the past (Moore, 1967) histological alterations in the tissue of vitamin A-deficient animals have been taken as a criterion to establish that tissue's need for vitamin A, we would now suggest that the presence of CRBP in a particular tissue may be a more sensitive indication that a tissue is influenced metabolically by retinol. A consequence of this suggestion (if it were indeed the case) is that more tissues and organs than previously thought would appear to be dependent on retinol supply.

IV. CHARACTERISTICS OF CELLULAR RETINOIC ACID-BINDING PROTEIN

A. Physicochemical Properties

CRABP has been purified to homogeneity from rat testes (Ong and Chytil, 1978b, 1980; Ross *et al.*, 1980) and bovine retina (Saari *et al.*, 1978a, 1980; Saari, 1982). There are reported purifications of CRABP from chick embryonic skin (Sani and Banerjee, 1978) and from mouse colon tumor (Sani *et al.*, 1980a) also.

CRABP migrates in sucrose gradients with a sedimentation constant of 2 S (Sani and Hill, 1974; Ong and Chytil, 1975a). The protein from rat testis is a single polypeptide chain of molecular weight 14,600 (Ong and Chytil, 1978b;

TABLE III

Presence of CRBP and CRABP in Tissues of Adult Rat

Tissue	CRBP	Reference[a]	CRABP	Reference
Adrenal	ND[b]	10	ND	10
Bladder	+	6	+	4
Brain (whole)	+	2	+	3
Cerebellum	+	10	+	10
Cerebral cortex	+	10	+	10
Hypothalamus	+	10	+	10
Colon	+	5	+	5
Epididymis	+	10	+	10
Eye	+	3	+	3
Fat pad, epididymal	ND	6	NT[c]	—
Heart	ND	2	ND	3
Small intestine (whole)	+	1	ND	3
Ileum, mucosa	ND	10	ND	10
Ileum, muscle	+	10	+	10
Jejunum, mucosa	+	10	+	10
Jejunum, muscle	+	10	+	10
Kidney	+	1	+	10
Liver	+	1	ND	3
Regenerating	+	8	+	8
Lung	+	1	+	10
Mammary gland	NT	—	+	4
Muscle (gastrocnemius)	ND	2	ND	3
Ovary	+	2	+	3
Pituitary	+	10	+	10
Prostate	+	10	+	4
Ventral	NT	—	ND	9
Dorsolateral	NT	—	+	9
Serum	ND	1	ND	3
Skin	ND	6	+	4
Spinal cord	+	10	+	10
Spleen	+	2	+	10
Testis	+	1	+	3
Sertoli cells	+	7	NT	—
Thymus	+	10	+	10
Trachea	NT	—	+	4
Uterus	+	2	+	3

[a] Reference is given for the first published report, detection may be by either binding ability for respective ligand or by radioimmunoassay (Refs. 6 and 10 only). Key to references: 1, Bashor *et al.* (1973); 2, Bashor and Chytil (1975); 3, Ong and Chytil (1975a); 4, Sani and Corbett (1977); 5, Ong *et al.* (1978); 6, Adachi *et al.* (1981); 7, Huggenvik and Griswold (1981); 8, Omori *et al.* (1981); 9, Gesell *et al.* (1982); 10, Ong *et al.* (1982b).

[b] ND, not detectable.

[c] NT, not tested.

Ross *et al.*, 1980). Human breast tissue CRABP was reported to have a molecular weight of 12,000 (Küng *et al.*, 1980). The proteins from chick embryo skin and from transplantable mouse tumor have reported molecular weights of 17,800 (Sani and Banerjee, 1978) and 14,600, respectively (Sani *et al.*, 1980a).

The binding protein from embryonic chick skin has a p*I* value of 4.6 (Sani and Hill, 1976); rat testis CRABP has a p*I* equal to 4.7 (Ross *et al.*, 1980). CRABP does not bind to plasma transthyretin (Saari and Futterman, 1976; Ross *et al.*, 1980).

B. Spectral Properties

The absorption spectrum of CRABP with bound retinoic acid purified from rat testis (Ong and Chytil, 1978b) is shown in Fig. 3. The spectrum shows typical protein absorbance centered at 280 nm but is dominated by the peak at 350 nm from the bound retinoic acid. The spectrum of retinoic acid bound to CRABP is very similar to free retinoic acid in organic solution. The molar extinction coefficient for retinoic acid bound to CRABP is about 50,000. The value of A_{350}/A_{280} for purified CRABP believed to be saturated with retinoic acid is 1.8. It is of interest that the absorption spectrum of the CRABP–retinoic acid complex is similar to that of the CRBP–retinol complex; these spectra differ substantially from that of the RBP–retinol complex (see Chapter 8).

The fluorescence excitation and emission spectra of retinoic acid in ethanol and bound to CRABP are shown in Fig. 4. All spectra are uncorrected. The

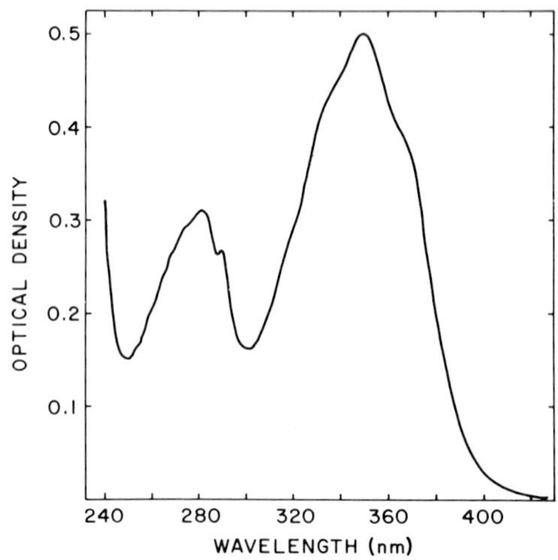

Fig. 3. Absorption spectrum of CRABP from rat testis, 1.0×10^{-5} *M* in 0.05 *M* Tris-HCl buffer, pH 8.3. (From Ong and Chytil, 1978b.)

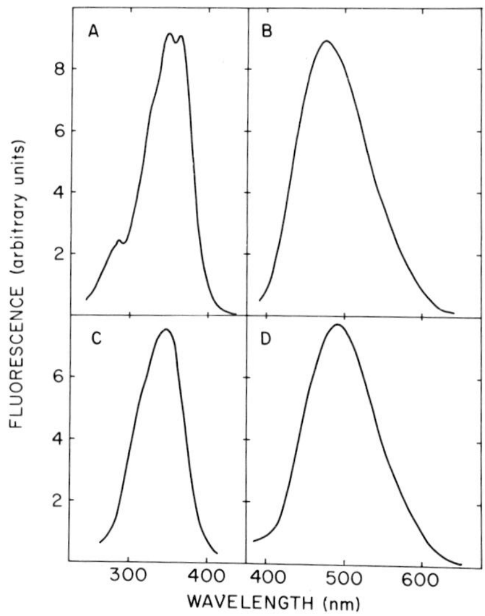

Fig. 4. Uncorrected fluorescence spectra of CRABP from rat testis, 3×10^{-7} *M* in 0.05 *M* Tris-HCl buffer, pH 7.5, and of retinoic acid, 1×10^{-5} *M* in ethanol. The excitation spectrum of CRABP (A) was determined measuring emission at 475 nm; the emission spectrum (B) was determined with excitation at 350 nm. For free retinoic acid, the excitation spectrum (C) was determined measuring emission at 490 nm; emission spectrum (D) with excitation at 350 nm. (From Ong and Chytil, 1978b.)

fluorescence spectra of retinoic acid bound to CRABP (shown in the top two panels) are about one order of magnitude greater in intensity than for retinoic acid in ethanol. In addition, the excitation spectrum of bound retinoic acid (top, left) shows a small peak at 290 nm, consistent with energy transfer from tryptophan residues to retinoic acid. The main excitation peak is almost a doublet, with maxima at 350 and 368 nm, in contrast to the smooth peak for retinoic acid in ethanol (bottom, left). A definite shift in the emission spectrum was also observed, with λ_{max} 475 m for bound retinoic acid in contrast to λ_{max} 490 nm for retinoic acid in ethanol.

C. Immunochemical Properties

The protein is immunochemically dissimilar from CRBP. Antibody to CRBP failed to react with CRABP (Ong and Chytil, 1979; Adachi *et al.*, 1981), and antibody to CRABP did not react with CRBP (Ong *et al.*, 1982b). In addition, antibody against plasma RBP did not recognize CRABP (Ross and Goodman, 1979; Ross *et al.*, 1980).

Like CRBP, CRABP is a very poor immunogen, although an antiserum against rat testis CRABP has been produced and a radioimmunoassay introduced (Ong *et al.*, 1982b). The antibody produced against testis CRABP could be used for determining the levels of the protein in other tissues, showing that there is no detectable tissue or sex differences in CRABP.

D. Amino Acid Sequences

The protein isolated from rat testis (Eriksson *et al.*, 1981) and from bovine retina (Crabb and Saari, 1981) has been partially sequenced. These partial sequences are identical for the first 30 amino acid residues as shown in Table IV. Thus it appears that CRABP may be even more conserved than CRBP. This conservation of primary structure may again help to explain the difficulties in obtaining antibody against this protein. The sequence of rat CRBP is also shown in Table IV. Significant homology can be seen between CRBP and CRABP, suggesting a common ancestor. Crabb and Saari (1981) and Takahashi *et al.* (1982) have also drawn attention to the extensive sequence homology of CRABP with the myelin protein P_2 and with rat liver Z-protein, respectively. The significance of that homology remains to be established.

E. Binding Specificity and Ligand Affinity

Displacement of label from the 2 S peak revealed after sucrose gradient centrifugation of tissue cytosols incubated with all-*trans*-[^{3}H]retinoic acid and test compounds has been used to determine the binding specificity of CRABP (Sani and Hill, 1974; Ong and Chytil, 1975a). The binding is quite specific for retinoic acid, as only excess unlabeled retinoic acid, but not retinol, retinaldehyde, or long-chain fatty acids, will compete with the radioactive ligand. Nonsaturable binding of retinoic acid has been observed in the 5 S region of these gradients. This has been shown to be binding to serum ablumin (Sani and Hill, 1976).

The ability of CRABP to bind synthetic analogs of retinoic acid has been explored. CRABP purified 1000-fold from rat testis, CRABP in an extract of human uterus and an extract of mouse skin papillomas (Chytil and Ong, 1976), and CRABP in an extract of chick embryo skin (Sani and Hill, 1976) have been examined. The binding affinity was assessed by competition experiments as previously described for CRBP. Again we find a striking correlation between the ability of analogs to interact with CRABP and their biological activity as determined by Sporn *et al.* (1975). The biological activity in question was the promotion of growth in cultured mouse skin epidermal cells (as indicated by increases in RNA, DNA, and protein levels after addition of the retinoid) or the ability to reverse the squamous metaplasia of tracheas explanted from retinoid-deficient hamsters (for description of these assay systems, see Chapter 5). Regardless of

TABLE IV

Comparison of the Amino Terminal Amino Acid Sequences of Cellular Retinoic Acid-Binding Proteins (CRABP) from Rat and Cow with Cellular Retinol-Binding Protein (CRBP) from Rat[a]

Protein	Source	Sequence (positions 5, 10, 15, 20, 25 marked above)
CRABP	Rat testis[b]	X - Asn-Phe-Ala-Gly-Thr-Trp-Lys-Met-Arg-Ser-Ser-Glu-Asn-Phe-Asp-Glu-Leu-Leu-Lys-Ala-Leu-Gly-Val-Asn-Ala-Met-Leu-Arg-
CRABP	Bovine retina[c]	Pro - Asn-Phe-Ala-Gly-Thr-Trp-Lys-Met-Arg-Ser-Ser-Glu-Asn-Phe-Asp-Glu-Leu-Leu-Lys-Ala-Leu-Gly-Val-Asn-Ala-Met-Leu-Arg-
CRBP	Rat liver[d]	Pro-Val-Arg-Phe-Asn-Gly-Tyr-Trp-Lys-Met-Leu-Ser-Asn-Glu-Asn-Phe-Glu-Glu-Tyr-Leu-Arg-Ala-Leu-Asp-Val-Asn-Val-Ala-Leu-Arg-

[a] Identical sequences are boxed.

[b] Eriksson *et al.* (1981).

[c] Crabb and Saari (1981).

[d] Rask *et al.* (1981).

its origin, the ability of CRABP to bind retinoic acid derivatives is very similar.

We have suggested that the binding ability of CRABP may be useful as a preliminary screening test to identify these derivatives or analogs of retinoic acid with potential biological activity (Chytil and Ong, 1976). Subsequent studies have extended the correlations between a compound's biological activity as defined by various test systems and its ability to bind to CRABP. Jetten and Jetten (1979) studied the induction of differentiation of embryonal carcinoma cells in tissue culture by several retinoic acid derivatives. The inducing activities of various compounds correlated well with the ability of these compounds to compete for the binding of retinoic acid to CRABP. This correlation was also observed by Trown *et al.* (1980), who conducted an extensive study in which the affinities of 18 retinoids for partially purified rat testis CRABP and carcinogen-induced rat mammary tumor CRABP were compared to the activity of those compounds in inhibiting the growth of a transplantable rat chondrosarcoma and of 3T6 cells. The compounds were also compared for activity against carcinogen-induced mouse skin papillomas as well as in their ability to induce differentiation in an embryonal carcinoma cell line.

The availability of a free carboxyl group at the C-15 position of such retinoids is necessary for ability to bind to CRABP (Sani and Hill, 1976; Lotan *et al.*, 1980a; Trown *et al.*, 1980). Consequently, if CRABP is involved in the action of retinoic acid and its derivatives, such compounds with a blocked C-15 carboxyl must first be hydrolyzed to the free acid in order to become biologically active. Evidence for such metabolic transformations have been reported (Bollag and Hanck, 1977).

Muto *et al.* (1981) have tested the ability of several synthetic polyprenoic acids to compete for binding of all-*trans*-[^{3}H] retinoic acid to rat testicular CRABP. They observed that the β-ionone rign of retinoic acid was not essential for binding to CRABP. The relative ability of these several compounds to bind to CRABP was similar to their relative activity in inhibiting the development of mouse skin papillomas.

CRABP from bovine retina has one binding site for retinoic acid (Saari *et al.*, 1978a). The apparent dissociation constant for a pure preparation of rat testicular CRABP has been estimated by fluorometric titration to be 4.1×10^{-9} M (Ong and Chytil, 1978b). Human breast tissue CRABP was reported to have an apparent dissociation constant of 2.5×10^{-7} M, when determined by polyacrylamide disc gel electrophoresis method (Küng *et al.*, 1980).

All-*trans*-retinoic acid has been identified by high-performance liquid chromatography as the endogenous ligand for CRABP (Saari, 1982; Saari *et al.*, 1982), suggesting a physiological role for retinoic acid and its binding protein in vitamin A metabolism. As isolated, the protein was about 20% saturated with retinoic acid (Saari *et al.*, 1982). Previously, CRABP isolated from testis was estimated to be about 40% saturated with endogenous ligand (Ross *et al.*, 1980).

F. Species and Tissue Distribution

Since the discovery of CRABP considerable energy has been spent in detecting this protein in tissues from various species. Table II shows that CRABP has been found in all species tested thus far. Human tissues that contain CRABP include breast, colon, kidney, testis, and uterus (see Chytil and Ong, 1983, for references). The recent introduction of a sensitive radioimmunoassay for rat CRABP (Ong *et al.*, 1982b) has contributed substantially to our knowledge of the occurrence of this protein in various tissue of the adult rat. Table III lists the organs which were tested; the great majority contain CRABP. Organs particularly rich in CRABP include eye, ovary, pituitary, prostate, testis, and uterus. It can be seen that most adult tissues contain both CRBP and CRABP but in several, only one binding protein was detected. Few tissues appear to be lacking in both of these proteins. The different tissue distribution of CRBP and CRABP suggests that the genes coding for these proteins are expressed differentially.

V. CELLULAR RETINOL- AND RETINOIC ACID-BINDING PROTEINS AND CANCER

There is considerable interest in the use of retinoids for cancer prevention or treatment (Bollag, 1979; Sporn and Newton, 1979; Ong and Chytil, 1983; see also Chapter 14). This has led to a number of studies designed to determine the presence and level of cellular retinoid-binding proteins in tumor tissue. The occurrence of CRABP in tumors was first reported for human lung and breast carcinomas (Ong *et al.*, 1975). A particularly interesting point was that CRABP was not detected in the histologically normal tissue adjacent to the tumor, suggesting that the tumor had either a dramatic increase in number of a cell type that normally contains this protein or an altered expression of the gene coding for this protein. This early communication did not determine the occurrence of CRBP in these tumors, a circumstance that may have focused more attention on CRABP rather than CRBP in neoplastic tissue. Subsequently however, it was found that both proteins could show dramatic changes in level because of malignancy, both in tumors in experimental animals (Ong *et al.*, 1978) as well as in humans (Ong *et al.*, 1982a). A better understanding of the role of binding proteins in tumors will require that the levels of both proteins be determined.

A. Human Cancer

Various laboratories have devoted their efforts to the detection and quantitation of CRABP and occasionally CRBP in human carcinomas (for review, see Chytil and Ong, 1982). These results are summarized in Table V. It was fre-

TABLE V

Cellular Retinol-Binding (CRBP) and Retinoic Acid-Binding (CRABP) Proteins in Human Carcinomas[a]

	CRBP			CRABP		
Tumor site	Control	Tumor	Reference	Control	Tumor	Reference
Breast	NT[b]	NT		ND[c]	+[d]	Ong *et al.* (1975)
				ND	+	Huber *et al.* (1978)
				+	10[e]	Palan and Romney (1980)
				+	14.85 + 12.0[f]	Küng *et al.* (1980)
				NT	1.5–8.0[f]	Mehta *et al.* (1982b)
Colon	ND–12.2[e] M[g]	14.4–24.7[e] M	Palan *et al.* (1980)	ND–37.4[e] M	31.4–50.2[e] M	Palan *et al.* (1980)
				NT	+	Sani *et al.* (1980b,c)
	3.9–10.0[e] F	ND–6.0[e] F		12.6–51.0[e] F	ND–35.3[e] F	
Endometrium	23.8–40.0[e]	3.0–11.1[e]	Palan and Romney (1980)	18.0–42.3[e]	86.0–159.0[e]	Palan and Romney (1980)
Kidney	NT	NT		NT	+	Chytil and Ong (1978a)
Liver	3.5–67.0[e]		Muto *et al.* (1979)	ND–4.1[e]	ND–6.6[e]	Muto *et al.* (1979)
Lung	NT	NT		ND	+	Ong *et al.* (1975)
				ND	0.048–0.426[f]	Clamon *et al.* (1981)
Oral cavity	ND–10.0[d]	45.0–175.0[d]	Ong *et al.* (1982a)	ND–60.0[d]	140.0–320.0[d]	Ong *et al.* (1982a)
		+		+	+	Bichler and Daxenbichler (1982)
Ovary	193.0–234.0[e]	48.0–154.0[e]	Palan and Romney (1980)	ND–1.4[e]	66.0–221.0[e]	Palan and Romney (1980)
Uterus Cervix	12.5–24.3[e]	ND–54.1[e]	Palan and Romney (1980)	10.1–22.3[e]	15.0–135.0[e]	Palan and Romney (1980)

[a] Detection and quantitation by binding ability.
[b] NT, not tested.
[c] ND, not detectable.
[d] +, present not quantitated.
[e] pmol/g wet tissue.
[f] pmol/mg cytosol protein.
[g] M, males; F, females.

quently observed that both proteins are at a high level in the tumor compared to normal tissue. However, the occurrence and amount of these proteins in tumors were quite variable, even when the organ site of the tumor was the same. At this point it seems safe to say that each tumor should be evaluated individually for binding-protein content. Sex differences in levels of these proteins have been reported in human colon carcinomas (Palan *et al.*, 1980). No correlation between the presence or absence of estrogen or progesterone recpetors with CRABP in human mammary tumors was observed (Mehta *et al.*, 1982a).

We suggest that one should not assume that an elevated level of either CRBP or CRABP in tumor tissue is indicative that the tumor tissue requires corresponding levels of retinol or retinoic acid. A number of proteins are expressed in higher levels in various malignant tissues, with no suggestion that they have utility to the tumor (Abelev, 1971; Shapira, 1973).

B. Experimental Tumors

The presence and level of the binding proteins has also been determined for a number of animal tumors, both transplantable or carcinogen-induced. The results of these determinations are summarized in Table VI. CRBP was frequently not determined, however. In several examples of carcinogen-induced tumors, it was observed, as for some "spontaneous" human tumors discussed above, that the level of binding protein in the tumor was considerably higher than in the adjacent, apparently normal tissue. Examples include elevated levels of CRABP in mouse papillomas (Chytil and Ong, 1976) and of CRBP in colon adenocarcinoma (Ong *et al.*, 1978).

CRABP from mouse papilloma and human breast tumor (Chytil and Ong, 1976) has been evaluated for its binding specificity and found to be quite similar, if not identical, with CRABP from normal tissues. Studies by Trown *et al.* (1980) have led to similar conclusions for CRABP partially purified from rat mammary tumors.

We have proposed that the presence of the binding proteins may be a necessary but not sufficient condition for a tumor to be sensitive to retinoid therapy (Ong and Chytil, 1976). If this proves to be true, the screening of tumors for the presence of the binding proteins will be a useful step to select those that might be candidates for retinoid treatment. Whether the levels of the binding proteins will correlate with the modulating effects of retinoids also remains to be established.

VI. CELLULAR RETINOL- AND RETINOIC ACID-BINDING PROTEINS IN CULTURED CELLS

Many lines of cultured cells have been examined by ligand-binding assays for the presence of the cellular retinoid-binding proteins. Many of these cell lines

TABLE VI

Cellular Retinol-Binding (CRBP) and Retinoic Acid-Binding (CRABP) Protein in Experimental Tumors[a,b]

Tumor	Host	Type	CRBP	Reference	CRABP	Reference
Carcinosarcoma	Rat	Walker 256	+	Ong and Chytil (1976b)	+[c]	Ong and Chytil (1976b)
Chondrosarcoma	Rat		ND[d]	Ong and Chytil (1976b)	+	Ong and Chytil (1976b)
Colon	Mice	26 and 57 metastatic	NT[e]		+	Sani and Corbett (1977) Sani and Titus (1977)
		36 and 58 (nonmetastatic)	NT		ND	Sani and Corbett (1977) Sani and Titus (1977)
Colon	Rat	DMH induced	+	Ong *et al.* (1978)	+	Ong *et al.* (1978)
Ehrlich carcinoma	Mice	Ascites	ND	Bashor and Chytil (1975)	ND	D. E. Ong (unpublished)
		Solid carcinoma	ND	Ong and Chytil (1976b)	ND	Ong and Chytil (1976b)
Hepatoma	Rat	AS—300 ascites	ND	Bashor and Chytil (1975)	NT	
		Novikoff	ND	Bashor and Chytil (1975)	NT	
Leukemia	Mice	L1210	ND	Ong and Chytil (1976b)	ND	Ong and Chytil (1976b)
Leukemia	Rat	Dunning	+	Ong and Chytil (1976b)	+	Ong and Chytil (1976b)
Lung	Mice	Lewis	NT		+	Sani and Corbett (1977) Sani and Titus (1977)
Melanoma	Mice	B16	NT		+	Sani and Corbett (1977) Sani and Titus (1977)
Mammary metastatic	Mice	C3H 13/C/24 C3H 04/A/64 C3H 16/C/13	NT		+	Sani and Corbett (1977) Sani and Titus (1977)

Mammary	Rat	MAC-1 adenocarcinoma	+	Ong and Chytil (1976b)	+	Ong and Chytil (1976b)
		CMBA induced	NT		+	Mehta *et al.* (1980, 1982b); Moon and Mehta (1981)
Prostatic carcinoma	Mice	Transplantable				
		R-3327H	NT		+	
		R-3327HI	NT		+	
		R-3327-AT	NT		+	Gesell *et al.* (1982)
		R-3327-Lylw	NT		+	
		MNU induced	NT		+	Mehta *et al.* (1980)
Sarcoma	Mice	180	ND	Ong and Chytil (1976b)	+	Ong and Chytil (1976b)
					ND	Sani and Titus (1977)
	Rat	Ridgeway osteogenic	NT		+	Sani and Corbett (1977)
Papilloma (skin)	Mice	DMBA induced	+	Ong and Chytil (1976b)	+	Ong and Chytil (1976b)
Papilloma (skin)	Rabbit	Shope	NT		+	Rattanapanone *et al.* (1981)
Urinary bladder	Mice	AC1/N	+	Kawamura and Hashimoto (1980)	NT	

[a] Detection by binding ability.

[b] Quantitative data available in the following references: Gesell *et al.* (1982); Kawamura and Hashimoto (1980); Mehta *et al.* (1982b); Ong and Chytil (1976b); Ong *et al.* (1978).

[c] Nondetectable in another sample.

[d] ND, not detectable.

[e] NT, not tested.

have been shown to be responsive to the presence of various retinoids in the culture media (Lotan, 1980). Tables VII and VIII review the results of such determinations for cells of human, mouse, and rat origin. A comment on the reported absence of a binding protein should be made. As discussed above, the ability to detect a particular protein by nonimmunochemical methods is limited by the specific radioactivity of the ligand. Consequently, it can not be concluded with certainty that a binding protein is absent, but only below detection by the method used.

Another caveat concerning the applicability of findings with cell cultures is necessary. It is quite possible that adaptation of cells to culture might result in an altered expression of the genes coding for the binding proteins when compared to the tissue of origin. It is also likely that different clones of the "same" cell line may contain different levels of either protein. The mode of cultivation of the cells may also influence the level of the binding proteins. These factors may explain discrepancies in findings from different laboratories. Last, it should be stressed that the cells to be tested for the presence of the binding proteins have usually been established and grown for many generations in a medium containing fetal calf serum with varying and undetermined levels of retinol bound to RBP. Consequently, the cell line may well be adapted to the presence of retinol, and the addition of exogenous retinoid to the medium may therefore have little effect, particularly if the effect under examination is a terminal differentiation. This "adaptation" could be accomplished by no longer expressing the gene for CRBP. It is of interest that although most animal tissues contain both binding proteins, of the cells in culture tested thus far, 86% contain CRABP while only 37% contain CRBP. The possible presence of endogenous retinoic acid in fetal calf serum should also be considered.

VII. REGULATION OF THE LEVELS OF CRBP AND CRABP

As discussed above, tissues of the adult animal vary with respect to the presence and level of both CRBP and CRABP. Moreover, it has been established that both proteins are present in all fetal organs studied (with the exception of serum; Ong and Chytil, 1976a) but at levels different than those observed for adult organs. This indicates that the expression of the genes for these proteins changes during development. In order to shed more light on this phenomenon, we examined rat tissues during perinatal development (Ong and Chytil, 1976a). Levels of CRBP and CRABP in liver and lung were determined as binding activity starting at 3 days before birth. As shown in Fig. 5, both CRBP and CRABP levels changed considerably during organ maturation. However, there was no correlation between the amount of CRBP and CRABP binding activity in either organ, suggesting that these proteins are regulated independently of each

TABLE VII

Cellular Retinol-Binding (CRBP) and Retinoic Acid-Binding (CRABP) Proteins in Human Cultured Cells[a]

Cell line	Type	CRBP	Reference	CRABP	Reference
Breast carcinoma	MCF-7	+	Lacroix and Lippman (1980)	+	Takenawa *et al.* (1980)
					Lacroix and Lippman (1980)
	Hs578T	+	Lacroix and Lippman (1980)	+	Lacroix and Lippman (1980)
	ZR-75-B	ND[b]	Lacroix and Lippman (1980)	+	Lacroix and Lippman (1980)
Cervical carcinoma	HeLa (clone M)	ND	Saari *et al.* (1978b)	ND	Saari *et al.* (1978b)
	HeLa (clone S3)	NT[c]		+	Lotan *et al.* (1980b)
Fibroblasts	Untransformed	ND	Swanson *et al.* (1976)	ND	Swanson *et al.* (1976)
		ND	Lacroix *et al.* (1981)	+	Lacroix *et al.* (1981)
Melanoma	Hs939	NT		+	Lotan (1980)
Promyelocytic leukemia	HL-60	ND	Takenaga *et al.* (1981)	+	Takenaga *et al.* (1981)
		NT		ND	Breitman *et al.* (1981)
Retinoblastoma	WERI-Rb1	ND	Saari *et al.* (1978b)	+	Saari *et al.* (1978b)
	Y-79	+	Wiggert *et al.* (1977b)	+	Wiggert *et al.* (1977b)
		+	Saari *et al.* (1978b)	+	Saari *et al.* (1978b)
		+	Russell *et al.* (1978)	+	Russell *et al.* (1978)

[a] Detection by binding ability.
[b] ND, not detectable.
[c] NT, not tested.

TABLE VIII

Cellular Retinol-Binding (CRBP) and Retinoic Acid-Binding (CRABP) Proteins in Cultured Cells[a]

Type	Origin	Line	CRBP	Reference	CRABP	Reference
Adenocarcinoma	Rat	DMBA No. 8	ND[b]	Lotan *et al.* (1980b)	+	Lotan *et al.* (1980b)
Embryonal carcinoma	Mice	PCC4. aza 1R	NT[c]		+	Jetten and Jetten (1979)
		PCC4. azaldif (RA)-1	NT		ND	Sherman *et al.* (1981); Schindler *et al.* (1981)
		PCC4. azaldif (RA)-3	NT		+	Sherman *et al.* (1981); Schindler *et al.* (1981)
		OC 15	NT		+	Jetten and Jetten (1979)
		Nulli SOC1	NT		+	Jetten and Jetten (1979)
Fibroblast	Mice	3T3 untransformed	NT		+	Jetten *et al.* (1979)
		3T6	NT		+	Jetten *et al.* (1979)
		SV3T3 transformed	ND	Chytil and Ong (1978a)	+	Chytil and Ong (1978a)
		MSV3T3-K 238 transformed	ND	Lotan *et al.* (1980b)	+	Lotan *et al.* (1980b)
		F361.2 transformed	ND	Lotan *et al.* (1980b)	+	Lotan *et al.* (1980b)
Glioma	Rat	Clone C6	ND	Saari *et al.* (1978b)	+	Saari *et al.* (1978b)
Leukemia	Mice	L1210-A5	ND	Lotan *et al.* (1980b)	ND	Lotan *et al.* (1980b)
		Myeloid M1	ND	Takenaga *et al.* (1981)	ND	Takenaga *et al.* (1981)

Lymphoma	Mice	BW 5147	ND	Lotan *et al.* (1980b)	+	Lotan *et al.* (1980b)
		BW 5147.Ric	ND	Lotan *et al.* (1980b)	+	Lotan *et al.* (1980b)
		RAW8	ND	Lotan *et al.* (1980b)	+	Lotan *et al.* (1980b)
McCoy	Mice	Heteroploid	+	Saari *et al.* (1978b)	+	Saari *et al.* (1978b)
Mammary	Mice	Mm5mT	+	Lotan *et al.* (1980b)	+	Lotan *et al.* (1980b)
Melanoma	Mice	S91 (clone 2)	+	Lotan *et al.* (1980b)	+	Lotan *et al.* (1980a,b)
		B16-F1	+	Lotan *et al.* (1980b)	+	Lotan *et al.* (1980b)
Myeloma	Mice	P3	ND	Lotan *et al.* (1980b)	+	Lotan *et al.* (1980b)
Neuroblastoma	Mice	C1300 (clone N 18)				
		Undifferentiated	+	Saari *et al.* (1978b)	+	Saari *et al.* (1978b)
		Differentiated	ND	Saari *et al.* (1978b)	+	Saari *et al.* (1978b)
Pituitary tumor	Mice	AtT-20	ND	Bashor and Chytil (1975)	+	Chytil and Ong (1978a)
Preadipocytes	Mice	ST 13	+	Sato *et al.* (1980)	+	Sato *et al.* (1980)
Testis	Rat	Sertoli	+	Huggenvik and Griswold (1981)	NT	
Trachea	Rat	Adenocarcinoma T8	NT	Marchok *et al.* (1981)	+	Marchok *et al.* (1981)
		1000-WT	NT	Marchok *et al.* (1981)	+	Marchok *et al.* (1981)
Urinary bladder	Mice	BES20B	+	Kawamura and Hashimoto (1980)	NT	
		BES3B, BC12-TC, BC47-TC, BC50-TC	ND	Kawamura and Hashimoto (1980)	NT	

[a] Detection by binding ability.
[b] ND, not detectable.
[c] NT, not tested.

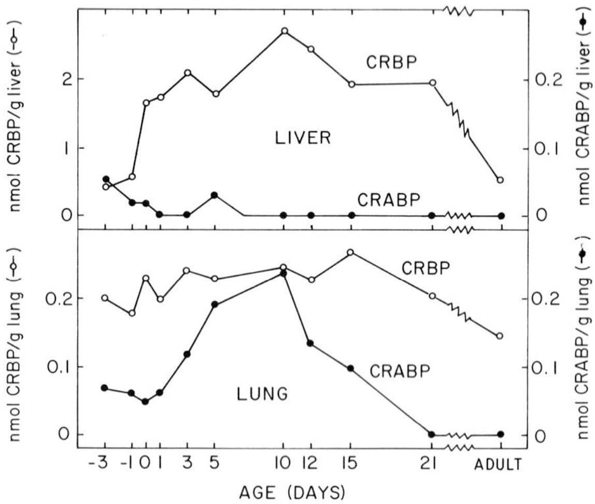

Fig. 5. Perinatal development of CRBP and CRABP of rat liver (top) and lung (bottom). Content of binding proteins was determined by specific binding of [^{3}H]retinol and [^{3}H]retinoic acid evaluated by sucrose gradient centrifugation. (From Ong and Chytil, 1976a.)

other. The changes in the levels of the binding proteins suggest that the requirements for retinol and retinoic acid during development of each organ also change dramatically. CRBP may be expressed differently within different parts of the same organ, as shown by Kylberg *et al.* (1981) who found CRBP is distributed differently in the jenunum compared to the ileum and underwent perinatal alternations in level (Kylberg *et al.*, 1981). It is not known which cells of a differentiated organ contain CRBP or CRABP. Consequently, localization of these proteins in various cell types will be of considerable interest.

Although we are unable to determine the significance of the changes occurring during the perinatal period, it is of interest that the disappearance of CRABP in the liver (Ong and Chytil, 1976a) generally follows the loss of the hematopoetic cells (Greengard *et al.*, 1972). Moreover, the highest level of CRABP in the developing rat lung correlates with the time period in which alveoli are formed (Burri *et al.*, 1974). These findings may be indicative of a role for retinoic acid and CRABP in the process of organ maturation (Ong and Chytil, 1976a; Chytil, 1984).

An important question is whether the amount of available ligand affects the level of its corresponding binding protein. Previous results have shown that binding-protein levels remain unchanged in the testis of mildly vitamin A-deficient rats (Ong *et al.*, 1976). The testes of these rats were smaller than control animals but did not have pronounced morphological changes. However, when determined by a sensitive radioimmunoassay, we have observed a considerable decrease in CRBP and CRABP in testis and lungs of severely deficient animals (D. E. Ong, G. Liau, and F. Chytil, unpublished). The testis of these animals had lost most of the germinal epithelium. The data agree with a preliminary report of Adachi *et al.* (1981). More work is needed to determine whether vitamin A status influences the levels of the binding proteins directly, or whether the lower level of binding proteins observed is a reflection of loss of specific cell types during deficiency.

The level of CRABP in mammary gland is affected by estrogen but not by progesterone (Mehta and Moon, 1981). Changes in the level of CRABP binding activity in the mammary gland during pregancy were also noted.

VIII. A ROLE FOR VITAMIN A AND THE CELLULAR RETINOID-BINDING PROTEINS IN DIFFERENTIATION

The pioneering studies of Wolbach and Howe (1925, 1933) have clearly shown that vitamin A is a nutritional component necessary for the maintenance of most, if not all, epithelia tissue. They observed in many tissues of the retinol-deficient rat a replacement of the normal columnar and transitional epithelium by squamous, frequently keratininzing, epithelial cells. Other changes were also observed. For example, in testis no keratinization occurred, but instead there was atrophy of the germinal epithelium in the seminiferous tubules. All such changes appeared to be fully reversible: Refeeding the animal with retinol resulted in the repopulation of the epithelia by normal cells and regeneration of testis (Wolbach and Howe, 1933). These observations indicate that a most important role of this nutrient will be found in the process of cellular differentiation (see also Chapter 12). Moreover, because the effects of vitamin A are tissue specific, it is likely that we will not be able to explain its action by a single, simple mechanism. Indeed, the observation that retinoic acid, when fed to retinol-deficient animals, was able to maintain proper differentiation of most epithelia but was unable to support vision (Dowling and Wald, 1960) or reproduction (Thompson *et al.*, 1964), suggested that there was more than one active physiological form of vitamin A.

A proposed hypothesis for vitamin A action should suggest experiments that will address not only the question of how vitamin A affects differentiation at the cellular level but also how the tissue specificity of this action might arise. In order to simplify our approach we have relied on two assumptions. First, because vitamin A affects cellular differentiation, we believe it should interact in some way with the nucleus, which is the site of genomic information. Second, we have assumed that both retinol and retinoic acid are active forms of vitamin A, with separable functions in differentiation.

A. Influence of Vitamin A on Genomic Expression

The influence of vitamin A-active compounds on genomic expression was previously only inferred from lesions in the deficient rat. No direct evidence was available. Such evidence now has been generated *in vivo* in the whole animal (Omori and Chytil, 1982). It was found that concurrently with the fall of the poly(A)$^+$-containing RNA levels in testis, i.e., 1 h after feeding retinyl acetate or retinoic acid to vitamin A-deficient rats, alterations of the *in vitro* translation products of poly(A)$^+$-containing RNA occurred. Some polypeptides disappeared and some became more numerous. Although the mechanisms underlying this phenomenon remain to be elucidated, it appears that retinol and retinoic acid simultaneously activate as well as suppress genomic expression *in vivo* in a rapid fashion. It appears also that retinoic acid acts on the genome in a manner that is similar to but not identical with that of retinol. Presently it is not known whether the observed effects are due to alteration in transcription or processing of primary transcripts or both (for more discussion, see Chapter 12).

There is also considerable evidence that retinoids influence expression of cells in culture. Several investigators have reported that retinoic acid added to the culture medium influences the synthesis of specific proteins. Lotan and Lotan (1981) and Lotan *et al.* (1980c) have found increased synthesis of melanin and tyrosinase when retinoic acid was added to the medium of human Hs 939 and mouse S91C2 melanoma cells. The authors also observed that after addition of retinoic acid, a cell-surface glycoprotein occurs in greater abundance in HeLa S_3 cells (Lotan and Lotan, 1980). On the other hand, few quantitative differences in protein pattern were detected when the growth of mouse S91C2 melanoma cells was inhibited by retinoic acid (Lotan *et al.*, 1982). Retinoic acid and to a smaller extent, retinol, stimulates activity of alkaline phosphatase in rat urinary bladder, prostate, and Dunning R 3327 transplantable rat prostatic adenocarcinoma in culture (Reese *et al.*, 1981; Reese and Politano, 1981). Furthermore, induction of vimentin and tropomyosin (Linder *et al.*, 1981) and laminin (Howe and Solter, 1981) in mouse F9 embroyonal carcinoma cell line by retinoic acid has been described. Fuchs and Green (1981) have shown that human keratinocytes in culture react to the addition of retinyl acetate to the culture medium by stimulation of synthesis of two specific keratins and their respective mRNAs. Moreover, these authors demonstrated that retinyl acetate led to inhibition of synthesis of a third specific keratin and its mRNA.

The results from these several systems indicate that the effect of retinoids on genomic expression is complex, with activation and suppression of specific genes occurring. In addition, the genes affected appear to be different for the different cells or tissues examined.

B. Are Cellular Retinoid-Binding Proteins Necessary for Vitamin A Action?

Some evidence for the involvement of the cellular retinoid-binding proteins in the effects of retinol and retinoic acid on growth and differentiation comes from work in cell culture. Frequently, cell lines are tested for responsiveness to retinoids using such parameters as growth stimulation or inhibition or induction of specific proteins (Lotan, 1980; Chapters 5 and 12). Here the question whether cellular retinoid-binding proteins are necessary for a cell to be responsive to retinoids can be examined. Striking correlations between the presence of the binding protein in the cell and its responsiveness to retinoids have been reported (Schindler *et al.*, 1981; Sherman *et al.*, 1981). Teratocarcinoma cells, in which CRABP was detected, differentiated when retinoic acid was added to the medium. However, mutant cells selected for failure to differentiate in response to the retinoic acid did not contain detectable CRABP (Schindler *et al.*, 1981; Sherman *et al.*, 1981). This is contrasted by the human myelocytic leukemia cell line HL-60, which is prompted to differentiate to mature granulocyts in the presence of retinoic acid (Breitman *et al.*, 1981), but in which CRABP could not be detected (Breitman *et al.*, 1981; Douer and Koeffler, 1982). In addition, Merriman and Bertram (1979) have reported that 3-methylcholanthrene-induced neoplastic transformation of C3H/10TP Clone 8 cells (from mouse embryo fibroblasts) can be reversibly inhibited by certain retinoids. Complete inhibition of transformation was achieved by the addition of retinyl acetate, retinol, or retinal dehyde but all-*trans*-retinoic acid and 13-*cis*-retinoic acid (E6) were inactive, in contrast to other systems (Bertram, 1980). Activity was observed for retinoic acid derivatives where the terminal carboxy group was modified [e.g., *N*-(4-hydroxyphenyl)retinamide (E16), retinylidene dimedone (C6), *N*-ethylretinamide (E11) and *N*-benzoylretinylamine (B2). No CRBP or CRABP was detected in this cell line (Libby and Bertram, 1982). Alternate mechanism(s) not involving the cellular retinoid-binding proteins should be taken into consideration to explain the inhibitory action of retinoids in this and perhaps other systems. The effects under study may be unrelated to any physiological function of vitamin A.

When growth inhibition by retinoids is the parameter under study, the level of binding protein in the responsive cell does not show good correlation with the percent inhibition observed (Lotan *et al.*, 1980b). Previous explanations for the differential response of malignant cells to anticancer drugs (Barranco *et al.*, 1978) or to the effect of glucocorticoids (McPartland *et al.*, 1977) may be applicable to this situation. Thus, cell lines may differ in the following properties: (a) uptake of the retinoid; (b) cellular metabolism of the retinoid; (c) presence of specific retinoid-binding protein(s); (d) presence of an altered binding protein that cannot transmit and localize the retinoid in the nucleus of the cell; (e) change in the nucleus that affects its ability to interact with cellular retinoid–retinoid-binding protein complexes; and (f) the existence of a mechanism that allows the cells to bypass a retinoid-inhibited step. These possible differences may explain why some cells that have a binding protein do not respond to retinoids; however, they do not explain the case of a cell that is inhibited but lacks detectable binding proteins. Possibly, such cells are extremely sensitive to retinoids and possess low but undetected levels of binding proteins. Alternatively, it cannot be excluded that certain cells may lack retinoid-binding proteins and be inhibited by a mechanism that does not call for the presence of binding proteins. Because inhibition of growth is not the only effect that the retinoids may have on these cells, possibly the level of the retinoid-binding proteins correlates with other effects of the retinoids on cellular metabolism.

C. Nuclear Interactions of Vitamin A and the Cellular Retinoid-Binding Proteins

The means by which regulation of the genome by retinol might be carried out was investigated in studies on the interaction of purified CRBP, retinol, and isolated nuclei. Radioactive retinol complexed with CRBP was found to bind to isolated rat liver nuclei in a specific and temperature-dependent manner (Takase *et al.*, 1979). CRBP charged with unlabeled retinol or apo-CRBP considerably diminished this binding of radioactivity, whereas free retinol did not. No specific binding sites could be detected if free retinol was used. Approximately 3×10^5 specific binding sites per nucleus was detected, similar to the number of CRBP molecules in the cytosol of a single liver cell (i.e., $5–15 \times 10^5$; Ong and Chytil, 1979). More binding sites were found in nuclei isolated from the liver of retinol-deficient rats than in nuclei from livers from retinol-sufficient rats, suggesting that the interaction of retinol with the nucleus is physiologically important. Furthermore, if nuclei were isolated 2 h after feeding retinol-deficient rats with retinyl acetate (a sufficient time to produce significant blood levels of retinol), the binding capacity of the isolated nuclei was reduced to values observed for retinol-sufficient animals. These results suggested that the liver nucleus reacts rapidly to vitamin A (Takase *et al.*, 1979).

The interaction of retinol–CRBP with the nucleus has been further studied (Liau *et al.*, 1981). Using CRBP labeled with tritium by reductive methylation, it was found that CRBP delivers retinol to the specific binding sites within the nucleus without itself remaining bound. The outer nuclear envelope is apparently not involved because Triton X-100-treated nuclei retained 70% of the retinol binding sites found in intact nuclei. Isolated liver chromatin also had the same number of binding sites per DNA as whole nuclei. CRBP did not remain bound after delivering retinol to the chromatin-binding sites. When the isolated nuclei were sectioned and autoradiographed after incubation with [^{3}H]retinol–CRBP, retinol was observed throughout the nucleus. In contrast, autoradiography after incubation with [^{3}H]retinol alone showed only a nonspecific association of retinol with the lipid-rich nuclear membrane (Liau *et al.*, 1981).

This evidence suggests that CRBP may carry retinol into the nucleus. The protein may recognize and transfer its ligand to a limited number of specific sites on the chromatin. The sites are not detectable and may not be accessible to retinol uncomplexed with CRBP. After the transfer, CRBP does not remain associated with the specific sites but leaves the nucleus. Whether this interaction of retinol with the chromatin then influences genomic expression is under investigation.

It is of interest that retinoic acid bound to CRABP can apparently be extracted from the nuclei of an embryonal carcinoma cell line (Jetten and Jetten, 1979), retinoblastoma cell line (Wiggert *et al.*, 1977b), and rat mammary carcinomas (Moon and Mehta, 1981). CRABP was also reported in the nuclei of embryonic chick skin (Sani and Donovan, 1979). This suggests there may be fundamental differences in the way that CRABP and retinoic acid interact with the nucleus compared with CRBP and retinol.

In any event, these demonstrated nuclear interactions suggest a means by which retinoids can direct differentiation. Much more work is under way and will be required to understand the mechanisms involved.

ACKNOWLEDGMENT

Supported by grants from the United States Public Health Service HD-09195, HL-14214, HL-1531, and CA-20850. The authors thank L. Chytil and D. Wadkins for help in preparation of the manuscript.

REFERENCES

Abelev, G. I. (1971). *Adv. Cancer Res.* **14,** 295–358.

Adachi, N., Smith, J. E., Sklan, D., and Goodman, DeW. S. (1981). *J. Biol. Chem.* **256,** 9471–9476.

Ames, S. R., Swanson, W. J., and Harris, P. L. (1955a). *J. Am. Chem. Soc.* **77,** 4136–4138.

Ames, S. R., Swanson, W. J., and Harris, P. L. (1955b). *J. Am. Chem. Soc.* **77,** 4134–4136.

Barranco, S. C., Haenelt, B. R., and Gee, E. L. (1978). *Cancer Res.* **38,** 656–660.

Bashor, M. M., and Chytil, F. (1975). *Biochim. Biophys. Acta* **411,** 87–96.

Bashor, M. M., Toft, D. O., and Chytil, F. (1973). *Proc. Natl. Acad. Sci. U.S.A.* **70,** 3483–3487.

Bertram, J. (1980). *Cancer Res.* **40,** 3141–3146.
Bichler, E., and Daxenbichler, G. (1982). *Cancer (Philadelphia)* **49,** 619–622.
Bollag, W. (1979). *Cancer Chemother. Pharmacol.* **3,** 207–215.
Bollag, W., and Hanck, A. (1977). *Acta Vitaminol. Enzymol.* **31,** 113–123.
Braude, E. A., Jones, E. R. H., Koch, H. P., Richardson, R. W., Sondheimer, F., and Toogood, J. B. (1949). *J. Chem. Soc.* pp. 1890–1897.
Breitman, T. R., Collins, S. J., and Keene, B. R. (1981). *Blood* **57,** 1000–1004.
Burri, P. H., Dbaly, J., and Weibel, E. R. (1974). *Anat. Rec.* **178,** 711–730.
Chytil, B. (1982). *Hepatology* **2,** 282–287.
Chytil, F. (1984). *Pediatrics* (in press).
Chytil, F., and Ong, D. E. (1976). *Nature (London)* **260,** 49–51.
Chytil, F., and Ong, D. E. (1978a). *Vitam. Horm.* **36,** 1–32.
Chytil, F., and Ong, D. E. (1978b). *In* "Receptors and Hormone Action" (B. W. O'Malley and L. Birnbaumer, eds.), pp. 573–591. Academic Press, New York.
Chytil, F., and Ong, D. E. (1979). *Fed. Proc., Fed. Am. Soc. Exp. Biol.* **38,** 2510–2513.
Chytil, F., and Ong, D. E. (1982). *In* "Molecular Interrelation of Nutrition and Cancer" (M. S. Arnott, J. van Eys, and Y. W. Wang, eds.), pp. 409–417. Raven, New York.
Chytil, F., and Ong, D. E. (1983). *In* "Advances in Nutritional Research" (H. H. Draper, ed.), Vol. 5, pp. 13–19. Plenum, New York.
Chytil, F., Page, D. L., and Ong, D. E. (1975). *J. Vitam. Nutr. Res.* **45,** 293–298.
Clamon, G. H., Sporn, M. B., Smith, J. M., and Saffiotti, U. (1974). *Nature (London)* **250,** 64–66.
Clamon, G. H., Nugent, K. M., and Rossi, N. P. (1981). *JNCI, J. Natl. Cancer Inst.* **67,** 61–63.
Crabb, J. W., and Saari, J. C. (1981). *FEBS Lett.* **130,** 15–18.
Das, R. C., Sarada, K., Murthy, S. K., and Ganguly, J. (1978). *Indian J. Biochem. Biophys.* **15,** 251–254.
Douer, D., and Koeffler, H. P. (1982). *Exp. Cell. Res.* **138,** 193–198.
Dowling, J. E., and Wald, G. (1960). *Proc. Natl. Acad. Sci. U.S.A.* **46,** 587–608.
Eriksson, U., Sundelin, J., Rask, L., and Peterson, P. A. (1981). *FEBS Lett.* **135,** 70–72.
Fex, G., and Johannesson, G. (1982). *Biochim. Biophys. Acta* **714,** 536–542.
Fuchs, E., and Green, H. (1981). *Cell* **25,** 617–625.
Futterman, S., and Heller, L. (1972). *J. Biol. Chem.* **247,** 5168–5172.
Futterman, S., Saari, J. C., and Swanson, D. E. (1976). *Exp. Eye Res.* **22,** 419–424.
Gesell, M. S., Brandes, M. J., Arnold, E. A., Isaacs, J. T., Ueda, H., Millan, J. C., and Brandes, D. (1982). *Prostate* **3,** 131–138.
Goodman, D. S. (1981). *Ann. N. Y. Acad. Sci.* **359,** 69–78.
Goodman, D. S. (1982). *J. Am. Acad. Dermatol.* **6,** 583–588.
Goodman, D. S., Smith, J. E., Hembry, R. M., and Dingle, J. T. (1974). *J. Lipid Res.* **15,** 406–414.
Greengard, O., Federman, M., and Knox, W. E. (1972). *J. Cell. Biol.* **52,** 261–272.
Howe, C. C., and Solter, D. (1981). *Dev. Biol.* **84,** 239–243.
Huber, P. R., Geyer, E., Küng, W., Matter, A., Torhorst, J., and Eppenberger, U. (1978). *JNCI, J. Natl. Cancer Inst.* **61,** 1375–1378.
Huggenvik, J., and Griswold, M. D. (1981). *J. Reprod. Fertil.* **61,** 403–408.
Jetten, A. M., and Jetten, M. E. R. (1979). *Nature (London)* **278,** 180–182.
Jetten, A. M., Jetten, M. E. R., Shapiro, S. S., and Poon, J. P. (1979). *Exp. Cell. Res.* **119,** 289–299.
Kawamura, H., and Hashimoto, Y. (1980). *Gann* **71,** 507–513.
Küng, W. M., Geyer, E., Eppenberger, U., and Huber, P. R. (1980). *Cancer Res.* **40,** 4265–4269.
Kylberg, H. K., Ong, D. E., and Chytil, F. (1981). *Biol. Neonate* **39,** 100–104.
Lacroix, A., and Lippman, M. E. (1980). *J. Clin. Invest.* **65,** 586–591.

Lacroix, A., Anderson, G. D., and Lippman, M. E. (1980). *Exp. Cell. Res.* **130,** 339–344.
Liau, G., Ong, D. E., and Chytil, F. (1981). *J. Cell Biol.* **91,** 63–68.
Libby, P. R., and Bertram, J. S. (1982). *Carcinogenesis* **3,** 481–484.
Linder, S., Krondahe, U., Sennerstam, R., and Ringertz, N. R. (1981). *Exp. Cell Res.* **132,** 453–460.
Liou, G. I., Fong, S. L., and Bridges, C. D. E. (1981). *J. Biol. Chem.* **256,** 3153–3155.
Lotan, R. (1980). *Biochim. Biophys. Acta* **605,** 33–91.
Lotan, R., and Lotan, D. (1980). *Cancer Res.* **40,** 3345–3350.
Lotan, R., and Lotan D. (1981). *J. Cell Physiol.* **106,** 179–189.
Lotan, R., Neumann, G., and Lotan, D. (1980a). *Cancer Res.* **40,** 1097–1102.
Lotan, R., Ong, D. E., and Chytil, F. (1980b). *JNCI, J. Natl. Cancer Inst.* **64,** 1259–1262.
Lotan, R., Kramer, R. H., Neuman, G., Lotan, D., and Nicolson, G. L. (1980c). *Exp. Cell. Res.* **130,** 401–414.
Lotan, R., Fischer, I., Meromsky, L., and Moldave, K. (1982). *J. Cell. Physiol.* **113,** 47–55.
McPartland, R. P., Milholland, R. J., Morris, H. P., and Rosen, F. (1977). *Cancer Res.* **37,** 1416–4165.
Marchok, A. C., Clark, J. N., and Klein-Szanto, A. (1981). *JNCI, J. Natl. Cancer Inst.* **66,** 1165–1174.
Mehta, R. G., and Moon, R. C. (1981). *Biochem. J.* **200,** 591–595.
Mehta, R. G., Cerny, W. L., and Moon, R. C. (1980). *Cancer Res.* **40,** 47–49.
Mehta, R. G., Kute, T. E., Hopkins, M., and Moon, R. C. (1982a). *Eur. J. Cancer Clin. Oncol.* **18,** 221–226.
Mehta, R. G., McCormick, D. L., Cerny, W. L., and Moon, R. C. (1982b). *Carcinogenesis* **3,** 89–91.
Merriman, D. L., and Bertram, J. S. (1979). Cancer Res. **39,** 1661–1666.
Moon, R. C., and Mehta, R. G. (1981). *Adv. Exp. Med. Biol.* **138,** 231–249.
Moore, T. (1967). *In* "The Vitamins: Chemistry, Physiology, Pathology, Methods: Second Edition" (W. H. Sebrell and R. S. Harris, eds.), Vol. 1, pp. 245–266. Academic Press, New York.
Muto, Y., and Omori, M. (1981). *Ann. N. Y. Acad. Sci.* **359,** 91–103.
Muto, Y., Omori, M., and Sugawara, K. (1979). *Gann* **70,** 215–222.
Muto, Y., Moriwaki, H., and Omori, M. (1981). *Gann* **72,** 974–977.
Newcomer, M. E., Liljas, A., Eriksson, U., Sundelin, J., Rask, L., and Peterson, P. A. (1981). *J. Biol. Chem.* **256,** 8162–8163.
Omori, M., and Chytil, F. (1982). *J. Biol. Chem.* **257,** 14370–14374.
Omori, M., Muto, Y., and Nagao, T. (1981). *J. Lipid. Res.* **22,** 899–904.
Ong, D. E. (1982). *Cancer Res.* **42,** 1033–1037.
Ong, D. E., and Chytil, F. (1974). *Biochem. Biophys. Res. Commun.* **59,** 221–229.
Ong, D. E., and Chytil, F. (1975a). *J. Biol. Chem.* **250,** 6113–6116.
Ong, D. E., and Chytil, F. (1975b). *Nature (London)* **255,** 74–75.
Ong, D. E., and Chytil, F. (1976a). *Proc. Natl. Acad. Sci. U.S.A.* **73,** 3976–3978.
Ong, D. E., and Chytil, F. (1976b). *Cancer Lett. (Shannon, Irel.)* **2,** 25–30.
Ong, D. E., and Chytil, F. (1978a). *J. Biol. Chem.* **253,** 828–832.
Ong, D. E., and Chytil, F. (1978b). *J. Biol. Chem.* **253,** 4551–4554.
Ong, D. E., and Chytil, F. (1979). *J. Biol. Chem.* **254,** 8733–8735.
Ong, D. E., and Chytil, F. (1980). *In* "Methods in Enzymology" (D. B. McCormick and L. D. Wright, eds.), Vol. 67, pp. 288–296. Academic Press, New York.
Ong, D. E., and Chytil, F. (1981). *Ann. N. Y. Acad. Sci.* **359,** 415–417.
Ong, D. E., and Chytil, F. (1983). *Vitam. Horm. (N. Y.)* **40,** 105–144.
Ong, D. E., Page, D. L., and Chytil, F. (1975). *Science* **190,** 60–61.
Ong, D. E., Tsai, C. H., and Chytil, F. (1976). *J. Nutr.* **106,** 204–211.

Ong, D. E., Markert, C., and Chiu, J. F. (1978). *Cancer Res.* **38,** 4422–4426.
Ong, D. E., Goodwin, W. J., Jesse, R. H., and Griffin, A. C. (1982a). *Cancer (Philadelphia)* **49,** 1409–1419.
Ong, D. E., Crow, J. A., and Chytil, F. (1982b). *J. Biol. Chem.* **257,** 13385–13389.
Oroshnik, W., Karmas, G., and Mebane, A. D. (1952). *J. Am. Chem. Soc.* **14,** 295–304.
Palan, P. R., and Romney, S. L. (1980). *Cancer Res.* **40,** 4221–4224.
Palan, P. R., Duttagupta, C., and Romney, S. L. (1980). *Cancer Lett. (Shannon Irel.)* **11,** 97–101.
Rao, M. R. S., Prasad, V. R., Padmanaban, G., and Ganguly, J. (1979). *Biochem. J.* **183,** 501–506.
Rask, L., Anundi, H., Bohme, J., Eriksson, U., Ronne, H., Sege, K., and Peterson, P. A. (1981). *Ann. N. Y. Acad. Sci.* **359,** 79–90.
Rattanapanone, V., Tashior, S., Tokuda, H., Rattanapanone, N., and Ito, Y. (1981). *Cancer Res.* **41,** 1483–1487.
Reese, D. H., and Politano, V. A. (1981). *Biochem. Biophys. Res. Commun.* **102,** 322–327.
Reese, D. H., Fiorentino, G. J., Claflin, A. J., Malinin, T. I., and Politano, V. A. (1981). *Biochem. Biophys. Res. Commun.* **102,** 315–321.
Ross, C. A., and Goodman, D. S. (1979). *Fed. Proc., Fed. Am. Soc. Exp. Biol.* **38,** 2515–2518.
Ross, A. C., Takahashi, Y. I., and Goodman, D. S. (1978). *J. Biol. Chem.* **253,** 6591–6598.
Ross, A. C., Adachi, N., and Goodman, D. S. (1980). *J. Lipid Res.* **21,** 100–109.
Russel, P., Wiggert, B., and Chader, G. J. (1978). *Biochim. Biophys. Acta* **543,** 586–589.
Saari, J. C. (1982). *In* "Methods in Enzymology" (L. Parker, ed.), Vol. 81, pp. 819–826. Academic Press, New York.
Saari, J. C., and Futterman, S. (1976). *Biochim. Biophys. Acta* **444,** 789–793.
Saari, J. C., Futterman, S., and Bredberg, L. (1978a). *J. Biol. Chem.* **253,** 6432–6436.
Saari, J. C., Futterman, S., Stubbs, G. W., Hefferman, J. T., Bredberg, L., Chan, K. Y., and Albert, D. M. (1978b). *Invest. Ophthalmol. Visual Sci.* **17,** 988–992.
Saari, J. C., Futterman, S., Stubbs, G. W., and Bredberg, L. (1980). *In* "Methods in Enzymology" (D. B. McCormick and L. D. Wright, eds.), Vol. 67, pp. 296–300. Academic Press, New York.
Saari, J. C., Bredberg, L., and Garwin, G. G. (1982). *J. Biol. Chem.* **257,** 13329–13333.
Sani, B. P., and Banerjee, C. K. (1978). *Biochem. J.* **173,** 643–694.
Sani, B. P., and Corbett, T. H. (1977). *Cancer Res.* **37,** 209–213.
Sani, B. P., and Donovan, M. K. (1979). *Cancer Res.* **39,** 2492–2496.
Sani, B. P., and Hill, D. L. (1974). *Biochem. Biophys. Res. Commun.* **61,** 1276–1281.
Sani, B. P., and Hill, D. L. (1976). *Cancer Res.* **36,** 409–413.
Sani, B. P., and Titus, B. C. (1977). *Cancer Res.* **37,** 4031–4034.
Sani, B. P., Titus, B. C., and Banerjee, C. K. (1978). *Biochem. J.* **171,** 711–717.
Sani, B. P., Condon, S. M., and Banerjee, C. K. (1980a). *Biochim. Biophys. Acta* **624,** 226–236.
Sani, B. P., Condon, S. M., Brockman, R. W., Weiland, L. H., and Schutt, A. J. (1980b). *Cancer (Philadelphia)* **45,** 1199–1206.
Sani, B. P., Banerjee, C. K., and Peckham, J. C. (1980c). *Cancer (Philadelphia)* **46,** 2421–2429.
Sato, M., Omori, M., and Muto, Y. (1979). *J. Nutr. Sci. Vitaminol.* **25,** 1–7.
Sato, M., Hiragun, A., and Mitsui, H. (1980). *Biochem. Biophys. Res. Commun.* **95,** 1839–1845.
Schindler, J., Matthaei, K. I., and Sherman, M. I. (1981). *Proc. Natl. Acad. Sci. U.S.A.* **78,** 1077–1080.
Shantz, E. M., and Brinkman, J. H. (1950). *J. Biol. Chem.* **183,** 467–471.
Shapira, F. (1973). *Adv. Cancer Res.* **18,** 77–153.
Sherman, M. I., Matthaei, K. I., and Schindler, J. (1981). *Ann. N. Y. Acad. Sci.* **359,** 192–199.
Shidoji, Y., and Muto, Y. (1978). *Comp. Biochem. Physiol. B* **61B,** 315–319.
Shinde, R., Das, R. C., Sarada, K., and Ganguly, J. (1980). *Indian J. Biochem. Biophys.* **17,** 135–138.

Sklan, D., Blaner, W. S., Adachi, N., Smith, J. E., and Goodman, D. S. (1982). *Arch. Biochem. Biophys.* **214,** 35–44.

Sporn, M. B., and Newton, D. L. (1979). *Fed. Proc., Fed. Am. Soc. Exp. Biol.* **38,** 2528–2534.

Sporn, M. B., Clamon, G. H., Dunlop, N. M., Newton, D. L., Smith, J. M., and Saffiotti, U. (1975). *Nature (London)* **253,** 47–50.

Swanson, D., Futterman, S., and Saari, J. C. (1976). *Invest. Opthalmol.* **15,** 1017–1020.

Takahashi, K., Odani, S., and Ono, T. (1982). *Biochem. Biophys. Res. Commun.* **106,** 1099–1105.

Takase, S., Ong, D. E., and Chytil, F. (1979). *Proc. Natl. Acad. Sci. U.S.A.* **76,** 2204–2208.

Takenaga, K., Honma, Y., and Hozumi, M. (1981). *Cancer Lett. (Shannon, Irel.)* **13,** 1–6.

Takenawa, T., Ueda, H., Millan, J. C., and Brandes, D. (1980). *Lab. Invest.* **42,** 490–494.

Thompson, J. N., Howell, J. McP., and Pitt, G. A. (1964). *Proc. R. Soc. London, Ser. B* **159B,** 510–535.

Trown, P. W., Palleroni, A. V., Bohoslawec, O., Richelo, B. N., Halpern, J. M., Gizzi, N., Geiger, R., Lewinski, C., Machlin, L. J., Jetten, A., and Jetten, M. E. R. (1980). *Cancer Res.* **40,** 212–220.

Wiggert, B., Bergsma, D. R., Lewis, M., Abe, T., and Chader, G. J. (1977a). *Biochim. Biophys. Acta* **498,** 366–364.

Wiggert, B., Russel, P., Lewis, M., and Chader, G. (1977b). *Biochem. Biophys. Res. Commun.* **79,** 218–226.

Wolbach, S. B., and Howe, P. R. (1925). *J. Exp. Med.* **43,** 753–777.

Wolbach, S. B., and Howe, P. R. (1933). *J. Exp. Med.* **57,** 511–526.

10

Retinoids in Photosensitive Systems

C. D. B. BRIDGES
Cullen Eye Institute
and Program in Neuroscience
Department of Ophthalmology
Baylor College of Medicine
Houston, Texas

THE RETINOIDS, VOL. 2

ISBN 0-12-658102-9

I. INTRODUCTION

It is remarkable that the photosensitivity of the insects, crustacea, arthropods, cephalopods, together with the vertebrates and even the halophilic bacteria, is based on a common chemistry. In each organism, the pigment responsible for absorption of light consists of a retinaldehyde (C1) prosthetic group coupled by a protonated aldimine bond to a membrane protein. The reactions following light absorption always involve cis–trans isomerization, a process that sometimes leads to deprotonation of the aldimine bond.

The biosynthesis and regeneration of these photosensitive pigments requires that retinaldehyde (or a retinoid precursor) is provided to the membranes in which they are sequestered. The delivery of retinoids to these membranes, which are generally found in specialized photoreceptor cells, has been investigated in some vertebrates. In these animals, proteins called *retinoid-binding proteins* serve as vehicles that transport retinoids in the serum, cell sap, and interstitial space of the eye. The critical role of these proteins in the normal functioning of the vertebrate visual system is one of the major topics of this chapter.

II. PROTEINS ASSOCIATED WITH RETINOIDS IN ANIMALS AND BACTERIA

A. Visual Pigments

Vision is initiated by the capture of quanta of radiation by integral transmembrane proteins called visual pigments. In all known visual pigments, the protein (opsin) is conjugated to a retinaldehyde prosthetic group. Two series of visual pigments can be extracted from dark-adapted retinas. The *rhodopsins* are based on 11-*cis*-retinaldehyde (C2), the *porphyropsins* on 11-*cis*-3,4-didehydroretinaldehyde (C3). Rhodopsins and porphyropsins have absorption spectra typified by those illustrated in Fig. 1. These spectra determine the spectral sensitivity curve of the dark-adapted eye. In vertebrate rhodopsins, the λ_{max} range from 440 to 562 nm. Invertebrate rhodopsins have λ_{max} that range down to 345 nm. While there is some overlap, the porphyropsins typically have absorption bands displaced to longer wavelengths, hence they are prevalent in organisms that live in freshwater environments where the ambient light is often reddish (Bridges, 1965, 1972; Lythgoe, 1972; Knowles and Dartnall, 1977). This is exemplified in Fig. 1. Frog and bovine rhodopsins (A, C) have λ_{max} close to 500 nm. Goldfish porphyropsin, on the other hand, has its λ_{max} shifted to 522 nm, a position more appropriate for a freshwater organism.

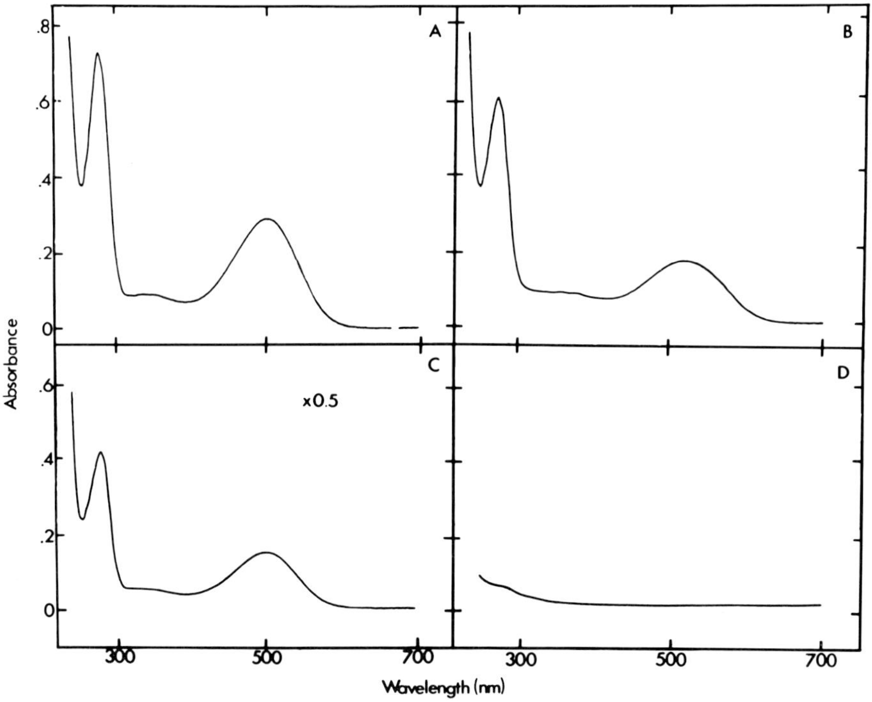

Fig. 1. Absorption spectra of visual pigments from (A) frog (*R. pipiens*; λ_{max} = 502 nm), (B) goldfish (*Carassius auratus*; λ_{max} = 522 nm), (C) cattle (λ_{max} = 498 nm). Solvent consists of 1% octyl glucoside in 50 m*M* sodium phosphate (pH 7.0); its spectrum is shown in (D). Concentrations of visual pigment were 7.1 μ*M* for frog and cattle rhodopsins, 5.5 μ*M* for goldfish porphyropsin. (Based on Fong *et al.*, 1982.)

1. Vertebrate Visual Pigments

Visual pigments are insoluble in water and must be solubilized with detergents such as CTAB (cetyl trimethylammonium bromide), Ammonyx LO, Emulphogene, digitonin, and octyl glucoside [see Fong *et al.* (1982) for references]. The majority of visual pigments extracted from vertebrate retinas are believed to originate in the *rod* photoreceptors. The visual pigments of *cones* probably do not survive detergent extraction. Because of its ready availability in large quantities, the best-characterized vertebrate visual pigment is bovine rhodopsin. The properties of bovine rhodopsin have been reviewed by Hargrave (1982) and Hargrave *et al.* (1982). The protein has M_r 39,000 (~ 350 amino acids), and Ovchinnikov *et al.* (1982) have determined the complete amino acid sequence. The polypeptide chain is believed to consist of seven hydrophobic transmembrane helices that range in length from 21 to 28 amino acids (Hargrave *et al.*, 1983). One molecule of 11-*cis*-retinaldehyde is attached to Lys-53′ by a Schiff base (aldimine) linkage (Wang *et al.*, 1980). The molar absorbance coefficient at 498 nm is 40,000–42,000 $M^{-1}\text{cm}^{-1}$ (in various detergents). The molecule contains two asparagine-linked oligosaccharides, each of the simple core structure $(\text{Man})_3$ $(\text{GlcNAc})_3$ (Fukuda *et al.*, 1979; Liang *et al.*, 1979). The biosynthesis of this oligosaccharide has been studied in several laboratories (e.g., Bok *et al.*, 1974; O'Brien, 1978; Kean, 1977).

When exposed to light, vertebrate visual pigments bleach with an overall quantum efficiency of about $\frac{2}{3}$ (Kropf, 1967, 1982; Dartnall, 1968), and the 11-*cis*-retinaldehyde prosthetic group isomerizes to all-trans (Wald, 1968). In rod outer segment membranes or in detergent solutions at room temperature and neutral pH, the bleached pigment hydrolyzes to opsin and all-*trans*-retinaldehyde. Low-temperature and picosecond spectroscopy have been used to identify a number of intermediate forms that appear between the initial absorption of radiation and the appearance of free all-*trans*-retinaldehyde (Yoshizawa and Shichida, 1982; Applebury and Rentzepis, 1982). As summarized in Fig. 2, these intermediates include hypsorhodopsin, bathorhodopsin, lumirhodopsin, metarhodopsins I and II, pararhodopsin, and *N*-retinylidene opsin. While there is basic agreement on the number of intermediates [although see, for example, Blazynski and Ostroy (1981), Sasaki *et al.* (1981), and Sasaki and Yoshizawa (1981)], their structure is still not well understood, neither has their order of

cattle
rhodopsin (498 nm)
↓ $h\nu$
hypsorhodopsin (430 nm)
↓ > − 251°C
bathorhodopsin (543 nm)
$\tau \cong$ 30 ns ↓ > − 140°C
lumirhodopsin (497 nm)
$\tau \cong$ 16 s ↓ > − 40°C
metarhodopsin I (478 nm)
$\tau \cong$ 0.2 s ↓ > − 15°C
metarhodopsin II (380 nm)
$\tau \cong$ 1 h ↓ > 0°C
pararhodopsin (465 nm)
(metarhodopsin III)
↓ > 5°C
(*N*-retinylideneopsin; 440 nm acid,
↓ 367 nm alkaline)
all-*trans*-retinaldehyde (387 nm)
+
opsin

squid
rhodopsin (480 nm)
↓ $h\nu$
hypsorhodopsin (446 nm)
↓ > − 238°C
bathorhodopsin (534 nm)
$\tau \cong$ 300 ns ↓ > − 160°C
lumirhodopsin (515 nm)
↓ > − 65°C
LM-rhodopsin (486 nm)
$\tau \cong$ 10 ms ↓ > − 20°C
acid metarhodopsin (482 nm)
⇃↾ H^+
alkaline metarhodopsin (367 nm)

Fig. 2. Photochemical reactions of cattle and squid rhodopsins. Simplified after Yoshizawa and Shichida (1982); this review, and others cited in the text, should be consulted for references to the original articles.

appearance been clarified, particularly during the final stages that precede hydrolysis of the aldimine bond (Baumann and Zeppenfeld, 1981; Blazynski and Ostroy, 1981; Brin and Ripps, 1977; Ernst *et al.*, 1978; Chabre and Breton, 1979). A problem that has not been completely resolved is the question of whether hypsorhodopsin exists at room temperature [e.g., compare Yoshizawa and Shichida (1982) with Applebury and Rentzepis (1982)]. A full discussion of the photochemistry of rhodopsin is beyond the scope of the present chapter: for reviews of this topic, the reader should consult Kropf (1977), Ostroy (1977), Knowles and Dartnall (1977), and Yoshizawa and Shichida (1982).

In intact rod outer segments, a membrane-bound retinol dehydrogenase reduces all-*trans*-retinaldehyde to all-*trans*-retinol (A1). This enzyme differs from the soluble liver alcohol dehydrogenase (which can also reduce retinaldehyde to retinol) in that it needs NADP rather than NAD and cannot use ethanol as a substrate (see Section III,F,1).

2. *Invertebrate Visual Pigments*

Rhodopsins from cephalopods, insects, crustacea, and arthropods have been investigated. In general, the visual pigments of invertebrates, like those of their vertebrate counterparts, appear to be based on 11-*cis*-retinaldehyde. Squid and octopus rhodopsins are glycoproteins that bind concanavalin A. However, octopus rhodopsin has a higher affinity for the lectin, which suggests that it has a different sugar complement (Kito *et al.*, 1982). The molecular weights of some invertebrate visual pigments have been determined, usually by sodium dodecyl sulfate polyacrylamide gel electrophoresis (SDS–PAGE). Squid rhodopsin has M_r 49,000 (Kito *et al.*, 1982) but easily loses an M_r 12,000 peptide by proteolysis of the C-terminus. Insect visual pigments, which range in λ_{max} from 345 to 610 nm (Schwemer and Langer, 1982), have M_r that are similar to those of vertebrate rhodopsins, ranging from 33,000 to 37,000.

As exemplified in Fig. 2, the visual pigments of invertebrates usually differ fundamentally from those of the vertebrates by forming stable metarhodopsins during light adaptation [see Bridges (1970), Ostroy (1977), Knowles and Dartnall (1977), Yoshizawa and Shichida (1982) for reviews]. Invertebrate metarhodopsins also differ from vertebrate metarhodopsins in their pH sensitivity. The long-wave absorbing variety of metarhodopsin is favored by low pH in invertebrates (acid metarhodopsin) and by high pH in vertebrates (metarhodopsin I).

B. Retinochrome

Retinochrome is an M_r 20,000 chromoprotein found in the retinas of cephalopods. Its λ_{max} is at 495 nm and its $\epsilon_{495} = 60{,}700\ M^{-1}\text{cm}^{-1}$ (Hara and Hara, 1982). Its prosthetic group consists of one molecule of retinaldehyde. In contrast to the prosthetic group of rhodopsin, the retinaldehyde is in the all-trans conformation and is converted to 11-cis during irradiation (Hara *et al.*, 1981). Retinochrome has been shown to be able to act catalytically, trace amounts being sufficient to insure that a preparation of all-*trans*-retinaldehyde is efficiently (> 98%) converted to 11-cis when irradiated with orange light. This finding has led to the suggestion that retinochrome plays a role in the regeneration of rhodopsin in the cephalopod retina (Seki *et al.*, 1982).

C. Retinaldehyde Proteins in Halobacteria

1. *Bacteriorhodopsin*

Bacteriorhodopsin is an integral transmembrane protein that is the only protein found in the purple membrane of *Halobacterium halobium* and other extremely halophilic bacteria [for reviews, see Stoeckenius *et al.* (1979), Stoeckenius (1980), and Stoeckenius and Bogomolni (1982) and also recent articles in Packer (1982)]. The purple membrane is a differentiated part of the plasma membrane of these organisms. It contains 25% lipid and approximately 75% protein (bacteriorhodopsin) by weight. Bacteriorhodopsin has M_r 26,000 and contains 248 amino acids, about 70% of which are grouped in hydrophobic segments. The amino acid sequence has been determined by protein sequencing (Ovchinnikov *et al.*, 1979; Khorana *et al.*, 1979) and by sequencing its gene (Dunn *et al.*, 1981).

As in rhodopsin, the polypeptide chain traverses the membrane seven times in the form of α-helical rods attached by linking sequences (Henderson and Unwin, 1975). Unlike rhodopsin, but in common with most bacterial proteins, it is not glycosylated. The molecule has λ_{max} = 568 nm and binds one molecule of retinaldehyde, which is linked covalently by a protonated Schiff base to the ϵ-amino group of Lys-216 (Rothschild *et al.*, 1982; Bayley *et al.*, 1981; Lemke and Oesterhelt, 1981; Mullen *et al.*, 1981). The 11-cis isomer is not present in significant amounts, the only isomers observed being 13-cis and all-trans in proportions that depend on the stage of the photocycle. Resonance Raman spectroscopy and retinaldehyde extraction have suggested that the "light-adapted" ground state (BR 570) contains all-trans, while the bleached intermediate with λ_{max} at 412 nm (MR 412) contains 13-cis (Stoeckenius *et al.*, 1979; Braiman and Mathies, 1982). A suggestion that retinaldehyde migrated between two attachment sites (Katre *et al.*, 1981; Ovchinnikov *et al.*, 1980) in the transition from BR 570 to MR 412 has not been substantiated (Rothschild *et al.*, 1982).

2. *Halorhodopsin*

Halorhodopsin is a retinaldehyde-containing protein with M_r close to that of bacteriorhodopsin (Schobert and Lanyi, 1982). It occurs in the cytoplasmic membranes of many halobacterial strains; its possible functions are discussed below.

3. *Contrasting Functions of Rhodopsin and the Retinaldehyde Proteins from Halobacteria*

There is now abundant evidence that bacteriorhodopsin is a light-activated vectorial ion pump that translocates protons across the bacterial plasma membrane. The mechanism of proton translocation is unknown, although the process is linked to a complex photocycle that involves a number of intermediates [Mowery and Stoeckenius (1981) and Stoeckenius and Bogomolni (1982); see also articles in Packer (1982)] in which the Schiff base is protonated in the BR 570 state and deprotonated in the MR 412 state (Lewis *et al.*, 1974; Aton *et al.*, 1977; Rothschild *et al.*, 1981). The electrochemical gradient generated in this way is used by the cell for ATP synthesis.

In addition to its function in using incident radiation for energy-requiring activity, bacteriorhodopsin may have a sensory role. Halobacteria are phototactic. A decrease in illumination at the long-wave end of the spectrum makes the organisms stop briefly and reverse direction. (An identical response is seen when the illumination is increased, but the effect is confined to the blue end of the spectrum.) The action spectrum closely fits the absorption spectrum of the purple membrane, which suggests that bacteriorhodopsin is the "visual pigment" of halobacteria (Stoeckenius *et al.*, 1979). The absolute sensitivity has not been determined, and it is not known whether the response is linked to the proton-pumping activity.

Less is known about halorhodopsin. Observations (Schobert and Lanyi, 1982) suggest that it is a light-activated ion pump that drives chloride inward into the cell. It may serve to regain intracellular chloride lost by membrane-potential-driven extrusion.

Rhodopsin, on the other hand, functions as a light sensor. Each rod photoreceptor in a vertebrate eye is exquisitely sensitive and responds to a single quantal absorption. The photoreceptor responds to light with a graded hyperpolarization caused by a reduction in the sodium conductance of its plasma membrane (Penn and Hagins, 1972; see Matthews and Baylor, 1981 for review and references therein). A single quantum absorbed blocks approximately 100 sodium channels and releases about 2×10^4 calcium ions into the medium surrounding the photoreceptors (Gold and Korenbrot, 1981).

The closure of sodium channels is believed to be effected by an internal transmitter, which has been linked to calcium and cGMP by various laboratories (Bownds, 1981; Liebman and Pugh, 1981, 1982; Stryer *et al.*, 1981; Bitensky *et al.*, 1981; Oakley and Pinto, 1981; Lolley and Racz, 1982; Hermolin *et al.*, 1982). Illuminated rhodopsin initiates a number of reactions. There is binding of GTPase to the disc membranes, coupled with its activation. There is also simultaneous activation of a cGMP phosphodiesterase, which is membrane-bound at physiologic ionic strength, irrespective of illumination. Additionally, rhodopsin is phosphorylated by a kinase that binds to the bleached membranes. For a review of other, light-dependent effects see Kühn (1981). To date, none of these findings has provided direct information on how the sodium conductance of the plasma membrane is modulated by light, and it is still not clear how rhodopsin acts in visual transduction beyond its initial, quantum-trapping function.

Even though rhodopsin and bacteriorhodopsin appear to differ fundamentally in their function, mode of action, structure, and photochemistry, it has been suggested (Stoeckenius *et al.*, 1979) that the two proteins did not evolve independently. One possibility is that the halobacteria may have acquired bacteriorhodopsin by gene transfer from a eukaryote (Stoeckenius *et al.*, 1979). However, a preliminary report by Hargrave *et al.* (1983) claims that no statistically significant sequence homology can be found. Thus, the resemblance between these proteins appears to be only superficial.

D. Analogs of Rhodopsin and Bacteriorhodopsin

Labeled retinaldehydes (14-^{13}C, ^{3}H, and ^{14}C) have been used for NMR (Shriver *et al.*, 1982) and metabolic studies in the eye (e.g., Bridges and Yoshikami, 1970; Hall and Bok, 1974). Additionally, a large variety of synthetic retinoids bind to opsin and bacterio-opsin [see Balogh-Nair and Nakanishi (1982a,b) for reviews]. A study of the analogs formed in this way can give specific information on the steric and electronic interactions between the protein and its prosthetic group and hence serve to shed light on the molecular events leading to visual excitation or to the pumping of ions across bacterial membranes. In general, the analogs have absorption spectra that are red-shifted in comparison with the free retinoid. The cattle rhodopsin analogs range in λ_{max} from 310 nm (9,10,11,12-tetrahydroretinaldehyde) to 520 nm (13-bromo-13-demethylretinaldehyde), close to that for porphyropsin (3,4-dehydroretinaldehyde). The λ_{max} of bacteriorhodopsin analogs may range from 400 nm (7,8-dihydroretinaldehyde) to 595 nm (13-bromo-13-demethylretinaldehyde). The latter is close to that of the analog obtained with 3,4-didehydroretinaldehyde.

A variety of isomers of retinaldehyde have been examined in relation to the rhodopsin system: these include 9-cis, 9,11-di-cis, 9,13-di-cis, 7-cis, 7,9-di-cis, 7,13-di-cis, 7,9,13-tri-cis, 7,11-di-cis, all-trans, and 13-cis. Only the latter two isomers fail to combine. It has been suggested that a distance of 9.6–10.0 Å from the center of the ring to C-15 may be the critical factor (Matsumoto *et al.*, 1980).

In the halobacterial system, the situation seems to be exactly opposite: only the all-trans and 13-cis isomers react with bacterio-opsin. In both bacteriorhodopsin and rhodopsin, the ability to react is comparatively insensitive to modifications of the β-ionone ring.

The following is a summary of the 9-*cis* or 11-*cis* (sometimes 7-*cis*) analogs that form pigments with bovine opsin.

1. Alkylated and Dealkylated Retinaldehydes

Methyl groups at 5, 9, and 13 are not necessary for pigment formation. Similarly, the addition of methyl groups at 10 and 14 or substituting ethyl at 9 has no adverse effects on the combination with opsin.

2. Halogenated Retinaldehydes

Substances that form pigments include the 9- and 13-bromo, demethyl analogs, and the 10-, 12-, and 14-fluoro and 10,14-difluoro analogs [see Liu and Matsumoto (1982)].

3. Allenic Retinaldehyde and Retinaldehydes with Modified Rings

Modified-ring retinaldehydes can be used to probe the chiroptical requirements of the binding site and the tolerance of the β-ionylidene ring binding area. The analogs tested contained a 6,7 allenic double bond in place of the normal (6*S*) bond. This served to immobilize the terminal ring at a fixed angle to the chain. In the 9-cis or 9,13-dicis forms, both antipodes bound to opsin, as did the adamantyl allenic retinaldehydes (Balogh-Nair and Nakanishi, 1982a,b).

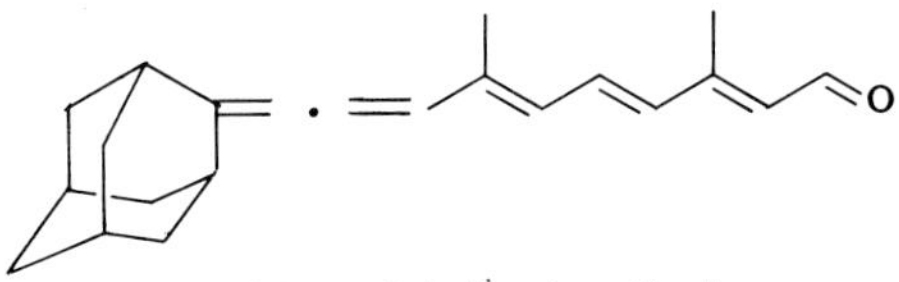

Adamantyl allenic retinal

This finding showed that the β-ionylidene ring binding area of rhodopsin was quite tolerant of large conformational variations. This is confirmed by the formation of pigments from retinaldehydes with oxygenated rings, aromatic rings, isomerized rings (e.g., γ-retroretinaldehyde, which has a 4,5 instead of a 5,6 double bond), a ring with an extra double bond (3,4-didehydroretinaldehyde), an open ring, or a five-membered ring.

In experiments on photoaffinity labels (Sen *et al.*, 1982), it was found that 9-*cis*-retinaldehyde with a diazoacetoxy group at C-3 formed a pigment. The opsin showed a preference for one of the C-3 enantiomers, indicating that the protein is capable of some chiroptical discrimination.

4. Dihydroretinaldehydes

These compounds were synthesized and examined by Nakanishi and his associates to test an external point-charge model to account for the spectrum of rhodopsin and the bathochromic shift when retinaldehyde is bound to the opsin molecule (Honig *et al.*, 1979). This model was an extension of the point-charge perturbation theory first advanced by Hubbard and Kropf (1965). The compounds tested were 5,6-, 7,8-, 9,10-, and 11,12-dihydroretinaldehydes, and 9,10,11,12-tetrahydroretinaldehyde. The largest bathochromic shift was seen in 11,12-dihydroretinaldehyde. Semi-empirical π-electron calculations indicated that the only plausible structure for rhodopsin included a counterion near the Schiff base nitrogen and a negative charge in the vicinity of C-12 and C-14. A full discussion of this model is presented by Balogh-Nair and Nakanishi (1982a).

5. *Blocked 11-cis Geometry*

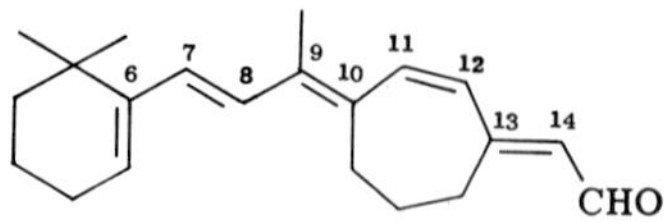

It is believed that in rhodopsin the prosthetic group is in the 11-cis-(12*S*)-trans conformation, which is nonplanar because of H-10, Me-13 interaction. The seven-membered ring in the cycloheptatrienylidene compound above forces it into a similar noncoplanar 9,11,13-triene conformation and has the additional property of being locked into the 11-cis conformation. When incubated with bleached bovine rod outer segment (ROS) membranes (Akita *et al.*, 1980), a pigment with λ_{max} = 488 is formed. This pigment is nonbleachable but has circular dichroic properties similar to rhodopsin. The 11,13- and 9,11-dicis compounds also formed nonbleachable pigments, as did the 9,11,13-tricis.

E. Proteins That Bind Endogenous Retinoids Noncovalently

In addition to their occurrence as covalently linked prosthetic groups in the light-sensitive pigments, retinoids are also found noncovalently associated with a number of soluble proteins known as retinoid-binding proteins, as summarized in Table I. They include serum retinol-binding protein (RBP) [Goodman (1980); also see Chapter 8], cellular retinol-binding protein (CRBP) [Ong and Chytil (1978a); see also Chapter 9], cellular retinaldehyde-binding protein (CRAlBP) [Futterman *et al.* (1977)], and cellular retinoic acid-binding protein (CRABP) Ong and Chytil (1978b); see Chapter 9]. Each of these proteins binds a single retinoid molecule. A glycoprotein capable of binding about two molecules of all-*trans*-retinol has been purified and characterized in the interstitial space between the retina and pigment epithelium of human and bovine eyes. It is a large protein [M_r 144,000; see also Adler and Martin (1982), Liou *et al.*, (1982c), and Fong *et al.*, (1983b, 1984).

To date, only a few vertebrates have been investigated: whether retinoid-binding proteins occur in all vertebrates and invertebrates remains to be determined. Further discussion of these proteins in relation to their possible functions in the vertebrate visual system is found in Section III.

TABLE I

Proteins That Bind Retinoids in Animals and Bacteria[a]

Protein	M_r	Location	Glycosylation	Retinoid	Moles ligand per mole binding protein	Linkage of retinoid to protein
Porphyropsin	39,000	Membrane	+	11-*cis*-3,4-didehydroretinaldehyde	1	Aldimine
Rhodopsin						
vertebrate	39,000	Membrane	+	11-*cis*-Retinaldehyde	1	Aldimine
invertebrate	33,000–49,000	Membrane	+ (?)	11-*cis*-Retinaldehyde (?)	1 (?)	Aldimine
Retinochrome	20,000	Membrane	−	All-*trans*-retinaldehyde	1	Aldimine
Bacteriorhodopsin	26,000	Membrane	−	All-*trans*-Retinaldehyde	1	Aldimine
Halorhodopsin	?	Membrane	?	Retinaldehyde	?	Aldimine (?)
CRBP	15,000	Cytosol	−	All-*trans*-retinol	1	Noncovalent
CRABP	15,000	Cytosol	−	All-*trans*-retinoic acid	1	Noncovalent
CRA1BP	33,000	Cytosol	−	11-*cis*-Retinol, 11-*cis*-retinaldehyde	1	Noncovalent
RBP	21,000	Serum	−	All-*trans*-retinol	1	Noncovalent
IRBP	144,000	Interstitial space (eye)	+	All-*trans*-retinol (11-*cis*-retinol)	~2	Noncovalent

[a] See text for references.

III. TRANSPORT AND UTILIZATION OF RETINOIDS IN THE VERTEBRATE VISUAL SYSTEM

A. Introduction

Because it is the precursor for the retinaldehyde prosthetic group of rhodopsin (Wald, 1968), retinol is needed to maintain the visual process. Additionally, retinal neurones may require retinoids for their normal functioning. Therefore, retinoids must be supplied to the eye from the circulation, transported between its layers, and stored in its tissues as a reserve against dietary deprivation.

The retinoids in dark-adapted eyes occur predominantly in the photoreceptor cell outer segments, which contain visual pigment, and in the adjacent retinal pigment epithelium (RPE) cells, which contain variable amounts of retinyl esters. These esters consist mainly of palmitate (A4) and stearate and are sometimes sequestered in lipid droplets (Bridges, 1975; Young and Bok, 1979; Alvarez *et al.*, 1981; Bridges *et al.*, 1982). The only major isomers present are all-trans and 11-cis. The 11-cis isomer is restricted to ocular tissues.

Retinoic acid (D1) is formed irreversibly from retinaldehyde (Futterman, 1962) and therefore is not a precursor for rhodopsin. Retinoic acid supports growth, promotes epithelial cell differentiation, and suppresses neoplastic development (e.g., Zile *et al.*, 1979; Roberts and Frolik, 1979; Lotan, 1980). It occurs in the retina (Saari *et al.*, 1982), but it is not known whether it has a specific retinal function.

Bleaching converts the 11-*cis*-retinaldehyde prosthetic group of visual pigments to all-*trans*-retinaldehyde (Wald, 1968). After hydrolysis of the Schiff base linkage, the free retinaldehyde is reduced to retinol by a membrane-bound retinol dehydrogenase (Futterman, 1963; DePont *et al.*, 1970; Bridges, 1977). In frogs and rats this retinol then flows into the pigment epithelium where it is esterified (Fig. 3) (Dowling, 1960; Zimmerman, 1974; Bridges, 1976b). In darkness, retinoid is returned to the retina where it is utilized by the photoreceptors for visual pigment regeneration (Fig. 3). Because retinoids are lipid-soluble compounds, specific binding proteins are required to transport them within the cytosol of the RPE and photoreceptor cells, as well as in the interstitial space between the rod outer segments and RPE.

Our present concern is primarily with those retinoids having a single double bond in the terminal ring, namely the "vitamin A_1" compounds, which are associated with retinas containing rhodopsin as the major visual pigment. However, many amphibians and fish have visual systems based on porphyropsin and "vitamin A_2", or 3,4-didehydroretinol (A9). The two systems may appear in temporal succession, as in amphibian metamorphosis, or may occur simultaneously, as in many freshwater and migratory fish and in adult bullfrogs (Wald, 1939; Reuter *et al.*, 1971; Bridges, 1965, 1972, 1974; Tsin and Beatty, 1979, 1980). Sometimes there is variation in the rhodopsin/porphyropsin ratio

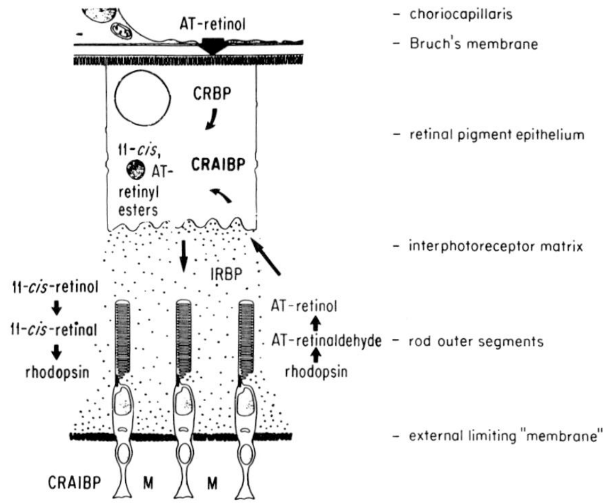

Fig. 3. Visual cycle retinoids and location of retinoid-binding proteins in the pigment epithelium, interphotoreceptor matrix and distal retina. Key: AT, all-trans configuration; M, Müller cell—a glial cell that spans the retina from the vitreous surface to the external limiting "membrane." This membrane is formed by the Müller cell ciliary processes that project into the space filled with interphotoreceptor matrix (dotted).

from dorsal to ventral retina, as in the bullfrog (Reuter *et al.*, 1971). Extensive investigations have been carried out on the environmental and hormonal factors that control the interconversions between rhodopsin and porphyropsin, and work on tadpoles and goldfish has shown that retinol can act as a precursor for 3,4-didehydroretinol (Tsin *et al.*, 1981). Although 3,4-didehydroretinol is less well studied than retinol, it is likely that its storage, transport, and delivery, as well as the enzymes implicated in its esterification, isomerization, and oxidation to 3,4-didehydroretinaldehyde are similar.

B. Stores of Retinoids in the Eye

1. *Quantity and Distribution*

Non-liver tissues such as heart, kidney, lung, muscle, show little tendency to accumulate retinoids. The highest concentrations in the body are found in the liver and RPE. The RPE plays an important role in governing the supply of nutrients and metabolites to the retina, particularly the photoreceptor cells. In most animals, the RPE is a single layer of hexagonal cells (each about 14 μm in diameter and 10–14 μm in thickness) that separates the choriocapillaris from the neural retina. On the choroidal side, the basal surfaces of the RPE cells are cemented to Bruch's membrane. On the retinal side, the apical plasma membrane gives rise to numerous microvilli, some of which form a sheath around the ends of the photoreceptor outer segments (Hogan *et al.*, 1971).

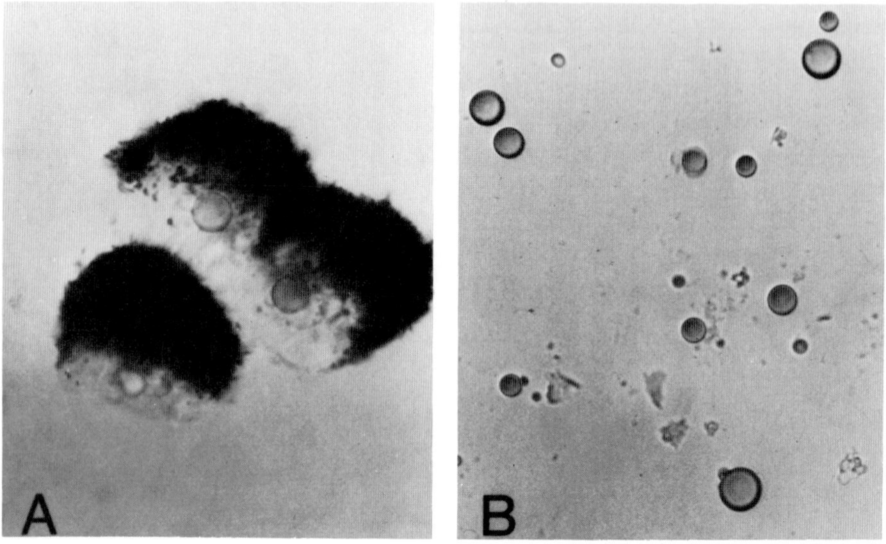

Fig. 4. Retinyl ester-bearing oil droplets in frog RPE cells. (A) Intact cells; (B) oil droplets collected from the surface of a centrifuged homogenate of frog RPE cells.

On average, the RPE from a pair of human eyes contains 4.5 ± 2.6 μg of vitamin A (retinol and retinyl esters) (15.8 ± 9.0 nmol), i.e., between 2 and 3 mol Eq of the rhodopsin in the retina (Bridges *et al.*, 1982). Excluding that bound in rhodopsin, this represents approximately 87% of the total retinoids in the eye. A similar situation is observed in frog RPE, which contains 96% of the total ocular vitamin A (Bridges, 1976b). The goldfish is exceptional in that its stores of vitamin A_2 in the RPE amount to as much as 40 mol Eq of the retinal porphyropsin (Bridges, 1975).

As shown in Fig. 4A, the frog RPE contains fluorescent yellow oil droplets (Hirosawa, 1979; Young and Bok, 1979) that may be isolated by centrifuging a homogenate of the tissue, causing them to accumulate as a thin layer on the surface (Bridges, 1975; Fig. 4B). Rabbit and guinea pig RPE cells also have large oil droplets. They contain 80–95% of the total nonvisual pigment ocular stores, a similar situation being applicable to the frog (Bridges, 1976b; Alvarez *et al.*, 1981). The droplets appear to augment the storage capacity of the cells by serving as reservoirs for retinoids. The total retinoid (vitamin A) stored per unit area of frog RPE ranges from 6.5 to 11.4 nmol/cm^2 [Bridges (1975); also Reuter *et al.* (1971)], compared with only 0.5–1.5 nmol/cm^2 in human and bovine tissues, which have much the same thickness but no oil droplets (Bridges and Alvarez, 1982a). Lipid droplets may accumulate in the RPE in response to high doses of vitamin A, as shown in mice by Robison and Kuwabara (1977). In this study, the appearance of droplets in the mouse RPE paralleled a dose-related increase in the numbers and sizes of lipid droplets in the stellate cells of the liver.

2. Composition

Except for that bound to visual pigment (and about 2% associated with cellular retinoid-binding proteins) most of the retinoid stored in the eye is in the form of retinyl ester. In the RPE, the proportion of total retinol that is esterified is 98–99%. The major esters are stearate and palmitate, the proportions varying according to species. In guinea pigs, rabbits, and frogs, the esters consist almost exclusively of retinyl palmitate. In humans, the ratio of palmitate to stearate is about 5:1, whereas in rats the two esters are present in nearly equivalent amounts, the ratio of palmitate to stearate being 1.3:1 (Alvarez *et al.*, 1981; Bridges *et al.*, 1982).

In some species of fish and in most tadpoles, mixtures of retinyl and 3,4-didehydroretinyl esters are found in the RPE (Wald, 1939; Bridges, 1972, 1975). Usually the compositions of these mixtures do not parallel the liver stores, which can contain predominantly retinyl ester whereas the RPE contains predominantly the 3,4-didehydro compound. In tadpoles, the 3,4-didehydroretinyl ester (mainly palmitate) disappears or diminishes sharply at metamorphic climax and is replaced by retinyl ester. Ambient levels of illumination also play a role in determining these compositions, which may also vary from the dorsal to the ventral half of the eye (Bridges, 1974; Reuter *et al.*, 1971).

Typical chromatograms obtained by high-performance liquid chromatography (HPLC) of the retinyl esters from human RPE are illustrated in Fig. 5. When conditions are optimized for isomer separation (normal phase, upper half of Fig. 5), there are two major peaks (2, 4), each with a shoulder on the leading edge. Peak 2 is 11-*cis*-retinyl palmitate; the shoulder corresponds to 11-*cis*-retinyl stearate. Peak 4 is all-*trans*-retinyl palmitate; the shoulder corresponds to all-*trans* retinyl stearate. Peak 5 is all-*trans*-retinyl stearate, and peak 3 (which is sometimes composite) is unidentified at present. Peaks 2 and 4 were collected separately and injected onto a reverse-phase column. Under these conditions the two component esters were completely resolved, as shown in the lower half of Fig. 5 (peaks 2.1, 2.2 and 4.1, 4.2).

3. Function of Stores

In the dark-adapted eyes of most animals, the retinoid stores in the RPE represent between 1 and 6 mol Eq of the retinal visual pigment. It is not known whether the magnitude of these stores depends on the vitamin A status of the animal, although it is probable that they serve as a reserve that protects the visual system from retinoid depletion under conditions of dietary deficiency. Since these stores typically contain up to 75% 11-*cis*-retinyl ester, the provision of 11-cis isomer for visual pigment synthesis may also be important under certain circumstances. However, the presence of this isomer in the RPE is not critical for regeneration, which can proceed efficiently in the absence of appreciable supplies of 11-*cis*-retinyl ester [see Bridges (1976b) for further discussion of this question].

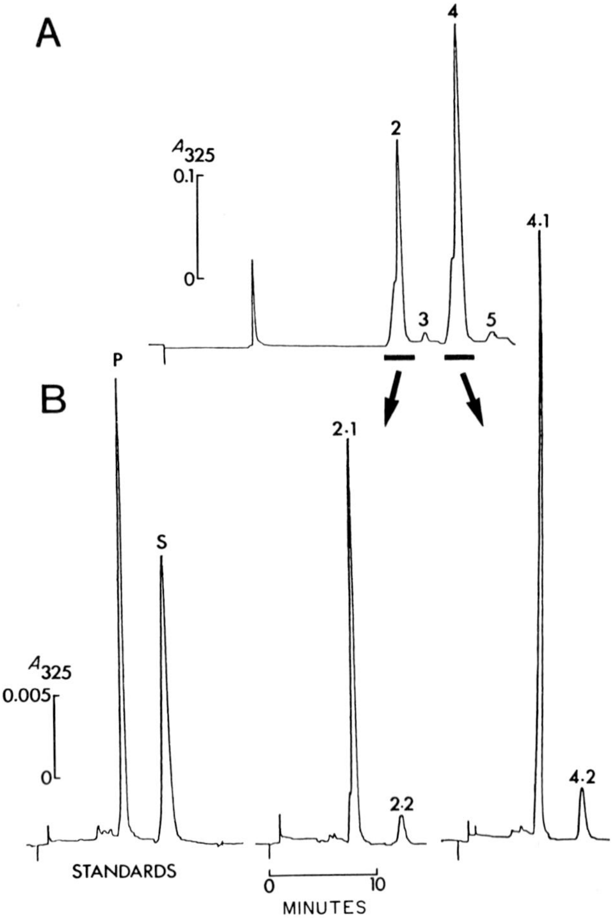

Fig. 5. Retinyl esters stored in human RPE. Analysis by high-performance liquid chromatography (HPLC). (A) Normal-phase column; peak 2, 11-*cis*-retinyl palmitate and stearate (shoulder); peak 3, unidentified; peak 4, all-*trans*-retinyl plamitate and stearate (shoulder); peak 5, all-*trans*-retinyl oleate. (B) Peaks 2 and 4 from the normal-phase HPLC eluent were collected and analyzed spearately in a reverse-phase system. The standards consisted of 11-*cis*-retinyl palmitate (P) and stearate (S). The 11-cis and all-trans isomers have the same retention time in this system. Peak 2.1, 11-cis-retinyl palmitate; peak 2.2, 11-*cis*-retinyl stearate; peak 4.1, all-*trans*-retinyl palmitate; peak 4.2, all-*trans* retinyl stearate. (From Bridges *et al.*, 1982.)

It has not been established whether the RPE supplies retinol exclusively for the visual cycle or whether it also provides retinoids for the remainder of the retina. In addition to the visual cycle retinoids, the retina appears to contain retinoic acid (cf. Saari *et al.*, 1982). However, although retinoic acid may be transported in the plasma bound to serum albumin in the same way as free fatty acids (see Chapter 8), its intracellular binding protein (CRABP) is virtually absent in the

RPE (Saari *et al.*, 1977). This observation may imply that the RPE has no role in delivering retinoic acid to the retina. Retinoic acid is often generated within the tissue where it presumably carries out its function (Roberts and DeLuca, 1967; Olson, 1967). The retinoic acid that may be needed by the retina, where CRABP is abundant (Saari *et al.*, 1977, 1982), may therefore be provided by oxidation of retinaldehyde.

C. Delivery of Retinoids to the Retinal Pigment Epithelium (RPE) from the Choriocapillaris

Retinoids enter the RPE by two routes. One is through the membrane that abuts the photoreceptor outer segments (Fig. 3). This is known as the apical membrane. The other is through the basal membrane adjacent to the choriocapillaris.

A two-way exchange of retinoid is known to take place through the apical membranes during the visual cycle. In contrast, it seems that the inward flow of vitamin A from the circulation through the basement membrane is not easily reversed. Consequently, the eyes of vitamin A-deficient rats retain their endogenous retinoids quite tenaciously (Dowling and Wald, 1958; Noell *et al.*, 1971).

Most of the retinol that is delivered to the eye originates in the liver, and is released into the circulation (mainly in the all-trans configuration) bound to an M_r 21,000 protein, serum retinol-binding protein (RBP). This protein, which contains no bound lipid or carbohydrate, circulates as a 1:1 complex with transthyretin (TTR, originally prealbumin) (see Chapter 8 for detailed discussion). In addition to solubilizing retinol and protecting it from oxidative degradation, RBP also provides the means for targeting the delivery of its ligand to cells with a special need for vitamin A (Goodman, 1980, 1981). This delivery process seems to involve cell-surface receptors for RBP. Data suggesting the existence of such receptors have been obtained with intestinal mucosa cells (Rask and Peterson, 1976), gonadal cells (Bhat and Cama, 1979), RPE cells (Heller, 1975; Bok and Heller, 1976), and corneal epithelial cells (Rask *et al.*, 1980).

Rask and Peterson (1976) concluded that uptake of retinol proceeded without concomitant uptake of RBP. The uptake was saturable and temperature-dependent and could be inhibited by retinol-depleted RBP and Fab′ fragments against RBP. Similarly, Bhat and Cama (1979) found that the binding of RBP to a chicken testicular cell membrane fraction was saturable and reversible. Saturable binding of RBP was also reported in interstitial cell preparations from rat testis (McGuire *et al.*, 1979). These authors also demonstrated that [^{3}H]retinol–RBP delivered its ligand specifically to the interstitial cells of the testis. Bhat and Cama (1979) claimed that the receptor level in their chicken preparations varied with the stage of differentiation. Testosterone treatment of 11-week-old birds led to a substantial increase in the receptor concentration, suggesting that the hormone induced synthesis of the receptor, producing a rise in the concentration of receptor before the formation of the acrosome. Because the acrosome constitutes a major reservoir for vitamin A (Gambhir and Ahluwahlia, 1974), this rise would have preceded a substantial augmentation of the cellular demand for vitamin A. A similar situation could occur in the RPE when the photoreceptor outer segments start to develop, but this has not been investigated.

The effect of vitamin A nutriture on the numbers of RBP receptors on the surfaces of target cells has not been studied extensively. Rask *et al.* (1981) compared the small intestinal mucosa cells from normal and hypovitaminotic A rats but found no differences.

The endothelial cells of the choroidal capillaries have numerous circular fenestrations that measure 800 Å in diameter (Hogan *et al.*, 1971). They probably facilitate the passage of the RBP–TTR complex, which has a combined M_r of 76,000. Heller, using about 10^{-3} of the physiological RBP concentration, demonstrated weak but significant binding of RBP to bovine RPE cells. The specific binding was temperature-dependent and saturable. Bound iodinated retinol–RBP was rapidly displaced by unlabeled retinol–RBP, and Heller suggested that uptake of ligand by the cell was not associated with internalization of the transport protein. However, it is exceedingly difficult to prepare RPE cells with intact plasma membranes. This is probably because the methods used have usually entailed brushing them away from Bruch's membrane, a procedure that leads to extensive tearing and possibly even some loss of the basal plasma membrane. Thus, although such preparations maintain a semblance of morphologic integrity when examined by light microscopy, electron microscopy reveals that the plasma membrane is fragmented along the apical and basal cell borders and the mitochondria appear ballooned and damaged (Saari *et al.*, 1977). Saari *et al.* concluded that as much as 97% of the CRBP was lost in these preparations. Consequently, some studies believed to have been carried out on "whole" RPE cells are difficult to interpret (e.g., Maraini *et al.*, 1975; Chen and Heller, 1977). For example, in Chen and Heller's gel-filtration analysis of cytosol from RPE cells incubated with [^{3}H]retinol–RBP, counts were observed only in the void and included volumes. Apparently, CRBP had not been retained in their preparations.

Studies on intact cultured RPE cells (Flood *et al.*, 1983) show that they are capable of taking up and esterifying retinol when bovine serum albumin is used as a carrier. This process presumably lacks the specificity associated with the RBP/receptor interaction.

As mentioned above, the eye loses its retinoids only very slowly under conditions of hypovitaminosis A. In rats, where the dark-adapted stores of retinyl esters are very small, the loss is virtually absent in darkness but is accelerated in the light (Noell *et al.*, 1971). In light, most of the retinoid is transferred from the rhodopsin in the ROS to the RPE (Dowling, 1960), where it resides as retinyl stearate and palmitate (Alvarez *et al.*, 1981). It returns to the ROS when rhodopsin regenerates. In the vitamin A-deficient animal, therefore, there must be losses during this cycling process, but it is not known whether retinol is lost by metabolic degradation or whether it leaves the cell by the same route implicated in its delivery. Apo-RBP–TTR (although present in the serum of vitamin A-deficient animals; Smith and Goodman, 1979) apparently has a markedly reduced affinity for its receptor and may not be able to remove retinol from the cell.

The mechanism of delivery of retinol to a target cell involves several steps. These probably include (1) binding of RBP to the receptors; (2) a simultaneous conformational change that reduces the affinity of RBP for its ligand; (3) transfer of retinol to the plasma membrane; (4) dissociation of RBP from the receptor. After delivery of its ligand, there is also a reduction in the affinity of apo-RBP for TTR (Smith and Goodman, 1979). Thus in vitamin A-deficient rats, the amount of apo-RBP in the free, uncomplexed RBP fraction is increased and is selectively filtered by the renal glomeruli (Muto *et al.*, 1972).

How retinol passes through the cell membrane and enters the cell interior is still not understood. The process does not seem to involve internalization of RBP (see Chapter 8). Because retinol is a lipid-soluble molecule, it would be free to diffuse through the hydrophobic region of the cell membrane, but in that case it would be immaterial whether the retinol had been delivered specifically by RBP or in some other way, such as bound to serum albumin. Consequently, it is difficult to see how delivery of retinol by RBP under physiological conditions would protect the cells from the adverse membranolytic effects of vitamin A toxicity (Dingle and Lucy, 1965) that are believed to arise when free retinol occurs in the plasma (Mallia *et al.*, 1975). An alternative possibility would implicate a more specific process, such as an intramembrane carrier protein (cf. Maraini *et al.*, 1977).

D. Intracellular Transport of Retinoids in the RPE—Cellular Retinoid-Binding Proteins

When retinoids pass from the plasma membrane to intracellular sites of utilization and storage they must be transported through the cytosol. Both human and bovine RPE cytosols contain retinol and retinaldehyde in amounts equivalent to ~ 1% of the retinyl esters in storage. Cytosol concentrations in these species are 3–7 μM for retinaldehyde and 2–5 μM for retinol.

A typical HPLC profile for the retinoid present in bovine RPE cytosol is shown in Fig. 6. After the initial "breakthrough" region at 8 min, only two major peaks are present. They correspond to 11-*cis*-retinaldehyde (peak 1) and all-*trans*-retinol (peak 6). Therefore, most of the retinaldehyde in the cytosol is 11-cis, while nearly all of the retinol is all-trans. This has been found to be generally true for bovine and human RPE cytosols.

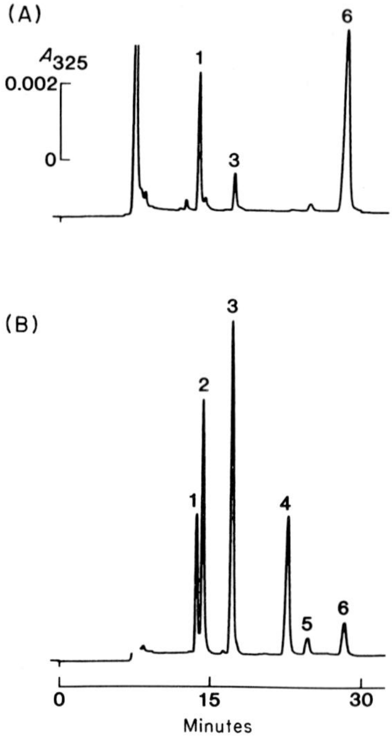

Fig. 6. Visual cycle retinoids extracted from the cytosol of bovine RPE cells (A) and standard retinoids (B). Normal-phase-high-performance liquid chromatography as described by Bridges and Alvarez (1982b). Peak 1, 11-*cis*-retinaldehyde; peak 2, 13-*cis*-retinaldehyde; peak 3, all-*trans*-retinaldehyde; peak 4, 11-*cis*-retinol; peak 5, 13-*cis*-retinol; peak 6, all-*trans*-retinol.

As Fig. 7 illustrates, gel-filtration analysis of RPE cytosol demonstrates that the 11-*cis*-retinaldehyde (open circles, continuous line) was bound to a protein of approximate M_r 33,000. In contrast, the all-*trans*- retinol (filled circles, dashed line) was bound to a protein of approximate M_r 15,000. These proteins are representatives of a group of intracellular proteins that bind retinoids in animal tissues [Chytil and Ong (1979); see also Chapter 9]. They include cellular retinol-binding protein [CRBP: Bashor *et al.* (1973), Ong and Chytil, (1978a), Saari *et al.* (1978), Liou *et al.* (1981), Rask *et al.* (1981), cellular retinoic acid-binding protein [CRABP: Ong and Chytil (1975, 1978b), Sani and Hill (1974), Sani and Banerjee (1978), Saari and Futterman (1976), Saari *et al.* (1978)] and cellular 11-*cis*-retinaldehyde-binding protein, which also binds 11-*cis*-retinol [CRAlBP: Futterman *et al.* (1977), Futterman and Saari (1977), Stubbs *et al.* (1979)]. The M_r 15,000 all-*trans*-retinol-binding protein in Fig. 7 probably corresponds to CRBP [CRBP is the 2 S "receptor" for exogenous [^{3}H]retinol described by Wiggert *et al.* (1978)]. Saari *et al.* (1982) have shown that CRBP isolated from RPE cytosol is saturated with all-*trans*-retinol and has no other ligand attached to it (CRBP from other sources is typically 40–70% saturated with retinol; Chytil and Ong, 1979). Therefore, it appears that, once it has arrived at the internal surface of the RPE plasma membrane, all-*trans*-retinol is transferred to CRBP.

Most of the CRBP in bovine eyes is found in the RPE: very little is associated with the retina (Saari *et al.*, 1977). CRBP is also found in the cytosol from other tissues (see Chapter 9).

CRBP is distinct from RBP, both immunologically (Ross and Goodman, 1979; Bashor and Chytil, 1975) and by a number of other criteria. It has a lower molecular weight and fails to complex with TTR (Ross and Goodman, 1979; Ong and Chytil, 1978a). CRBP will not bind retinaldehyde, retinoic acid, and retinyl acetate, although these ligands will bind to RBP (Horwitz and Heller, 1974). CRBP has spectroscopic properties that differ from those of RBP. Like free retinol, holo-RBP has a smooth absorption band centered at 330 nm. In CRBP, however, the retinol absorption band is shifted bathochromically to 350

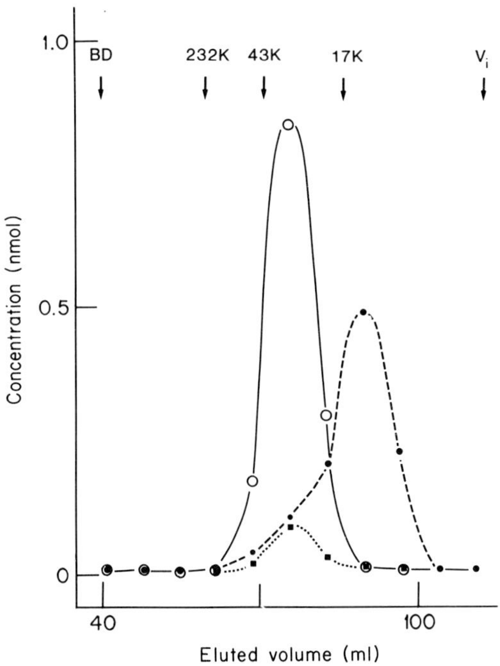

Fig. 7. Gel-filtration chromatography of bovine RPE cytosol; analysis of eluted fractions for endogenous retinol and retinaldehyde. Cytosol was applied to a column of Sephadex G-100 overlaid with Sepharose CL-4B. Retinoids were extracted from successive fractions and analyzed by HPLC as described in Fig. 5. Key: ○, 11-*cis*-retinaldehyde; ●, all-*trans*-retinol; ■, 11-*cis*-retinol. Molecular weight calibration: BD = blue dextran 2000; 232 K = catalase; 43 K = ovalbumin; 17 K = myoglobin; V_i = 3-nitrotyrosine.

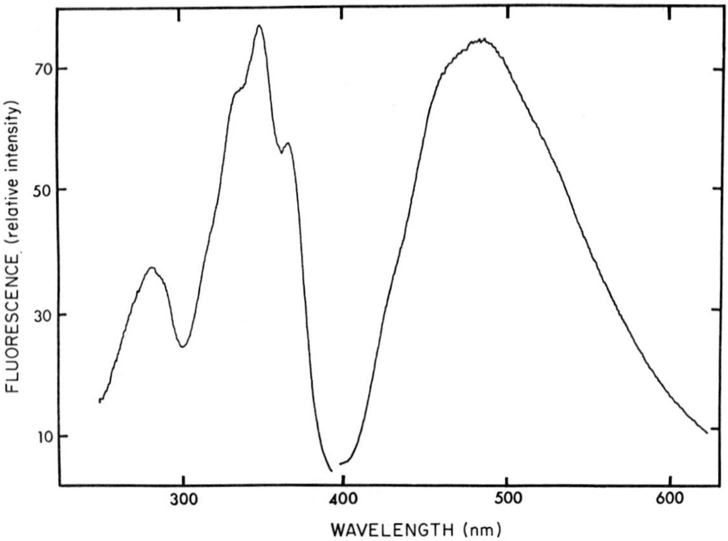

Fig. 8. Uncorrected fluorescence excitation (left) and emission spectra (right) for cellular retinol-binding protein (dog liver; Liou *et al.*, 1981). Excitation at 350 nm; emission at 475 nm.

nm and displays fine structure that is evidenced by a subsidiary peak at 367 nm and a shoulder at 335 nm. The excitation spectrum for fluorescence emitted at 475 nm displays similar features (Fig. 8). This excitation spectrum is also seen in the M_r 15,000 fractions from RPE cytosol (Liou *et al.*, 1982c). Vibrational fine structure is also observed in the spectra of complexes between all-*trans*-retinol and β-lactoglobulin. It may arise from fixation of the β-ionylidene ring and tetraene side-chain torsional angle. Such fixation also occurs in the retro-retinoids, leading to selection of one particular conformation out of the many that contribute to the diffuseness of the normal spectrum in solution (Hemley *et al.*, 1979).

CRBP seems to be a structurally conserved protein. CRBPs from different species and tissues, including the retina, show a striking degree of identity in their fluorescence and absorption spectra, molecular weights, and amino acid composition (Liou *et al.*, 1981). Liou *et al.* found that CRBPs from rat liver, dog liver, and bovine retina were immunologically identical when tested against their rabbit anti-rat CRBP antiserum, although immunochemical species differences have been reported by Ong and Chytil (1981; see Chapter 9). Immunological identity between rat liver and rat testis CRBP, however, was reported by Ong and Chytil (1979).

The function of CRBP within cells is not understood. Presumably it serves to transport retinol to its sites of action and utilization. One point of action may be the nucleus, where it may modify gene expression (Takase *et al.*, 1979; Liau *et *al.*, 1981). One function of CRBP in the RPE is probably to serve as a transport system between the apical and basal plasma membrane and the sites of esterification and storage.

Saari *et al.* (1982) have shown (and Fig. 7 illustrates) that CRBP is not involved in the transport of 11-*cis*-retinol. This observation raises an important question. The stores of vitamin A in the RPE of most animals consist of both 11-*cis*- and all-*trans*-retinyl esters. In the frog, the 11-cis isomer is consumed and replaced by all-trans during rhodopsin bleaching and regeneration (Bridges, 1976b). It is likely, although not established, that this occurs in other species as well. A transport system for 11-*cis*-retinoid is therefore a necessity. This need is possibly met within the RPE by the M_r 33,000 11-*cis*-retinaldehyde-binding protein of Fig. 7, which apparently corresponds to CRAlBP (cf. Saari *et al.*, 1982). CRAlBP occurs in both the RPE and the retina; it is not present in other tissues (Futterman *et al.*, 1977; Futterman and Saari, 1977). Work by Bunt-Milam and Saari (1983) has shown that retinal CRAlBP occurs in the Müller cells (Fig. 3; see Section III,G,4,e).

E. Esterification of Retinol in the RPE

Krinsky (1958) showed that retinol is readily esterified by RPE cells and their homogenates. Figure 9 illustrates the HPLC profile of radioactivity in extracts from homogenized frog choroid, homogenized RPE, as well as the RPE pellet and cytosol fractions following 2 h incubation with all-*trans*-[11,12-^{3}H]retinol (Fong *et al.*, 1983a).

No significant radioactivity attributable to retinyl esters is present in the choroid and RPE cytosol. In contrast, a large peak of radioactivity coincident with all-*trans*-retinyl palmitate (AT) can be seen in both the homogenate and the pellet. Although no radioactivity was present in the position of 11-*cis*-retinyl palmitate (peak 11 of Fig. 9), a peak that varied considerably in height from experiment to experiment coincided with the corresponding 13-cis isomer (see below, also Fong *et al.*, 1983a). These results demonstrate that the esterifying enzyme is restricted to the particulate fraction and that it is associated with an endogenous palmityl donor.

Similar findings have been reported for bovine RPE (Krinsky, 1958; Berman *et al.*, 1980). According to Berman *et al.* (1980), the most active esterifying fraction of bovine pigment epithelium cells is the microsomes. The pH optimum is 7.5, and there are no clear requirements for ATP, CoA, or free fatty acids. As in frog, it appears that these membrane preparations have an adequate supply of endogenous acyl donors (possibly acyl-CoA).

When provided with exogenous retinol, rat, human, and frog RPE tissues synthesize retinyl esters in proportions that appear to be identical with those found in endogenous stores (Alvarez *et al.*, 1981; Bridges *et al.*, 1982; Fong *et al.*, 1983a). In human RPE cells in culture, this situation is changed. During the first few days in culture, the cells lose their endogenous supplies (Flood *et al.*, 1983). This is coupled with a sharp decrease in esterifying activity (Oka *et al.*, 1983). Exogenous retinol (provided bound to bovine serum albumin) is esterified by these cells to some extent, but the proportions of stearate and oleate relative to palmitate increase gradually with increasing time in culture (Flood *et al.*, 1982, 1983).

The esterifying enzyme has not been characterized, mainly because it is inactivated by solubilizing agents such as deoxycholate, digitonin, and octyl glucoside (Krinsky, 1958; S.-L. Fong, unpublished data). In cattle, it appears to be specific for retinol and will not act on cholesterol (Berman *et al.*, 1980). Regarding its specificity for different isomers of retinol, Krinsky (1958) reported that the esterification of 13-cis and 9-cis isomers by bovine RPE did not proceed as effectively as for 11-cis and all-trans. However, owing to limitations in his method (reverse-phase paper chromatography), Krinsky did not identify the isomeric conformation of the esters formed during the reaction. In another study (Flood *et al.*, 1983), HPLC analysis has demonstrated that human RPE homoge-

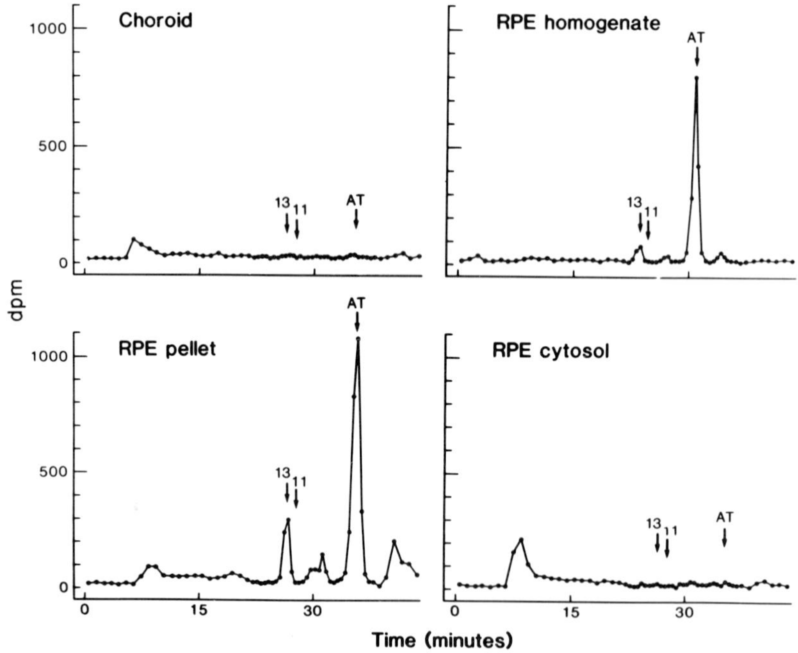

Fig. 9. Normal-phase HPLC of extracts from homogenized frog choroid, a homogenate of the corresponding RPE and the RPE pellet and cytosol prepared from this homogenate, all incubated with [^{3}H]all-*trans* retinol. Labeled retinyl esters are formed only by the RPE homogenate and its particulate component. Peak 13, 13-*cis*-retinyl palmitate; peak 11, 11-*cis*-retinyl palmitate; peak AT, all-*trans*-retinyl palmitate. (From Fong *et al.*, 1983a.)

nates rapidly synthesize retinyl stearate and palmitate when provided with exogenous 11-*cis*-retinol, but the situation is complicated by the simultaneous isomerization of the substrate to 13-cis and all-trans. Similar results have been obtained with frog RPE (C. D. B. Bridges, unpublished data). Therefore, although all three retinol isomers can be esterified by RPE cells, any preference that may exist for one particular isomer over another is obscured by parallel, probably nonenzymatic isomerizing reactions. This may have happened because the substrate was delivered in a drop of ethanol. Future experiments utilizing a more "physiological" system, i.e., using an appropriate binding protein as a carrier, will hopefully provide more information.

F. The Visual Cycle—Bleaching

1. Formation of all-trans-Retinol in the Photoreceptors

In addition to entering the RPE from the circulation, all-*trans*-retinol passes into the RPE cells through their apical surfaces during bleaching of rhodopsin in the ROS. This phase of the visual cycle is better characterized than the reverse step, namely the reverse flow of retinoid that culminates in visual pigment regeneration.

When rhodopsin is exposed to light it forms a series of intermediates, most of them transients at physiological temperatures (Section II,A,1). The last intermediate, possibly *N*-retinylideneopsin, is unstable at physiological pH and hydrolyzes to opsin and all-*trans*-retinaldehyde. In the human retina at 36°C, free retinaldehyde has a half-life of 23 s (Baumann and Bender, 1973). This is because it is reduced to all-*trans*-retinol by a membrane-bound dehydrogenase that may also act on the retinylideneimine compound (DePont *et al.*, 1970).

Frog and bovine retinol dehydrogenases require NADPH as their cofactor (Futterman, 1963; DePont *et al.*, 1970; Bridges, 1977). Additionally, the enzyme has a strong preference for the all-trans isomer in bovine (Lion *et al.*, 1975) and frog rod outer segments (Bridges, 1977).

The visual pigments are strongly dichroic, showing that their chromophores are oriented within the membrane. When bleaching occurs in isolated rod outer segments, the retinol that accumulates also exhibits a high dichroic ratio but in the opposite sense (Harosi and MacNichol, 1974).

2. Transport of all-trans-retinol to the Apical Surfaces of the RPE Cells—Interstitial Retinol-Binding Protein, a Glycoprotein Synthesized by the Retina

In rats and frogs exposed to high levels of illumination, retinoid passes from the retina to the RPE: it accumulates in the RPE in the form of all-*trans*-retinyl esters. In the case of the frog, these are found in the oil droplets (Fig. 10). The end result is therefore similar to that observed when retinol is delivered to the

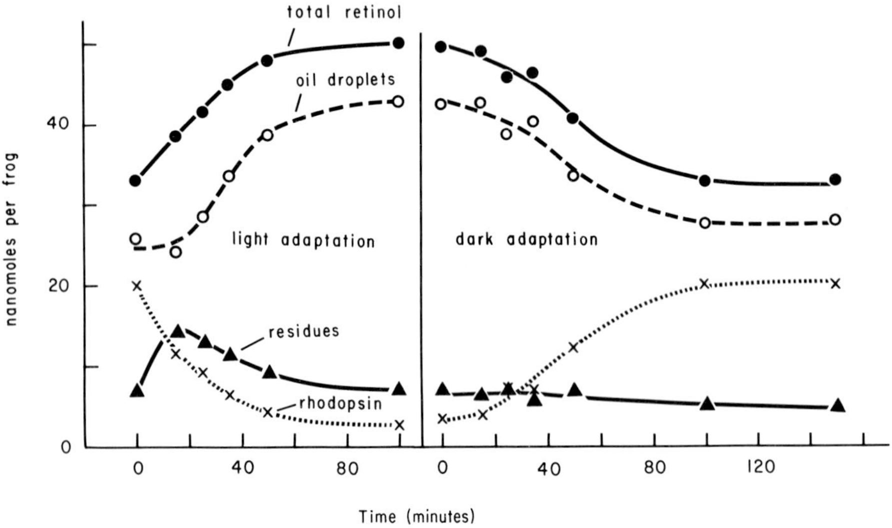

Fig. 10. Distribution of vitamin A and rhodopsin during the course of light and dark adaptation in the frog. The frogs in each experiment were matched as closely as possible for eye size, and each point represents the averaged results from two groups of four frogs in two separate identical experiments carried out on animals from the same batch on 2 consecutive days. The curve labeled "residues" represents the particulate fraction from RPE cells and whole retina combined. (From Bridges, 1976b.)

RPE from the choriocapillaris. The evidence strongly suggests that the transport form is retinol, which is esterified on arrival in the RPE as described in Section III,E. An esterifying enzyme has not been demonstrated in frog rod outer segments. Consequently, when an isolated frog retina is bleached, all-*trans*-retinol accumulates in the retina and remains unchanged for many hours (Bridges, 1976b). Mammalian tissue is probably similar (Andrews and Futterman, 1964), but this has been disputed by Berman *et al.* (1980).

Although there is often close contact between the RPE processes and the rod outer segment surfaces, much of the retinol is probably delivered to the RPE via the subretinal space rather than by direct membrane-membrane transfer (but see Rando and Bangerter, 1982). It has been estimated that the subretinal space occupies 0.01–0.02 ml in human eyes and 0.03–0.10 ml in cattle eyes. It is filled with an interphotoreceptor matrix (IPM) (Feeney, 1973; Berman 1982) that contains proteoglycans (Adler and Severin, 1981; Adler and Klucznik, 1982). The ratio of protein to proteoglycans (measured as glycosaminoglycans) is greater than 50:1. Over 90% of the IPM remains with the retina when this tissue is separated from the RPE.

Other workers (Liou *et al.*, 1982a,b,c; Adler and Martin, 1982) have sought to determine whether retinoids were present in IPM, and if so, whether they were associated with binding proteins. These studies led to the discovery in bovine, rat, monkey, and human interphotoreceptor matrix preparations of a high-molecular-weight glycoprotein that carries endogenous retinol and is confined to the IPM. The name interphotoreceptor matrix or *interstitial retinol-binding protein* (IRBP) was coined by Liou *et al.* (1982b,c), who showed that its M_r was 144,000 on SDS–PAGE (bovine). The presence of this protein and its endogenous ligand in IPM preparations is illustrated in Fig. 11, where IPM from light-adapted cattle eyes was subjected to liquid chromatography on high-performance size-exclusion columns. The effluent was monitored by its absorbance at 280 nm and its fluorescence due to endogenous bound retinol. The absorbance peak (arrowed) corresponds to M_r 260,000, and coincides with the peak of retinol fluorescence in the lower trace. Fractions were collected and analyzed by SDS–PAGE. Those that corresponded to the fluorescent peak showed only one protein band that corresponded to M_r 144,000. Unlike the other known retinoid-binding proteins, IRBP is a glycoprotein that contains about 8% by weight of carbohydrate. Lectin-binding studies by the procedure of Irimura and Nicolson (1983), direct carbohydrate analysis and sequencing, and biosynthetic studies show that it has

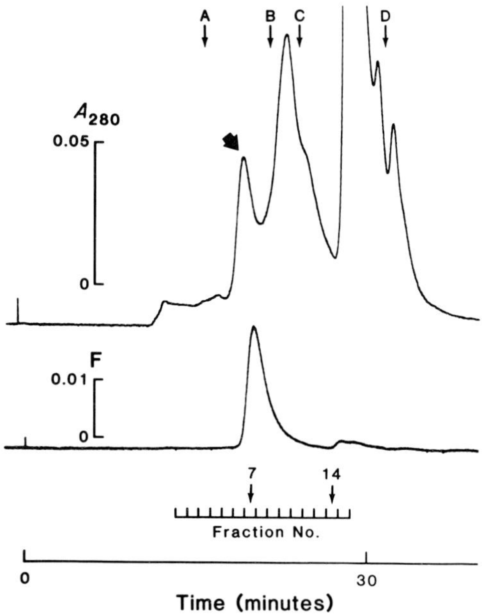

Fig. 11. High-performance size-exclusion chromatography of bovine interphotoreceptor matrix on Bio-Rad TSK-400 and TSK-250 columns connected in series. The effluent was monitored by its absorbance at 280 nm (A_{280}, upper trace) or the fluorescence due to its endogenous retinol (F, lower trace). Because fluorescence was monitored after absorbance, the fluorescence peak was slightly displaced with respect to the corresponding absorbance peak (arrowed). Protein standards: A, thyroglobulin (M_r 669,000); B, IgG (M_r 150,000); C, ovalbumin (M_r 43,000); D, cyanocobalamin (M_r 1,355). Fractions were collected, numbered, and analyzed by SDS-PAGE. (From Liou *et al.*, 1982c.)

4–5 N-linked oligosaccharides of the biantennary complex-type. They contain sialic acid, galactose, glucosamine, mannose, and fucose (Fong *et al.*, 1983b, 1984; Bridges *et al.*, 1983b).

The disparate M_r values observed by gel filtration and SDS–PAGE may arise from the fact that glycoproteins behave anomalously in these systems. The K_d of IRBP for all-*trans*-retinol is 10^{-6} *M*, higher than for RBP and CRBP. One molecule of IRBP binds two to three molecules of all-*trans*-retinol, giving an ϵ_{330} of 1.14×10^5 M^{-1}cm^{-1}. The absorption spectrum of the holoprotein is illustrated in Fig. 12. When subjected to isoelectric focusing on agarose slabs, the purified protein shows at least four bands between pH 4.4 and 4.8. All these bands are glycosylated, as judged by their ability to bind the lectin concanavalin A. The amino acid analysis shows that about 50% of the residues are hydrophobic. In experiments where isolated bovine, monkey, or rat retinas were incubated with radiolabeled glucosamine, fucose or leucine, it has been demonstrated that IRBP is synthesized and secreted by the neural retina (Bridges *et al.*, 1983b; Fong *et al.*, 1984). Although the cell of origin has not been determined, some evidence suggests that it is the photoreceptors (Gonzalez-Fernandez *et al.*, 1983b). No synthesis of IRBP by the cells of the pigment epithelium or choroid could be detected.

Although it has been shown to be capable of binding retinaldehyde and retinoic acid (Liou *et al.*, 1982c), as well as α-tocopherol (Liou *et al.*, 1983), the major endogenous ligand of interstitial retinol-binding protein appears to be

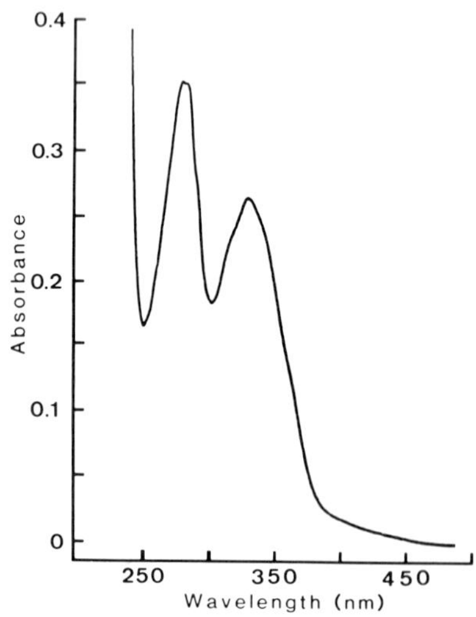

Fig. 12. Absorption spectrum of interstitial retinol-binding protein. $\epsilon_{330} = 1.14 \times 10^5$ M^{-1}cm^{-1}corresponding to between 2 and 3 mol of all-*trans*-retinol per mol of protein. (From Fong *et al.*, 1983b.)

all-*trans*-retinol, together with variable lesser amounts of 11-*cis*-retinol. The quantity of all-*trans*-retinol bound to interstitial retinol-binding protein was found to increase when the rhodopsin in the retina was bleached (Liou *et al.*, 1982c). There was no change, however, in the quantity of binding protein.

The high-M_r protein carrying endogenous retinol was also demonstrated by Adler and Martin (1982), who used gel-filtration columns. In agreement with Liou *et al.* (1982c), the amount of ligand increased when the eyes were exposed to light. Although Adler and Severin (1981) and Adler and Klucznik (1982) had previously observed this glycoprotein by SDS–PAGE of IPM preparations, they did not identify it as a retinoid-binding protein.

Interstitial retinol-binding protein may correspond to a 7–8 S "receptor" for exogenous [^{3}H]retinol assigned to bovine retina cytosol and brain by Wiggert *et al.* (1978), to bovine ROS by Lee and Wiggert (1982), and to the interphotoreceptor matrix by Lai *et al.* (1982). For further discussion, see Liou *et al.* (1982c).

IRBP has the capacity to bind retinol formed when eyes are exposed to light. Therefore, provided it can diffuse freely through the matrix (which contains proteoglycans), it is a possible candidate for a retinol transport protein between the retina and the RPE (Liou *et al.*, 1982c). If so, there may be receptors for this protein on the plasma membranes of cells that line the space between the retina and RPE. The glycoprotein may also be internalized together with its ligand: thus Orzalesi *et al.* (1982) have described coated pits and vesicles in RPE that may be associated with this activity.

If IRBP is hindered from diffusing by the presence of proteoglycans in the matrix, however, it is possible that its ligand could be transferred from one IRBP molecule to the next, thus traversing the space between the retina and RPE by a process of facilitated diffusion.

G. The Visual Cycle—Rhodopsin Regeneration

The retinoid lost to the RPE during bleaching must be restored during dark adaptation, or while a steady-state level of visual pigment is maintained when illumination conditions are more moderate. Although it is possible that the visual cycle may be restricted to the retina (and may not even involve hydrolysis of the aldimine bond) when bleaching levels are low, the evidence (Zimmerman *et al.*, 1974, 1975) is subject to several interpretations (see Bridges, 1976b).

It has been suggested that retinoids may also continue to cycle between a small pool of retinoid in the ROS and the RPE after rhodopsin regeneration has been completed (Bridges, 1976a). A recent finding, however, has a critical bearing on our knowledge of the dynamics of retinoid movements in darkness. Autoradiographic studies on frogs and mice injected with [^{3}H]retinol–RBP (with or without TTR) by Defoe and Bok (1983) have demonstrated an apparent exchange of retinoid between rhodopsin molecules and the RPE in the absence of bleaching.

Nearly all studies on the visual cycle have been carried out on frogs and rats, where substantial (> 90%) bleaching leads to accumulation of all-*trans*-retinyl esters in the RPE. When dark-adapted, rats have only very small stores of retinyl esters. In frogs, any preexisting supplies of 11-*cis*-retinyl esters are consumed over several cycles of bleaching and regeneration and are only restored during prolonged dark adaptation (24 h or more; Hubbard and Dowling, 1962; Bridges, 1976b). Therefore, at the end of a period of strong light adaptation, both rats and frogs are faced with the undertaking of carrying out the overall conversion of all-*trans*-retinyl ester to 11-*cis*-retinaldehyde, coupled with the transfer of one of these retinoids, or an intermediate, from the RPE to the ROS. The retinoid that is transported and isomerized, and where this isomerization occurs has not been determined. It should also be pointed out that there may be differences between frogs and rats, and that these should not be overlooked when comparing results obtained from the two animals (see below).

The following distinct steps are probably involved in rhodopsin regeneration after a massive bleach: (1) mobilization of retinoid in the RPE; (2) transport and delivery of retinoid to the retina; (3) isomerization of retinoid and its utilization in the ROS. Steps (2) and (3) do not necessarily occur in that order.

1. Mobilization of Stored Retinoid

As in the liver, the mobilization of retinol in the RPE almost certainly entails the initial hydrolysis of retinyl esters. Although a hydrolase acting on the esters of long-chain fatty acids should be present in the RPE, none has been convincingly demonstrated. Lest this be considered an indication that the vitamin A is mobilized as ester, it should be noted that until quite recently compelling evidence for such a hydrolase in liver had also been lacking. Hepatic uptake of vitamin A involves hydrolysis of chylomicron retinyl esters followed by intrahepatic reesterification and storage (Goodman, 1980, 1981; see Chapter 7). During mobilization, the stored retinyl esters (mainly palmitate, as in the eye) must be hydrolyzed to free retinol that can then be bound to RBP. An unusual cholate-stimulated and cholate-soluble hydrolase has been shown to occur in rat livers, although it varies in amount by as much as 50-fold between individual animals (Prystowsky *et al.*, 1981). It correlates strongly with hydrolytic activity against cholesteryl oleate and triolein. According to Prystowsky *et al.* (1981), the hydrolase may be associated with lipid droplets or lipid–protein aggregates in the cell cytoplasm.

Are the various isomers mobilized and transported in different ways in the RPE? This appears to be the case. The retinyl esters stored in the RPE contain 11-cis as well as all-trans isomers. As discussed in Section III,D, 11-*cis*- and all-*trans*-retinoids are "streamed" separately in the RPE cytosol, the 11-cis isomer as *retinaldehyde* and the all-trans isomer as *retinol*, each bound to its own, separate transport protein. Whether both retinoids are transported through the apical plasma membrane into the subretinal space and thence into the retina has not been established (see Section IV,6,2, below).

The 11-*cis*-retinaldehyde in bovine RPE cytosol may be derived directly from 11-*cis*-retinol by a specific 11-*cis*-retinol dehydrogenase (see also Section III,G,4,d). Zimmerman *et al.* (1975) found that the highest total and specific activity of 11-*cis*-retinol dehydrogenase was in bovine RPE [bovine rod outer segments, unlike frog rod outer segments (Bridges, 1977) lack 11-*cis* retinol dehydrogenase]. Because the bovine RPE enzyme is confined to the microsomes, it may be in close proximity to the esterase.

2. *Transport and Delivery of Retinoid to the Photoreceptors*

All-*trans*-retinol, 11-*cis*-retinol (A2) and 11-*cis*-retinaldehyde are found in the IPM (Liou *et al.*, 1982c). At present, it is not clear which retinoid, or which isomer, is in transit from the RPE to the ROS. With regard to carrier proteins, the evidence suggests that CRBP and CRAlBP are present in the IPM (Liou *et al.*, 1982c), but it is probable that they have leached from the RPE and retina cytosols where both proteins occur. On the other hand, interstitial retinol-binding protein is believed to be a true IPM protein because it is absent from the RPE and retina cytosols. This conclusion has been confirmed by immunocytochemical observations in the author's laboratory (Fong *et al.*, 1984) and by Bunt-Milam and Saari (1983). Although IRBP binds the all-*trans*-retinol that is released when rhodopsin is bleached, it also carries some 11-cis isomer, and it has not been established whether it may be implicated in the two-way transport of retinoids.

3. *Observations on the Utilization of Exogenous Retinoids by the Retina*

Experiments that investigate the regeneration of rhodopsin by bleached retinas or ROS supplied with exogenous retinoids may provide important clues concerning the form delivered to them. The means of delivery in the experimental situation may also be important, although this question has not been seriously addressed. Bridges (1976b, 1977), Pepperberg *et al.* (1976, 1978), Pepperberg and Masland (1978), and Pepperberg (1982) have used aqueous colloidal dispersions, while Yoshikami and Nöll (1978) delivered their retinoid in liposomes. Both methods gave similar results.

Rhodopsin regeneration has been determined by extracting it directly or by measuring the spectrum of the intact retina. The recovery of ERG sensitivity has also been used as an index of rhodopsin regeneration by Pepperberg and his colleagues.

All-*trans*-retinoids are not able to support regeneration in frogs and mammalian retinas. The 11-cis retinoids that have been tested are retinaldehyde, retinol, and retinyl palmitate.

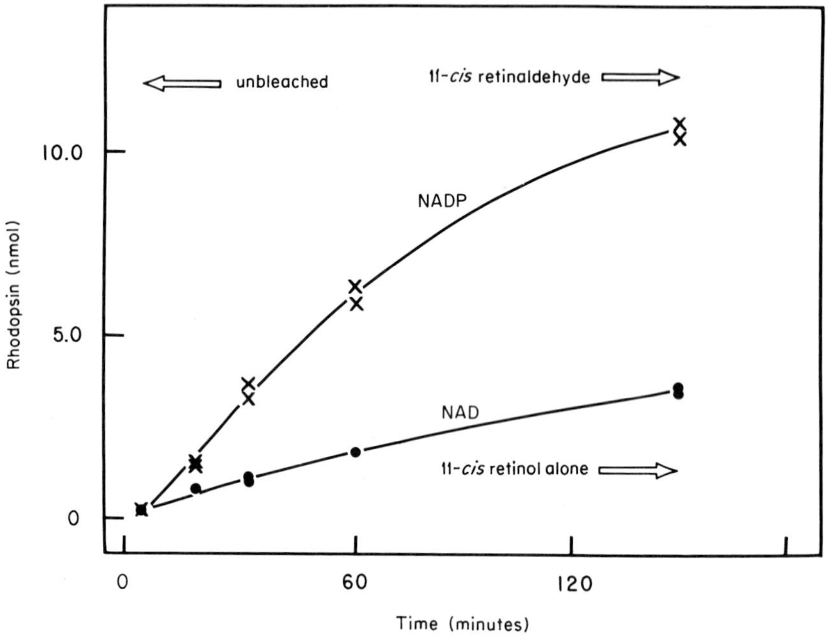

Fig. 13. Regeneration of rhodopsin in suspensions of frog ROS membranes. Enzyme and opsin source: bleached ROS membranes in Ringer's (1 ml), pH 7.4, temperature 25°. Additives were as follows: 11-*cis*-retinol, 100 nmol; 11-*cis*-retinaldehyde, 50 nmol; NAD^+ (●) or $NADP^+$ (X), 200 nmol. Retinoids were added in about 10 μl ethanol. (From Bridges, 1976b.)

Frog and mammalian retinas regenerate when provided with 11-*cis*-retinaldehyde. They differ in that frogs can regenerate rhodopsin from 11-*cis*-retinol (Fig. 13), while the mammals investigated cannot (e.g., Pepperberg and Masland, 1978; Yoshikami and Nöll, 1978). This finding is consistent with the observation that 11-*cis*-retinol dehydrogenase is present in frog ROS (Bridges, 1976b, 1977) but not in cattle (Zimmerman *et al.*, 1975, 1976). The 11-*cis*-retinol dehydrogenase in frog ROS will also oxidize 9-*cis*-retinol to 9-*cis*-retinaldehyde (Bridges, 1977). Since the method of assay consisted of measuring rhodopsin or isorhodopsin formation in frog ROS membrane preparations supplemented with NADP and substrate (Fig. 13), it could not be determined whether this enzyme can also act on the all-trans isomer. It differs from the membrane-bound retinol dehydrogenase, which is responsible for reducing all-*trans*-retinaldehyde to all-*trans*-retinol, in that it is more sensitive to detergents and to storage and washing of the membranes (Bridges, 1977). In frogs, reduction of all-*trans*-retinaldehyde and the simultaneous oxidation of 11-*cis*-retinol—as might occur during moderate light adaptation—have been experimentally coupled through their common NADP cofactor (Bridges, 1977).

At present, there is no evidence that either frog or mammalian retinas can use

11-*cis*-retinyl palmitate for rhodopsin regeneration (Perlman *et al.*, 1983). This finding indicates that frog ROS lack an ester hydrolase because any 11-*cis*-retinol released would have been used to regenerate rhodopsin.

With the exception of one early report on rat tissue (Cone and Brown, 1969), appreciable regeneration has not been observed in excised retinas. Since the Cone and Brown experiment was done in a small volume of perfusate and involved the bleaching and regeneration of only a restricted patch of retina, it is possible that regeneration occurred from an endogenous store of 11-*cis*-retinoid (bound to CRAlBP or present in the ROS membranes) and did not represent true isomerization. Therefore, it would seem logical to suppose that the retina lacks an isomerizing enzyme and must be provided with 11-*cis*-retinaldehyde (mammals) or 11-*cis*-retinaldehyde or retinol (frog) from another tissue. This has been a role frequently suggested for the RPE. Evidence in favor of the RPE having a role in visual pigment regeneration includes (1) with few exceptions, it contains preformed 11-*cis*-retinoid; (2) regeneration in isolated frog retinas occurs only if they are placed in contact with the RPE (e.g., Reuter, 1966; Bridges, 1973; Ratzlaff, 1975). Evidence against this view is that it has not proved possible to demonstrate the existance of an isomerase in RPE cells, either freshly isolated or in culture.

4. Isomerizing Reactions

Since the key to understanding visual pigment regeneration lies in obtaining an insight into how 11-*cis*-retinoid is formed from all-*trans*-retinoid, this section is devoted to this difficult as yet unsolved problem. Under what circumstances do retinoids isomerize?

a. PHOTOISOMERIZATION. Retinaldehyde is a conjugated isoprenoid that exists in a number of configurations owing to the occurrence of cis–trans isomerism at the double bonds in its tetraene side chain. Although these isomers are usually comparatively stable when stored in dry solvents in darkness, irradiation of their solutions converts them into mixtures of isomers. Irradiation of any isomer yields the same photosteady state (Kropf and Hubbard, 1970). The polarity of the solvent plays an important role in determining the composition of this mixture. In hexane, photoisomerization proceeds preferentially from cis to trans (Hubbard and Wald, 1952; Hubbard *et al.*, 1953), whereas in more polar solvents (such as ethanol or acetonitrile), photoisomerization favors cis isomers (Brown and Wald, 1956; Denny *et al.*, 1981; Denny and Liu, 1977; Tsukida *et al.*, 1977; Liu and Asato, 1982).

If the negligible contribution of di-cis isomers is ignored, the composition of a typical photoisomerate (incandescent lamp) in ethanol is 22% 13-cis, 29% 11-cis, 11% 9-cis, 0.7% 7-cis, and 37% all-trans (Bridges and Alvarez, 1982b). If the 11-*cis*-retinaldehyde from this reaction is collected from the column effluent, transferred to hexane, and irradiated in the presence of a trace amount of iodine, most of it is converted back to the original all-trans configuration.

Strong light adaptation causes photoisomerization in living frog eyes as indicated by the formation of small amounts of isorhodopsin (i.e., 9-*cis*-rhodopsin) in the retina (Reuter, 1966; Bridges, 1976b). Photoisomerization of all-*trans*- to 11-*cis*-retinaldehyde has been observed in bovine ROS membranes (Shichi and Somers, 1974), and flash photolysis has also been shown to generate isorhodopsin in suspensions of frog ROS membranes (Bridges, 1962). Photoisomerization also occurs in isolated retinas and probably involves the irradiation of pararhodopsin (Azuma and Azuma, 1980; Baumann, 1972). The effect is probably unimportant under moderate illumination, particularly if the light is lacking shorter wavelengths (as would occur in human eyes where the lens has appreciable absorption at the blue end of the spectrum). In any event, there is no evidence that photoisomerization influences the course of subsequent dark adaptation (Lewis, 1957; Rushton, 1957; Reuter, 1966). Further, light does not induce accumulation of 11-*cis*-retinoid in the retina or RPE; on the contrary, preformed stores of 11-*cis*-retinyl palmitate in the frog are steadily consumed during light and dark adaptation (Hubbard and Dowling, 1962; Bridges, 1976b).

In contrast, photoisomerization as a means for visual pigment regeneration appears to be the rule in invertebrates (Schwemer *et al.*, 1971; Hara *et al.*, 1981).

b. ISOMERIZATION IN DARKNESS IN CELL-FREE MEDIA. Since photoisomerization of all-*trans*-retinoid does not appear to play a role in the physiological process of dark adaptation in vertebrates, it is clear that a dark process must be sought.

The 9-cis and 13-cis isomers of retinaldehyde are often formed thermally in aqueous detergent solutions or dispersions of all-*trans*-retinaldehyde (Hubbard, 1956; Groenendijk *et al.*, 1980; Futterman and Rollins, 1973). The presence of nucleophiles ($FADH_2$, $FMNH_2$), bacteria, and reducing conditions seems to be important for obtaining appreciable yields of 9-*cis*-retinaldehyde (Futterman and Rollins, 1973; Rotmans *et al.*, 1972). With respect to retinol, which is not acted upon by nucleophiles, our own observations show that in RPE homogenates, all-*trans*-retinol is partially isomerized to 13-*cis*-retinol (see Fig. 9).

In no case has the formation of an 11-*cis*-retinoid from all-*trans*-retinoid been confirmed *in vitro* in darkness. However, numerous studies have shown that 11-cis isomers of retinaldehyde, retinol, and retinaldehyde oxime are easily converted to all-trans in a variety of solvents (Hubbard, 1956, 1966; Futterman and Rollins, 1973; Futterman and Futterman, 1974; Rotmans *et al.*, 1972), particularly under conditions favoring Schiff base formation or interaction with phospholipids such as phosphatidylethanolamine (Groenendijk *et al.*, 1980). Various studies of this process have been carried out with the rationale that important clues concerning the mechanism of the reverse reaction would be obtained.

Many authors lean toward the belief that the *in vivo* form isomerized to 11-cis is retinaldehyde. A possible mechanism suggested by Futterman and Rollins (1973) is that a nucleophilic attack could be directed specifically to the 11,12-*trans* double bond if some degree of immobilization in a defined hydrophobic environment were achieved (cf. Groenendijk *et al.*, 1980). Stubbs *et al.* (1979) have gone further and have suggested that CRAlBP might act as a template during isomerization. While this protein may be involved in the mobilization, transport, and delivery of 11-*cis*-retinoid, a role for CRAlBP in isomerizing all-*trans*- to 11-*cis*-retinoid has not been established but will be discussed further in Section III,G,4,e. At present, the mechanism of this process is still almost wholly a matter for conjecture.

c. ISOLATED RPE CELLS. As shown by Fong *et al.* (1983a), frog RPE cells that have been separated mechanically from the eyecups esterify exogenous all-*trans*-retinol in essentially the same way as homogenates (Section III,E). The bottom half of Fig. 14 shows that when these cells were incubated with all-*trans*-[11,12-^{3}H]retinol, the only isomers that were labeled corresponded to all-trans and 13-cis. The latter was probably formed by esterification of nonenzymatically produced 13-*cis*-retinol. In order to determine whether these preparations were able to esterify 11-*cis*-retinol, RPE cells were incubated with a mixture of ^{3}H-labeled 13-*cis*-, 11-*cis*-, and all-*trans*-retinol. In contrast to the situation where pure all-*trans*-[^{3}H]retinol was used, Fig. 14 shows that there was an additional peak of radioactivity attributable to 11-*cis*-retinyl palmitate. Therefore, if 11-*cis*-retinol had been formed from all-*trans*-retinol in these preparations, it would have been esterified and a peak of radioactivity would have been observed in the position of 11-*cis*-retinyl palmitate.

Similar results were obtained with cultured human RPE cells (Flood *et al.*, 1982). Even after 18-h incubation with exogenous all-*trans*-retinol, no 11-*cis*-retinyl esters could be found.

d. FORMATION OF 11-*cis*-RETINOID IN THE LIVING ANIMAL. While there has been repeated failure to form 11-*cis*-retinoids from all-*trans*-retinoids in *in vitro* situations, it is clear that this conversion can be effected by the living organism. Aside from the occurrence of small amounts of 11-*cis*-retinoids in the cytosol and ROS membranes, dark-adapted eyes have 11-*cis*-retinaldehyde bound in the visual pigments in the ROS and (in most animals) 11-*cis*-retinyl ester in the RPE storage depots. The net effect of strong and repeated light adaptation is the loss of the visual pigment 11-cis prosthetic group and its appearance as all-*trans*-retinyl ester (mainly palmitate and stearate) in the RPE. In frogs, any preexisting 11-*cis*-retinyl ester is progressively replaced by all-*trans*. It has not been established whether this is also applicable to mammals.

In darkness, all-*trans*-retinoid is converted to 11-*cis*-retinoid in the living

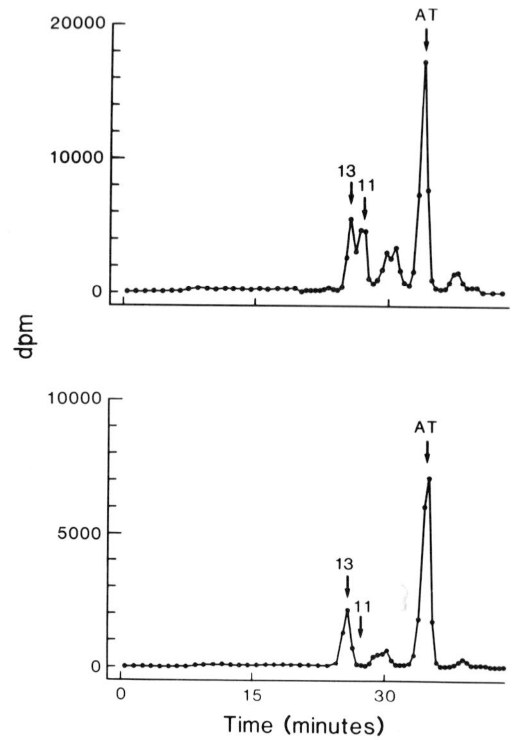

Fig. 14. Labeled isomeric retinyl esters synthesized by frog RPE cell suspensions (not homogenized) incubated with [^{3}H]retinol. Top: 2-h incubation; the substrate consisted of labeled 11-*cis*-, 13-*cis*-, and all-*trans*-retinol isomers. Bottom: 18 h- incubation; the substrate was purified all-*trans*-[^{3}H]retinol. HPLC under normal phase conditions; peak 13, 13-*cis*-retinyl palmitate; peak 11, 11-*cis*-retinyl palmitate; peak AT, all-*trans*-retinyl palmitate. (From Fong *et al.*, 1983a.)

animal. In frogs, the reaction is manifested by (1) rhodopsin regeneration, and (2) the replacement of half of the all-*trans*-retinyl palmitate by 11-*cis*-retinyl palmitate in the RPE. The time course of these two processes is different. Rhodopsin regeneration is complete in about 1 h, whereas a roughly equivalent amount of 11-*cis*-retinyl ester is formed over a period of 24 h (Bridges, 1976b).

A slow appearance of labeled 11-*cis*-retinyl palmitate in the RPE is demonstrable after injection of all-*trans*-[11,12-^{3}H]retinol directly into the eyes of dark-adapted frogs (Fong *et al.*, 1983a). Initially, the all-*trans*-retinol is taken up by the RPE and converted to all-*trans*-retinyl palmitate. After several hours, the all-*trans*-retinyl palmitate is found to have a fairly constant specific activity, as shown in Fig. 15. No labeled 11-cis isomer is detectable at first, but between 2 and 7 h, radioactivity starts appearing in the 11-*cis*-retinyl palmitate. After 15 h, however, Fig. 15 shows that its specific activity is still only 20% of that of the all-trans, even though the endogenous pools of the two isomers are about equal in size, each representing about 1 mol Eq of the rhodopsin in the retina.

Since the rate of accumulation of 11-*cis*-retinoid in the RPE is not fast enough to support rhodopsin regeneration, it has been suggested (Bridges, 1976a,b; Fong *et al.*, 1983a) that the RPE is not the primary site of isomerization. According to this view, 11-*cis*-retinoid is rapidly generated in the retina and is slowly transferred to the esterifying site in the RPE. It should be emphasized that the presumptive release of 11-*cis*-retinaldehyde by proteolysis of rhodopsin in phagocytosed shed ROS discs (Young and Bok, 1969) could account for only 6% of the observed increase in stored retinyl ester in frog RPE over a 24-h period [see Fong *et al.* (1983a)], and therefore plays an insignificant role in the process.

e. CONCLUDING REMARKS. The view that the retina is the site of isomerization must be reconciled with the limited regeneration capacity of isolated retinas. It had been suggested that the isomerase required retinyl ester as a substrate, and that in frogs this could only be supplied by the RPE (Bridges, 1976a,b). Studies by Perlman *et al.* (1983), however, have indicated that isolated retinas cannot utilize 11-*cis*-retinyl palmitate for rhodopsin regeneration and hence lack the necessary hydrolase capable of acting on retinyl esters of long-chain fatty acids.

Of particular interest is the fact that 11-*cis*-retinoid in bovine retina cytosol consists of 11-*cis*-retinol as well as 11-*cis* retinaldehyde. In contrast, only 11-*cis*-retinaldehyde is present in the RPE cytosol [Saari (1982); Saari *et al.* (1982); see also Liou *et al.* (1982c)]. Since bovine and other mammalian retinas cannot utilize 11-*cis*-retinol for rhodopsin regeneration (see Section III,G,3), its

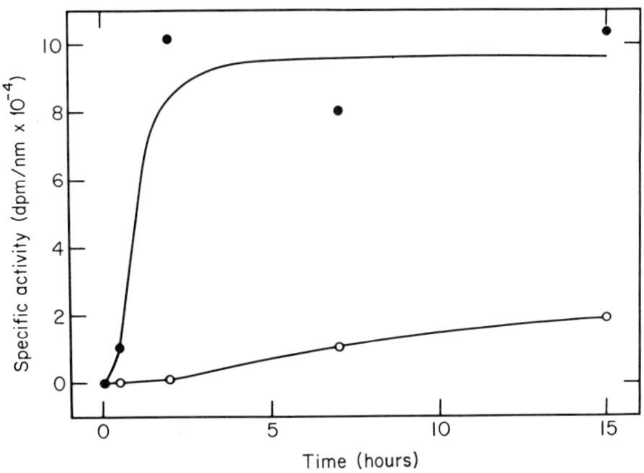

Fig. 15. Specific activity of retinyl palmitate isomers in the RPE of dark-adapted frogs injected intraocularly with all-*trans*-[^{3}H]retinol and sacrificed at intervals afterwards. Key: ●, all-*trans*-retinyl palmitate; ○, 11-*cis*-retinyl palmitate. Each point represents the injected eyes from one pair of frogs. (From Fong *et al.*, 1983a.)

presence in the retina and virtual absence in the RPE is surprising. In retina and RPE, 11-*cis*-retinoid is bound to CRAlBP (Saari *et al.*, 1982). An important new dimension to our concept of the visual cycle has been provided by Bunt-Milam and Saari (1983). Using immunocytochemistry, they have localized CRAlBP not only in the RPE but also in the Müller cells of the retina. These are glial cells that span the full depth of the retina from the external limiting membrane (which is formed partly from their ciliary processes; Fig. 3) to its vitreous surface. A highly speculative possibility is that this third cell type participates in the visual cycle by isomerizing all-*trans*- to 11-*cis*-retinol, but that the 11-*cis*-retinol must be oxidized to 11-*cis*-retinaldehyde (Section III,G,1) in the RPE cell before it can be utilized for rhodopsin regeneration by the rod outer segment. The esterifying system could therefore be completely by-passed under conditions of moderate illumination. Any surplus 11-*cis*-retinol would be esterified and stored, perhaps accounting for the slow formation of 11-*cis*-retinyl ester in darkness (Section III,G,4,d).

IV. RETINOIDS AND HUMAN RETINAL DISEASE

A. Night Blindness and Vitamin A Deficiency

The association of night blindness with vitamin A deficiency is well documented (e.g., McClaren, 1980; Berson, 1982), and the effects of maintaining rats on vitamin A-deficient diets, with or without retinoic acid supplements, have been the subject of several studies (Dowling and Gibbons, 1961; Dowling, 1964; Noell *et al.*, 1971; Carter-Dawson *et al.*, 1979; Delmelle *et al.*, 1975). In its initial stages, nutritional night blindness appears to be due to a loss of rhodopsin, which in rats is then followed by degeneration of the outer segments and loss of the photoreceptors. The cause of this degeneration has not been satisfactorily explained. One possibility is that opsin is less stable than rhodopsin [as observed in detergent extracts, Radding and Wald (1956)] thus leading to destruction of the outer segment membranes of which it is the major protein constituent. This still fails to account for the subsequent complete degeneration of the remainder of the visual cell, even when the animals are supplemented with retinoic acid. The possibility that retinol has another function in the visual cell that cannot be replaced by retinoic acid is unlikely because vitamin A-deficient, retinoic acid-supplemented guinea pigs lose their rhodopsin but keep their rod outer segments and the remainder of their visual cells (Howell *et al.*, 1967). *Xenopus laevis* toads appear to be similar (Witkovsky *et al.*, 1981). This suggests that the survival of the photoreceptor cell is critically dependent on the presence of an outer segment.

B. Retinitis Pigmentosa

The visual cell degeneration seen in vitamin A-deficient, retinoic acid-supplemented rats is similar to the histological picture of retinitis pigmentosa (Kolb and Gouras, 1974; Szamier and Berson, 1977; Szamier *et al.*, 1979). Retinitis pigmentosa, one of the most prevalent of hereditary degenerations involving the eye, is characterized by a ring of blindness that slowly and progressively invades the central areas of vision. Loss of vision is due to photoreceptor degeneration, which is accompanied by irregular deposits of melanin pigmentation as the deterioration progresses (Merin and Auerbach, 1976). It is a complex disease that is probably associated with more than one gene. Inheritance can be by the autosomal recessive, autosomal dominant, or sex-linked modes. Sporadic cases are probably the result of mutations that manifest themselves as dominant or sex-linked genes (Merin and Auerbach, 1976).

The potential magnitude of the problem can be appreciated when it is realized that 1 in 50 people carry the gene(s) for this disorder, even though they enjoy normal vision (Boughman *et al.*, 1980). An estimated 50,000–100,000 people in the United States alone are actually afflicted, and some 1,450 new cases of legal blindness due to retinitis pigmentosa are reported each year (National Society to Prevent Blindness, 1980).

Because of the resemblance between the histopathology of the retina in retinitis pigmentosa and in vitamin A deficiency, it is not surprising that a link between the disease and some failure in the absorption, transport, delivery, or utilization of vitamin A has been sought. Thus a generalized retinal degeneration coupled with widespread loss of visual cells, pigment proliferation, and visual fields characteristic of retinitis pigmentosa is observed in abetalipoproteinemia (Bassen-Kornzweig syndrome: Gouras *et al.*, 1971; Carr, 1976; von Sallmann *et al.*, 1969), where there is an inability to form chylomicrons and where the liver and serum become depleted in vitamin A. Patients in the early stages of the disease have been shown to respond to vitamin A [see Berson (1982) for review]. However, it has now been established that "true" retinitis pigmentosa patients have normal serum retinol and RBP levels, and that the RBP molecule itself is not abnormal (Futterman *et al.*, 1974; Maraini *et al.*, 1975). Therefore, it is unlikely that retinitis pigmentosa arises from a defect in the absorption of vitamin A or in its delivery to the eye. Additionally, negative findings were reported in trials where an attempt was made to arrest or reverse the progress of the disease by vitamin A therapy (Bergsma and Wolf, 1976).

What is the likelihood that retinitis pigmentosa is associated with a defect in the transport of vitamin A *within* the ocular tissues or in an enzyme responsible for its isomerization or conversion to one of its derivatives? Some of the evidence against this view depends on comparisons between retinitis pigmentosa patients

TABLE II

Reactions of the Visual Cycle

Reaction	Site
Rhodopsin —($h\nu$, thermal)→ All-*trans*-retinaldehyde + opsin	ROS (all vertebrates)
All-*trans*-retinaldehyde —(NADPH / dehydrogenase)→ All-*trans*-retinol	ROS (cattle, frog)
Retinol (all isomers) —(acyl donor / ester synthetase)→ retinyl palmitate, stearate, etc.	RPE (frog, cattle, human)
Retinyl palmitate (11-*cis*, all-*trans*) —(ester hydrolase)→ retinol	Probably RPE
All-*trans*-retinoid (All-*trans*-retinol?) —(isomerase)→ 11-*cis* retinoid (11-*cis* retinol?)	Müller cell? (see Section III G,4e)
11-*cis*-Retinol —(NADP, NAD / dehydrogenase)→ 11-*cis* retinaldehyde	ROS (frog), RPE (cattle)
11-*cis*-Retinaldehyde + opsin —(thermal)→ rhodopsin	ROS (all vertebrates)

and patients with vitamin A deficiency (Berson, 1982). However, since the causes of inadequate vitamin A nutriture were complex (in this instance, it was associated with chronic alcoholism), many diverse bodily processes could have been affected. In any event, systemic hypovitaminosis A may have effects on the eye that are very different from a local, chronic, and possibly progressive deficiency in vitamin A delivery and utilization.

Other observations relating to this question have compared retinal rhodopsin content with visual sensitivity in retinitis pigmentosa patients. When the amount of rhodopsin in the living eye is reduced by bleaching or in experimental hypovitaminosis A, a graph of the logarithm of visual sensitivity against the amount of rhodopsin is roughly linear. That is, loss of sensitivity is severe for a comparatively minor reduction in the amount of rhodopsin (Dowling, 1964). The effect, which is fundamental to visual physiology, has not been explained. It may be due to a depression of rod sensitivity caused by the presence of free opsin. In early studies on retinitis pigmentosa patients, it seemed that the loss of rhodopsin observed in this disease did not affect visual sensitivity in the same way (Highman and Weale, 1973; Ripps *et al.*, 1978). Rather, the observations were explicable by assuming a progressive shortening of the ROS with a concomitant reduction in the "quantum catch" but no accumulation of free opsin [see also Ripps (1982)]. However, it now appears that not all retinitis pigmentosa patients fall into this category. In 8 out of 11 patients examined by Perlman and Auerbach (1981), there was severe loss of visual sensitivity coupled with rhodopsin depletion, the relationship being similar to that seen in partial bleaching or hypovitaminosis A.

What defects in vitamin A transport or utilization are likely to occur in the diseased eye? The cytosol and IPM binding proteins and their possible receptors are clearly essential components of the visual system, as are the specific enzymes that are implicated at certain stages of the visual cycle (Table II). Failure to isomerize all-*trans*- to 11-*cis*-retinoid is one obvious possibility, and a clinical trial was carried out by Chatzinoff *et al.* (1968) in an attempt to answer this question. The compound 11-*cis*- "vitamin A" (the retinoid used was not reported) was injected intramuscularly into retinitis pigmentosa patients over a 3-year period. This group was then compared with a parallel group of patients that had received the all-trans isomer. No beneficial effect of 11-*cis*-retinoid was found. However, the ease with which the 11-cis isomer isomerizes back to all-trans when dispersed in tissue preparations (see Section III,G,4,b) casts doubt on the likelihood that in this study any 11-*cis*-retinol would have survived to be delivered to the RPE.

Another possible defect would be failure to esterify retinol by the RPE. Reading (1966) suggested that this could lead to an accumulation of free retinol which might cause breakdown of lysosomes and release of their potentially damaging contents. Although there has been some support for this view in studies on the

RCS (Royal College of Surgeons) rat model of retinal dystrophy (Berman *et al.*, 1981), observations on human tissue have indicated that this is not the problem in retinitis pigmentosa. These data are discussed in the final section of this chapter.

C. Studies on Tissues from Eyes with Hereditary Chorioretinal Degenerations

Tissues from the diseased eyes of two donors (RP 137 and 179) have been examined (Bridges and Alvarez, 1982a; Bridges *et al.*, 1983a). In RP 137 the disease appeared to be a hereditary chorioretinal degeneration that resembled retinitis pigmentosa but involved primarily the choroidal circulation (Cope *et al.*, 1976). The disease of RP179 was sporadic retinitis pigmentosa. The comparative

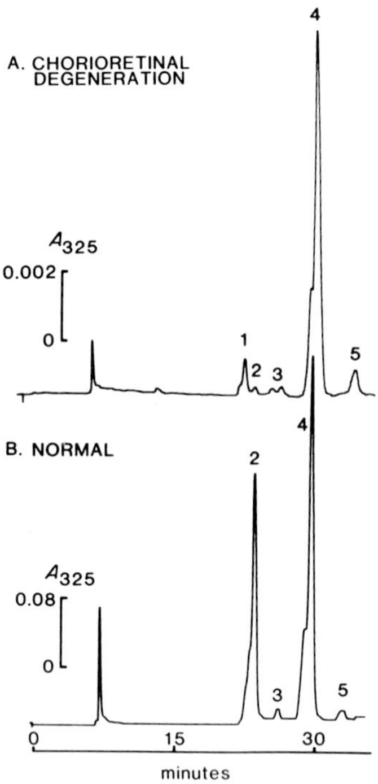

Fig. 16. Selective loss of 11-*cis*-retinyl ester (peak 2) in the RPE storage depots of eyes from a donor with dominantly inherited bilateral chorioretinal degeneration similar to retinitis pigmentosa (RP137). (A) HPLC of retinyl esters from the affected tissue; (B) HPLC of retinyl esters from normal tissue. Approximate amounts injected were 0.8 nmol for trace (A) and 19.5 nmol for trace (B). Peak 1, 13-*cis*-retinyl palmitate and stearate; peak 2, 11-*cis*-retinyl palmitate and stearate; peak 3, not identified; peak 4, all-*trans*-retinyl palmitate and stearate; peak 5, all-*trans*-retinyl oleate. (From Bridges and Alvarez, 1982a.)

rarity of this tissue meant that only portions could be used for the few selected studies described below.

1. Storage of Retinoids

In both cases, there was no evidence for depletion of the retinoid stored in the RPE.

2. Esterification of Retinol

The retinoid stored in the RPE from these eyes was predominantly retinyl ester. No accumulation of free retinol was observed. As in normal eyes (Bridges *et al.*, 1982), the esters consisted mainly of palmitate with a smaller proportion of stearate. The rate of esterification of all-*trans*-[^{3}H]retinol was measured in RPE homogenates and was not considered to be significantly lower than normals.

3. 11-cis-Retinyl Esters in the RPE

The RPE in the eye with retinitis pigmentosa contained 11-*cis*-retinyl esters in proportions that were close to normal. In RP137, however, there was selective loss of the 11-cis isomer, as shown in Fig. 16. Clearly, some impairment of the isomerizing system had occurred, possibly as a consequence of photoreceptor loss (Bridges and Alvarez, 1982a).

4. Interstitial Retinol-Binding Protein (IRBP)

Interstitial retinol-binding protein could not be detected in the eye with retinitis pigmentosa (Bridges *et al.*, 1983a). This is illustrated in the SDS gel of Fig. 17. Tracks 3 and 6 are interphotoreceptor matrix from normal human eyes, while tracks 1,2 and 7,8 are from the retina and RPE cytosols of normal human eyes. These data clearly show that the interstitial binding protein (arrow) is restricted to the interphotoreceptor matrix. Track 4 represents the combined interphotoreceptor matrix, retina cytosol, and RPE cytosol from the patient with sporadic retinitis pigmentosa. There is no band corresponding to the interstitial binding protein. The loss of a 7–8 S "receptor" for retinol in an eye with advanced retinitis pigmentosa was reported by Bergsma *et al.* (1977). Recent reports (Gonzalez-Fernandez *et al.*, 1983a,b) describe the loss of IRBP during the active phase of photoreceptor degeneration in the RCS (Royal College of Surgeons) rat model of retinal dystrophy. This loss is associated with a decline in the ability of the retina to synthesize and secrete IRBP *in vitro*. Clearly, partial or outright loss of interstitial binding protein could mean a malfunctioning in the transfer of vitamin A between RPE and retina, either causing an impairment of rhodopsin synthesis or possibly permitting a damaging accumulation of free retinol in the ROS in the light.

Interstitial retinol-binding protein may also carry other lipid-soluble ligands; vitamin E is one possibility (Section II,F,2). This compound is present in bovine

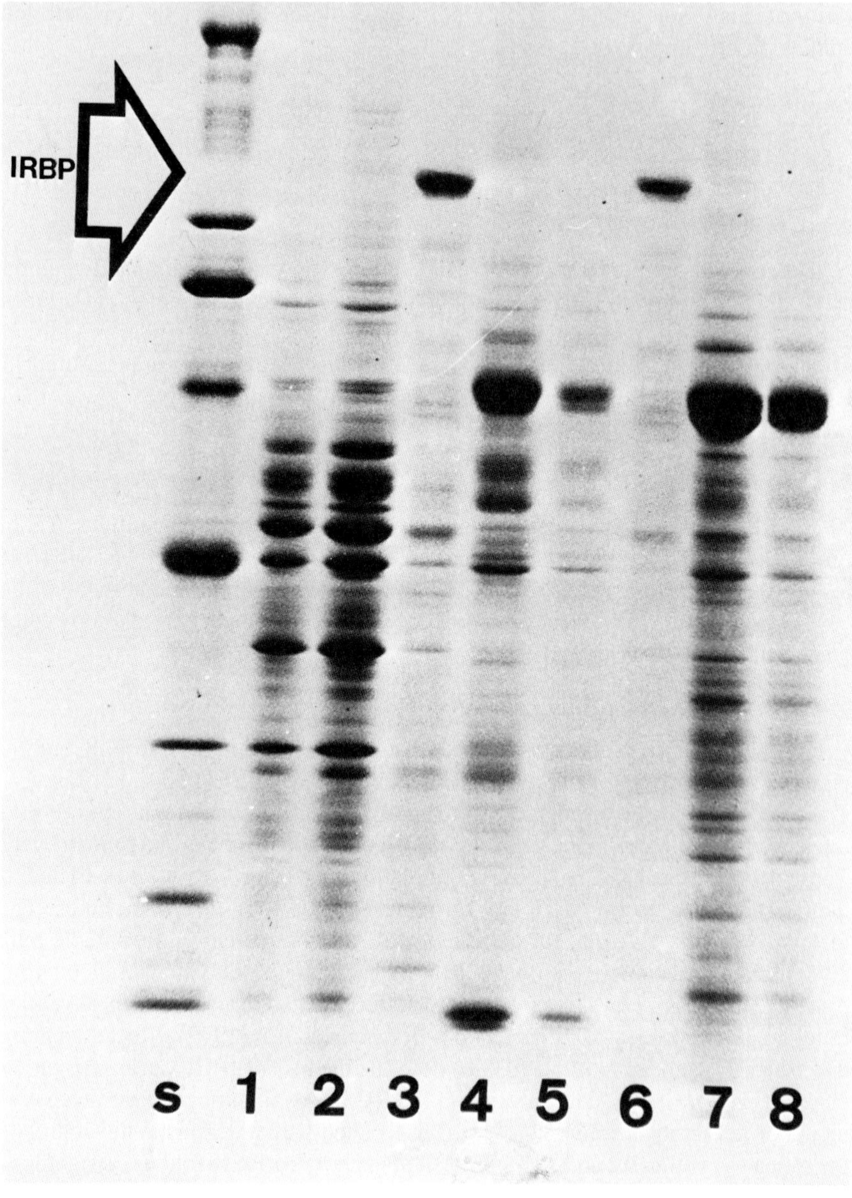

Fig. 17. SDS–PAGE analysis of proteins from retinitis pigmentosa-affected and normal ocular tissue. Tracks 1 and 2, normal retina cytosol (two loading levels); track 3, normal interphotoreceptor matrix; track 4, combined retina cytosol, interphotoreceptor matrix, RPE, and choroid cytosol from RP179; track 5, choroid cytosol from RP179; track 6, normal interphotoreceptor matrix; tracks 7 and 8, normal RPE cytosol (two loading levels); track s, protein standards (from the top, they correspond to M_r = 210,000, 130,000, 94,000, 68,000, 43,000, 29,000, 21,000, and 14,300, respectively). See Liou *et al.* (1982c) for methods.

rod outer segment membranes, where it is believed to act as an antioxidant (Dilley and McConnell, 1970). Failure to provide the ROS with α-tocopherol would also have severe effects on their structural integrity (Robison *et al.*, 1980, 1982; Hayes, 1974).

5. *Conclusions*

The cause or causes of retinitis pigmentosa are yet to be found and at present a possible role for vitamin A in this condition has not been established. However, the above studies have enabled us to eliminate some possibilities and to identify others. The retinitis pigmentosa eye was normal in its ability to store retinoid, to esterify retinol, and to convert the all-trans isomer to 11-cis. A major abnormality was the absence of interstitial retinol-binding protein. It has not been established whether this is a selective loss, or whether other components of the interphotoreceptor matrix are also involved. Clearly, much more work is needed before we arrive at an insight into the causes of chorioretinal binding diseases.

ACKNOWLEDGMENTS

This work was supported by grants from the Retina Research Foundation of Houston, and the National Institutes of Health EY 02489 and EY 02520.

I would like to express my appreciation to Ms. Vicki L. Burch for assistance in preparation of this chapter.

REFERENCES

Adler, A. J., and Klucznik, K. M. (1982). *Exp. Eye Res.* **34,** 423–434.

Adler, A. J., and Martin, K. J. (1982). *Biochem. Biophys. Res. Comm.* **108,** 1601–1608.

Adler, A. J., and Severin, K. M. (1981). *Exp. Eye Res.* **32,** 755–769.

Akita, H., Tanis, S. P., Adams, M., Balogh-Nair, V., and Nakanishi, K. (1980). *J. Am. Chem. Soc.* **102,** 6370–6372.

Alvarez, R. A., Bridges, C. D. B., and Fong, S.-L. (1981). *Invest. Ophthalmol. Visual Sci.* **20,** 304–313.

Andrews, J. S., and Futterman, S. (1964). *J. Biol. Chem.* **239,** 4073–4076.

Applebury, M. L., and Rentzepis, P. M. (1982). *In* "Methods in Enzymology, Vol. 81, Part H: Visual Pigments and Purple Membranes" (L. Packer, ed.), pp. 354–368. Academic Press, New York.

Aton, B., Doukas, A. G., Callender, R. H., Becher, R. H., and Ebrey, T. (1977). *Biochemistry* **16,** 2995–2999.

Azuma, K., and Azuma, M. (1980). *Photochem. Photobiol.* **32,** 529–538.

Balogh-Nair, V., and Nakanishi, K. (1982a). *In* "New Comprehensive Biochemistry" (Ch. Tamm, ed.), Vol. 3, pp. 283–334. Elsevier, Amsterdam.

Balogh-Nair, V., and Nakanishi, K. (1982b). *In* "Methods in Enzymology, Vol. 88: Biomembranes, Part I" (L. Packer, ed.), pp. 496–506. Academic Press, New York.

Bashor, M. M., and Chytil, F. (1975). *Biochim. Biophys. Acta* **411,** 87–96.

Bashor, M. M., Toft, D. O., and Chytil, F. (1973). *Proc. Natl. Acad. Sci. U.S.A.* **70,** 3483–3487.

Baumann, C. (1972). *In* "Handbook of Sensory Physiology, VII/I: Photochemistry in Vision" (H. J. A. Dartnall, ed.), pp. 395–416. Springer-Verlag, Berlin and New York.
Baumann, C., and Bender, S. (1973). *J. Physiol. (London)* **235,** 761–773.
Baumann, C., and Zeppenfeld, W. (1981). *J. Physiol. (London)* **317,** 347–364.
Bayley, H., Huang, K.-S., Radhakrishnan, R., Ross, A. H., Takagaki, Y., and Khorana, H. G. (1981). *Proc. Natl. Acad. Sci. U.S.A.* **78,** 2225–2229.
Bergsma, D. R., and Wolf, M. L. (1976). *In* "Retinitis Pigmentosa" (M. A. Landers, M. L. Wolbarsht, A. M. Laties, and J. E. Dowling, eds.), pp. 197–209. Plenum, New York.
Bergsma, D. R., Wiggert, B., Funahashi, M., Kuwabara, T., and Chader, G. (1977). *Nature (London)* **265,** 66–67.
Berman, E. R. (1982). *In* "Methods in Enzymology, Vol. 81, Part H: Visual Pigments and Purple Membranes, I" (L. Packer, ed.), pp. 77–85. Academic Press, New York.
Berman, E. R., Horowitz, J., Segal, N., Fisher, S., and Feeney-Burns, L. (1980). *Biochim. Biophys. Acta* **630,** 36–46.
Berman, E. R., Segal, N., Photiou, S., Rothman, H., and Feeney-Burns, L. (1981). *Nature (London)* **293,** 217–220.
Berson, E. L. (1982). *Retina* **2,** 236–255.
Bhat, M. K., and Cama, H. R. (1979). *Biochim. Biophys. Acta* **587,** 273–281.
Bitensky, M. W., Wheeler, G. L., Yamazaki, A., Rasenick, M. M., and Stein, P. J. (1981). *Curr. Top. Membr. Transp.* **15,** 237–271.
Blazynski, C., and Ostroy, S. E. (1981). *Vision Res.* **21,** 833–841.
Bok, D., and Heller, J. (1976). *Exp. Eye Res.* **22,** 395–402.
Bok, D., Basinger, S. F., and Hall, M. O. (1974). *Exp. Eye Res.* **18,** 225–240.
Boughman, J. A., Conneally, P. M., and Nance, W. E. (1980). *Am. J. Hum. Genet.* **32,** 223–235.
Bownds, M. D. (1981). *Am. J. Med. Genet.* **15,** 203–214.
Braiman, M., and Mathies, R. (1982). *Proc. Natl. Acad. Sci., U.S.A.* **79,** 403–407.
Bridges, C. D. B. (1962). *Vision Res.* **2,** 215–232.
Bridges, C. D. B. (1965). *Cold Spring Harbor Symp. Quant. Biol.* **30,** 317–334.
Bridges, C. D. B. (1970). *In* "Biochemistry of the Eye" (C. N. Graymore, ed.) pp. 563–644. Academic Press, New York.
Bridges, C. D. B. (1972). *In* "Handbook of Sensory Physiology 7/1A" (H. J. A. Dartnall, ed.), pp. 418–480. Springer-Verlag, Berlin and New York.
Bridges, C. D. B. (1973). *In* "Biochemistry and Physiology of Visual Pigments" (H. Lauger, ed.), pp. 115–121. Springer-Verlag, Berlin and New York.
Bridges, C. D. B. (1974). *Vision Res.* **14,** 779–793.
Bridges, C. D. B. (1975). *Vision Res.* **15,** 1311–1323.
Bridges, C. D. B. (1976a). *Nature (London)* **259,** 247–248.
Bridges, C. D. B. (1976b). *Exp. Eye Res.* **22,** 435–455.
Bridges, C. D. B. (1977). *Exp. Eye Res.* **24,** 571–580.
Bridges, C. D. B., and Alvarez, R. A. (1982a). *Retina* **2,** 256–260.
Bridges, C. D. B., and Alvarez, R. A. (1982b). *In* "Methods in Enzymology, Vol. 81, Part H: Visual Pigments and Purple Membranes, I" (L. Packer, ed.), pp. 463–485. Academic Press, New York.
Bridges, C. D. B., and Yoshikami, S. (1970). *Nature (London)* **221,** 275–276.
Bridges, C. D. B., Alvarez, R. A., and Fong, S.-L. (1982). *Invest. Ophthalmol & Vis. Sci.* **22,** 706–714.
Bridges, C. D. B., Alvarez, R. A., and Fong, S.-L. (1983a). *Invest Ophthalmol. & Vis. Sci., Suppl.* **24,** 141.
Bridges, C. D. B., Fong, S.-L., Liou, G. I., Landers, R. A., Gonzalez-Fernandez, F., Glazebrook, P., and Lam, D. M. K. (1983b). *Soc. Neurosci. Abstr.* **9,** 345.

Brin, K. P., and Ripps, H. (1977). *J. Gen. Physiol.* **69,** 97–120.
Brown, P. K., and Wald, G. (1956). *J. Biol. Chem.* **222,** 865–877.
Bunt-Milam, A. H., and Saari, J. C. (1983). *J. Cell Biol.* **97,** 703–712.
Carr, R. E. (1976). *In* "The Eye and Inborn Errors of Metabolism" (D. Bergsma, A. J. Bron, and E. Cotlier, eds.), pp. 385–399. Alan Liss Inc., New York.
Carter-Dawson, L., Kuwabara, T., O'Brien, P. J., and Bieri, J. G. (1979). *Invest. Ophthalmol. Vis. Sci.* **18,** 437–446.
Chabre, M., and Breton, J. (1979). *Vision Res.* **19,** 1005–1018.
Chatzinoff, A., Nelson, E., Stahl, N., and Clahane, A. (1968). *Arch. Ophthalmol* **80,** 417–419.
Chen, C.-C., and Heller, J. (1977). *J. Biol. Chem.* **252,** 5216–5221.
Chytil, F., and Ong, D. E. (1979). *Fed. Proc., Fed. Am. Soc. Exp. Biol.* **38,** 2510–2514.
Cone, R. A., and Brown, P. K. (1969). *Nature (London)* **221,** 818–820.
Cope, L. A., Teeters, V. W., Borda, R. P., and McCrary, J. A. (1976). *Doc. Ophthal. Proc. Ser. Int. Symp. Fluorescein Angiography* pp. 431–437.
Dartnall, H. J. A. (1968). *Vision Res.* **8,** 339–358.
Defoe, D. M., and Bok, D. (1983). *Invest. Ophthalmol Vis. Sci.* **24,** 1211–1226.
Delmelle, M., Noell, W. K., and Organisciak, D. T. (1975). *Exp. Eye Res.* **21,** 369–380.
Denny, M., and Liu, R. S. H. (1977). *J. Am. Chem. Soc.* **99,** 4865–4867.
Denny, M., Chun, M., and Liu, R. S. H. (1981). *Photochem. Photobiol.* **33,** 267–269.
DePont, J. J. H. H. M., Daemen, F. J. M., and Bonting, S. L. (1970). *Arch. Biochem. Biophys.* **140,** 275–285.
Dilley, R. A., and McConnell, D. G. (1970). *J. Memb. Biol.* **2,** 317–323.
Dingle, J. T., and Lucy, J. A. (1965). *Biol. Rev.* **40,** 422–461.
Dowling, J. E. (1960). *Nature (London)* **188,** 114–118.
Dowling, J. E. (1964). *Exp. Eye Res.* **3,** 348–356.
Dowling, J. E., and Gibbons, I. R. (1961). *In* "The Structure of the Eye" (G. K. Smelser, ed.), pp. 85–99. Academic Press, New York.
Dowling, J. E., and Wald, G. (1958). *Proc. Natl. Acad. Sci. U.S.A.* **44,** 648–661.
Dunn, R., McCoy, J., Simsek, M., Majumdar, A., Chang, S. H., RajBhandary, V. L., and Khorana, H. G. (1981). *Proc. Natl. Acad. Sci. U.S.A.* **78,** 6744–6748.
Ernst, W., Kemp, C. M., and White, H. A. (1978). *Exp. Eye Res.* **26,** 337–350.
Feeney, L. (1973). *Dev. Biol.* **32,** 115–128.
Flood, M. T., Bridges, C. D. B., Alvarez, R. A., Gouras, P., and Blaner, W. S. (1982). *Invest. Ophthalmol. Vis. Sci., Suppl.* **22,** 185.
Flood, M. T., Bridges, C. D. B., Alvarez, R. A., Blaner, W. S., and Gouras, P. (1983). *Invest. Ophthalmol. Vis. Sci.* **24,** 1227–1235.
Fong, S.-L., Tsin, A. T. C., Bridges, C. D. B., and Liou, G. I. (1982). *In* "Methods in Enzymology, Vol. 81, Part H: Visual Pigments and Purple Membranes, I" (L. Packer, ed.), pp. 133–140. Academic Press, New York.
Fong, S.-L., Bridges, C. D. B., and Alvarez, R. A. (1983a). *Vision Res.* **23,** 47–52.
Fong, S.-L., Liou, G. I., Landers, R. A., Gonzalez-Fernandez, F., and Bridges, C. D. B. (1983b). *Fed. Proc., Fed. Am. Soc. Exp. Biol.* **42,** 2179.
Fong, S.-L., Liou, G. I., Landers, R. A., Alvarez, R. A., Glazebrook, P. A., Lam, D. M. K., and Bridges, C. D. B. (1984). *J. Neurochem.* (in press).
Fukuda, M. N., Papermaster, D. S., and Hargrave, P. A. (1979). *J. Biol. Chem.* **254,** 8201–8207.
Futterman, S. (1962). *J. Biol. Chem.* **237,** 677–680.
Futterman, S. (1963). *J. Biol. Chem.* **238,** 1145–1150.
Futterman, S., and Futterman, A. (1974). *Biochim. Biophys. Acta* **337,** 390–394.
Futterman, S., and Rollins, M. (1973). *J. Biol. Chem.* **248,** 7773–7779.
Futterman, S., and Saari, J. C. (1977). *Invest. Ophthalmol. Vis. Sci.* **16,** 768–771.

Futterman, S., Swanson, D., and Kalina, R. E. (1974). *Invest. Ophthalmol. Vis. Sci.* **13,** 798–801.
Futterman, S., Saari, J. C., and Blair, S. (1977). *J. Biol. Chem.* **252,** 3267–3271.
Gambhir, K. K., and Ahluwahlia, B. S. (1974). *Biochem. Biophys. Res. Commun.* **61,** 551–558.
Gold, G. H., and Korenbrot, J. I. (1981). *Curr. Top. Membr. Transp.* **15,** 307–330.
Gonzalez-Fernandez, F., Bridges, C. D. B., Liou, G. I., and Fong, S.-L. (1983a). *Invest. Ophthalmol. Vis. Sci., Suppl.* **24,** 42.
Gonzalez-Fernandez, F., Fong, S.-L., Liou, G. I., Landers, R. A., Lam, D. M. K., Glazebrook, P., and Bridges, C. D. B. (1983b). *J. Cell. Biol.* **97,** 4549.
Goodman, D. S. (1980). Plasma retinol-binding protein. *Ann. N. Y. Acad. Sci.* **348,** 378–388.
Goodman, D. S. (1981). *Ann. N. Y. Acad. Sci.* **359,** 69–78.
Gouras, P., Carr, R. E., and Gunkel, R. D. (1971). *Invest. Ophthalmol. Vis. Sci.* **10,** 784–793.
Groenendijk, G. W. T., Jacobs, C. W. M., Bonting, S. L., and Daemen, F. J. M. (1980). *Eur. J. Biochem.* **106,** 119–128.
Hall, M. O., and Bok, D. (1974). *Exp. Eye Res.* **18,** 105–117.
Hara, T., and Hara, R. (1982). *In* "Methods in Enzymology, Vol. 81, Part H: Visual Pigments and Purple Membranes, I" (L. Packer, ed.), pp. 827–833. Academic Press, New York.
Hara, R., Hara, T., Tokunaga, F., and Yoshizawa, T. (1981). *Photochem. Photobiol.* **33,** 883–891.
Hargrave, P. A. (1982). *Prog. Retinal Res.* **1,** 1–51.
Hargrave, P. A., McDowell, J. H., Siemiatkowski-Juszczak, E. C., Fong, S.-L., Kühn, H., Wang, J. K., Curtis, D. R., Rao, J. K. M., Argos, P., and Feldmann, R. J. (1982). *Vision Res.* **22,** 1429–1438.
Hargrave, P. A., McDowell, J. H., Curtis, D. R., Wang, J., Juszczak, E., Fong, S.-L., Rao, J. K. M., and Argos, P. (1983). *Biophys. Struct. Mech.* **9,** 235–244.
Harosi, F. I., and MacNichol, E. (1974). *J. Opt. Soc. Am.* **64,** 903–910.
Hayes, K. C. (1974). *Invest. Ophthalmol. Vis. Sci.* **13,** 499–510.
Heller, J. (1975). *J. Biol. Chem.* **250,** 3613–3619.
Hemley, R., Kohler, B. E., and Siviski, P. (1979). *Biophys. J.* **28,** 447–455.
Henderson, R., and Unwin, P. N. T. (1975). *Nature (London)* **257,** 28–32.
Hermolin, J., Karell, M. A., Hamm, H. E., and Bownds, M. D. (1982). *J. Gen. Physiol.* **79,** 633–655.
Highman, V. N., and Weale, R. A. (1973). *Am. J. Ophthalmol.* **75,** 822–832.
Hirosawa, K. (1979). *Exp. Eye Res.* **29,** 169–179.
Hogan, M. J., Alvarado, J. A., and Weddell, J. E. (1971). "History of the Human Eye." Saunders, Philadelphia, Pennsylvania.
Honig, B., Dinur, U., Nakanishi, K., Balogh-Nair, V., Gawinowicz, M. A., Arnaboldi, M., and Motto, M. G. (1979). *J. Am. Chem. Soc.* **101,** 7084–7086.
Horwitz, J., and Heller, J. (1974). *J. Biol. Chem.* **249,** 4712–4719.
Howell, J. McC., Thompson, J. N., and Pitt, G. A. J. (1967). *Br. J. Nutr.* **21,** 37–45.
Hubbard, R. (1956). *J. Am. Chem. Soc.* **78,** 4662–4667.
Hubbard, R. (1966). *J. Biol. Chem.* **241,** 1814–1818.
Hubbard, R., and Dowling, J. E. (1962). *Nature (London)* **193,** 341–343.
Hubbard, R., and Kropf, A. (1965). *J. Gen. Physiol.* **49,** 381–385.
Hubbard, R., and Wald, G. (1952). *J. Gen. Physiol.* **36,** 269–315.
Hubbard, R., Gregerman, R. I., and Wald, G. (1953). *J. Gen. Physiol.* **36,** 415–429.
Irimura, T., and Nicolson, G. L. (1983). *Carbohyd. Res.* **115,** 209–220.
Katre, N. V., Wolber, P. K., Stoeckenius, W., and Stroud, D. M. (1981). *Proc. Natl. Acad. Sci. U.S.A.* **78,** 4068–4072.
Kean, E. L. (1977). *J. Biol. Chem.* **252,** 5622–5629.
Khorana, H. G., Gerber, G. E., Herlihy, W. C., Gray, C. P., Anderegg, R. J., Nihei, K., and Biemann, K. (1979). *Proc. Natl. Acad. Sci. U.S.A.* **76,** 5046–5050.

Kito, Y., Naito, T., and Nashima, K. (1982). *In* "Methods in Enzymology, Vol. 81, Part H: Visual Pigments and Purple Membranes, I" (L. Packer, ed.), pp. 167–171. Academic Press, New York.

Knowles, A., and Dartnall, H. J. A. (1977). *In* "The Eye, Vol. 2B: The Photobiology of Vision" (H. Davson, ed.), Academic Press, New York.

Kolb, H., and Gouras, P. (1974). *Invest. Ophthalmol. Vis. Sci.* **13,** 487–498.

Krinsky, N. I. (1958). *J. Biol. Chem.* **232,** 881–894.

Kropf, A. (1967). *Vision Res.* **7,** 811–818.

Kropf, A. (1977). *In* "Vertebrate Photoreception" (H. B. Barlow and P. Fatt, eds.), pp. 15–18. Academic Press, New York.

Kropf, A. (1982). *In* "Methods in Enzymology, Vol. 81, Part H: Visual Pigments and Purple Membranes, I" (L. Packer, ed.), pp. 384–392. Academic Press, New York.

Kropf, A., and Hubbard, R. (1970). *Photochem. Photobiol.* **12,** 249–260.

Kühn, H. (1981). *Curr. Top. Membr. Transp.* **15,** 172–201.

Laemmli, U. K. (1970). *Nature (London)* **227,** 680–685.

Lai, Y. L., Wiggert, B., Liu, Y. P., and Chader, G. J. (1982). *Nature (London)* **298,** 848–849.

Lee, L., and Wiggert, B. (1982). *Invest. Ophthalmol. Vis. Sci. Suppl.* **22,** 186.

Lemke, H. D., and Oesterhelt, D. (1981). *FEBS Lett.* **128,** 255–260.

Lewis, A., Spoonhower, J., Bogomolni, R. A., Lozier, R. H., and Stoeckenius, W. (1974). *Proc. Natl. Acad. Sci. U.S.A.* **71,** 4462–4466.

Lewis, D. M. (1957). *J. Physiol. (London)* **136,** 624–631.

Liang, C.-J., Yamashita, K., Muellenberg, C. G., Shichi, H., and Kobata, A. (1979). *J. Biol. Chem.* **254,** 6414–6418.

Liau, G., Ong, D. E., and Chytil, F. (1981). *J. Cell Biol.* **91,** 63–68.

Liebman, P. A., and Pugh, E. N., Jr. (1981). *Curr. Top. Membr. Transp.* **15,** 157–170.

Liebman, P. A., and Pugh, E. N., Jr., (1982). *Vision Res.* **22,** 1475–1480.

Lion, F., Rotmans, J. P., Daemen, F. J. M., and Bonting, S. L. (1975). *Biochim. Biophys. Acta* **384,** 283–292.

Liou, G. I., Fong, S.-L., and Bridges, C. D. B. (1981). *J. Biol. Chem.* **256,** 3153–3155.

Liou, G. I., Fong, S.-L., and Bridges, C. D. B. (1982a). *Fed. Proc., Fed. Am. Soc. Exp. Biol.* **41,** 683.

Liou, G. I., Bridges, C. D. B., and Fong, S.-L. (1982b). *Invest. Ophthalmol. Vis. Sci., Suppl.* **22,** 65.

Liou, G. I., Bridges, C. D. B., and Fong, S.-L. (1982c). *Vision Res.* **22,** 1457–1468.

Liou, G. I., Bridges, C. D. B., Alvarez, R. A., and Fong, S.-L. (1983). *Invest. Ophthalmol. Vis. Sci., Suppl.* **24,** 42.

Liu, R. S. H., and Asato, A. E. (1982). *In* "Methods in Enzymology, Vol. 88: Biomembranes, Part I" (L. Packer, ed.), pp. 506–516. Academic Press, New York.

Liu, R. S. H., and Matsumoto, H. (1982). *In* "Methods in Enzymology, Vol. 81, Part H: Visual Pigments and Purple Membranes, I" (L. Packer, ed.), pp. 694–698. Academic Press, New York.

Lolley, R. N., and Racz, E. (1982). *Vision Res.* **22,** 1481–1486.

Lotan, R. (1980). *Biochim. Biophys. Acta* **605,** 33–91.

Lythgoe, J. N. (1972). *In* "Handbook of Sensory Physiology, Vol. VII/I: Photochemistry of Vision" (H. J. A. Dartnall, ed.), pp. 566–603. Springer-Verlag, Berlin and New York.

McClaren, D. S. (1980). "Nutritional Ophthalmology." Academic Press, New York.

McGuire, B. W., Orgebin-Crist, M. C., and Chytil, F. (1981). *Endocrinology (Baltimore)* **108,** 658–667.

Mallia, A. K., Smith, J. E., and Goodman, D. S. (1975). *J. Lipid Res.* **16,** 180–188.

Maraini, G., Fadda, G., and Gozzoli, F. (1975). *Invest. Ophthalmol. Vis. Sci.* **14,** 236–237.

Maraini, G., Ottonello, S., Gozzoli, F., and Merli, A. (1977). *Nature (London)* **265,** 68–69.
Matsumoto, H., Liu, R. S. H., Simmons, C. J., and Seff, K. (1980). *J. Am. Chem. Soc.* **102,** 4259–4262.
Matthews, G., and Baylor, D. A. (1981). *Curr. Top. Membr. Transp.* **15,** 3–18.
Merin, S., and Auerbach, E. (1976). *Surv. Ophthalmol.* **20,** 303–346.
Mowery, P. C., and Stoeckenius, W. (1981). *Biochemistry* **20,** 2302–2306.
Mullen, E., Johnson, A. H., and Akhtar, M. (1981). *FEBS Lett.* **130,** 187–193.
Muto, Y., Smith, J. E., Milch, P. O., and Goodman, D. S. (1972). *J. Biol. Chem.* **247,** 2542–2550.
National Society to Prevent Blindness (1980). Vision Problems in the United States.
Noell, W. K., Delmelle, M. C., and Albrecht, R. (1971). *Science* **172,** 72–76.
Oakley, B., and Pinto, L. H. (1981). *Curr. Top. Membr. Transp.* **15,** 405–416.
O'Brien, P. J. (1978). *Recept. Recognition Ser. A* **6,** 109–150.
Oka, M. S., Landers, R. A., Alvarez, R. A., and Bridges, C. D. B. (1983). *Invest. Ophthalmol. Vis. Sci. Suppl.* **24,** 159.
Olson, J. A. (1967). *Pharmacol. Rev.* **19** 559–596.
Ong, D. E., and Chytil, F. (1975). *J. Biol. Chem.* **250,** 6113–6117.
Ong, D. E., and Chytil, F. (1978a). *J. Biol. Chem.* **253,** 828–832.
Ong, D. E., and Chytil, F. (1978b). *J. Biol. Chem.* **253,** 4551–4554.
Ong, D. E., and Chytil, F. (1979). *J. Biol. Chem.* **254,** 8733–8735.
Ong, D. E., and Chytil, F. (1981). *Ann. N. Y. Acad. Sci.* **359,** 415–417.
Orzalesi, N., Fossarello, M., Carta, S., Del Fiacco, G., and Diaz, G. (1982). *Invest. Ophthalmol. Vis. Sci.* **23,** 689–714.
Ostroy, S. E. (1977). *Biochim. Biophys. Acta* **112,** 265–277.
Ovchinnikov, Y. A., Abdulaev, N. G., Feigina, M. Y., Kiselev, A. V., and Lobanov, N. A. (1979). *FEBS Lett.* **100,** 219–224.
Ovchinnikov, Y. A., Abdulaev, N. G., Tseltin, V. I., Kiselev, A. V., and Zakis, V. I. (1980). *Bioorg. Khim.* **6,** 1427–1429.
Ovchinnikov, Y. A., Abdulaev, N. G., Feigina, M. Y., Artamonov, I. D., Zolotarev, A. S., Kostina, M. B., Bogachuk, A. S., Miroshnikov, A. I., Martinov, V. I., and Kudelin, A. B. (1982). *Bioorg. Khim.* **8,** 1011–1014.
Packer, L., ed. (1982). "Methods in Enzymology, Vol. 88, Biomembranes, Part I," pp. 836. Academic Press, New York.
Penn, R. D., and Hagins, W. A. (1972). *Biophys. J.* **12,** 1073–1094.
Pepperberg, D. R. (1982). *In* "Methods in Enzymology, Vol. 81, Part H: Visual Pigments and Purple Membranes, I" (L. Packer, ed.), pp. 452–459. Academic Press, New York.
Pepperberg, D. R., and Masland, R. H. (1978). *Brain Res.* **151,** 194–200.
Pepperberg, D. R., Lurie, M., Brown, P. K., and Dowling, J. E. (1976). *Science* **191,** 394–396.
Pepperberg, D. R., Brown, P. K., Lurie, M., and Dowling, J. E. (1978). *J. Gen. Physiol.* **71,** 369–396.
Perlman, I., and Auerbach, E. (1981). *Invest. Ophthalmol. Vis. Sci.* **20,** 758–765.
Perlman, J. I., Nodes, B. R., and Pepperberg, D. R. (1983). *J. Gen. Physiol.* **80,** 885–913.
Prystowsky, J. H., Smith, J. E., and Goodman, D. S. (1981). *J. Biol. Chem.* **256,** 4498–4503.
Radding, C. M., and Wald, G. (1956). *J. Gen. Physiol.* **39,** 923–933.
Rando, R. R., and Bangerter, F. W. (1982). *Biochim. Biophys. Res. Commun.* **104,** 430–436.
Rask, L., and Peterson, P. (1976). *J. Biol. Chem.* **251,** 6360–6366.
Rask, L., Geijer, C., Bill, A., and Peterson, P. A. (1980). *Exp. Eye Res.* **31,** 201–211.
Rask, L., Anundi, H., Bohme, J., Eriksson, U., Ronne, H., Sege, K., and Peterson, P. A. (1981). *Ann. N. Y. Acad. Sci.* **359,** 79–90.
Ratzlaff, K. O. (1975). *Vision Res.* **15,** 73–77.
Reading, H. W. (1966). *Biochem. J.* **100,** 34. (Abstr.)
Reuter, T. (1966). *Nature* **202,** 1119–1120.

Reuter, T. E., White, R., and Wald, G. (1971). *J. Gen. Physiol.* **58,** 351–371.
Ripps, H. (1982). *Invest. Ophthalmol. Vis. Sci.* **23,** 588–609.
Ripps, H., Brin, K. P., and Weale, R. A. (1978). *Invest. Ophthalmol. Vis. Sci.* **17,** 735–745.
Roberts, A. B., and DeLuca, H. F. (1967). *Biochem. J.* **102,** 606–611.
Roberts, A., and Frolik, C. (1979). *Fed. Proc., Fed. Am. Soc. Exp. Biol.* **38,** 2524–2527.
Robison, W. G., and Kuwabara, T. (1977). *Invest. Ophthalmol. Vis. Sci.* **16,** 1110–1117.
Robison, W. G., Kuwabara, T., and Bieri, J. G. (1980). *Invest. Ophthalmol. Vis. Sci.* **19,** 1030–1037.
Robison, W. G., Kuwabara, T., and Bieri, J. G. (1982). *Retina* **2,** 263–281.
Ross, A. C., and Goodman, D. S. (1979). *Fed. Proc., Fed. Am. Soc. Exp. Biol.* **38,** 2515–2523.
Rothschild, K. J., Zagaeski, M., and Cantore, W. (1981). *Biochem. Biophys. Res. Commun.* **103,** 483–489.
Rothschild, K. J., Argade, P. V., Earnest, T. N., Huang, K.-S., London, E., Liao, M.-J., Bayley, H., Khorana, H. G., and Herzfeld, J. (1982). *J. Biol. Chem.* **257,** 8592–8595.
Rotmans, J. P., Daemen, F. J. M., and Bonting, S. L. (1972). *Biochim. Biophys. Acta* **267,** 583–587.
Rushton, W. A. H. (1957). *J. Gen. Physiol.* **41,** 419–428.
Saari, J. C. (1982). *In* "Methods in Enzymology, Vol. 81, Part H: Visual Pigments and Purple Membranes, I" (L. Packer, ed.), pp. 819–826. Academic Press, New York.
Saari, J. C., and Futterman, S. (1976). *Biochim. Biophys. Acta* **444,** 789–793.
Saari, J. C., Bunt, A. H., Futterman, S., and Berman, E. R. (1977). *Invest. Ophthalmol. Vis. Sci.* **16,** 797–806.
Saari, J. C., Futterman, S., and Bredberg, L. (1978). *J. Biol. Chem.* **253,** 6432–6436.
Saari, J. C., Bredberg, L., and Garwin, G. C. (1982). *J. Biol. Chem.* **257,** 13329–13333.
Sani, B. P., and Banerjee, C. K. (1978). *Biochem. J.* **173,** 643–649.
Sani, B. P., and Hill, D. L. (1974). *Biochem. Biophys. Res. Commun.* **61,** 1276–1282.
Sasaki, N., and Yoshizawa, T. (1981). *Photobiochem. Photobiophys.* **2,** 365.
Sasaki, N., Tokunaga, F., and Yoshizawa, T. (1981). *FEBS Lett.* **114,** 1–3.
Schobert, B., and Lanyi, J. K. (1982). *J. Biol. Chem.* **257,** 10306–10313.
Schwemer, J., and Langer, H. (1982). *In* "Methods in Enzymology, Vol. 81, Part H: Visual Pigments and Purple Membranes, I" (L. Parker, ed.), pp. 182–190. Academic Press, New York.
Schwemer, J., Gogala, M., and Hamdorf, K. (1971). *Z. Vgl. Physiol.* **75,** 174–188.
Seki, T., Hara, R., and Hara, T. (1982). *Exp. Eye Res.* **34,** 608–621.
Sen, R., Carriker, J. D., Balogh-Nair, V., and Nakanishi, K. (1982). *J. Am. Chem. Soc.* **104,** 3214–3216.
Shichi, H., and Somers, R. L. (1974). *J. Biol. Chem.* **249,** 6570–6577.
Shriver, J. W., Mateescu, G. D., and Abrahamson, E. W. (1982). *In* "Methods in Enzymology, Vol. 81, Part H: Visual Pigments and Purple Membranes, I" (L. Packer, ed.), pp. 698–703. Academic Press, New York.
Smith, J. E., and Goodman, D. S. (1979). *Fed. Proc., Fed. Am. Soc. Exp. Biol.* **38,** 2504–2509.
Stoeckenius, W. (1980). *Acc. Chem. Res.* **13,** 337–344.
Stoeckenius, W., and Bogomolni, R. A. (1982). *Annu. Rev. Biochem.* **51,** 587–616.
Stoeckenius, W. Lozier, R. H., and Bogomolni, R. A. (1979). *Biochim. Biophys. Acta* **505,** 215–278.
Stryer, L., Hurley, J. B., and Fung, B. K.-K. (1981). *Curr. Top. Membr. Transp.* **15,** 93–108.
Stubbs, G. W., Saari, J. C., and Futterman, S. (1979). *J. Biol. Chem.* **254,** 8529–8533.
Szamier, R. B., and Berson, E. L. (1977). *Invest. Ophthalmol. Vis. Sci.* **16,** 947–962.
Szamier, R. B., Berson, E. L., Klein, R., and Meyers, S. (1979). *Invest. Ophthalmol. Vis. Sci.* **18,** 145–160.
Takase, S., Ong, D., and Chytil, F. (1979). *Proc. Natl. Acad. Sci.* **76,** 2204–2208.

Tsin, A. T. C., and Beatty, D. D. (1979). *Exp. Eye Res.* **29,** 15–26.
Tsin, A. T. C., and Beatty, D. D. (1980). *Exp. Eye Res.* **30,** 143–153.
Tsin, A. T. C., Bridges, C. D. B., and Alvarez, R. A. (1981). *Invest. Ophthalmol. Vis. Sci., Suppl.* **20,** 112.
Tsukida, K., Kodama, A., and Ito, M. (1977). *J. Chromatogr.* **134,** 331–336.
von Sallmann, L., Gelderman, A. H., and Laster, L. (1969). *Doc. Ophthalmol.* **26,** 451–460.
Wald, G. (1939). *J. Gen. Physiol.* **22,** 775–794.
Wald, G. (1968). *Nature (London)* **219,** 800–807.
Wang, J. K., McDowell, J. H., and Hargrave, P. A. (1980). *Biochemistry* **19,** 5111–5117.
Wiggert, B., Mizukawa, A., Kuwabara, T., and Chader, G. J. (1978). *J. Neurochem.* **30,** 653–659.
Witkovsky, P., Levine, J. S., Engbrestson, G. A., Hassin, G., and MacNichol, E. F., Jr. (1981). *Vision Res.* **21,** 867–874.
Yoshikami, S., and Nöll, G. N. (1978). *Science* **200,** 1393–1395.
Yoshizawa, T., and Shichida, Y. (1982). *In* "Methods in Enzymology, Vol. 81, Part H: Visual Pigments and Purple Membranes, I" (L. Packer, ed.), pp. 333–354. Academic Press, New York.
Young, R. W., and Bok, D. (1969). *J. Cell Biol.* **42,** 392–403.
Young, R. W., and Bok, D. (1979). *In* "The Retinal Pigment Epithelium" (K. M. Zinn and F. Marmor, eds.), pp. 103–123. Harvard Univ. Press, Cambridge, Massachusetts.
Zile, M., Bunge, E., and DeLuca, H. (1979). *J. Nutr.* **109,** 1787–1796.
Zimmerman, W. F. (1974). *Vision Res.* **14,** 795–802.
Zimmerman, W. F., Yost, M. T., and Daemen, F. J. M. (1974). *Nature (London)* **250,** 66–67.
Zimmerman, W. F., Lion, R., Daemen, F. J. M., and Bonting, S. L. (1975). *Exp. Eye Res.* **21,** 325–332.
Zimmerman, W. F., Daemen, F. J. M., and Bonting, S. L. (1976). *J. Biol. Chem.* **251,** 4700–4705.

11

Metabolism of Retinoids

CHARLES A. FROLIK
Laboratory of Chemoprevention
National Cancer Institute
National Institutes of Health
Bethesda, Maryland

I. INTRODUCTION

Although the presence of a "fat-soluble A" factor in butter fat, egg yolk, and cod liver oil was described in 1913 by McCollum and Davis, it was not until 1931 that the structure of this factor was first elucidated as retinol (A1; Karrer *et al.*, 1931). Since then, there have been numerous reports concerning the metabolism of retinol and other natural retinoids. Many of these studies have been concerned with the question as to what is the final physiologically active form of retinol involved in carrying out its functions in vision, reproduction, and differentiation. In the course of these experiments, it has become evident that in the visual process, the active retinoid is 11-*cis*-retinaldehyde (C2; Wald, 1968). However, it is still not clear as to what form of retinol is involved in the other

THE RETINOIDS, VOL. 2

ISBN 0-12-658102-9

biological functions of retinoids (growth promotion, reproduction, differentiation, and maintenance of epithelial tissue). Retinoic acid (D1) has been shown by many investigators to be a biologically active retinoid, and it has been suggested by some to be perhaps metabolically closer to the retinoid that acts within cells in supporting growth (Zile and DeLuca, 1968) and differentiation (Sporn *et al.*, 1976). This retinoid is, however, unable to completely fulfill the function of retinol in maintaining reproduction (Thompson *et al.*, 1961; Thompson, 1969) and in the visual cycle (Dowling and Wald, 1960). It therefore appears that no single natural retinoid is able to function by itself in all the varied aspects of vitamin A activity. This naturally leads to the suggestion that different metabolites of retinol are responsible for different vitamin A activities.

Although the conversion of retinol to retinaldehyde (C1) and retinoic acid has been under investigation for many years, it has only been with the introduction of high-pressure liquid chromatography (HPLC) that studies into the further metabolism of these retinoids have begun to yield actual structural data. An overall pathway for this metabolism is summarized in Figs. 1 and 2. In the course of this chapter, each of these metabolic conversions are described in greater detail with reference, where possible, to other chapters in this book that also discuss these reactions. Several reviews covering various aspects of the metabolism of the natural retinoids include Olson (1969a,b), DeLuca (1979), and Roberts and Frolik (1979).

In addition to the natural retinoids, the metabolism of synthetic retinoids has been the focus of renewed interest over the past few years because of their possible use in the prevention of cancer (see Chapter 14). In the final pages of this chapter the metabolism of several of these synthetic retinoids is discussed.

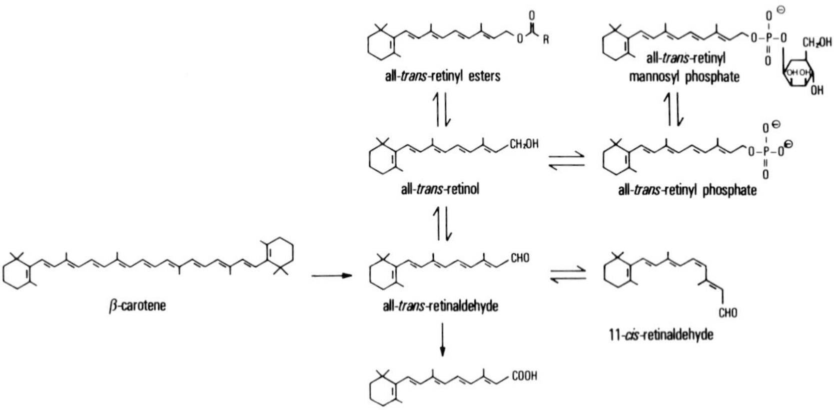

Fig. 1. Metabolic pathway for β-carotene, retinol, and retinaldehyde.

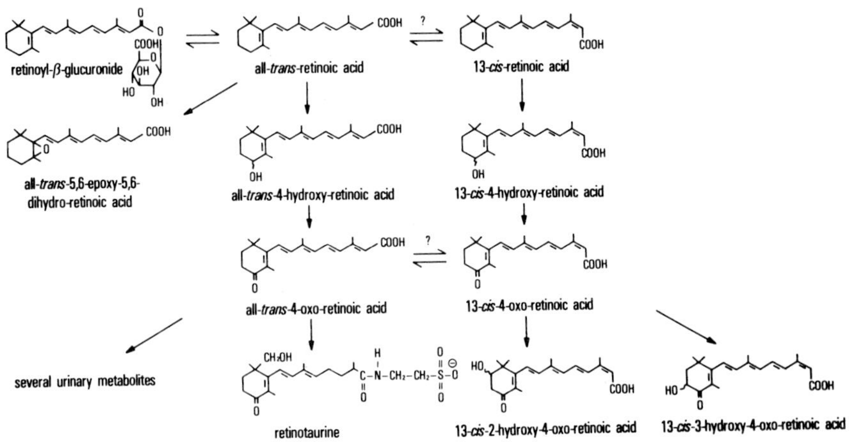

Fig. 2. Metabolic pathway for retinoic acid.

II. METABOLISM OF THE NATURAL RETINOIDS

A. Conversion of β-Carotene to Retinol

Until the introduction of chemically synthesized retinoids into food, β-carotene, a natural plant product, was the major dietary source of retinol for man. This compound is cleaved by a soluble enzyme (β-carotene 15,15′-oxygenase) found in the intestine and liver to two molecules of retinaldehyde (Fig. 3) (for a

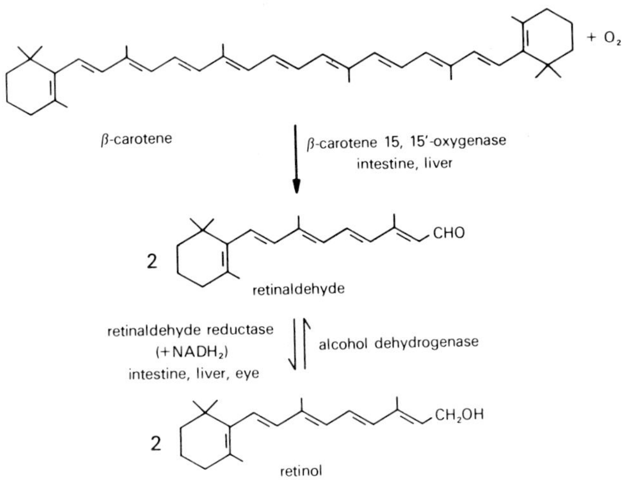

Fig. 3. Conversion of β-carotene to retinol.

review, see Goodman and Olson, 1969 and Chapter 7, this volume). The reaction requires molecular oxygen and is inhibited by sulfhydryl group inhibitors and by chelators of ferrous ion. Once the retinaldehyde is formed, it is reduced in the intestinal mucosa to retinol (Glover *et al.*, 1948; Olson, 1961; Huang and Goodman, 1965; Fidge *et al.*, 1968). The enzyme involved in this conversion (retinaldehyde reductase) can be found not only in the intestine but also in the liver (Zachman and Olson, 1961) and eye (Wald, 1950; Blaner and Churchich, 1980). The reduction requires a reduced pyridine nucleotide as a cofactor. Alcohol dehydrogenase can also convert retinol to retinaldehyde (Bliss, 1951; Zachman and Olson, 1961). It has been postulated that in rat liver both the retinaldehyde reductase and the alcohol dehydrogenase activity reside in the same enzyme (Zachman and Olson, 1961). For a further discussion of the interconversion of retinol and retinaldehyde the reader is referred to Chapters 7 and 10.

B. Nonoxidative Metabolism of Retinol

1. Formation of Retinyl Esters

Dietary retinol is quickly esterified in the intestinal mucosa with long-chain fatty acids (Fig. 4) (Olson, 1961; Murthy *et al.*, 1961; Huang and Goodman, 1965; Goodman *et al.*, 1966) and is incorporated into chylomicra and transported via the lymph to the liver. In the liver, the retinyl esters are hydrolyzed and then reesterified for storage (Goodman *et al.*, 1965; Lawrence *et al.*, 1966). Retinyl palmitate is the major ester in both the lymph (Huang and Goodman, 1965; Goodman *et al.*, 1966) and in the liver (Mahadevan and Ganguly, 1961), with

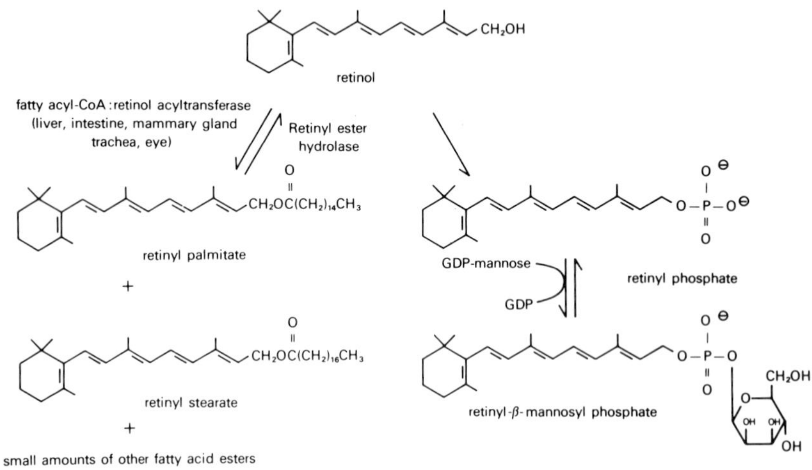

Fig. 4. Nonoxidative metabolism of retinol.

saturated esters (retinyl palmitate plus stearate) comprising two-thirds to three-fourths of the retinyl esters. The enzyme responsible for retinol esterification in the liver is found in the microsomal fraction (Andrews and Futterman, 1964; Ross, 1982a) and appears to qualify as a fatty acyl-CoA:retinol acyltransferase (Ross, 1982a). A similar enzyme has been demonstrated in mammary gland microsomes (Ross, 1982b) and in the intestine (Helgerud *et al.*, 1982). In addition, esterification of retinol has been reported to occur in the eye (Krinsky, 1958; Andrews and Futterman, 1964) and in the tracheal epithelium (Frolik *et al.*, 1981a).

The enzymatic hydrolysis of retinyl esters in liver has been examined in retinol-depleted, retinoic acid-supplemented rats (Harrison *et al.*, 1979) as well as in normal rats (Prystowsky *et al.*, 1981). In both studies, the enzyme retinyl palmitate hydrolase showed an unusual subcellular distribution, being found in the nuclear, mitochondrial, and supernatant fractions and displayed a large variability in activity from rat to rat. For additional information regarding the esterification and hydrolysis of retinol the reader is again referred to Chapters 7 and 10.

2. *Formation of Retinyl Phosphate*

It has been postulated that retinol may be involved in the synthesis of glycoproteins (for a recent review, see De Luca *et al.*, 1979a,b; Wolf *et al.*, 1979) in a manner similar to dolichyl mannosyl phosphate (Behrens and Leloir, 1970). Indeed, retinyl phosphate (A12; Fig. 4) can be synthesized *in vitro* using either cell extracts or cells from liver (De Luca *et al.*, 1973), intestinal epithelium (Frot-Coutaz *et al.*, 1976), and epidermis (Adamo *et al.*, 1979a). This formation of retinyl phosphate shows an apparent dependence on the presence of ATP (De Luca *et al.*, 1973). The *in vitro* synthesis of retinyl pyrophosphate by rat thyroid homogenates has also been reported (Gaede and Rodriguez, 1973). The further conversion of retinyl phosphate to retinyl mannosyl phosphate (A13; Fig. 4) has likewise been demonstrated *in vitro* in rat liver cells and requires guanosine diphosphate mannose (De Luca *et al.*, 1973; Rosso *et al.*, 1975). The enzyme responsible for this conversion is found in the microsomal fraction of rat liver (Shidoji and De Luca, 1981). This fraction is also involved in the synthesis of dolichyl mannosyl phosphate and, although different enzyme activities have been reported for the two substrates (Shidoji *et al.*, 1981), two different enzymes have not yet been isolated and purified. Finally, [^{3}H,^{14}C]retinyl mannosyl phosphate has been isolated directly from the liver and intestine of rats given [^{3}H]retinol and [^{14}C]mannose (De Luca *et al.*, 1973; Masushige *et al.*, 1978). In addition to retinyl mannosyl phosphate, retinyl galactosyl phosphate has been reported to be synthesized from retinol and UDP-galactose by a particulate enzyme fraction from mouse mastocytoma (Helting and Peterson, 1972; Peterson *et al.*, 1976). The physiological importance of these glycosylated derivatives of retinol is at this time still unknown (see Chapter 12). Only a very small percentage of the retinol present in the liver of a rat appears as retinyl mannosyl phosphate (Masushige *et al.*, 1978). This therefore does not appear to be a quantitatively major pathway for the further metabolism of retinol.

3. *Dehydration of Retinol*

When spontaneously transformed mouse fibroblasts (BALB/c 3T12-3) are grown in the presence of certain retinoids, the cells display an increased adhesion and a change in their morphology to that more characteristic of a normal phenotype (Adamo *et al.*, 1979b). These cells will metabolize [^{14}C]retinol to an as yet unidentified metabolite that is then converted to anhydroretinol (A8; Bhat *et al.*, 1979). The anhydroretinol appears to be the end product of retinol metabolism in this system. While cultured BALB/c 3T3 mouse fibroblasts, as well as primary cultures of mouse dermal fibroblasts, also form anhydroretinol from retinol, hamster intestinal mucosa cells do not. The significance of this reaction *in vivo* still remains to be determined.

C. Oxidative Metabolism of Retinol

1. *Conversion to Retinoic Acid*

As mentioned in Section II,A, retinol can be reversibly oxidized to retinaldehyde. In the eye, the all-*trans*-retinaldehyde can isomerize to 11-*cis*-retinaldehyde as part of the visual cycle (see Chapter 10). In many other tissues, the further irreversible oxidation of retinaldehyde to retinoic acid can occur (Fig. 2). Retinoic acid was first chemically synthesized in 1946 by Arens and van Dorp (1946a), who also demonstrated that this retinoid was active in supporting growth of retinoid-deficient rats (Arens and van Dorp, 1946a; van Dorp and Arens, 1946) and could not be converted back to retinol (Arens and van Dorp, 1946b). It took more than 15 years, however, to demonstrate that retinoic acid was a natural metabolite of retinol. In 1961 Dmitrovskii demonstrated that retinaldehyde could be irreversibly oxidized to retinoic acid by a hog liver aldehyde oxidase preparation and that a compound which behaved as retinoic acid on an alumina column could be found in rat intestine (but not liver) 24 h after the animals had been given an oral dose of 10–60 mg of retinaldehyde. The compound, however, was not well characterized. This initial study was quickly followed by three additional reports concerning the *in vitro* oxidation of retinaldehyde to retinoic acid by an NAD-linked enzyme found in the liver, kidney, and intestine of rats (Mahadevan *et al.*, 1962) and rabbits (Futterman, 1962; Elder and Topper, 1962).

Demonstration of the occurrence of retinoic acid in the living animal, however, remained elusive. Krishnamurthy *et al.* (1963) failed to detect the acid in the chick intestine after oral administration of retinaldehyde, and Wright (1961) also was unable to find retinoic acid in rat tissues following an intraperitoneal injection of [^{14}C]retinyl acetate (A3). In 1964 Dunagin *et al.* convincingly demonstrated the presence of retinoic acid in rat bile, collected 6 h after an intravenous 3-mg dose of [^{14}C]retinaldehyde. During the next few years several manuscripts were published describing the *in vivo* production of retinoic acid after milligram doses of retinaldehyde (Deshmukh *et al.*, 1965; Crain *et al.*, 1967). Additionally, retinoic acid has been shown to be a natural metabolite of both retinol (Emerick *et al.*, 1967) and retinyl acetate (Kleiner-Bössaler and DeLuca, 1971; Ito *et al.*, 1974a) after the administration of physiological levels of the parent compound. Retinoic acid has also been shown to be detectable at nanogram levels in normal human plasma (De Ruyter *et al.*, 1979; Vane *et al.*, 1982b; De Leenheer *et al.*, 1982). Finally, it has been demonstrated that retinoic acid can be synthesized not only in the liver, kidney, and intestine, but it can be produced also from retinyl acetate in an *in vitro* vitamin A target tissue system the hamster trachea (Frolik *et al.*, 1981a). Therefore, from the data that have accumulated over the past 22 years, it is clearly evident that retinoic acid is a natural physiological metabolite of retinol and retinaldehyde.

Although it is now clear that retinoic acid is a naturally occurring retinoid, the question still remains as to whether this compound is an obligatory metabolite of retinol. Initial experiments seemed to indicate that the acid was not as active as the alcohol in supporting growth of retinoid-deficient rats (Ames, 1965). However, Zile and DeLuca (1968) demonstrated that this lower biological activity probably was due to the rapid metabolism of retinoic acid and that when retinoic acid was administered to the rat in frequent small doses, it was just as active as retinyl acetate in supporting growth. A number of *in vitro* and *in vivo* assay systems have indicated that retinoic acid is more active than retinyl acetate or retinol in controlling differentiation and in reversing changes induced in various experimental models by carcinogens (see Chapter 5, Vol. 1). Whether retinoic acid is closer metabolically than retinol to the biologically active form of vitamin A and whether retinol itself can also function as an active retinoid in differentiation without being metabolized to retinoic acid, are questions that still require further investigation.

2. *Urinary Metabolites*

The overall metabolism of retinol has been examined by several investigators with emphasis placed on the urinary and fecal or biliary excretion of radioactive metabolites. During the first 24–48 h after administration to rats of either [^{14}C]retinol (Wolf *et al.*, 1957; Garbers *et al.*, 1960; Goodman *et al.*, 1965; Sundaresan and Sundaresan, 1975; Sundaresan, 1977) or [^{3}H]retinyl acetate (Varma and Beaton, 1972), 4–15% of the dose was recovered in the urine with the majority being excreted during the first 24 h. The amount of the dose found in the urine depends on the position of the radioactive label in the starting retinoid.

Roberts and DeLuca (1967) have reported that during a 4-day period, 27% of a dose of [6,7-^{14}C]retinyl acetate was recovered in the urine of rats while during the same time, only 8–15% of a [15-^{14}C]retinyl acetate dose was found. This suggests that decarboxylation of retinol is occurring, most likely passing through retinoic acid as an intermediate. In support of this hypothesis, 9–10% of an administered dose of [15-^{14}C]retinyl acetate was recovered as expired $^{14}CO_2$. Although some of the urinary retinol metabolites have been partially characterized by Wolf *et al.* (1957), none has yet been ascribed a chemical structure. The urinary metabolites of retinol are mainly water-soluble (Goodman *et al.*, 1965; Varma and Beaton, 1972) and no free retinol or retinyl esters have been detected (Wolf *et al.* 1957; Varma and Beaton, 1972). A protein-bound form of retinol, or of a metabolite of retinol that is different from retinoic acid, was reported to be excreted into the urine of rats administered a large (23.1 mg) dose of retinol (Clark *et al.*, 1975). Again, however, this compound has not been further characterized. Finally, it has been suggested that determination of the urinary metabolites of retinol might be a potential index for assessment of the vitamin A nutritional status in humans (Varma and Beaton, 1972). However, Rietz *et al.* (1974) did not find this to be a suitable parameter for measuring vitamin A status.

3. *Biliary and Fecal Metabolites*

As is the case with the urinary metabolites, the biliary retinol metabolites also have not been extensively characterized. Within 24 h after administration, 18–25% of a 10–26 μg dose of retinol can be found in the bile of rats (Zachman and Olson, 1964; Zachman *et al.*, 1966a; Fidge *et al.*, 1968). Similarly for retinaldehyde, 23–35% of a 2- to 3.5-mg dose (Dunagin *et al.*, 1964, 1966; Zachman *et al.*, 1966a) and 15–20% of an 8-μg dose (Fidge *et al.*, 1968) is excreted into rat bile 6–24 h after administration. As mentioned previously, a small percentage of the biliary retinol or retinaldehyde metabolites have been identified as free or conjugated retinoic acid (0.4–6% of the administered dose) (Dunagin *et al.*, 1964, 1966; Emerick *et al.*, 1967), while little (0.2% of the administered dose, Emerick *et al.*, 1967) or no (Zachman and Olson, 1964, 1965) free retinol can be detected. Lippel and Olson (1968a) have reported the presence of an *O*-ether derivative of retinol, retinyl β-glucosiduronate, in the bile of rats given a 3-mg dose of retinol. They postulated that at physiological levels of retinol, a mixture of retinoyl and retinyl conjugates are normally excreted. The biliary metabolites have often been separated into three groups based on their chromatographic properties on anion-exchange columns: those less polar than retinoic acid, those with properties similar to retinoic acid, and those that are more polar than retinoic acid (Zachman and Olson, 1964; Zachman *et al.*, 1966a,b; Dunagin *et al.*, 1964, 1966). However, except for the presence of retinol, retinoic acid, and their conjugates, the structures of these compounds have not been clarified.

The majority of the fecal metabolites probably originate from the bile. As is also the case with urinary and biliary metabolites, the amount of the administered dose found in the feces varies depending on the position of the label, the mode of administration, and on the amount administered. Values from 0.3% of a 10-μg dose in 24 h (Sundaresan and Sundaresan, 1975) to 26% of a 2-μg dose in 5 days (Roberts and DeLuca, 1967) have been reported. There appears to be an initial delay in the fecal excretion of metabolites after injection of retinol (Olson, 1968; Varma and Beaton, 1972). Again, however, little has been done to obtain structural data on these compounds.

It should be mentioned that most structural studies on retinoid metabolites have been performed after administration of retinoic acid to the animal. Because it is likely that at least some retinol is metabolized through retinoic acid, many of the metabolites identified for retinoic acid are probably also metabolites of retinol. These compounds, derived from retinoic acid, are discussed in Section II,D,2.

D. Metabolism of Retinoic Acid

1. *Biologically Active Metabolites*

Much of the research into the metabolism of the natural retinoids has centered on the search for new metabolites of retinol or retinoic acid that are more active than the parent compound in controlling epithelial differentiation. Retinoic acid itself has been shown to be active in supporting growth in the chick (Krishnamurthy *et al.*, 1963; DeMan *et al.*, 1964) and in the rat (Arens and van Dorp, 1946a; van Dorp and Arens, 1946; Sharman, 1949; Dowling and Wald, 1960; Redfearn, 1960; Dowling, 1961; Malathi *et al.*, 1963; Zile and DeLuca, 1968). However, because it cannot be converted back to retinaldehyde or retinol (Arens and van Dorp, 1946b; Sharman, 1949; Dowlings and Wald, 1960; Redfearn, 1960; Dowling, 1961), retinoic acid fails to satisfy the visual and reproductive requirements for retinol (Dowling and Wald, 1960; Thompson *et al.*, 1964).

The existence of biologically active metabolites of retinoic acid has been reported. Krishnamurthy *et al.* (1963) detected a fat-soluble metabolite of retinoic acid, which displayed biological activity in a rat curative assay, in the liver of chicks administered a 10-mg oral dose of the parent retinoid. Similarly, Wolf *et al.* (1963) reported that an intestinal metabolite, isolated from retinoid-deficient rats injected with labeled retinoic acid, was active in restoring to normal levels the mucopolysaccharide biosynthesis in retinoid-deficient rat intestinal cell-free particle suspensions. The same laboratory also described a decarboxylated metabolite of both retinol and retinoic acid that was isolated from the intestine of retinoid-deficient rats administered retinol or retinoic acid. This compound, which appeared to have both carboxyl and hydroxyl functional groups, was biologically active in a growth assay (Yagishita *et al.*, 1964). In

1965 Zile and DeLuca reported the existence of a metabolite of retinoic acid that was found in rat liver and that was able to support the growth of retinoid-deficient rats. This metabolite was later identified as 13-*cis*-retinoic acid (E6) and most, if not all, of it appeared to be generated from all-*trans*-retinoic acid by the extraction procedure (Emerick *et al.*, 1967). Sundaresan (1966) has likewise reported that an acidic metabolite of retinoic acid extracted from normal rat liver was able to restore to near normal the activity of a sulfur-activating enzyme (ATP-sulfurylase) whose activity was depressed in retinoid-deficient rat liver supernatants. Except for the 13-*cis*-retinoic acid, none of these metabolites have been characterized further.

The compound 5,6-epoxy-5,6-dihydroretinoic acid (D6) has been isolated from the intestinal mucosa of retinoid-deficient rats given retinoic acid (McCormick *et al.*, 1978; Napoli *et al.*, 1978). This compound was claimed to be highly biologically active (McCormick *et al.*, 1978). However, both earlier work and subsequent findings have not supported this idea. 5,6-Epoxy-5,6-dihydroretinoic acid was synthesized in the mid-1960s (Morgan and Thompson, 1966; John *et al.*, 1967), and it was reported to be 157% as active as retinyl acetate in supporting growth in rats (John *et al.*, 1967). This value was later revised to 80% (Mallia *et al.*, 1970). Later data, however, showed even a lower biological activity for this retinoid. In a hamster tracheal organ culture assay, 5,6-epoxy-5,6-dihydroretinoic acid had only 1% the activity of all-*trans*-retinoic acid in maintaining epithelial differentiation (Newton *et al.*, 1980). Similarly, when the retinoid was tested for its growth-promoting activity in retinoid-deficient rats, it was found to be only 0.5% as active as all-*trans*-retinoic acid (Zile *et al.*, 1980b). Finally, in a vaginal-smear assay, which measures reversal of keratinization in the vaginal epithelium of retinoid-deficient ovariectomized rats, the 5,6-epoxy compound was about 200 times less active then all-*trans*-retinoic acid (Sietsema and DeLuca, 1982b). Because in all these assays the 5,6-epoxy-5,6-dihydroretinoic acid was introduced extracellularly, the possibility remains that in order for the epoxide to be active, it must be synthesized within the cell. That this is not the case is apparent from results published by Sietsema and DeLuca (1982a). They have demonstrated that blocking the epoxidation of retinoic acid *in vivo* does not inhibit the function of retinoic acid in maintaining differentiation of the vaginal epithelium or in promoting growth of retinoid-deficient rats. It can therefore be stated that epoxidation of retinoic acid is not necessary for expression of the biological activity of the parent acid.

Finally, the synthesis of a mannosyl retinoid phosphate compound from retinoic acid in spontaneously transformed mouse 3T12 cells has been reported (Bhat and De Luca, 1981). The structure of this metabolite, its physiological significance, and its biological activity are still unknown.

In conclusion, although much activity has centered around the search for an active metabolite of retinoic acid, no such metabolite has yet been detected and characterized.

2. *Inactivation Reactions*

a. INTRODUCTION. The metabolism of retinoic acid has been shown to be much more rapid than that observed for retinol, with little or no storage taking place in the animal (Roberts and DeLuca, 1967; Zachman *et al.*, 1966a; Sundaresan and Bhagavan, 1971; Smith *et al.*, 1973; Ito *et al.*, 1974b). The initial investigations into the metabolism of retinoic acid concentrated on those compounds that were found in the bile and urine. Within 24 h after administration of a physiological or pharmacological dose of retinoic acid to either retinoid-deficient or normal rats, 55–85% of the dose was excreted into the bile (Zachman *et al.*, 1966a; Zile *et al.*, 1980a, 1982a; Swanson *et al.*, 1981b; Skare *et al.*, 1982; Skare and DeLuca, 1983) while during a 5-day period, 82–95% of the dose was found in the bile or feces (Zachman *et al.*, 1966a; Roberts and DeLuca, 1967; Hänni and Bigler, 1977). Although the total percent of the dose detected in the bile during the first 24 h is independent of the retinoid status of the animal, the rate of excretion of these metabolites has been found to vary (Skare and DeLuca, 1983). For example, it takes only 3 h for vitamin A-normal rats to excrete 30% of a physiological (2 μg) dose of all-*trans*-retinoic acid into the bile, whereas retinoid-deficient rats take 9 h to excrete this same percentage. By 24 h, however, the bile from both groups of animals contains an identical percentage of the dose (Skare and DeLuca, 1983).

In the urine the amount of retinoic acid metabolites detected after a 10- to 17-μg dose of retinoic acid has been shown to vary between 30 and 60% of the dose (Geison and Johnson, 1969; Sundaresan and Bhagavan, 1971; Sundaresan and Sundaresan, 1973; Ito *et al.*, 1974b; Swanson *et al.*, 1981b). Because some of the urinary metabolites appear to be derived from biliary metabolites (Swanson *et al.*, 1981b), apparently through the process of enterohepatic circulation (Zachman *et al.*, 1966a, Swanson *et al.*, 1981b), those experiments in which urine was collected from biliary cannulated rats gave a lower percentage of the dose excreted as urinary metabolites (10–22% of the dose; Swanson *et al.*, 1981b; Skare *et al.*, 1982) than in those experiments where the urine was collected from intact animals (19–60% of the dose; Emerick *et al.*, 1967; Geison and Johnson, 1969; Sundaresan and Bhagavan, 1971; Sundaresan and Sundaresan, 1973; Ito *et al.*, 1974b; Swanson *et al.*, 1981b). The percentage of the dose appearing as urinary metabolites has also been found to decrease with an increase in dose (Swanson *et al.*, 1981b) and to vary depending on the position of the radioactive label in the parent compound. With [6,7-^{14}C]retinoic acid, 38–50% of the dose appeared in the urine, whereas with [15-^{14}C]retinoic acid only 20–30% of the dose was found in the urine (Roberts and DeLuca, 1967; Geison and Johnson 1969; Sundaresan and Bhagavan, 1971; Sundaresan and Sundaresan, 1973).

A number of studies (Redfern, 1960; Geison and Johnson, 1969; Ito *et al.*, 1974b; Wang *et al.*, 1980; Kalin *et al.*, 1981; Swanson *et al.*, 1981b; Napoli and

McCormick, 1981; Silva and DeLuca, 1982) have examined the overall tissue distribution and metabolism of retinoic acid in the rat and mouse. Each laboratory has utilized different extraction and separation methods, making comparison of metabolic profiles extremely difficult. Several general statements, however, emerge from these experiments. First, the liver, kidney, and small intestine contain the highest levels of retinoic acid and its metabolites after administration of the labeled acid, although radioactive compounds can be found in all tissues studied. Second, at early times after administration of the dose, free retinoic acid is the major retinoid present in most tissues studied except for the small intestine where polar metabolites predominate [although many of these intestinal metabolites may originate from the bile (Zile *et al.*, 1982b)]. At later times, the amount of free retinoic acid in all tissues decreases and the proportion of polar products increases. Finally, the elimination of retinoic acid from tissues and serum follows non-first-order kinetics. The degree of deviation from first-order kinetics is dose-dependent and can be diminished by lowering the dose.

Retinoic acid undergoes a variety of reactions including conjugation with glucuronic acid (Dunagin *et al.*, 1964, 1965; Zile, *et al.*, 1980a; Swanson *et al.*, 1981b), decarboxylation (Roberts and De Luca, 1967; Sundaresan and Theriault, 1968; Lippel *et al.*, 1970; Nelson *et al.*, 1971), oxidation at the 4 position of the cyclohexenyl ring (Dmitrovskii *et al.*, 1977; Hänni and Bigler, 1977; Frolik *et al.*, 1979), epoxidation (McCormick *et al.*, 1978), isomerization (Zile *et al.*, 1967; Frolik, 1981), and esterification (Fidge *et al.*, 1968; Smith *et al.*, 1973; Lippel and Olson, 1968b; Ito *et al.*, 1974b). In the following text, these reactions are examined in greater detail.

b. CONJUGATION. The first metabolite of retinoic acid to be identified was retinoyl β-glucuronide (E24; Fig. 2) (Dunagin *et al.*, 1965, 1966; Zachman *et al.*, 1966a). It was first detected in the bile of rats but has since been shown to occur in the urine (Nath and Olson, 1967) and intestinal mucosa (Zile *et al.*, 1982b) as well. In the intestine of a bile duct-cannulated, retinoid-deficient rat, retinoyl β-glucuronide represents up to 48% of the metabolites present 2 h after a 12.7-μg dose of all-*trans*-retinoic acid. Whether this compound has an important role in the action of retinoic acid or whether it is merely an excretory product is still unclear. In this regard, the glucuronide has been shown to be as active as retinoic acid in promoting growth of retinoid-deficient rats (Nath and Olson, 1967) and in causing epithelial differentiation in a vaginal-smear assay (Sietsema and DeLuca, 1982b), but it is not certain whether it first has to be hydrolyzed to the free acid before it can act. Although the glucuronide was originally reported to be the major biliary metabolite of retinoic acid (Dunagin *et al.*, 1965; Lippel and Olson, 1968b), it has been demonstrated that at a physiological dose, this compound represents only 10–15% of the retinoic acid metabolites present in bile. This percentage, however, increases in proportion to the dosage level of the retinoic acid (Zile *et al.*, 1980a; Swanson *et al.*, 1981b). The rate of excretion of retinoyl β-glucuronide into the bile of rats given a small dose of retinoic acid varies depending on the retinoid status of the animal. In vitamin A-normal rats, the glucuronide reaches a maximum concentration 2 h after administration of the dose and then declines rapidly, while in retinoid-deficient rats, the concentration of retinoyl β-glucuronide is relatively constant over an 11-h period (Skare and DeLuca, 1983).

Other retinoids identified in the bile include retinoic acid itself (0.2–10% of the biliary metabolites; Zile *et al.*, 1980a, 1982a; Swanson *et al.*, 1981b; Skare and DeLuca, 1983) and retinotaurine (Fig. 2; 10% of the biliary metabolites; Skare *et al.*, 1982). Like retinoyl β-glucuronide, the concentration of retinotaurine in the bile differs between vitamin A-normal and retinoid-deficient rats, being approximately equal to the glucuronide levels in normal rats but about fivefold lower then the glucuronide in deficient animals (Skare and DeLuca, 1983). The remaining 65–80% of the dose comprises metabolites that are more polar than retinoyl β-glucuronide and that still remain to be identified. The appearance of these more polar metabolites in the bile is again dependent on the retinoid status of the animal as well as on the isomer of retinoic acid injected into the animal. At early time points after administration of the acid, the concentration of the polar metabolites is much lower for retinoid-deficient rats than that observed for normal rats (Skare and DeLuca, 1983). Similarly, a greater portion of a dose of all-*trans*-retinoic acid is metabolized to these more-polar compounds when compared to 13-*cis*-retinoic acid (Frolik *et al.*, 1981b).

Several metabolites of retinoic acid have been isolated from the feces of normal rats administered a large dose of the parent acid (Hänni and Bigler, 1977). Presumably these compounds originated from the bile; their structures are shown in Fig. 5. The importance of the appearance of the hydroxylated compounds after a physiological dose of retinoic acid is still uncertain. However, more is said about the 4-oxoretinoic acid (D4) in a later part of this section.

The urinary metabolites have not been studied as extensively. It is apparent that little if any free retinoic acid occurs in the urine (Emerick *et al.*, 1967; Ito *et al.*, 1974b; Hänni *et al.*, 1976; Frolik *et al.*, 1978a; Kalin *et al.*, 1981). Retinoyl β-glucuronide is present in the urine (Nath and Olson, 1967; Sundaresan and Bhagavan, 1971), although whether it occurs under physiological conditions is uncertain. Rietz *et al.* (1974) have postulated the structure of several metabolites that were isolated from the urine of rats administered high doses of retinoic acid (Fig. 6A), while Hänni and co-workers (1976) have structurally characterized three urinary metabolites (Fig. 6B) that were again obtained after the administration of pharmacological (27 mg) doses of retinoic acid to rats. It is interesting to note that in all seven proposed metabolites, there is a keto group at C-4 of the cyclohexenyl ring. The reactions that lead to oxidation at this position are discussed in greater detail later.

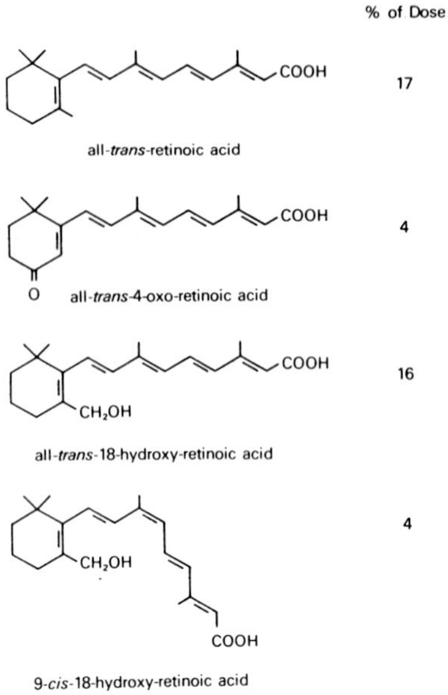

Fig. 5. Metabolites of retinoic acid in rat feces after a pharmacological dose of retinoic acid. Vitamin A-normal rats were administered an intraperitoneal dose of 27.2 mg all-*trans*-retinoic acid. Feces were collected over a 5-day period and extracted. The retinoid metabolites were purified to homogeneity and identified by mass spectrometry and proton magnetic resonance spectroscopy (Hänni and Bigler, 1977).

c. DECARBOXYLATION. It should be pointed out that several of the biliary and urinary metabolites already mentioned have a shortened side chain when compared to the parent acid. That retinoic acid may undergo decarboxylation was suggested by Yagishita *et al.* (1964), although it had been demonstrated 7 years earlier that 5% of a dose of [14-^{14}C]retinol appeared in expired $^{14}CO_2$ (Wolf *et al.*, 1957). Roberts and DeLuca (1967) later reported that 35% of a 14.5-μg dose of [15-^{14}C]retinoic acid given to retinoic acid supplemented rats was recovered as $^{14}CO_2$, while Sundaresan and Therriault (1968) found that 14% of an injected dose of the same ^{14}C-labeled compound given to retinoid-deficient rats appeared as $^{14}CO_2$. This decarboxylation reaction has been investigated in several *in vitro* systems. Roberts and DeLuca (1968a,b, 1969) observed that tissue slices and microsomes prepared from the liver and kidney of rats were able to decarboxylate retinoic acid by a free-radical mechanism similar to lipid peroxidation. However, the *in vivo* reaction apparently does not occur, to any great extent, via this mechanism (Roberts and DeLuca, 1969; DeLuca and Roberts, 1969). Lippel *et al.* (1970) reported that rat liver mitochondria do not extensively decarboxylate retinoic acid. Finally, the decarboxylation of retinoic acid has been shown to occur in a horseradish peroxidase system (Nelson *et al.*, 1971;

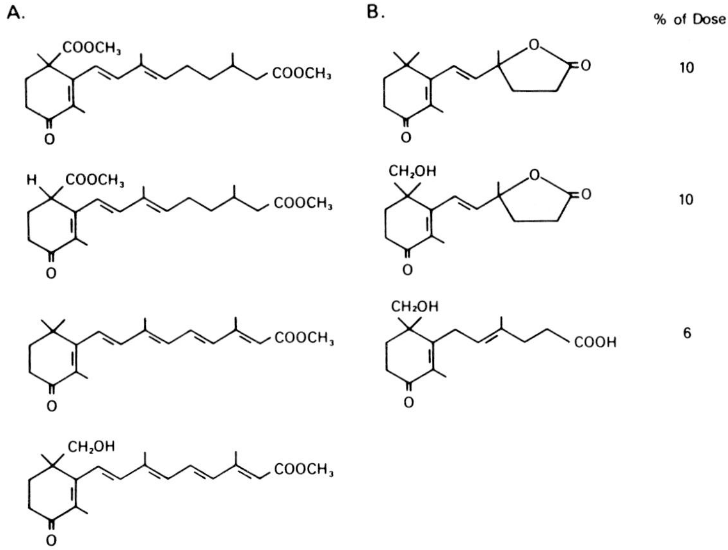

Fig. 6. Urinary metabolites of all-*trans*-retinoic acid. (A) Metabolites proposed by Rietz *et al.* (1974) to be present in the urine of rats after a pharmacological dose of all-*trans*-retinoic acid. (Metabolites were methylated with diazomethane prior to partial characterization.) (B) Metabolites identified by Hänni *et al.* (1976) in the urine of rats administered an intraperitoneal dose (27 mg) of all-*trans*-retinoic acid.

McKenzie and Nelson, 1979) yielding a 4-oxo-C_{19} aldehyde (Rockley *et al.*, 1980) (Fig. 7). However, the physiological significance of this retinoid remains unknown. The most convincing evidence that decarboxylation of retinoic acid does indeed occur *in vivo* is the structural elucidation of several chain-shortened biliary (Skare *et al.*, 1982) and urinary metabolites (Hänni *et al.*, 1976).

d. OXIDATION. Several of the compounds already discussed have been oxidized at C-4 of the cyclohexenyl ring to yield 4-oxoretinoic acid or its derivatives (Figs. 2, 5–7). Similar to retinoic acid, 4-oxoretinoic acid was chemically synthesized (Surekha Rao *et al.*, 1972) several years before it was found to occur naturally *in vivo*, (Hänni and Bigler, 1977; Dmitrovskii *et al.*, 1977; Frolik *et*

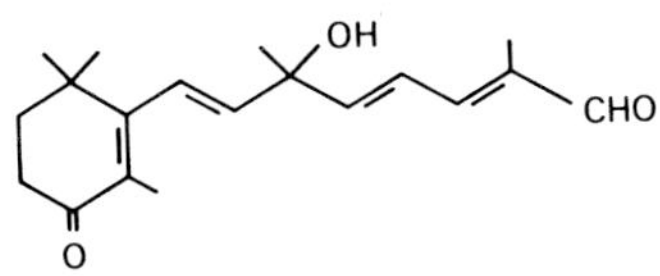

Fig. 7. Decarboxylated metabolite of all-*trans*-retinoic acid synthesized *in vitro* in a horseradish peroxidase system (Rockley *et al.*, 1980).

al., 1979). In 1977 Dmitrovskii *et al.* reported the characterization of 4-oxoretinyl palmitate that had been purified from the livers of rats administered a large (25 mg) dose of retinol plus retinyl acetate. They also demonstrated the *in vitro* conversion of 4-oxoretinol to 4-oxoretinoic acid. In the same year, Hänni and Bigler (1977) detected the presence of 4-oxoretinoic acid in the feces of normal rats administered a pharmacological (27.2 mg) dose of retinoic acid. This compound represented 4% of the injected dose. Finally, 4-oxoretinoic acid has been shown to be a metabolite of low (1.5–15 μg) levels of retinoic acid in the hamster (Frolik *et al.*, 1978c, 1980). It is also interesting to note that in *Phycomyces,* β-carotene and retinaldehyde are intermediates in the biosynthesis of trisporic acids, compounds that have a keto moity at the 4 position similar to 4-oxoretinoic acid (Austin *et al.*, 1970).

The formation and metabolism of 4-hydroxyretinoic acid (D3) and 4-oxoretinoic acid have been studied using several *in vitro* techniques (Roberts and Frolik, 1979; Frolik, 1981; Roberts, 1981). These two retinoids have been identified as *in vitro* metabolites of retinoic acid in hamster tracheal organ cultures (Frolik *et al.*, 1978c, 1979), a system which is responsive to nanomolar concentrations of retinoids *in vitro* (Clamon *et al.*, 1974; Sporn *et al.*, 1974). Using several different high-pressure liquid chromatography methods, these *in vitro*-formed metabolites were shown to cochromatograph with *in vivo* metabolites obtained from hamsters given a 1.5-μg dose of retinoic acid (Frolik *et al.*, 1978c). The *in vitro* conversion of retinoic acid to the 4-hydroxy and 4-oxo compounds has also been demonstrated in hamster liver and intestinal homogenates and microsomal preparations (Frolik *et al.*, 1979; Roberts *et al.*, 1979a).

The enzymatic activity required for the synthesis of 4-hydroxyretinoic acid from retinoic acid is located in the microsomal fraction of the hamster liver, intestinal mucosa, and other tissues, requires NADPH and oxygen, and is inhibited by carbon monoxide, all suggesting that the enzyme involved belongs to a class of mixed function oxidases containing cytochrome *P*-450 (Roberts *et al.*, 1979a, 1980). Like other drug-hydroxylating enzymes, the activity of this enzyme could be induced by its substrate retinoic acid. This induction appeared to be fairly specific for retinoids since other classical *P*-450 inducers, such as phenobarbital and 3-methycholanthrene, did not cause a significant increase in retinoid metabolism (Roberts *et al.*, 1979a,b, 1980). The further oxidation of 4-hydroxy- to 4-oxoretinoic acid has been characterized as a dehydrogenase reaction because it required NAD, was not inhibited by carbon monoxide, and did not need oxygen. This reaction was not inducible by retinoid pretreatment (Roberts *et al.*, 1980). Finally, the metabolism of 4-oxoretinoic acid to more polar compounds again appeared to utilize a *P*-450-type enzyme since it required NADPH and oxygen and was inhibited by carbon monoxide (Roberts *et al.*, 1980). This suggests that these more polar metabolites may be hydroxylated derivatives of 4-oxoretinoic acid. It is interesting to note that the induction of retinoic acid metabolism was observed both *in vivo* and *in vitro* and that two target tissues, the testis and trachea, were not inducible.

Although the *in vitro* and *in vivo* metabolism of all-*trans*-retinoic acid to 4-hydroxy- and 4-oxoretinoic acid has been shown to occur in the rat as well as the hamster (Hänni and Bigler, 1977; Vane *et al.*, 1982a; Sato and Lieber, 1982; Silva and DeLuca, 1982; Zile *et al.*, 1982b), the relative amounts of the two metabolites were significantly lower in the rat than that observed for the hamster. Whether this is due to a slower conversion of retinoic acid to 4-hydroxyretinoic acid or to a faster metabolism of the 4-oxoretinoic acid to more polar compounds is not yet known. It is clear, however, that oxidation at C-4 of the cyclohexenyl ring does occur in rats since several urinary, fecal, and biliary metabolites have been isolated from this species with a keto group at that position (Hänni *et al.*, 1976; Hänni and Bigler, 1977; Skare *et al.*, 1982). Isomerization of retinoic acid to yield a mixture of all-*trans*- and 13-*cis*-4-oxoretinoic acid may also be a physiologically important reaction and is discussed further in Section II,D,3.

Several lines of evidence indicate that hydroxylation at C-4 may be the first step in the elimination pathway of retinoic acid. First, 4-hydroxy- and 4-oxo-retinoic acid have less than one-tenth the biological activity of the parent acid in causing epithelial differentiation in a tracheal organ-culture assay (Frolik *et al.*, 1979) and in promoting growth in retinoid-deficient rats (Surekha Rao *et al.*, 1972). Second, the formation of these two retinoids is maximal in both retinoic acid-induced and vitamin A-normal animals and is accompanied by an increase in the rate of elimination of retinoic acid metabolites from the plasma and tissues (Roberts *et al.*, 1979b). Finally, as mentioned previously, several urinary, biliary, and fecal metabolites have been identified that contain the 4-oxo moiety as a common element (Hänni *et al.*, 1976; Hänni and Bigler, 1977; Skare *et al.*, 1982). Therefore, oxidative attack at the 4 position of the cyclohexenyl ring of retinoic acid is probably an early step in a series of reactions that lead to the elimination of retinoic acid from the body.

e. EPOXIDATION. Epoxidation of retinoic acid at the 5,6 position of the cyclohexenyl ring was discussed earlier. Although it had been suggested that this retinoid had greater biological activity than the parent compound, all evidence now indicates that this is not the case. The questions remain as to whether 5,6-epoxy-5,6-dihydroretinoic acid is a metabolite that is normally formed in the body and whether it is important in the elimination pathway of retinoic acid. It has been reported that low concentrations of 5,6-epoxy-5,6-dihydroretinoic acid are present in the small intestine, kidney, liver, testes, and serum after pharmacological or physiological doses of either retinoic acid (McCormick *et al.*, 1978, 1979, 1980) or retinol (McCormick and Napoli, 1982). In retinoid-deficient rats, the concentration of the epoxide in the intestine is maximal at 3 h and is not detected by 7 h after administration of the dose (McCormick *et al.*, 1980).

However, in retinoic acid-repleted animals the 5,6-epoxide could not be detected at 3 h after injection of [^{3}H]retinoic acid, possibly owing to the induction of retinoic acid metabolism that lowered the amount of available retinoic acid for subsequent epoxidation (Roberts *et al.*, 1979b; Napoli and McCormick, 1981) or to the repressed synthesis of the [^{3}H]5,6-epoxide by endogenous 5,6-epoxide.

In vitro studies have demonstrated the presence of a retinoic acid epoxidase in homogenates of rat kidney, small intestine, liver, and several other tissues (Sietsema and DeLuca, 1979). In the rat kidney, this epoxidation reaction is located in the particulate fraction, requires NADPH, ATP, and oxygen, and is strongly inhibited by *N,N'*-diphenyl-*p*-phenylenediamine, a free-radical scavenger (DeLuca *et al.*, 1981; Sietsema and DeLuca, 1982a). The epoxidase activity is not affected by the retinoid status of the animal and is not induced by high levels of retinoic acid. This inability to be induced together with the low levels of epoxidation that occur *in vivo,* strongly suggest that epoxidation of retinoic acid is not a major pathway for deactivation and elimination. Furthermore, when *in vivo* epoxidation in retinoid-deficient rats is blocked by administration of *N,N'*-diphenyl-*p*-phenylenediamine, retinoic acid is still able to promote growth and to maintain differentiation of the vaginal epithelium (Sietsema and DeLuca, 1982a), indicating that epoxidation is not necessary for these functions of retinoic acid. In conclusion, then, it appears that retinoic acid epoxidation does not play a major if in fact any, role in either retinoic acid function or elimination.

3. *Isomerization of Retinoic Acid*

Isomerization of all-*trans*-retinaldehyde to 11-*cis*-retinaldehyde has been shown to be an obligatory event in the function of retinaldehyde in the visual process (Hubbard and Wald, 1952; Wald, 1968). Isomerization of 11-*cis*- and 13-*cis*-retinol has also been reported to occur in rats both *in vivo* (Murray *et al.*, 1959; Stainer *et al.*, 1960) and *in vitro* (Stainer and Murray, 1960). In addition, 13-*cis*-retinol accounts for approximately 35% of the retinol found in fish liver oils (Robeson and Baxter, 1945, 1947; Hayes and Petitpierre, 1952). Except for the visual cycle, the physiological significance of these isomerization reactions is not yet known. It has been suggested that the occurrence of 13-*cis*-retinol is not necessarily indicative of a requirement of animals for that isomer, but rather that retinol may naturally occur as a mixture of the cis and trans compounds (Cawley *et al.*, 1948).

It has been demonstrated that isomerization of retinoic acid may also be a naturally occurring event. In 1967 Zile *et al.* reported the identification of 13-*cis*-retinoic acid in tissue extracts of rats given all-*trans*-retinoic acid. They found, however, that the isolation methods employed resulted in extensive isomerization of the all-trans-retinoid to the 13-cis compound, and therefore the biological significance of this reaction was in doubt. Lippel and Olson (1968b) also tentatively identified 13-*cis*-retinoic acid in rat bile, although they concluded that this compound was being produced from retinoyl β-glucuronide during the isolation procedure. That extraction and chromatographic procedures can artifactually convert the all-trans to the 13-cis isomer is important to remember. Because of the ease of this artifactual conversion, all work dealing with the possible involvement of isomers in the physiological mechanism of action of retinoids must be properly controlled to prevent incorrect interpretation of the data obtained.

It was not until 1980 that the possible importance of 13-*cis*-retinoic acid in the overall metabolism scheme of all-*trans*-retinoic acid again became apparent. Frolik *et al.* (1980) reported that at physiological doses, both all-*trans*- and 13-*cis*-retinoic acid are metabolized *in vivo* to a common metabolite 13-*cis*-4-oxoretinoic acid. It was also demonstrated that in an *in vitro* liver-metabolizing system, the isomeric distribution of 4-oxoretinoic acid is concentration-dependent. At high initial concentrations of all-*trans*-retinoic acid, the all-trans-4-oxo compound predominates, whereas the 13-cis isomer is the major form at lower concentrations (Roberts and Frolik, 1979; Table I). Interestingly, in two target tissues of retinoid function, the testes and the intestinal mucosa, the *in vitro* metabolism of retinoic acid yielded mainly the 13-cis isomer, regardless of the starting retinoid concentration (Table I). The physiological importance of this observation is not yet clear. From the basis of these *in vivo* and *in vitro* studies, it can be postulated (Fig. 8) that at low concentrations of all-*trans*-retinoic acid, the formation of 13-*cis*-4-oxoretinoic acid is favored, while at higher concentrations of retinoic acid the all-trans-4-oxo isomer becomes increasingly important. Indeed, 2 h after the injection of a physiological (5 μg) dose of all-*trans*-retinoic acid into vitamin A-normal hamsters, only 13-*cis*-4-oxoretinoic acid and no all-trans isomer is detected in the plasma (Frolik *et al.*, 1980). However, after a large pharmacological (27.5 mg)

TABLE I

Effect of Retinoic Acid Concentration on the *in Vitro* Conversion of all-*trans*-Retinoic Acid to 4-Oxoretinoic Acid

Source of microsomes[a]	Retinoic acid concentration (*M*)	Percent of 4-oxoretinoic acid appearing as: 13-*cis*	all-*trans*
Liver	10^{-7}	27	74
	10^{-8}	59	41
Intestine	10^{-7}	86	14
	10^{-8}	87	13
Testis	10^{-6}	76	24
	10^{-7}	66	34

[a] Retinoid-deficient hamsters were pretreated with all-*trans*-retinoic acid or retinyl acetate 3 days prior to sacrifice. Microsomes were prepared and *in vitro* metabolism of retinoic acid was performed as described by Roberts *et al.* (1979a,b).

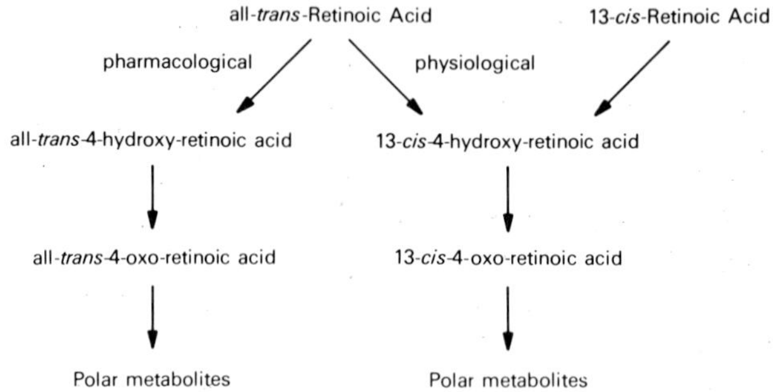

Fig. 8. Postulated metabolic pathway for all-*trans*- and 13-*cis*-retinoic acid after the administration of physiological or pharmacological doses to hamsters (Frolik, 1981).

dose of all-*trans*-retinoic acid to rats, all-*trans*-4-oxoretinoic acid has been shown to be excreted in the feces (Hänni and Bigler, 1977).

The concept that 13-*cis*-retinoic acid may play an important role physiologically is further supported with data from several investigators. Zile *et al.* (1982a) has reported detecting both all-*trans*- and 13-*cis*-retinoyl β-glucuronide in the bile of retinoid-deficient rats administered all-*trans*-retinoic acid. Similarly, in other carefully controlled experiments, 13-*cis*- and all-*trans*-retinoic acid as well as the isomer of the glucuronide were shown to be the major metabolites present in the small intestinal mucosa of bile duct-cannulated, retinoid-deficient rats 2 h after a dose of all-*trans*-retinoic acid (Zile *et al.*, 1982b). Isomerization of retinoic acid has also been reported to occur in mice (Kalin *et al.*, 1981), vitamin A-normal rats (Sundaresan and Bhat, 1982), and dogs (Patel *et al.*, 1982).

Although it appears that the isomerization of all-*trans*-retinoic acid to the 13-cis isomer is a physiological event, the importance of this step is not known. The 13-*cis*-retinoic acid is equivalent in biological activity to all-*trans*-retinoic acid, both *in vivo* (Zile and DeLuca, 1968) and *in vitro* (Newton *et al.*, 1980). More work is needed, however, before it can be determined whether isomerization of retinoids might be an important event in the control of epithelial differentiation as has already been shown to be the case for the visual cycle.

E. Metabolism of Other Natural Retinoids

The retinoids discussed so far in this chapter, all-*trans*-retinol, retinaldehyde, and retinoic acid, together comprise the major portion of the naturally occuring retinoids. However, several isomers of retinol have been detected in fish liver oil, including 13-*cis*-, 9-*cis*-, and 9,13-di-*cis*-retinol (Roberson and Baxter, 1945, 1947; Brown *et al.*, 1959), although the metabolism of these compounds has not been investigated. 3,4-Didehydroretinol (A9; vitamin A_2) is also a naturally occurring form of retinol that has been found in freshwater fish (Heilbron *et al.*, 1931; Edisbury *et al.*, 1937; Lederer *et al.*, 1937; Goswami and Barua, 1981). Although this retinoid can be synthesized *in vivo* and *in vitro* from all-*trans*-retinol (Naito and Wilt, 1962), the major biosynthetic pathway appears to be through the conversion of lutein to anhydrolutein followed by cleavage to 3-hydroxyretinol and 3,4-didehydroretinol (Fig. 9; Baura *et al.*, 1973, 1977; Barua and Das, 1975; Barua and Goswanni, 1977). The 3-hydroxyretinol thus formed can be further metabolized to 3-hydroxyanhydroretinol, which is then converted to a second molecule of 3,4-didehydroretinol (Barua *et al.*, 1979). This conversion of anhydroleutin to 3,4-didehydroretinol has also been shown to take place in retinoid-deficient chicks (Budowski *et al.*, 1963, 1966) and rats (Savithry *et al.*, 1972) administered anhydroleutin. However, 3,4-didehydroretinol is not a physiologically significant retinoid in mammals. The biopotency of anhydroleutin in supporting the growth of retinoid-deficient rats is 21% of that observed for β-carotene (Savithry *et al.*, 1972), whereas 3,4-didehydroretinol has about 40% of the activity of retinol (Shantz and Brinkman, 1950). Part, if not all of this activity, may be due to the conversion of 3,4-didehydroretinol to retinol, which has been shown to occur in the rat (Yoshikami *et al.*, 1969). The further metabolism of 3,4-didehydoretinol has not been stud-

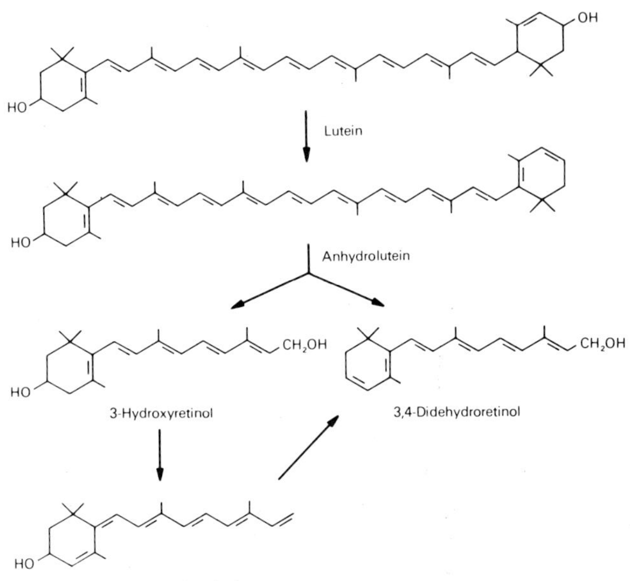

Fig. 9. Biosynthesis of 3,4-didehydroretinol from lutein in fish (see Barua *et al.*, 1973, 1979).

ied extensively, although it has been shown both in the rat and the fish that the alcohol can be oxidized to the aldehyde (Wald, 1937; Cama *et al.*, 1952). As yet, 3,4-didehydroretinoic acid has not been detected (Barua, 1978).

III. METABOLISM OF THE SYNTHETIC RETINOIDS

The use of natural retinoids in dermatology and in the prevention of epithelial cancer has been an area of extreme interest during the past few years (for reviews, see Sporn and Newton, 1979; Frolik and Roller, 1981; Chapters 14 and 16, this volume). However, because of the toxicity of the pharmacological levels of the natural retinoids employed in these studies, less-toxic analogs were sought. Although several effective compounds have been found, only four have undergone extensive metabolic studies, the details of which are presented below. These include 13-*cis*-retinoic acid, *N*-(4-hydroxyphenyl)retinamide (E16), *N*-(2-hydroxyethyl)retinamide (E12), and the TMMP analog of retinoic acid ethyl ester (F1; etretinate).

The metabolism of several other synthetic retinoids has been briefly examined but not analyzed to any great extent. Included are studies in rats of the hydrolysis of retinoic acid anhydride to retinoic acid (Premachandra and Cama, 1975), the conversion in hamsters of retinaldehyde acetylhydrazone (C8) and *N*-acetyl-retinylamine (B1) to retinol and retinyl esters (Roberts *et al.*, 1978), the metabolism of 15-methyl retinone, 15-methylretinol, and 15,15-dimethylretinol (A10) and the excretion of their metabolites in the bile and urine of rats (Tosukhowong and Olson, 1978), and the metabolism of 5,6-epoxy-5,6-dihydroretinol and retinaldehyde both in rats (Jungalwala and Cama, 1965; Lakshmanan *et al.*, 1965) and in cultured, spontaneously transformed mouse fibroblasts (Bhat *et al.*, 1981). Because of the lack of extensive metabolic studies, these reactions are not discussed further here.

As indicated in Section II,D,3, 13-*cis*-retinoic acid could possibly be considered a natural retinoid and a physiological metabolite of all-*trans*-retinoic acid. However, owing to the preliminary nature of this observation, 13-*cis*-retinoic acid will be considered a synthetic retinoid and its metabolism is presented in this section. Initial studies with this retinoid were concerned with detection of the 13-cis compound in the plasma and its rate of absorption and disappearance. Frolik *et al.* (1978b), employing a high-pressure liquid chromatography assay for quantitative determination of 13-*cis*-retinoic acid in the presence of all-*trans*-retinoic acid and retinol, observed that after a single 80- to 100-mg oral dose of the 13-cis isomer to humans, the plasma levels of 13-*cis*-retinoic acid rose rapidly to a peak concentration of about 400 ng/ml at 4–8 h after dosage followed by a gradual decline. This time course has been verified by a number of investigators using similar techniques (Besner *et al.*, 1980; Vane *et al.*, 1982b; Goodman *et al.*,

1982). In mice, after either an intravenous (Wang *et al.*, 1980) or oral (Kalin *et al.*, 1982) dose of 13-*cis*-retinoic acid, the starting compound disappeared from the serum in an initial distribution phase, followed by a slower exponential elimination phase. Similar results were also obtained in rats (Shelley *et al.*, 1982). This is in contrast to the nonexponential elimination of all-*trans*-retinoic acid from serum (Wang *et al.*, 1980; Swanson *et al.*, 1981b; Kalin *et al.*, 1981; Shelley *et al.*, 1982).

The further metabolism of 13-*cis*-retinoic acid has been investigated in the hamster, the rat, and the human. In the normal hamster, 13-*cis*-4-oxoretinoic acid has been detected in the plasma 2 h after an intravenous 5.3-μg dose of ^{3}H-labeled 13-*cis*-retinoic acid (Frolik *et al.*, 1980). As mentioned earlier, when 5.1 μg of all-*trans*-retinoic acid was injected into hamsters, the 13-*cis*-4-oxo metabolite and not the corresponding all-trans analog was detected. It has therefore been concluded that this metabolite is common to both isomers of retinoic acid. A basic difference, however, is that 13-*cis*-4-oxoretinoic acid is a major metabolite in the 13-*cis*-retinoic acid-dosed hamster, representing 21% of the radioactive metabolites in the plasma compared to 2.2% in the all-*trans*-retinoic acid-dosed animals (Frolik *et al.*, 1980). In the same study, 13-*cis*-4-hydroxyretinoic acid was also tentatively identified.

When 13-*cis*-retinoic acid was administered intravenously to vitamin A-normal rats, 69% of the dose appeared in the bile after 24 h while 9% was found in the urine (Frolik *et al.*, 1981b). This suggests that, similar to all-*trans*-retinoic acid (see Section II,D,2), 13-*cis*-retinoic acid and its metabolites are rapidly eliminated from the body and that the major route of excretion is via the bile. Analysis of the bile by high-pressure liquid chromatography indicated the presence of several polar metabolites with the two major compounds being identified as 13-*cis*-4-oxoretinoyl β-glucuronide (8.1% of the dose during the first 4 h) and 13-*cis*-retinoyl β-glucuronide (7.0% of the dose). Several other unidentified biliary metabolites also appeared to be conjugated with glucuronic acid. A comparison of the chromatographic profiles of bile from rats administered 13-*cis*- versus all-*trans*-retinoic acid indicated that a greater proportion of the all-*trans*-retinoic acid is metabolized to compounds that elute in the more polar regions of the column effluent when compared to the metabolism of 13-*cis*-retinoic acid (Frolik *et al.*, 1981b).

These initial observations have been verified and expanded by Vane and coworkers. Using an *in vitro* rat liver supernatant-metabolizing system, these workers identified four different metabolites of 13-*cis*-retinoic acid (Vane and Buggé, 1981a; Vane *et al.*, 1982a). Similar to the metabolic profile observed in the hamster, the major metabolite was 13-*cis*-4-oxoretinoic acid with a minor amount of its precursor 13-*cis*-4-hydroxyretinoic acid also detected. The 4-oxo compound was further metabolized to 13-*cis*-2-hydroxy-4-oxoretinoic acid and 13-*cis*-3-hydroxy-4-oxoretinoic acid (Fig. 2).

Metabolite	% OF DOSE IN HUMAN URINE: Free	CONJUGATE: a	CONJUGATE: b
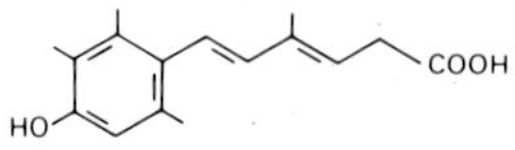	0.5	1.5	0.5
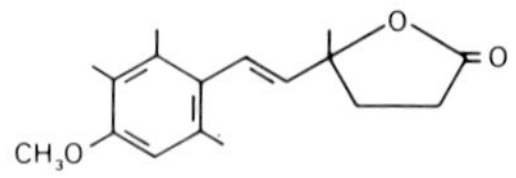	0.5	–	–
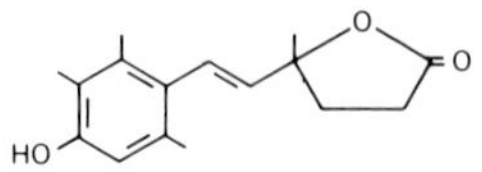	1.5	0.5	–
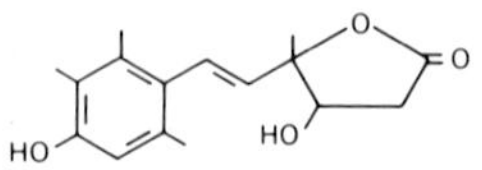	0.5	–	–
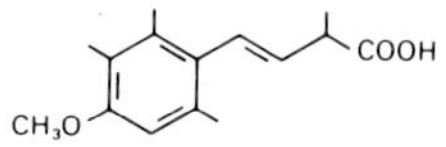	0.5	–	–
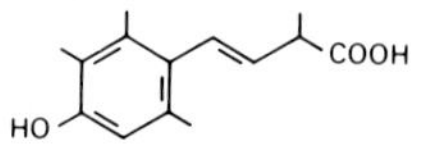	0.5	1.5	0.5
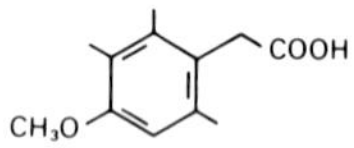	–	0.5	–
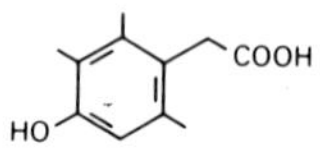	0.5	1.5	–

Fig. 10. Metabolites found in the urine of humans after administration of etretinate. The "a" conjugates have not been identified, while the "b" conjugates are glycine conjugates of the free metabolites (see Hänni *et al.*, 1977).

Investigation of the metabolism of 13-*cis*-retinoic acid in humans has shown that the 13-*cis*-4-oxo compound is the major metabolite found in the blood of patients on chronic 13-*cis*-retinoic acid therapy (Vane and Buggé, 1981b). A small amount of 13-*cis*-4-hydroxyretinoic acid and the dioxygenated metabolites were also detected. In general, it appears that during the first 6 h after a single oral dose of 13-*cis*-retinoic acid to patients, 13-*cis*-retinoic acid itself is the major drug-derived compound in the blood while, after longer intervals of time or after chronic dosing, 13-*cis*-4-oxoretinoic acid becomes the major drug-derived compound (Vane *et al.*, 1982b; Kerr *et al.*, 1982). In conclusion, analogous to the metabolism of the all-trans isomer, a major metabolic pathway of 13-*cis*-retinoic acid in hamsters, rats, and humans is oxidation at C-4 of the cyclohexenyl ring.

A second synthetic retinoid with an extensively investigated metabolism is etretinate, the TMMP analog of retinoic acid ethyl ester. After an oral 100-mg dose of [^{3}H]etretinate to psoriasis patients, 75% of the dose was found in the feces within the first 5 days, while 15% occured in the urine (Hänni *et al.*, 1977; Hänni, 1978). The metabolites present in human urine were divided into two groups, free metabolites (6% of dose) and conjugated metabolites (8% of dose), and purified by thin-layer and high-pressure liquid chromatography. These purified metabolites were then identified by NMR and mass spectroscopy (Hänni *et al.*, 1977). As is shown in Fig. 10, all of the urinary metabolites contained a shortened tetraene side chain. Except for the glycine conjugates, the water-soluble residues (conjugated with glucuronic acid or sulfuric acid) were not identified. In the feces, the major portion of the radioactive material (60% of the dose) was found in the unaltered parent compound. Most of this apparently resulted from nonabsorbed material. The remainder of the metabolites in the feces were unidentified. In the plasma, the parent drug and its free acid were the major compounds detected. Because the free acid has been shown to be a biologically active compound and is most likely the active form of the parent drug, it is important to measure levels of this metabolite as well as the administered drug when analyzing plasma levels in humans. Hänni and co-workers (1979) have shown that after an oral 100-mg dose of etretinate to the adult male, levels of both the parent drug and its free acid reach a maximum at 3–4 h. By 24 h after administration, the concentrations of both metabolites have declined to unmeasurable amounts. The metabolism of etretinate to the free acid has also been shown to occur *in vitro* in human epidermal keratinocytes (Kitano *et al.*, 1982).

After an intravenous injection of etretinate to rats, 70% of the dose was excreted into the bile within the first 48 h (Hänni *et al.*, 1977). Of this material, 70% was a conjugate of the free acid, 10% was the free acid that had a hydroxyl

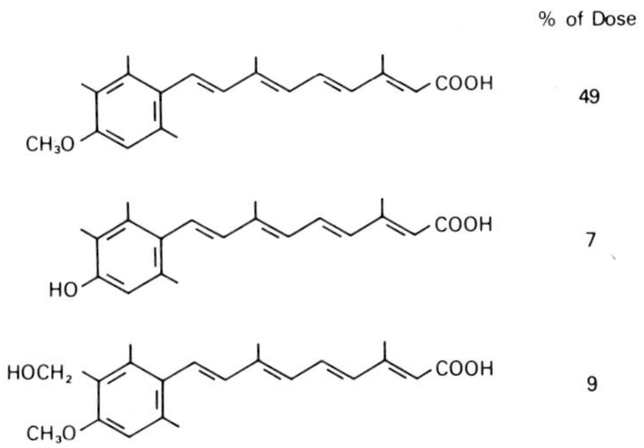

Fig. 11. Metabolites found in the bile of rats administered etretinate. The metabolites were found in the bile as conjugates that were cleaved by enzymatic hydrolysis prior to identification (see Hänni *et al.*, 1977).

group at position 4 in the aromatic ring in place of the methoxy moiety, while 13% was found as the free acid hydroxylated at the 3-methyl group (Fig. 11). No unaltered parent drug was detected. Some enterohepatic circulation of these biliary metabolites appeared to occur (Hänni *et al.*, 1977). In the mouse, esterases that hydrolyze the ethyl ester to the free acid have been found to be widely distributed in all tissues examined (Wang and Hill, 1977) with the brain, liver, and ovary containing the highest activity and plasma the lowest.

The last group of synthetic retinoids with metabolisms that have been investigated to some extent are various analogs of retinamide, including *N*-(2-hydroxyethyl)retinamide (E12) and *N*-(4-hydroxyphenyl)retinamide (E16). The disposition of the *N*-(2-hydroxyethyl)retinamide has been examined in mice after either an oral (Kalin *et al.*, 1982) or an intravenous (Wang *et al.*, 1980) 10-mg/kg dose. After either method of administration, the drug disappeared from the serum in an initial distribution phase, followed by an exponential elimination phase. Tissue levels of the *N*-(2-hydroxethyl)retinamide were in most cases higher than serum levels and also declined exponentially. After an intravenous dose, the parent drug persisted in the tissues for a considerable length of time with measurable levels still being found in the kidney, liver, and testis 18 h after administration. The bioavailability of the *N*-(2-hydroxyethyl)retinamide after oral dosing was 60% of that observed after an intravenous dose, while elimination of the parent drug and its metabolites proceeded at rates independent of the route of administration. Seventy-two hours after an oral dose of ^{3}H-labeled *N*-(2-hydroxyethyl)retinamide to rats, 64% of the dose was found in the feces and 11% in the urine. Although polar metabolites were observed in the bile, feces, and

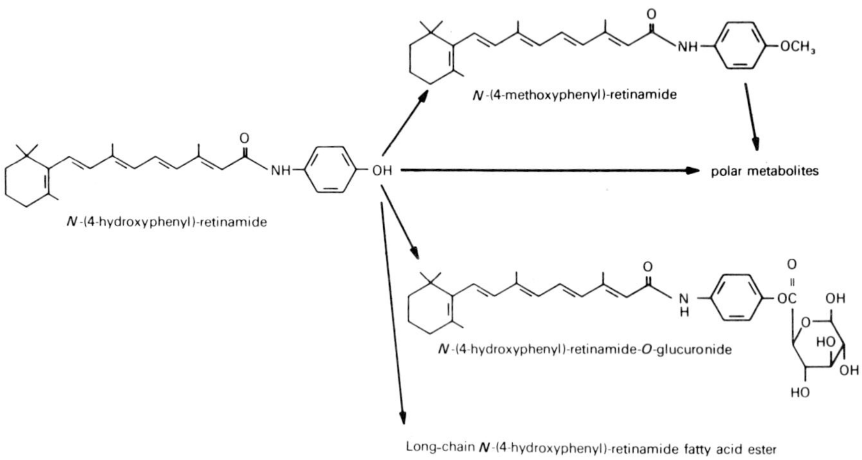

Fig. 12. Metabolism of *N*-(4-hydroxyphenyl)retinamide in the rat (see Swanson *et al.*, 1980, 1981a).

urine, none of these compounds have been structurally identified. Neither the parent drug nor its free acid derivative could be detected in the urine or bile.

Finally, some initial metabolic studies have also been performed with *N*-(4-hydroxyphenyl)retinamide (Swanson *et al.*, 1980, 1981a). When a single intravenous dose (5 mg/kg) of ^{3}H-labeled *N*-(4-hydroxyphenyl)retinamide was administered to vitamin A-normal rats, 64% of the dose appeared in the feces after 5 days and 13% in the urine. Less than 2% of these excreted compounds appeared as the unchanged parent retinoid. Although several polar metabolites of this retinoid have been observed in the bile of rats, only one, *N*-(4-hydroxyphenyl)retinamide *O*-glucuronide (Fig. 12), has been identified. Metabolites of the initial retinoid were also detected in many tissues and by 24 h after an intravenous dose, these metabolites were reported to be more abundant than the intact drug. Two of these tissue metabolites, *N*-(4-methoxyphenyl)retinamide and a fatty acid ester of *N*-(4-hydroxyphenyl)retinamide (Fig. 12), have been identified.

In conclusion, with the advances in extraction and purification techniques, the further characterization of metabolites of both natural and synthetic retinoids should be rapidly forthcoming. Once these metabolic pathways have been elucidated, the roles of the various metabolites in the expression of the biological activity of the different retinoids should become more defined.

REFERENCES

Adamo, S., De Luca, L. M., Silverman-Jones, C. S. and Yuspa, S. H. (1979a). *J. Biol. Chem.* **254,** 3279–3287.

Adamo, S., De Luca, L. M., Akalovsky, I., and Bhat, P. V. (1979b). *JNCI, J. Natl. Cancer Inst.* **62,** 1473–1478.

Ames, S. R. (1965). *Fed. Proc., Fed. Am. Soc. Exp. Biol.* **24,** 917–923.

Andrews, J. S., and Futterman, S. (1964). *J. Biol. Chem.* **239,** 4073–4076.

Arens, J. F., and van Dorp, D. A. (1946a). *Nature (London)* **157,** 190–191.

Arens, J. F., and van Dorp, D. A. (1946b). *Nature (London)* **158,** 622–623.

Austin, D. J., Bu'Lock, J. D., and Drake, D. (1970). *Experientia* **26,** 348–349.

Barua, A. B. (1978). *World Rev. Nutr. Diet.* **31,** 89–94.

Barua, A. B., and Das, R. C. (1975). *Br. J. Nutr.* **33,** 319–327.

Barua, A. B., and Goswami, U. S. (1977). *Biochem. J.* **164,** 133–136.

Barua, A. B., Singh, H. T., and Das, R. C. (1973). *Br. J. Nutr.* **30,** 1–12.

Barua, A. B., Das, R. C., and Verma, K. (1977). *Biochem. J.* **168,** 557–564.

Barua, A. B., Verma, K., and Das, S. R., (1979). *Biochem. J.* **177,** 791–799.

Behrens, N. H., and Leloir, L. F. (1970). *Proc. Natl. Acad. Sci. U.S.A.* **66,** 153–159.

Besner, J. G., Leclaire, R., and Band, P. R. (1980). *J. Chromatogr.* **183,** 346–351.

Bhat, P. V., and De Luca, L. M. (1981). *Ann. N. Y. Acad. Sci.* **359,** 135–149.

Bhat, P. V., De Luca, L. M., Adamo, S., Akalovsky, J., Silverman-Jones, C. F., and Peck, G. L. (1979). *J. Lipid Res.* **20,** 357–362.

Bhat, P. V., Roller, P. P., and De Luca, L. M. (1981). *J. Lipid Res.* **22,** 1069–1078.

Blaner, W. S., and Churchich, J. E. (1980). *Biochem. Biophys. Res. Commun.* **94,** 820–826.
Bliss, A. F. (1951). *Arch. Biochem. Biophys.* **31,** 197–204.
Brown, P. S., Blum, W. P., and Stern, M. H. (1959). *Nature (London)* **184,** 1377–1379.
Budowski, P., Ascarelli, I., Gross, J., and Nir, I. (1963). *Science* **142,** 969–971.
Budowski, P., Auerbach, E., and Rowe, H. (1966). *Nature (London)* **211,** 77–78.
Cama, H. R., Dalvi, P. D., Morton, R. A., and Salah, M. K., (1952). *Biochem. J.* **52,** 542–547.
Cawley, J. D., Robeson, C. D., Weisler, L., Shantz, E. M., Embree, N. D., and Baxter, J. G. (1948). *Science* **107,** 346.
Clamon, G. H., Sporn, M. B., Smith, J. M., and Saffiotti, U. (1974). *Nature (London)* **250,** 64–66.
Clark, J. N., Nathaniel, R. S., Kleinman, H. K., and Wolf, G. (1975). *Biochem. Biophys. Res. Commun.* **62,** 233–240.
Crain, F. D., Lotspeich, F. J., and Krause, R. F. (1967). *J. Lipid Res.* **8,** 249–254.
De Leenheer, A. P., Lambert, W. E., and Claeys, I. (1982). *J. Lipid Res.* **23,** 1362–1367.
DeLuca, H. F. (1979). *Fed. Proc., Fed. Am. Soc. Exp. Biol.* **38,** 2519–2523.
DeLuca, H. F., and Roberts, A. B. (1969). *Am. J. Clin. Nutr.* **22,** 945–952.
DeLuca, H. F., Zile, M., and Sietsema, W. K. (1981). *Ann. N. Y. Acad. Sci.* **359,** 25–36.
De Luca, L. M., Maestri, N., Rosso, G., and Wolf, G. (1973). *J. Biol. Chem.* **248,** 641–648.
De Luca, L. M., Adamo, S., Bhat, P. V., Wlodzimierz, S., Silverman-Jones, C. S., Akalovsky, J., Frot-Coutaz, J. P., Fletcher, T. R., and Chader, G. J. (1979a). *Pure Appl. Chem.* **51,** 581–591.
De Luca, L. M., Bhat, P. V., Sasak, W., and Adamo, S. (1979b). *Fed. Proc., Fed. Am. Soc. Exp. Biol.* **38,** 2535–2539.
DeMan, T. J., van Leeuwen, P. H., and Roborgh, J. R. (1964). *Nature (London)* **201,** 77–78.
De Ruyter, M. G., Lambert, W. E., and De Leenheer, A. P. (1979). *Anal. Biochem.* **98,** 402–409.
Deshmukh, D. S., Malathi, P., and Ganguly, J. (1965). *Biochim. Biophys. Acta* **107,** 120–122.
Dmitrovskii, A. A. (1961). *Biokhimiya* **26,** 126–131.
Dmitrovskii, A. A., Soloveva, N. B., Dmitrovskaya, T. A., Yermilova, L. K., and Rozynov, B. V. (1977). *Prikl. Biokhim. Mikrobiol.* **13,** 205–212.
Dowling, J. E. (1961). *Am. J. Clin. Nutr.* **9,** 23–26.
Dowling, J. E., and Wald, G. (1960). *Proc. Natl. Acad. Sci. U.S.A.* **46,** 587–608.
Dunagin, P. E., Zachman, R. D., and Olson, J. A. (1964). *Biochim. Biophys. Acta* **90,** 432–434.
Dunagin, P. E., Meadows, E. H., and Olson, J. A. (1965). *Science* **148,** 86–87.
Dunagin, P. E., Zachman, R. D., and Olson, J. A. (1966). *Biochim. Biophys. Acta* **124,** 71–85.
Edisbury, J. R., Morton, R. A., and Simpkins, G. W. (1937). *Nature (London)* **140,** 234.
Elder, T. D., and Topper, Y. J. (1962). *Biochim. Biophys. Acta* **64,** 430–437.
Emerick, R. J., Zile, M., and DeLuca, H. F. (1967). *Biochem. J.* **102,** 606–611.
Fidge, N. H., Shiratori, T., Ganguly, J., and Goodman, D. S. (1968). *J. Lipid Res.* **9,** 103–109.
Frolik, C. A. (1981). *Ann. N. Y. Acad. Sci.* **359,** 37–44.
Frolik, C. A., and Roller, P. P. (1981). *In* "Antitumor Compounds of Natural Origin: Chemistry and Biochemistry" (A. Aszalos, ed.), Vol. II, pp. 87–128. CRC Press, Boca Rotan, Florida.
Frolik, C. A., Tavela, T. E., and Sporn, M. B. (1978a). *J. Lipid Res.* **19,** 32–37.
Frolik, C. A., Tavela, T. E., Peck, G. L., and Sporn, M. B. (1978b). *Anal. Biochem.* **86,** 743–750.
Frolik, C. A., Tavela, T. E., Newton, D. L., and Sporn, M. B. (1978c). *J. Biol. Chem.* **253,** 7319–7324.
Frolik, C. A., Roberts, A. B., Tavela, T. E., Roller, P. P., Newton, D. L., and Sporn, M. B. (1979). *Biochemistry* **18,** 2092–2097.
Frolik, C. A., Roller, P. P., Roberts, A. B., and Sporn, M. B. (1980). *J. Biol. Chem.* **255,** 8057–8062.
Frolik, C. A., Dart, L. L., and Sporn, M. B. (1981a). *Biochim. Biophys. Acta* **663,** 329–335.

Frolik, C. A., Swanson, B. N., Dart, L. L., and Sporn, M. B. (1981b). *Arch. Biochem. Biophys.* **208,** 344–352.

Frot-Coutaz, J. P., Silverman-Jones, C. S., and De Luca, L. M. (1976). *J. Lipid Res.* **17,** 220–230.

Futterman, S. (1962). *J. Biol. Chem.* **237,** 677–680.

Gaede, K., and Rodriquez, P. (1973). *Biochem. Biophys. Res. Commun.* **54,** 76–81.

Garbers, C. F., Gillman, J., and Peisach, M. (1960). *Biochem. J.* **75,** 124–132.

Geison, R. L., and Johnson, B. C. (1969). *Lipids* **5,** 371–378.

Glover, J., Goodwin, T. W., and Morton, R. A. (1948). *Biochem. J.* **43,** 109–114.

Goodman, D. S., and Olson, J. A. (1969). *In* "Methods in Enzymology, Vol. 15: Steroids and Terpenoids" (B. Raymond, ed.), pp. 462–475. Academic Press, New York.

Goodman, D. S., Huang, H. S., and Shiratori, T. (1965). *J. Lipid Res.* **6,** 390–396.

Goodman, D. S., Blomstrand, R., Werner, B., Huang, H. S., and Shiratori, T. (1966). *J. Clin. Invest.* **45,** 1615–1623.

Goodman, G. E., Einspahr, J. G., Alberts, D. S., Davies, T. P., Leigh, S. A., Chen, H. S. G., and Meyskens, F. L. (1982). *Cancer Res.* **42,** 2087–2091.

Goswami, U. C., and Barua, A. B. (1981). *Ind. J. Biochem. Biophys.* **18,** 383–385.

Hänni, R. (1978). *Dermatologica* **157,** 5–10.

Hänni, R., and Bigler, F. (1977). *Helv. Chim. Acta* **60,** 881–887.

Hänni, R., Bigler, F., Meister, W., and Englert, G. (1976). *Helv. Chim. Acta* **59,** 2221–2227.

Hänni, R., Bigler, F., Vetter, W., Englert, G., and Loeliger, P. (1977). *Helv. Chim. Acta* **60,** 2309–2325.

Hänni, R., Hervouet, D., and Bussinger, A. (1979). *J. Chromatog.* **162,** 615–621.

Harrison, E. H., Smith, J. E., and Goodman, D. S. (1979). *J. Lipid Res.* **20,** 760–771.

Hayes, E., and Petitpierre, M. A. (1952). *J. Pharm. Pharmacol.* **4,** 879–885.

Heilbron, I. M., Gillam, A. E., and Morton, R. A. (1931). *Biochem. J.* **25,** 1352–1366.

Helgerud, P., Peterson, L. B., and Norum, K. R. (1982). *J. Lipid Res.* **23,** 609–618.

Helting, T., and Peterson, P. A. (1972). *Biochem. Biophys. Res. Commun.* **46,** 429–436.

Huang, H. S., and Goodman, D. S. (1965). *J. Biol. Chem.* **240,** 2839–2844.

Hubbard, R., and Wald, G. (1952). *Science* **115,** 60–63.

Ito, Y. L., Zile, M., Ahrens, H., and DeLuca, H. F. (1974a). *J. Lipid Res.* **15,** 517–524.

Ito, Y., Zile, M., DeLuca, H. F., and Ahrens, H. M. (1974b). *Biochim. Biophys. Acta* **369,** 338–350.

John, K. V., Lakshmanan, M. R., and Cama, H. R. (1967). *Biochem. J.* **103,** 539–543.

Jungalwala, F. B., and Cama, H. R. (1965). *Biochem. J.* **95,** 17–26.

Kalin, J. R., Starling, M. E., and Hill, D. L. (1981). *Drug Metab. Dispos.* **9,** 196–201.

Kalin, J. R., Wells, J., and Hill, D. L. (1982). *Drug Metab. Dispos.* **10,** 391–398.

Karrer, P., Morf, R., and Schopp, K. (1931). *Helv. Chem. Acta* **14,** 1431–1436.

Kerr, J. G., Lippman, M. E., Jenkens, J., and Myers, C. E. (1982). *Cancer Res.* **42,** 2069–2073.

Kitano, Y., Okada, N., and Nagase, N. (1982). *Arch. Dermatol. Res.* **273,** 327–331.

Kleiner-Bössaler, A., and DeLuca, H. F. (1971). *Arch. Biochem. Biophys.* **142,** 371–377.

Krinsky, N. I. (1958). *J. Biol. Chem.* **232,** 881–894.

Krishnamurthy, S., Bieri, J. G., and Andrews, E. L. (1963). *J. Nutr.* **79,** 503–510.

Lakshmanan, M. R., Jungalwala, F. B., and Cama, H. R. (1965). *Biochem. J.* **95,** 27–34.

Lawrence, C. W., Crane, F. D., Lotspeich, F. J., and Krause, R. F. (1966). *J. Lipid Res.* **7,** 226–229.

Lederer, E., Gillam, A. E., Rosanova, V., and Heilbron, I. M. (1937). *Nature (London)* **140,** 233.

Lippel, K., and Olson, J. A. (1968a). *J. Lipid Res.* **9,** 168–175.

Lippel, K., and Olson, J. A. (1968b). *J. Lipid Res.* **9,** 580–586.

Lippel, K., Manyan, D. R., and Llewellyn, A. (1970). *Arch. Biochem. Biophys.* **139,** 421–424.

McCollum, E. V., and Davis, M. (1913). *J. Biol. Chem.* **15,** 167–175.
McCormick, A. M., and Napoli, J. L. (1982). *J. Biol. Chem.* **257,** 1730–1735.
McCormick, A. M., Napoli, J. L., Schnoes, H. K., and DeLuca, H. F. (1978). *Biochemistry* **17,** 4085–4090.
McCormick, A. M., Napoli, J. L., Schnoes, H. K., and DeLuca, H. F. (1979). *Arch. Biochem. Biophys.* **192,** 577–583.
McCormick, A. M., Napoli, J. L., Yoshizawa, S., and DeLuca, H. F. (1980). *Biochem. J.* **186,** 475–481.
McKenzie, R. M., and Nelson, E. C. (1979). *Biochim. Biophys. Acta* **574,** 1–7.
Mahadevan, S., and Ganguly, J. (1961). *Biochem. J.* **81,** 53–58.
Mahadevan, S., Murthy, S. K., and Ganguly, J. (1962). *Biochem. J.* **85,** 326–331.
Malathi, P., Rao, K. S., Sastry, P. S., and Ganguly, J. (1963). *Biochem. J.* **87,** 305–311.
Mallia, K. A., John, J., John, K. V., Lakshmanan, M. R., Jungalwala, F. G., and Cama, H. R. (1970). *Indian J. Biochem.* **7,** 102–103.
Masushige, S., Schreiber, J. B., and Wolf, G. (1978). *J. Lipid Res.* **19,** 619–627.
Morgan, B., and Thompson, J. N. (1966). *Biochem. J.* **101,** 835–842.
Murray, T. K., Stainer, D. W., and Cambell, J. A. (1959). *Can. J. Biochem. Physiol.* **37,** 1469–1474.
Murthy, S. K., Mahadevan, S., Seshadri Sastry, P., and Ganguly, J. (1961). *Nature (London)* **189,** 482.
Naito, K., and Wilt, F. H. (1962). *J. Biol. Chem.* **237,** 3060–3064.
Napoli, J. L., and McCormick, A. M. (1981). *Biochim. Biophys. Acta* **666,** 165–175.
Napoli, J. L., McCormick, A. M., Schnoes, H. K., and DeLuca, H. F. (1978). *Proc. Natl. Acad. Sci. U.S.A.* **75,** 2603–2605.
Nath, K., and Olson, J. A. (1967). *J. Nutr.* **93,** 461–469.
Nelson, E. C., Mayberry, M., Reid, R., and John, K. V. (1971). *Biochem. J.* **121,** 731–733.
Newton, D. L., Henderson, W. R., and Sporn, M. B. (1980). *Cancer Res.* **40,** 3413–3425.
Olson, J. A. (1961). *J. Biol. Chem.* **236,** 349–356.
Olson, J. A. (1968). *Vitam. Horm. (N. Y.)* **26,** 1–55.
Olson, J. A. (1969a). *Am. J. Clin. Nutr.* **22,** 953–962.
Olson, J. A. (1969b). *Fed. Proc., Fed. Am. Soc. Exp. Biol.* **28,** 1670–1677.
Patel, I. H., Khoo, K.-C., and Colburn, W. A. (1982). *Drug Metab. Dispos.* **10,** 387–390.
Peterson, P. A., Rask, L., Helting, T., Ostberg, L., and Fernstedt, Y. (1976). *J. Biol. Chem.* **251,** 4986–4995.
Premachandra, B. R., and Cama, H. R. (1975). *Int. J. Vitam. Nutr. Res.* **45,** 305–316.
Prystowsky, J. H., Smith, J. E., and Goodman, D. S. (1981). *J. Biol. Chem.* **256,** 4498–4503.
Redfearn, E. R. (1960). *Arch. Biochem. Biophys.* **91,** 226–229.
Rietz, P., Wiss, O., and Weber, F. (1974). *Vitam. Horm. (N. Y.)* **32,** 237–249.
Roberts, A. B. (1981). *Ann. N. Y. Acad. Sci.* **359,** 45–53.
Roberts, A. B., and DeLuca, H. F. (1967). *Biochem. J.* **102,** 600–605.
Roberts, A. B., and DeLuca, H. F. (1968a). *Arch. Biochem. Biophys.* **123,** 279–285.
Roberts, A. B., and DeLuca, H. F. (1968b). *J. Lipid Res.* **9,** 501–508.
Roberts, A. B., and DeLuca, H. F. (1969). *Arch. Biochem. Biophys.* **129,** 290–295.
Roberts, A. B., and Frolik, C. A. (1979). *Fed. Proc., Fed. Am. Soc. Exp. Biol.* **38,** 2524–2527.
Roberts, A. B., Nichols, M. D., Frolik, C. A., Newton, D. L., and Sporn, M. B. (1978). *Cancer Res.* **38,** 3327–3332.
Roberts, A. B., Nichols, M. D., Newton, D. L., and Sporn, M. B. (1979a). *J. Biol. Chem.* **254,** 6296–6302.
Roberts, A. B., Frolik, C. A., Nichols, M. D., and Sporn, M. B. (1979b). *J. BIol. Chem.* **254,** 6303–6309.

Roberts, A. B., Lamb, L. C., and Sporn, M. B. (1980). *Arch. Biochem. Biophys.* **199,** 374–383.

Robeson, C. D., and Baxter, J. G. (1945). *Nature (London)* **155,** 300.

Robeson, C. D., and Baxter, J. G. (1947). *J. Am. Chem. Soc.* **69,** 136–141.

Rockley, N. L., Halley, B. A., and Nelson, E. C. (1980). *Biochim. Biophys. Acta* **627,** 270–275.

Ross, A. C. (1982a). *J. Biol. Chem.* **257,** 2453–2459.

Ross, A. C. (1982b). *J. Lipid Res.* **23,** 133–144.

Rosso, G. C., De Luca, L., Warren, C. D., and Wolf, G. (1975). *J. Lipid Res.* **16,** 235–243.

Sato, M., and Lieber, C. S. (1982). *Arch. Biochem. Biophys.* **213,** 557–564.

Savithry, K. N., Mallia, K. A., and Cama, H. R. (1972). *Ind. J. Biochem. Biophys.* **9,** 325–327.

Shantz, E. M., and Brinkman, J. H. (1950). *J. Biol. Chem.* **183,** 467–471.

Sharman, I. M. (1949). *Br. J. Nutr.* **3,** viii.

Shelley, R. S., Jun, H. W., Price, J. C., and Cadwallader, D. E. (1982). *J. Pharm. Sci.* **71,** 904–906.

Shidoji, Y., and De Luca, L. M. (1981). *Biochem. J.* **200,** 529–538.

Shidoji, Y., Sasak, W., Silverman-Jones, C. S., and De Luca, L. M. (1981). *Ann. N. Y. Acad. Sci.* **359,** 345–357.

Sietsema, W. K., and DeLuca, H. F. (1979). *Biochem. Biophys. Res. Commun.* **90,** 1091–1097.

Sietsema, W. K., and DeLuca, H. F. (1982a). *J. Biol. Chem.* **25,** 4265–4270.

Sietsema, W. K., and DeLuca, H. F. (1982b). *J. Nutr.* **112,** 1481–1489.

Silva, D. P., and DeLuca, H. F. (1982). *Biochem. J.* **206,** 33–41.

Skare, K. L., and DeLuca, H. F. (1983). *Arch. Biochem. Biophys.* **224,** 13–18.

Skare, K. L., Schnoes, H. K., and DeLuca, H. F. (1982). *Biochemistry* **21,** 3308–3317.

Smith, J. E., Milch, P. O., Muto, Y., and Goodman, D. S. (1973). *Biochem. J.* **132,** 821–827.

Sporn, M. B., and Newton, D. L. (1979). *Fed. Proc., Fed. Am. Soc. Exp. Biol.* **38,** 2528–2534.

Sporn, M. B., Clamon, G. H., Smith, J. M., Dunlop, N. M., Newton, D. L., and Saffiotti, U. (1974). *In* "Carcinogenesis and Bioassays" (E. Karbe and J. F. Park, eds.), pp. 575–582. Springer-Verlag, Berlin and New York.

Sporn, M. B., Dunlop, N. M., Newton, D. L., and Henderson, W. R. (1976). *Nature (London)* **263,** 110–113.

Stainer, D. W., and Murray, T. K. (1960). *Can. J. Biochem. Physiol.* **38,** 1467–1470.

Stainer, D. W., Murray, T. K., and Cambell, J. A. (1960). *Can. J. Biochem. Physiol.* **38,** 1219–1222.

Sundaresan, P. R. (1966). *Biochim. Biophys. Acta* **113,** 95–109.

Sundaresan, P. R. (1977). *J. Nutr.* **107,** 70–78.

Sundaresan, P. R., and Bhagavan, H. N. (1971). *Biochem. J.* **122,** 1–4.

Sundaresan, P. R., and Bhat, P. V. (1982). *J. Lipid Res.* **23,** 448–455.

Sundaresan, P. R., and Sundaresan, G. M. (1973). *Int. J. Vitam. Nutr. Res.* **43,** 61–69.

Sundaresan, P. R., and Sundaresan, G. M. (1975). *Biochem. J.* **152,** 99–104.

Sundaresan, P. R., and Therriault, D. G. (1968). *Biochim. Biophys. Acta* **158,** 92–97.

Surekha Rao, M. S., John, J., and Cama, H. R. (1972). *Int. J. Vitam. Nutr. Res.* **42,** 368–378.

Swanson, B. N., Zaharevitz, D. W., and Sporn, M. B. (1980). *Drug Metab. Dispos.* **8,** 168–172.

Swanson, B. N., Newton, D. L., Roller, P. P., and Sporn, M. B. (1981a). *J. Pharmacol. Exp. Ther.* **219,** 632–637.

Swanson, B. N., Frolik, C. A., Zaharevitz, D. W., Roller, P. P., and Sporn, M. B. (1981b). *Biochem. Pharmacol.* **30,** 107–113.

Thompson, J. N. (1969). *In* "The Fat-Soluble Vitamins" (H. F. DeLuca and J. W. Suttie, eds.), pp. 267–281. Univ. of Wisconsin Press, Madison.

Thompson, J. N., Howell, J. McC., and Pitt, G. A. J. (1961). *Biochem. J.* **80,** 25p–26p.

Thompson, J. N., Howell, J. McC., and Pitt, G. A. J. (1964). *Proc. R. Soc. London, Ser. B* **159,** 510–535.

Tosukhowong, P., and Olson, J. A. (1978). *Biochim. Biophys. Acta* **529,** 438–453.
van Dorp, D. A., and Arens, J. F. (1946). *Nature (London)* **158,** 60.
Vane, F. M., and Buggé, C. J. L. (1981a). *Ann. N. Y. Acad. Sci.* **359,** 424–426.
Vane, F. M., and Buggé, C. J. L. (1981b). *Drug Metab. Dispos.* **9,** 515–520.
Vane, F. M., Buggé, C. J. L., and Williams, T. H. (1982a). *Drug Metab. Dispos.* **10,** 212–219.
Vane, F. M., Stoltenborg, J. K., and Buggé, C. J. L. (1982b). *J. Chromatog.* **227,** 471–484.
Varma, R. N., and Beaton, G. H. (1972). *Can. J. Physiol. Pharmacol.* **50,** 1026–1037.
Wald, G. (1937). *Nature (London)* **139,** 1017–1018.
Wald, G. (1950). *Biochim. Biophys. Acta* **4,** 215–228.
Wald, G. (1968). *Science* **162,** 230–239.
Wang, C.-C., and Hill, D. L. (1977). *Biochem. Pharmacol.* **26,** 947–950.
Wang, C.-C., Cambell, S., Furner, R. L., and Hill, D. L. (1980). *Drug Metab. Dispos.* **8,** 8–11.
Wolf, G., Kahn, S. G., and Johnson, B. C. (1957). *J. Am. Chem. Soc.* **79,** 1208–1212.
Wolf, G., Bergan, J. G., and Sundaresan, P. R. (1963). *Biochim. Biophys. Acta* **69,** 524–532.
Wolf, G., Kiorpes, T. C., Masushige, S., Schreiber, J. B., Smith, M. J., and Anderson, R. S. (1979). *Fed. Proc., Fed. Am. Soc. Exp. Biol.* **38,** 2540–2543.
Wright, G. J. (1961). *Am. J. Clin. Nutr.* **9,** 26.
Yagishita, K., Sundaresan, P. R., and Wolf, G. (1964). *Nature (London)* **203,** 410–412.
Yoshikami, S., Pearlman, J. T., and Crescitelli, F. (1969). *Vision Res.* **9,** 633–646.
Zachman, R. D., and Olson, J. A. (1961). *J. Biol. Chem.* **236,** 2309–2313.
Zachman, R. D., and Olson, J. A. (1964). *Nature (London)* **201,** 1222–1223.
Zachman, R. D., and Olson, J. A. (1965). *J. Lipid Res.* **6,** 27–32.
Zachman, R. D., Dunagin, P. E., and Olson, J. A. (1966a). *J. Lipid Res.* **7,** 3–9.
Zachman, R. D., Singer, M. B., and Olson, J. A. (1966b). *J. Nutr.* **88,** 137–142.
Zile, M., and DeLuca, H. F. (1965). *Biochem. J.* **97,** 180–186.
Zile, M., and DeLuca, H. F. (1968). *J. Nutr.* **94,** 302–308.
Zile, M. H., Emerick, R. J., and DeLuca, H. F. (1967). *Biochim. Biophys. Acta* **141,** 639–641.
Zile, M. H., Schnoes, H. K., and DeLuca, H. F. (1980a). *Proc. Natl. Acad. Sci. U.S.A.* **77,** 3230–3233.
Zile, M. H., Inhorn, R. C., and DeLuca, H. F. (1980b). *J. Nutr.* **110,** 2225–2230.
Zile, M. H., Inhorn, R. C., and DeLuca, H. F. (1982a). *J. Biol. Chem.* **257,** 3537–3543.
Zile, M. H., Inhorn, R. C., and DeLuca, H. F. (1982b). *J. Biol. Chem.* **257,** 3544–3550.

12

Cellular Biology and Biochemistry of the Retinoids

ANITA B. ROBERTS AND MICHAEL B. SPORN
Laboratory of Chemoprevention
National Cancer Institute
National Institues of Health
Bethesda, Maryland

THE RETINOIDS, VOL. 2

ISBN 0-12-658102-9

I. INTRODUCTION

This chapter reviews the cellular mechanism of action of retinoids, which is the central unsolved problem in retinoid biology and biochemistry. The problem of the role of retinoids in maintaining the normal physiology of the body as a whole, and of its various constituent organs, has been essentially solved. However, at the cellular and molecular levels of investigation, it is not yet possible to present a consistent, unifying analysis of the mechanism of action of retinoids which provides a comprehensive explanation of the experimental data.

The many advances in *in vitro* culture of embryos, of differentiating epithelia, and especially of isolated cell types have provided new impetus to the search for a molecular mechanism of action of the retinoids. However, as will become readily apparent to the reader, the cellular effects of retinoids in the induction and control of various biochemical processes are diverse. In fact, if one picks and chooses appropriate cellular systems, appropriate concentrations of retinoids, and appropriate time points, then one can eventually find a set of data that will corroborate almost any preconceived notion relating to the effects of retinoids on cell proliferation and differentiation. For example, the demonstrated antipromoter effects of retinoids in carcinogenesis (see Chapter 14) have prompted numerous investigations comparing cellular effects of retinoids with those of tumor promoters such as the phorbol esters. These effects are so profound and so diverse that it is not difficult to obtain data in a single system that will substantiate particular prejudices about the desirability or undesirability of retinoids for therapeutic purposes. Often even within the same experimental system the results obtained with retinoids are biphasic; a low concentration of retinoid will produce one effect, and a very high concentration the opposite effect. Results obtained at an early time point after administration of the retinoid may be entirely the opposite of those obtained at a later time.

Some of the confusion resulting from this diversity of cell-specific effects can be resolved by considering a retinoid to be merely an effector that closes a cellular switch. It is likely that there are cellular receptors (yet to be defined) for retinoids somewhere within the cell or on the cell surface. More than one set of critical receptors may exist for biologically active retinoids. These hypothetical receptors may be considered to be the equivalent of cellular switches that become "closed" when a retinoid interacts with them. The data accumulated from many studies suggest a remarkable similarity in the specificity of the molecular fit between active retinoids and the receptor system in all sorts of different cells, be they epithelial, mesenchymal, neoplastic, or nonneoplastic. However, once this cellular switch has been closed by an active retinoid, a large number of different responses occur in different cells, ranging from enhancement of proliferation to suppression of proliferation, and from enhancement of differentiation to suppression of differentiation. The information for these varied responses is not in the signal (the effector substance); it is in the constellation of cellular events that the signal can influence. Thus, one cannot speak of the generic activity of a retinoid; one can only consider its particular activity in a particular defined cellular system. This is hardly a new concept and is one that is well-known in classical pharmacology. As an example, it has long been known that the results of the interaction of the effector epinephrine with the β-adrenergic receptor vary greatly, depending on the type of cell in which the receptor is located. Epinephrine is a powerful stimulant of contraction of cardiac muscle and of arterial smooth muscle; in other cell types, such as the smooth muscle of the bronchi, the gastrointestinal tract, and the bladder, interaction of epinephrine with the β-adrenergic receptor leads to relaxation of muscle (Weiner, 1980). Thus, one cannot state in general whether epinephrine makes muscle contract or relax. The information for the nature of the response is clearly *not* in the effector substance but in the cellular context of the receptor (Roth, 1981). This point has not been appreciated adequately in studies on the cell biology of retinoids and unfortunately has led to some rather meaningless polemics.

During the past 10 years, there has been an intense level of investigation of the ability of retinoids to modify the processes of cell differentiation and proliferation. Many of these investigations have been based on the observations that retinoids can interfere with the progression of neoplastic transformation and that they can, in certain instances, control the differentiation and proliferation of fully neoplastic cells. The availability of a large selection of transformed cell lines and the well-studied *in vitro* effects of tumor promoters have facilitated this experimental approach. Thus, as will become apparent in the development of this chapter, it is impossible to review the *in vitro* effects of retinoids or to discuss the mechanism of action of retinoids without including a review of their effects in neoplastic systems. It is hoped that an understanding of the mechanisms operative in control of "abnormal" growth and differentiation will lead to an understanding of the "normal" processes. Since carcinogenesis is essentially a disorder of cell growth and differentiation, the overall scientific problem of the role of retinoids in either differentiation or carcinogenesis is essentially the same problem (Sporn and Roberts, 1983).

The results of these investigations of the effects of retinoids on both nonneoplastic and neoplastic cells are reviewed in detail in the following pages, with emphasis on the most recent literature (as of this writing); nearly 75% of the references cited have appeared since 1978. In addition, the reader may wish to consult reviews on the subject by Lotan (1980), Schroder *et al.* (1983), Sporn and Roberts (1983), Wolf (1980, 1983), and Zile and Cullum (1983). However, it will become apparent that the data obtained from these studies of retinoid effects on enzymes, effectors, structural proteins, and glycoconjugates do not yet lead to a satisfactory unifying hypothesis. Toward this end, we attempt, in the conclusion to the chapter, to suggest new experimental approaches that might allow more direct investigation of the interaction of retinoids with the genetic apparatus of the cell.

II. EFFECTS OF RETINOIDS ON EMBRYONIC DEVELOPMENT *IN VITRO*

Before considering the effects of retinoids on isolated cellular systems, either in organ culture or cell culture, we briefly note some of the marked effects that retinoid deficiency or excess has on the developing embryo. For many years, retinoids have been known to be potent teratogens and an extensive summary of their teratogenic effects can be found in review articles (Kalter and Warkany, 1961; Kochhar, 1967, 1968; Shenefelt, 1972; Sporn *et al.*, 1981) and in Chapter 13. Apart from the clinical importance of these phenomena, they may also provide valuable clues relating to the mechanism of action.

The most striking teratogenic effects that have been described in experimental studies in rodents are the production of various craniofacial abnormalities that resemble those which also occur in human embryos. These abnormalities include malformation of many parts of the cartilaginous and bony facial and oral skeleton such as the mandible, maxilla, palate, as well as the external ear (Kalter and Warkany, 1961; Shenefelt, 1972). The development of defined systems for isolated explant culture of both mammalian and avian embryos (in which essentially normal development takes place during a brief period *in vitro*) has allowed investigation of the mechanism of teratogenesis by retinoids (New, 1966; New *et al.*, 1976). The levels of retinoids that are required to produce these teratogenic effects *in vitro* are extremely small and suggest that these responses are quite selective, rather than being the reflection of a generalized toxic action of the retinoids on the embryo as a whole. The effects of both retinol (A1) and retinoic acid (D1) on cells of mesenchymal origin are particularly noteworthy. In isolated 9-day-old rat embryos, retinoic acid (1.7×10^{-6} *M*) prevents the formation of the pharyngeal arches, which are derived from cephalic mesenchyme (Morriss and Steele, 1968); these structures later form the maxilla and the mandible. Another mesenchymal derivative whose formation is markedly suppressed by retinoic acid in rat embryos is the yolk sac circulation (Morriss and Steele, 1977); similar observations have been made in isolated chick embryos (M. B. Sporn and D. L. Newton, unpublished observations).

The mechanism of the selective inhibition of the formation of these structures derived from mesenchyme is not known. It has been shown that retinoids can inhibit the migration of primitive mesenchymal cells, either within the embryo itself (Kwasigroch and Kochhar, 1975) or on glass coverslips (Morriss, 1976). Alternatively, retinoic acid may have some selective effect on differentiation or proliferation of the mesenchymal cells which are the precursors of the structures that do not form. It is of interest that retinoid deficiency as well as excess may also cause similar selective malformation of structures derived from mesenchyme in the early embryo. Thus, Thompson *et al.* (1969) have shown that retinoid deficiency causes a conspicuous failure to form a functional yolk sac circulation in chick embryos that have been incubated approximately 48 h; thus, both retinoid excess and retinoid deficiency yield the same final result with respect to the developing circulatory system. No molecular interpretation can yet be made of any of the above observations. They have been noted only to suggest that they may provide useful systems in the future for molecular studies, particularly if one wishes to develop a genetic analysis of retinoid action. As it now appears that "oncogenes" may have some critical role in the process of embryological development (Mintz and Fleischman, 1981; Müller *et al.*, 1982), one may anticipate that the study of the selective effects of retinoids on the developing embryo may provide a useful approach to future studies of molecular mechanisms of retinoid action.

III. EFFECTS OF RETINOIDS ON DIFFERENTIATING EPITHELIA IN ORGAN CULTURE

The effects of retinoids on epithelial differentiation have been discussed at length in Chapter 5, Vol. 1. It is clear that retinoids can exert a powerful influence on cell differentiation in many different types of epithelia. The original experiments were those of Fell and Mellanby (1953), who showed that retinol or retinyl acetate (A3) could change keratinized skin epithelium into one that secreted mucus and also had ciliated cells. In 1955 Lasnitzki showed that the premalignant phenotype of mouse prostate glands that had been treated with the carcinogen 3-methylcholanthrene could be altered by retinoid treatment in organ culture. The atypical epithelial cells that were induced by the carcinogen disappeared, and they were replaced by cells with more normal morphology. The effects of the retinoids were thus to suppress abnormal cellular differentiation that had been induced by the carcinogen and to restore a more normal pattern of epithelial differentiation.

Details of these two experimental systems and the further use of organ cultures for studying epithelial differentiation are presented in Chapter 5, Vol. 1. Although organ culture studies were of major significance in making the initial observations on the critical role of retinoids in control of differentiation, organ culture methodology has now been replaced by cell culture techniques for molecular analysis of mechanism, as discussed in the next section.

IV. EFFECTS OF RETINOIDS ON PROLIFERATION AND DIFFERENTIATION OF CELLS IN CULTURE

The effects of retinoids on differentiation of epithelia in organ culture undoubtedly result from a combination of complex cellular responses and interactions of different cell types in the explant. These systems therefore have definite limitations for analysis of the molecular mechanism of action of the retinoids. The introduction of cell culture methodology to such studies was therefore of great importance and now is allowing molecular investigation of the role of retinoids in proliferation and differentiation.

Culture of many cells, both neoplastic and nonneoplastic, has become routine, and many investigators have utilized these more defined systems for the study of retinoids. Many studies have shown that retinoids play a role in controlling both proliferation and differentiation of cells in culture. In certain systems retinoids act to inhibit cell growth, while in others retinoids can be shown to stimulate cell growth. In more specialized cell culture systems retinoids have been shown to promote differentiation of cells, while in others retinoids act to suppress differentiation. Depending on the cell types studied, certain investigators have chosen to emphasize the role of retinoids in control of proliferation, whereas others have emphasized the role in control of differentiation. The actual separation of these effects may be more semantic than real; this may become apparent when complete genetic analyses using recombinant DNA techniques are available.

In this section we first discuss the effects of retinoids on nonneoplastic cells in culture and then the effects of retinoids on cultured neoplastic cells. As discussed earlier, the cell types affected are not limited to epithelial cells but include also cells of mesenchymal origin and cells derived from neural ectoderm. Since most of these cell culture systems require serum, they inherently include retinoids. However, it is not known whether all of these cell types have a *requirement* for retinoids. Therefore, one should always attempt to relate studies of retinoid action *in vitro* to the function of those particular cell types *in vivo*.

In this discussion, the following points will be addressed whenever possible.

1. The concentration of retinoid employed in the study is extremely important. Only those effects demonstrated to be noncytotoxic are discussed in this

chapter. Tolerances of various cell types to retinoids appear to vary over several logs of concentration, but it is clear that effects at physiological ($< 10^{-6}$ *M*) rather than pharmacological ($> 10^{-6}$ *M*) concentrations should be given greater consideration. In this regard, studies showing the concentration dependence of the effects of retinoids are most useful. The quantity of serum used in the culture must be considered as well. In certain instances the effect of retinoids on cells is diminished by the addition of serum, which acts to reduce the concentration of unbound retinoid. In other situations, the presence of serum proteins such as retinol-binding protein may be necessary for the appropriate effect of a retinoid in certain cells that have specific cell-surface receptors for retinol-binding protein (see Chapter 8).

2. Selectivity of response to retinoids will be discussed, keeping in mind that certain isolated cell types have restricted metabolic activity. Particularly revealing are those studies that have employed a variety of retinoids.

3. Whether the cells are derived from an established cell line or from primary cultures is another element that must be considered. While primary cultures most closely resemble the parent cell type *in vivo*, most cell culture experiments have been carried out using established cell lines, many of which are aneuploid. The ready availability of these established cell lines, many of which are neoplastic, and their reproducible response pattern have made them the subject of numerous investigations of retinoid action. However, their physiological relevance may be questionable; therefore it is important, whenever possible, to relate the observed *in vitro* effects to similar effects of retinoids *in vivo*.

4. Whenever available, data concerning the presence of specific intracellular retinoid-binding proteins are included, especially as they might relate to the mechanism of action of the retinoids (for a detailed discussion, see Chapter 9).

A. Effects on Nonneoplastic Cells

Retinoids have been shown to exert effects on differentiation of both epithelial cells (keratinocytes) and mesenchymal cells (chondrocytes, adipocytes, and hematopoetic cells) in culture, much as had been observed both *in vivo* (Wolbach and Howe, 1925) and in organ culture *in vitro* (see previous section). In certain other cells, only growth and not differentiation is affected. Whether both types of effects have a common basis will have to await mechanistic studies.

1. Effects on Epithelial Cells

a. KERATINOCYTES. It has clearly been shown both *in vivo* (Wolbach and Howe, 1925) and in organ culture *in vitro* (see Chapter 5, Vol. 1) that the expression of differentiation in keratinizing stratified squamous epithelium is enhanced in retinoid deficiency and suppressed in retinoid excess. Experiments using primary cultures of keratinocytes elegantly demonstrate these effects at the cellular and molecular level. Green and co-workers have characterized the effects of retinoid deficiency on various subtypes of human keratinocytes, whereas Yuspa and co-workers have characterized the effects of added retinoids on both normal and Ca^{2+}-induced differentiation of keratinocytes derived from newborn mice. Green (1977) showed that although human epidermal keratinocytes cultured in the presence of 10–20% fetal bovine serum (estimated to contain 4- to 8×10^{-8} *M* retinol equivalents) undergo terminal differentiation and stratification, they fail to form a typical stratum corneum, even after several weeks in culture. In addition, these cultures did not synthesize the large (> 60-K*) keratins previously found to be characteristic of the strateum corneum but rather continued to synthesize smaller keratins (46–58K) typical of deeper cells (Fuchs and Green, 1980). However, if the cells were grown in 10% delipidized serum [serum from which the retinoid has been removed by solvent extraction; Rothblat *et al.* (1976)], the pattern of keratin synthesis changed (Fuchs and Green, 1981). After 2–3 weeks in the retinoid-deficient medium, synthesis of the 67-K keratin typical of epidermis was elevated and synthesis of the 40-K and 52-K keratins was suppressed. This effect could be reversed by addition of retinyl acetate to the medium. Interestingly, conjunctival keratinocytes, which typically express the 40-K and 52-K keratins (Sun and Green, 1978) could also be induced to express the larger 67-K keratin when grown in the retinoid deficient medium, thus mimicking the results of vitamin A deficiency *in vivo* (Wolbach and Howe, 1925). *In vitro* translation of isolated keratin-specific poly(A)$^+$RNA demonstrated that the retinoid-dependent control of keratin synthesis occurred at the transcriptional level; thus, for example, a 67-K keratin was translated from mRNA of cells grown in the absence of retinoid, and a 52-K keratin was translated from mRNA of cells grown in the presence of retinyl acetate. Further investigations by Green and Watt (1982) have shown that the addition of retinyl acetate to human keratinocyte cultures maintained in retinoid-deficient delipidized serum can suppress the formation of cornified envelopes, even though retinoids had little effect on the concentration of an envelope precursor protein, involucrin. The degree of inhibition varied for keratinocytes derived from conjunctiva, esophagus, vagina, or epidermis; synthesis of cornified envelopes by conjunctival keratinocytes was completely inhibited at 10^{-8} *M* retinyl acetate while synthesis by epidermal keratinocytes was only slightly inhibited at 10^{-6} *M*. Thus the sensitivity of these cultured keratinocytes to retinoids accurately

*Abbreviations used are: K, kilodaltons; TPA, 12-*O*-tetradecanoylphorbol-13-acetate; TGF, transforming growth factor; EGF, epidermal growth factor; SGF, sarcoma growth factor; ODC, ornithine decarboxylase; TG, transglutaminase; GAG, glycosaminoglycan; RetP, retinyl phosphate; DolP, dolichyl phosphate; MRP, β-mannosyl retinyl phosphate; MDP, β-mannosyl dolichyl phosphate; CRBP, cellular retinol-binding protein; CRABP, cellular retinoic acid-binding protein; TMMP, trimethylmethoxyphenyl; DFMO, α-difluoromethylornithine.

reflects the degree of cornification of these tissues *in vivo*; the conjunctiva is not cornified, whereas the epidermis forms a thick cornified layer.

Yuspa and co-workers have taken a different approach to the problem of differentiation of keratinocytes in culture. Using primary cultures of newborn mouse epidermal cells (Yuspa and Harris, 1974), they showed that retinyl acetate (10^{-5} *M*) prevented the appearance of a keratin-like product and increased glycoprotein synthesis in these cells; this response in some ways resembles the mucus metaplasia observed in organ culture. Later, they showed that extracellular Ca^{2+} is a critical regulator of epidermal cellular proliferation and differentiation (Hennings *et al.*, 1980). At concentrations of Ca^{2+} in the medium of 0.02–0.1 m*M*, epidermal basal cells proliferate rapidly and do not stratify or differentiate to cornified cells; elevation of the Ca^{2+} concentration to 1.2 m*M* blocks proliferation of the cells and induces terminal differentiation with sloughing of cornified sheets of cells into the medium. Addition of retinoic acid (10^{-6} *M*) to the low Ca^{2+} medium did not significantly affect cellular behavior, but its continuous presence upon shifting to the high Ca^{2+} medium reduced stratification and the formation of sheets of squamous cells and maintained a more undifferentiated morphology (Yuspa *et al.*, 1981). Retinoic acid, however, could only delay the Ca^{2+}-induced differentiation but not prevent it.

Accompanying the suppression of differentiation, it could be shown that retinoic acid treatment of these cells suppressed cornified envelope production (Yuspa *et al.*, 1982), in agreement with the data of Green and Watt (1982). This observation is particularly interesting in light of the finding that retinoic acid treatment actually elevated the amount of a Ca^{2+}-dependent epidermal transglutaminase that is known to participate in the production of cornified envelopes (Yuspa *et al.*, 1981, 1982). Retinoids must therefore interfere with the intracellular activity of this enzyme via either a direct or an indirect mechanism. One mechanism that has been proposed is the stabilization of the cell membrane by retinoids, thus preventing the Ca^{2+} influx assumed to be necessary for cornified envelope production. In support of this concept, Yaar *et al.* (1981) using suspension cultures of guinea pig ear keratinocytes, and Green and Watt (1982) using cultures of human conjunctival keratinocytes, have shown that treatment of cells with membrane-active agents, such as the ionophore X537A or the detergent Triton X-100, reverses the inhibitory effect of retinoids on cornified envelope production.

Several other experiments on cultured epidermal cells deserve mention. Newborn mouse epidermal cells cultured in serum-free defined medium respond to retinoid treatment in a dose-dependent manner, with an increase in both DNA and RNA synthesis (Sporn *et al.*, 1973). The response is extremely sensitive with an ED_{50} of 4×10^{-10} *M* for all-*trans*-retinoic acid (Sporn *et al.*, 1976) and has been standardized in a sensitive assay for measurement of structure–function relationships of a diverse series of retinoids (Sporn *et al.*, 1976; see also Chapter 5, Vol. 1). Both Christophers (1974) and Wilkinson (1978) using guinea pig ear keratinocytes, and Chopra and Flaxman (1975) using human keratinocytes, showed a dramatic (6- to 10-fold) increase in DNA synthesis in the first 48 h after plating in the presence of retinoic acid (10^{-5} *M*). However, Yuspa *et al.* (1977) using mouse epidermal cells, showed that retinyl acetate decreased the rate of cellular proliferation (DNA synthesis), although it increased cell survival. Further experiments by Wilkinson and Orenberg (1980) demonstrated that the mitogenic effect of retinoic acid is observed only when it is added at the time of plating, and not if it is added either 14 h or 14 days after plating. Whether or not these differences derive from the use of different retinoids or from experimental design has not been resolved.

Cell-surface changes also accompany the effects of retinoids on epidermal cells. Christophers and Wolff (1975) using cultured epidermal cells derived from adult guinea pig ear skin fragments, showed that addition of retinoic acid (10^{-5} *M*) increased the number of hemidesmosomes formed by the plasma membrane; stratification and differentiation, on the other hand, is characterized by the formation of desmosomes. Hemidesmosome formation was accompanied by decreased cell–cell adhesion and increased cell–substratum adhesion, as had been noted earlier by Yuspa and Harris (1974).

b. INTESTINAL CELLS. Results of Shapiro and Poon (1979) using an established human embryonic intestinal cell line are in striking contrast to those just discussed. Retinoic acid treatment of these cells brought about a reversible density-dependent inhibition of growth that was most evident at lower cell densities. This was accompanied by morphological change to a more fibroblast-like appearance; trypsinization brought about an earlier reversal of these effects than did scraping of the cells, suggesting retinoid-induced cell surface changes. However, until more detailed biochemical studies are undertaken, the relationships between the effects of retinoids on these cells and on cultures of stratified keratinizing epithelial cells will remain unclear.

c. TESTICULAR CELLS. Two clonal testicular cell lines of Leydig (TM3) and Sertoli (TM4) origin have been isolated and cultured in serum-free medium supplemented with insulin, transferrin, and epidermal growth factor (Mather, 1981). Interestingly, the TM4 cell line responds to retinoic acid or retinol (10^{-7} *M*) treatment with a significant increase in growth rate, whereas the growth of the TM3 cell line is inhibited. Growth of a transformed variant of the TM4 cell line was also inhibited by retinoic acid at concentrations as low as 10^{-8} *M*. Retinoid treatment also changes the gonadotropin responsiveness of these cells, as discussed in Section IV,E.

2. *Effects on Cells of Mesenchymal Origin*

a. CHONDROCYTES. Retinoic acid (10^{-5} *M*) inhibits chondrogenesis in mesenchymal cells derived from prechondrogenic limb buds of mouse embryos; confluent cell layers of retinoic acid-treated cells remain mesenchymal in appearance, do not form cartilagenous nodules, and do not synthesize the sulfated glycosaminoglycans characteristic of cartilage (Hassell *et al.*, 1978a; Lewis *et al.*, 1978; Pennypacker *et al.*, 1978). Retinoids also affect the mature chondrocyte, inducing transformation into fibroblastic cells 1–2 days after their addition (Takigawa *et al.*, 1980, 1982) and inhibiting glycosaminoglycan synthesis (Vasan and Lash, 1975; Shapiro and Poon, 1976). Trechsel *et al.* (1982) have shown that dedifferentiation of rabbit articular chondrocytes induced by retinol (10^{-6} *M*) treatment is accompanied by a switch in collagen synthesis from the cartilage-specific type II in favor of type I collagen. These effects of retinoids on chondrocytes in culture are probably related to certain of the effects of retinoids on embryonic development in culture (see Section II) and the reported teratogenic effects of retinoids on limb formation in the developing embryo (Kochhar, 1973; see also Chapter 13).

b. ADIPOCYTES. Another differentiating system derived from mesenchyme is that of adipocyte conversion. Both Sato *et al.* (1980) and Murray and Russell (1980), using different cloned lines of murine preadipocytes, have shown that retinoic acid can block terminal differentiation into mature adipocytes in a dose-dependent reversible fashion. In these experiments, in contrast to many of the foregoing, retinoic acid at concentrations up to 10^{-5} *M* had no effect on the morphology, growth rate, or saturation density of the preadipocytes (Sato *et al.*, 1980), yet triglyceride formation and increase in the number of insulin receptors (two markers of differentiation) were significantly inhibited at 10^{-8} *M* retinoic acid; differentiation progressed on removal of the retinoid. In the experiments of Murray and Russell (1980), conversion of preadipocytes to adipocytes was triggered by exposure of the cells to 1-methyl-3-isobutylxanthine plus dexamethasone; in order to block the conversion, retinoic acid (10^{-11}–10^{-5} *M*) had to be added simultaneously with these drugs. Treatment of the differentiated adipocytes with 10^{-6} *M* retinoic acid had no effect as assessed by measurement of the enzymatic activity of fatty acid synthetase. Thus, in this system as well as in the chondrogenic differentiation just discussed, the timing of retinoic acid treatment is critical, suggesting that physiological regulation of retinoid concentration could be important in controlling the early events in cytodifferentiation.

c. FIBROBLASTS. From other experiments, it is clear that retinoids have effects on nondifferentiating mesenchymal cells as well. Several murine fibroblastic cell lines including 3T3, 3T6, MB4-7 (cloned from a mouse blastocyst-derived cell line), as well as primary cultures of murine dermal fibroblasts respond to retinoic acid treatment (10^{-8}– 10^{-5} *M*) with an increase in doubling time, decrease in growth rate, a general flattening of cellular morphology, and a reduction in saturation density (Jetten *et al.*, 1979b; Schroder *et al.*, 1982). However, in certain situations, a retinoid-dependent increase in growth of 3T3 cells has been observed (Dicker and Rozengurt, 1979; Schroder *et al.*, 1982). In 3T6 cells (Jetten *et al.*, 1979b) and in C3H/10T1/2 mouse fibroblasts (Bertram, 1980), retinoid treatment results in a reversible increase in cell-to-substratum adhesion. These changes appear to result from an alteration in the pattern of glycoprotein synthesis by these cells, but the structure–function (adhesion) relationships of a series of retinoids are somewhat unusual and suggest a mechanism unrelated to that of growth regulation (Bertram, 1980; Bertram *et al.*, 1981; see also Section VI and Chapter 5, Vol. 1).

The response of primary cultures of human fibroblasts to retinoid treatment has also been examined (Lacroix *et al.*, 1981). Consistent with the results on murine fibroblast cell lines, those cells responded to retinoic acid (10^{-8}–10^{-5} *M*) or to retinol with a decrease in exponential growth rate and a decrease in saturation density. However, cellular morphology and cell-to-substratum adhesiveness were unchanged by retinoid treatment. Cytoplasmic retinoic acid-binding protein has been identified in both the murine 3T3 and 3T6 cell lines (Jetten *et al.*, 1979b) and the human fibroblast cultures (Lacroix *et al.*, 1981); no cytoplasmic retinol-binding protein could be detected in the human fibroblasts.

d. HEMATOPOIETIC CELLS. Effects of retinoids in the control of growth or differentiation can also be observed in hematopoietic cells, which are of mesenchymal origin. Early morphological studies demonstrated a reduction of hematopoietic cells in the bone marrow of vitamin A-deficient rats (Findlay and McKenzie, 1922; Wolbach and Howe, 1925), and an association has been made between anemia and vitamin A deficiency in a group of otherwise healthy men (Hodges *et al.*, 1978). Two studies, one on the effects of retinoic acid on the growth of erythroid progenitor cells in culture (Douer and Koeffler, 1982a) and the other on the effects of retinoic acid on the clonal growth of myeloid progenitor cells (Douer and Koeffler, 1982b), indicate that these results can be reproduced *in vitro* in cell culture. Normal human peripheral blood cells form colonies in methyl cellulose (burst-forming units-erythroid, BFU-E) when plated in the presence of erythropoietin. Colony formation is increased over twofold in a dose-dependent manner by addition of either 13-*cis*- (E6) or all-*trans*-retinoic acid (D1); a concentration of 3×10^{-8} *M* gave a maximal response. Granulopoiesis, in culture, can be promoted by the addition of colony-stimulating factor to normal human bone marrow stem cells suspended in soft agar (Douer and Koeffler, 1982b). The number of colonies of granulocytes and macrophages so formed was increased twofold by the addition of 3×10^{-7} *M* retinoic acid (13-*cis* or all-*trans*). A 10-fold higher concentration of either retinaldehyde (C1) or retinyl acetate was required to enhance colony formation maximally, but this was only 50% of the stimulation seen with retinoic acid. Retinol was inactive in both of these systems.

The normal levels of retinoic acid in human plasma have been determined to be approximately 10^{-8} *M* (De Ruyter *et al.*, 1979; De Leenheer *et al.*, 1982). However, clinical studies with 13-*cis*-retinoic acid have demonstrated that plasma concentrations as high as 10^{-6} *M* are attainable in man following oral dosing with the retinoid (Frolik *et al.*, 1978). Therefore it is reasonable to assume that 13-*cis*-retinoic acid may have useful applications in promoting erythropoiesis or granulopoiesis in clinical situations.

e. NEUTROPHILS. Using a different approach, Carmisa *et al.* (1982) surveyed the effects of a 10-min exposure of freshly isolated human neutrophils to a series of retinoids. They found all-*trans*-retinoic acid to be 100 times more effective than its 13-cis isomer in the inhibition of superoxide anion production ($ED_{50} = 3 \times 10^{-6}$ *M*), whereas 13-*cis*-retinoic acid was six times more effective than the all-trans isomer in the inhibition of lysosomal enzyme release ($ED_{50} = 1.5 \times 10^{-5}$ *M*). These results are in agreement with those of Witz *et al.* (1980) who found that retinoids inhibited the generation of superoxide anion by neutrophils stimulated by phorbol ester tumor promoters. These observations may be relevant to certain aspects of the effectiveness of retinoids in suppression of tumor promotion or in treatment of cutaneous diseases in which neutrophils are the predominant cell type in the accompanying inflammatory response (Plewig and Wagner, 1981; also see Chapter 16). However, the unusual structure–activity relationships, the short exposure time, as well as the reversal of the effects by serum, all suggest that these effects are unrelated to the role of retinoids in control of cell growth or differentiation of hematopoietic cells (see, for example, the role of retinoids in promotion of differentiation of neoplastic promyelocytic leukemia cells, Section IV,B,2).

f. MACROPHAGES. In agreement with early descriptions of changes in the reticuloendothelial system associated with vitamin A deficiency (De Ruyter, 1934; Frank, 1935), studies have shown that macrophage function can be modulated *in vitro* by treatment with retinoids (Rhodes and Oliver, 1980). Retinoic acid (10^{-7} *M*), or to a lesser extent retinol, suppressed the expression by human peripheral blood monocytes of both cellular receptors for antibody (Fc receptors) and of associated phagocytic functions; in the absence of retinoids, Fc receptors increased significantly in culture as the monocytes acquired a morphology characteristic of adherent macrophages. In cultures of guinea pig peritoneal macrophages, retinoic acid (10^{-9}–10^{-7} *M*) increased intracellular levels of the tumoricidal enzyme arginase. The involvement of macrophages in induction of the immune reponse, in resistance to infection, and in antitumor host defense suggests that further investigations utilizing these cells might help to clarify the role of retinoids in the immune system (see Chapter 15).

B. Effects on Neoplastic Cells

Retinoids have been shown to have striking effects on the growth and differentiation of many fully neoplastic cell types growing in culture, in some instances appearing to make these cells responsive to control mechanisms operative on their untransformed counterparts (for a review, see Lotan *et al.*, 1983c). Neoplastic cell types classified as epithelial (tracheal, bladder, breast, or cervical carcinoma), mesenchymal (myeloid leukemia, chondrosarcoma, osteosarcoma, L cells), as well as cells derived from neural ectoderm (melanoma, neuroblastoma) and primitive ectoderm (embryonal carcinoma) have all been shown to respond to treatment with retinoids. Most likely, the changes in growth potential of such responsive cells result from a block or retardation of the progression of cells through the cell cycle or from enhanced differentiation or maturation of the cells (Sachs, 1978, 1980). Observed effects of retinoids include inhibition of growth in monolayer, inhibition of anchorage-independent growth in semi-solid medium [generally thought to be the best *in vitro* correlate of tumorigenicity (Kahn and Shin, 1979; Cifone and Fidler, 1980)] and promotion of the terminal differentiation of certain fully neoplastic embryonal carcinoma and promyelocytic leukemia cells to nonneoplastic differentiated cell types. While many of the effects of retinoids on growth of transformed cells are reversible (and thus probably represent suppression of certain aspects of the transformed phenotype), the latter effects of retinoids on differentiation represent a stable alteration of the transformed phenotype to a nonneoplastic phenotype. Such effects may in themselves be able to induce remission of disease, whereas the inhibitory effects of retinoids on the growth of neoplastic cells may prove a valuable adjunct to other therapeutic modes. Discussion of the effects of retinoids on carcinogenesis is deferred to Chapter 14.

1. Embryonal Carcinoma Cells

One of the most fascinating and potentially most illuminating examples of the ability of retinoids to promote differentiation is the effect of retinoids on mouse embryonal carcinoma cells. These undifferentiated stem cells of teratocarcinomas are multipotential, that is, they can differentiate into a multiplicity of somatic cell types (Kleinsmith and Pierce, 1964). When injected into mouse blastocysts, these malignant cells can differentiate and contribute to the normal tissues of the developing mouse (Brinster, 1974; Mintz and Illmensee, 1975). In this way, these cells resemble primordial germ cells and embryonic ectoderm.

However, in certain teratocarcinoma cell lines, the ability to differentiate has been greatly reduced or restricted; such "nullipotent" cells can be propagated as undifferentiated stem cells. The discovery by Strickland and Mahdavi (1978) that retinoic acid at very low concentrations (10^{-7} *M*) could promote the differentiation of the nullipotent F9 embryonal carcinoma cell line to a new, stable, non-neoplastic phenotype opened the way to exploration of this system both as a paradigm for normal embryogenesis and as a model in which to investigate the mechanism of action of retinoids in control of cellular differentiation. The ability of retinoic acid to induce differentiation of teratocarcinoma stem cells has been confirmed in the F9 system (Segal and Khoury, 1979; Kuff and Fewell, 1980; Rizzino and Crowley, 1980; Linder *et al.*, 1981; Howe and Solter, 1981) as well as with other cell lines (Jetten *et al.*, 1979a; Rees *et al.*, 1979). The development by Rizzino and Crowley (1980) of a serum-free culture medium for F9 cells (containing fibronectin, insulin, and transferrin in place of serum) demonstrated that retinoic acid alone is sufficient to induce differentiation in these cells; the sensitivity of the differentiation to retinoic acid was 100-fold greater in the absence of serum.

The key to the study of the effects of retinoids on teratocarcinoma cells was the discovery that primitive endoderm cells produce the serine protease plasminogen activator (Strickland *et al.*, 1976). These are one of the first cell types to differentiate from embryonal carcinoma cells. Using this enzyme as a marker of differentiation, Strickland and Mahdavi (1978) were able to quantitate the degree of differentiation of F9 cells and show that optimal differentiation requires 72 h and depends on the concentration of retinoic acid; a detectable response could be observed at 10^{-9} *M* retinoic acid. Interestingly, these primitive cells utilize retinaldehyde or retinol 1000-fold less effectively than retinoic acid, suggesting that they are devoid of retinoid-metabolizing enzymes. The potential use of this system as an assay for screening of new retinoids for biological activity is discussed in Chapter 5, Vol. 1. In an extension of these initial observations, Strickland *et al.* (1980) have shown that retinoic acid treatment of the F9 cells converts the cells to an intermediate phenotype that can further differentiate to either visceral or parietal endoderm. Aggregation of the retinoic acid-treated cells gives rise to visceral endoderm (Hogan *et al.*, 1981), whereas exposure of the retinoic acid-treated cells to compounds that elevate intracellular cyclic AMP gives rise to cells characteristic of parietal endoderm (Strickland *et al.*, 1980; Kuff and Fewell, 1980). Cyclic AMP acts only on retinoic acid-treated cells; it has no effect on the F9 cells themselves. The final differentiated phenotypes are stable on removal of the inducing agents. Accompanying the retinoic acid-induced differentiation are changes in morphology and in the pattern of protein synthesis of the cells (Strickland *et al.*, 1980; Linder *et al.*, 1981; Howe, and Solter, 1981); thus retinoic acid induces F9 cells to synthesize the type IV collagen, laminin, and vimentin characteristic of parietal endoderm cells.

These studies have been extended to other murine embryonal carcinoma cell lines by Jetten *et al.* (1979a) and by Sherman *et al.* (1981), who have compared the differentiated phenotype induced by retinoic acid with that induced by simple aggregation of the cells. They made the important observation that retinoic acid promotes but does not influence the direction of differentiation of these cells. In addition, Jetten and Jetten (1979) were able to clone from the cell line PCC4-aza 1, for which retinoic acid was cytotoxic, a derivative PCC4-aza 1 R, which was tolerant to retinoic acid and which could be induced to differentiate by retinoic acid. They found a correlation between the ability of various retinoids to compete for binding to the cellular retinoic acid-binding protein (CRABP) and their ability to promote differentiation of these cells. In further support of a role of the cellular retinoic acid-binding protein in this differentiation, Schindler *et al.* (1981) were able to develop two mutants of the PCC4-aza 1 R cell line that failed to differentiate in response to retinoic acid. The uptake of retinoic acid by these mutants was not different from that of the parental clone, but these mutants [dif(RA)$^-$] were distinguished by the *absence* of the specific cytoplasmic retinoic acid-binding protein, CRABP.

In contrast to the murine embryonal carcinoma cells, several human teratocarcinoma-derived cell lines fail to differentiate when treated with retinoic acid (Matthaei *et al.*, 1983). Yet unlike the mutant murine cells that are refractory to retinoic acid, it has been shown that the human cell lines contain high levels of CRABP. These results and other data presented in the ensuing portions of this chapter make it unclear whether the retinoic acid–CRABP complex plays an obligatory role in the mechanism of promotion of differentiation.

These effects of retinoic acid on the differentiation of murine embryonal carcinoma cells are not an artifact of *in vitro* tissue culture. Both Strickland and Sawey (1980) and Speers (1982) have shown that the differentiation of embryonal carcinoma cells to benign derivatives can be induced *in vivo* either by feeding mice with retinoic acid or by intratumor injection of a combination of retinoic acid and another differentiation-inducing chemical dimethylacetamide. Both treatments increased the survival time of the tumor-bearing host. Using a different approach, Sherman *et al.* (1981) injected mice with an inoculum of the stem cell clone PCC4-aza 1 R or of the same cells that first had been treated *in vitro* with retinoic acid. Whereas mice injected with the parental clone all developed palpable tumors in 10–12 days, six or eight mice injected with the retinoic acid-treated cells had no palpable tumors after 100 days. Thus, in each of these systems, retinoic acid greatly reduced the malignancy of the cells by promoting their differentiation.

2. *Leukemia Cells*

Another example of the ability of retinoic acid to promote terminal differentiation of neoplastic cells to nonneoplastic cell types is the effect of retinoic acid on human promyelocytic leukemia cells. Acute myelogenous leukemia has been viewed as a disease involving an arrest in the normal maturation sequence (Clarkson, 1972; Sachs, 1978, 1980; Lotem and Sachs, 1982). Analogous to the experiments just discussed with teratocarcinoma cells, Gootwine *et al.* (1982) have shown that mouse myeloid leukemia cells injected into embryos can participate in normal development of the hematopoietic system. Various leukemia cell lines can also be induced to differentiate *in vitro* in response to a variety of agents including butyrate and dimethyl sulfoxide; however, as first reported by Breitman *et al.* (1980b) and Honma *et al.* (1980) for suspension cultures of the human promyelocytic leukemia cell line HL-60, retinoic acid is 500–160,000 times as potent as these agents on a molar basis, in inducing terminal differentiation to morphologically and functionally mature granulocytes (Collins *et al.*, 1978). In this system, differentiation can be quantitated by measuring the reduction of a water-soluble dye nitroblue tetrazolium to an insoluble formazan by stimulated mature granulocytes (Collins *et al.*, 1979). Quantitation of this response as well as development of serum-free culture conditions (Breitman *et al.*, 1980a) have made this system amenable to biological assay of various retinoids (see Chapter 5, Vol. 1); all-*trans*- and 13-*cis*-retinoic acid are equally effective, but retinaldehyde (C1), retinol, and retinyl acetate are less than 1/1000 as potent (Breitman *et al.*, 1980b). Newer studies have shown that this system is exquisitely sensitive to the synthetic aromatic series of retinoidal benzoic acid derivatives (G1–G18) as well (Strickland *et al.*, 1983).

In studies employing fresh human myelocytic leukemia cells, Breitman *et al.* (1981) found that the cells of only two patients with promyelocytic leukemia could be induced to differentiate in response to retinoic acid treatment; less-differentiated myeloid leukemia cells from 19 other patients showed no granulocytic differentiation on exposure to retinoic acid. Studies with established cell lines yielded similar results: HL-60 could be induced to differentiate, whereas the human myeloid leukemia cell line KG-1 and the human blast cell line K-562 could not. On the other hand, Douer and Koeffler (1982c) showed that the clonal growth in soft agar of not only HL-60 but also of KG-1 and of fresh cells from five of seven patients with acute myeloid leukemia and acute myelomonocytic leukemia was inhibited by retinoic acid, although differentiation (except for HL-60) was not induced. Inhibition of clonal growth was irreversible but was not due to a cytotoxic effect. Clonal growth of the human K-562 cells or of the murine myeloid leukemia M-1 cells was not affected by retinoic acid. Thus retinoic acid appears to have a stage-specific effect on differentiation and growth of myeloid leukemia cells that is not yet well understood. Interestingly, it has been found that 1,25-dihydroxyvitamin D_3 can induce differentiation of HL-60 cells into monocytes (Miyaura *et al.*, 1981) and differentiation of murine M-1 cells into macrophages (Abe *et al.*, 1981). Elucidation of the specific mechanistic relationship of the action of these two compounds on differentiation of leukemia cells remains an exciting problem for future research that will most likely require unraveling of the genetic events involved in the differentiation process itself. The implication of a role of the active metabolite of vitamin D in the differentiation of these cells is suggestive of a Ca^{2+}-mediated response; we discuss the putative role of Ca^{2+} in the retinoid-mediated differentiation of keratinocytes in Section IV,A,1,a.

Other experiments also relate to the mechanism of differentiation of these leukemia cells. In contrast to the situation of the mouse teratocarcinoma cell lines, where a correlation appeared to exist between responsiveness of cell lines to retinoic acid and the presence of a cellular retinoic acid-binding protein (Schindler *et al.*, 1981), neither the HL-60 cell line (Brietman *et al.*, 1981) nor the retinoic acid-responsive leukemia cells examined by Douer and Koeffler (1982c) have detectable levels of CRABP binding activity. At the very minimum, these data suggest that CRABP is not universally required in the mechanism of retinoid acid-induced effects on cellular differentiation and proliferation. However, Sachs (1978) has suggested that control mechanisms operative on differentiation of hematopoietic cells may not necessarily be identical to those operating on nonmigrating cells that normally maintain strong cell–cell contacts within a tissue.

In another study possibly relating to the mechanism of action of retinoic acid on these myeloid cells, Takenaga *et al.* (1980) using the murine myeloid leukemia cell line M-1 have shown that retinoic acid is able to inhibit in a reversible manner the dexamethasone-dependent differentiation of these cells, and that retinoic acid induces the M-1 cells to synthesize and release into the medium a protease-sensitive inhibitor of differentiation (Takenaga *et al.*, 1981).

Also relevant to the mechanism of action of retinoic acid on these cells are provocative new experiments by Olsson *et al.* (1982), which show that the granulocytic differentiation of HL-60 cells and the differentiation of human monoblast-like lymphoma cells (U-937) to phagocytic monocyte-like cells (Olsson and Breitman, 1982) can be primed by a relatively short (20-h) exposure to retinoic acid, followed by exposure to a cyclic AMP-inducing agent or to a T lymphocyte-derived differentiation-inducing factor. In both cases, the cells must *first* be primed with retinoic acid; the reverse procedure was ineffective. These observations strengthen the analogy of the mechanism of action of retinoic acid in the induction of differentiation in this system and in the teratocarcinoma stem cell system. Possible implications of these results, particularly as they might relate to a retinoic acid-induced cyclic AMP-dependent protein kinase, are discussed in the section on mechanism of action (see Section V,B,1 and 2).

A report by Garg and Brown (1983) shows that retinoids can also promote differentiation of certain cell lines of Friend erythroleukemia cells, inducing cells to synthesize hemoglobin and to increase acetylcholinesterase activity analogous to the response of the cells to the classic inducer dimethyl sulfoxide. Not all Friend leukemic cell lines respond to retinoic acid (Breitman *et al.*, 1980b; Garg and Brown, 1983); the basis for the variability is not understood at the present time.

3. Melanoma Cells

The response of malignant melanoma cells to retinoic acid has also been well characterized. Unlike embryonal carcinoma cells and promyelocytic leukemia cells, in which retinoic acid could induce terminal (irreversible) differentiation to nontumorigenic cell types, melanoma cells treated with retinoic acid respond with a *reverisble* inhibition of growth accompanied by an increase in melanogenesis (for a review, see Lotan *et al.*, 1981). Also, in contrast to the marked *qualitative* changes in protein synthesis that accompany the retinoic acid-induced differentiation of murine embryomal carcinoma cells (Strickland *et al.*, 1980; Linder *et al.*, 1981; Howe and Solter, 1981), retinoic acid treatment of the murine melanoma cell line S91 brings about only minor *quantitative* changes in the proteins expressed (Lotan *et al.*, 1982a).

The response of both murine melanoma cell lines S91 and B16 as well as several human melanoma cell lines to retinoids has been investigated in detail (Lotan *et al.*, 1978, 1981; Lotan, 1979; Meyskens and Fuller, 1980; Maniglia and Sartorelli, 1981). In every case where inhibition of growth was observed, it was dependent on the concentration of retinoic acid and independent of cell density; 48- to 72-h treatment was required before a decrease in cell growth rate could be detected. Both anchorage-dependent and anchorage-independent growth (see Section IV,B,4,c) of these cells was affected by continuous exposure to retinoic acid. It has been suggested that retinoic acid may affect cells in the G_1 phase of growth and slow their entry into S phase (Lotan *et al.*, 1981; Lotan *et al.*, 1982a; see also Section IV,D). Retinoic acid was not cytotoxic to the cells; growth rate was restored to control levels 48–72 h after removal of the retinoic acid (Lotan *et al.*, 1978).

The response of various melanoma cell lines and even of subclones derived from the same melanoma cell line (B16) was heterogeneous (Lotan *et al.*, 1978; Lotan and Nicolson, 1979; Lotan, 1979; Meyskens and Fuller, 1980; Hoal *et al.*, 1982). Among the murine melanoma cell lines, S91 was more responsive to retinoic acid inhibition than was B16 (Lotan *et al.*, 1978). More striking however, was the variable response of clonal derivatives of B16; at 10^{-9} *M* retinoic acid, the growth of the parental clone was inhibited 40%, whereas several cloned derivatives showed no growth inhibition (Lotan and Nicolson, 1979). Of six human melanoma cell lines studied, the growth of two of them was inhibited about 60% by 10^{-6} *M* retinoic acid, three were unaffected, and the growth of one, Hs294, was stimulated approximately two-fold (Lotan, 1979). The basis of this variability is not understood at the present time.

Retinoic acid stimulated melanogenesis, as measured by increased tyrosinase activity and increased melanin production, in certain murine and human melanoma cell lines (Lotan and Lotan, 1980, 1981; Meyskens and Fuller, 1980). In other human melanoma cell lines (UCT-Me12), exposure to retinoids inhibited melanogenesis (Hoal *et al.*, 1982). Melanogenesis induced by retinoic acid could be distinguished from that induced by α-melanocyte stimulating hormone in that the former treatment, unlike the latter, was not accompanied by increased cyclic AMP levels (Lotan and Lotan, 1980, 1981). The human melanoma cell line Hs939, which was not responsive to the hormone, nonetheless could be stimulated to increase melanin production by treatment with retinoic acid. Retinoic acid inhibition of growth was not a result of increased melanogensis; inhibition of tyrosinase activity did not interfere with the growth inhibition (Lotan and Lotan, 1981).

Extensive investigation has been made of the efficacy of different retinoids in the suppression of growth of S91 melanoma cells (see also Chapter 5, Vol. 1). In general, all-*trans*- and 13-*cis*-retinoic acid were the most active retinoids; retinyl acetate, retinaldehyde, and retinol were approximately 1/100 as active (Lotan *et al.*, 1980b). As had been described for differentiation of an embryonal carcinoma cell line (Jetten and Jetten, 1979a), a positive correlation could be made between the ability of retinoids possessing a carboxyl group at C-15 to bind to the cellular retinoic acid-binding protein and to inhibit the growth of S91 melanoma cells. However, the correlation between the actual levels of CRABP (assessed by ligand binding activity) and the responsiveness of cells to retinoic acid is less clear. The S91 melanoma cells had approximately four times the amount of CRABP as did human Hs939 melanoma cells (Lotan, 1980), yet they were nearly 30-fold more sensitive to retinoic acid in the inhibition of growth and the induction of melanogenesis (Lotan and Lotan, 1980, 1981).

Mutants of the wild type S91–C2 melanoma cells have been isolated that are resistant to the antiproliferative action of retinoic acid (Lotan *et al.*, 1983b). The mutant cells were also resistant to growth inhibition by retinol and a retinoidal benzoic acid derivative, suggesting a single growth-inhibitory mechanism common to all of these retinoids. Uptake of the retinoids was unimpaired in the mutant cells, and the cells had normal levels of CRABP. These results contrast with those of the retinoid-resistant embryonal carcinoma cell line mutants that were deficient in CRABP (Schindler *et al.*, 1981). Eventually a mechanism of action of the retinoids will have to be described that can explain these diverse results, and mutants such as these should prove invaluable to such studies.

4. *Other Neoplastic Cells*

In addition to the extensively studied models examined thus far, there are numerous other examples of the effects of retinoids on both growth and differentiation of neoplastic cells. Also, the important observation has been made that for certain neoplastic cells, anchorage-independent growth in semisolid medium is several orders of magnitude more sensitive to retinoid inhibition than is anchorage-dependent growth in monolayer. These examples are discussed below.

a. EFFECTS ON DIFFERENTIATION. Preliminary results indicate that retinoids may have an effect on differentiation of the rat bladder tumor cell line NBT II similar to the effects on nonneoplastic keratinocytes discussed earlier (Section IV,A,1,a). Treatment of dense cultures of NBT II cells with an aqueous-soluble preparation of 10^{-5} *M* retinyl palmitate (A4) (Aquasol A) suppressed keratinization of these cells in a reversible manner (Toyoshima and Leighton, 1975) and enhanced the number of proliferating cells (Tchao, 1980). Another example of retinoic acid-induced differentiation of neoplastic cells is that of the human neuroblastoma cell line LA-N-1 (Sidell, 1982). These cells, like the melanoma cells, are derived from the neural crest and undergo irreversible inhibition of growth and form long neurites in response to retinoic acid treatment. Unlike other cell types discussed below, the anchorage-dependent monolayer growth of the neuroblastoma cells was more sensitive to inhibition by retinoic acid ($ED_{50} = 10^{-8}$ *M*), than the anchorage-independent growth of the cells in soft agar ($ED_{50} = 10^{-7}$ *M*). There was a lag of 48–72 h before the inhibitory effects of retinoic acid were evident; the degree of inhibition was concentration dependent.

Two rat tracheal epithelial cell lines, one cloned from a mucus-secreting adenocarcinoma and the other from a keratinizing squamous cell carcinoma, also responded to retinoid treatment (10^{-6} *M* retinyl acetate) with a reduction in growth rate and saturation density (Marchok *et al.*, 1981). Accompanying these changes, was a shift in differentiation toward the secretory phenotype, as evidenced by stimulation of the synthesis and secretion of high-molecular-weight glycoproteins. Both of these cell lines contained CRABP. Another tracheal epithelial cell line also responded to retinoic acid (10^{-5}–10^{-7} *M*) treatment with a significant decrease in keratinization (Marchok *et al.*, 1978), but unlike the other two cell lines showed 25- and 35-fold increases in thymidine uptake and labeling index (Klann and Marchok, 1982).

b. EFFECTS ON GROWTH. Although effects of retinoids on differentiation are highly specialized, the ability of retinoids to suppress the growth of neoplastic cells is a more general phenomenon. Lotan and Nicolson (1977) laid the groundwork for future investigations and examined the effects of retinoic acid and retinyl acetate (10^{-5} *M* retinoid) on the growth of 31 different cell lines. While they observed a broad variation in response, ranging from no inhibition to cells whose growth was inhibited 75% by 10^{-5} *M* retinoic acid, the growth of over 60% of the cell lines they tested was inhibited by retinoic acid. Those most strongly inhibited included murine melanoma, mammary adenocarcinoma, lymphosarcoma, and myeloma cell lines, as well as a human lymphoid cell line and a Burkitts' lymphoma cell line.

In similar experiments, Lacroix and Lippman (1980) reported a reversible inhibitory effect of retinoic acid (and to a lesser extent retinol) on four human breast cancer cell lines. As had been noted before in comparisons of human melanoma and breast carcinoma cell lines (Lotan, 1979; Section IV,B,3), these cell lines varied widely in their susceptability to retinoid inhibition; the growth of ZR-75-B was inhibited 50% by 6×10^{-9} *M* retinoic acid, whereas 100 times more was required for equivalent inhibition of MCF-7 or Hs578T. Somewhat surprisingly all three cell lines contained approximately equal amounts of CRABP. On the other hand, ZR-75-B contained no measurable intracellular retinol-binding protein (CRBP), even though the growth inhibition of all three cell lines by retinol was equivalent. Results such as these must ultimately be explained to arrive at a full understanding of the putative role of these binding proteins in the biochemical mechanism of action of the retinoids.

Reports on murine neuroblastoma and rat glioma cells (Chapman, 1980), on human rectal adenocarcinoma cells (Tsao *et al.*, 1982), and on herpesvirus-transformed marmoset lymphoblastoid cell lines (Abb and Deinhart, 1981) also demonstrate retinoid-dependent inhibition of tumor cell growth. In a more detailed study of cell lines derived from human osteosarcomas and chrondrosarcomas, Thein and Lotan (1982) showed that exposure to 10^{-5} *M* retinoic acid brought about a flattened morphology, decrease in exponential growth rate, and a reduced saturation density. The cell lines that were capable of anchorage-independent growth lost their ability to grow in semisolid medium after retinoic acid treatment. Like previously discussed effects of retinoids on growth of neoplastic cells, these effects were reversible. As already demonstrated with clones of human melanoma cells (Lotan, 1979) and with human breast cancer cell lines (Lotan, 1979; Lacroix and Lippman, 1980) marked differences in response were detected in these cell lines, with the ED_{50} for growth inhibition by retinoic acid ranging from 10^{-9} *M* for the Hs705 chondrosarcoma cell line to 10^{-7} *M* for the Hs819 cell line. There is evidence in this system that these *in vitro* results are predictive for inhibitory effects of retinoid treatment on growth of chondrosarcomas *in vivo*. Trown *et al.* (1976) and Oegema and Parzych (1981) have both demonstrated that retinoids not only inhibited the growth of rat chondrosarcomas but also brought about regression of established tumors at tolerated doses.

c. EFFECTS ON ANCHORAGE-INDEPENDENT GROWTH OF CELLS. Anchorage-independent growth of cells in semisolid medium has been shown to have a high correlation with the tumorgenicity of cells (Kahn and Shin, 1979; Cifone and Fidler, 1980). Therefore, the observation that the anchorage-independent growth of some tumor cells is more sensitive to retinoids than anchorage-dependent growth may be important in terms of both mechanism of action of retinoids and clinical application. Dion *et al.* (1977) in their studies on the L-929 mouse fibroblast cell line observed that of a wide spectrum of retinoids, retinoic acid (10^{-5} *M*) was most effective in the inhibition of monolayer growth of the cells. Furthermore, it did not act as a cell cycle block or as a cytotoxic agent but rather appeared to restore contact inhibition, a parameter of normalcy, to these cells. Extending their work, they were able to make the important observation that retinoic acid could also restore anchorage-dependent growth of these L-929 cells, and that inhibition of anchorage-independent growth of these cells in agarose was 10,000 times as sensitive to retinoic acid treatment as inhibition of growth in monolayer culture (Dion *et al.*, 1978). They again found wide variability in the susceptibility of cell lines to inhibition of anchorage-independent growth; the growth of L-929 cells, B16 melanoma cells, and HeLa cells was inhibited, but the growth of four other cell lines was not. Like the effects described in monolayer culture, the inhibition was reversible.

Lotan *et al.* (1982b) have compared the effects of retinoic acid on the anchorage-dependent and the anchorage-independent growth of several other cultured tumor cell lines. As in the studies on the L-929 cells, they showed that the anchorage-independent growth of these cells was more sensitive to retinoic acid inhibition than anchorage-dependent growth. More importantly, they demonstrated inhibition of the anchorage-independent growth of certain cell lines that were resistant to retinoic acid under anchorage-dependent conditions (human melanoma cell lines Hs852 and SH4, and murine B16 melanoma sublines and UV-2237P fibrosarcoma). Meyskens and Salmon (1979) reported that a 1-h treatment of fresh human melanoma cells with several retinoids, followed by a wash, was sufficient to reduce significantly colony formation in soft agar. However, Lotan *et al.* (1981) could not reproduce these results with S91 melanoma cells. Finally, in virally transformed cells it was reported that retinoic acid (10^{-5} *M*) could restore both density-dependent inhibition of growth and anchorage-dependent growth to a Kirsten sarcoma virus-transformed 3T3 cell line (Mukherjee *et al.*, 1982).

5. *Summary*

It should be kept in mind that examples discussed here represent interactions of retinoids with specific cell types and cannot necessarily be generalized (for a review, see Lotan, 1980). In particular, it should be noted that:

1. Not all transformed cells respond to retinoic acid treatment with an inhibition of growth, nor are inhibitory effects on growth limited to neoplastic cells.
2. The inhibitory effects of retinoids on the growth of neoplastic cells are not restricted to cells growing in monolayer but can be demonstrated in suspension culture or soft agar culture as well. Anchorage-independent growth of cells may be considerably more sensitive to retinoids than anchorage-dependent growth of the same cells. Understanding of the mechanistic basis of this increased sensitivity might contribute to our understanding of the suppressive effects of retinoids on tumor growth and metastasis.
3. There seems to be no correlation between susceptibility to growth inhibitory effects of retinoids and the doubling time of the cells.
4. Not only is there a wide range of response patterns in different cell types,

but even cell lines derived from tumors of similar histopathological types, or more striking, subclones of a transformed cell line, differ widely in their response to retinoid treatment.

5. Although the effects of retinoid treatment in certain terminally differentiating systems are nonreversible, the effects of retinoids on cell growth are usually reversible, taking approximately 2–3 days for the effect to take place and an equal time for the effect to be reversed.

6. In most cases examined, retinoic acid and its analogs are considerably more active than retinol, retinyl esters, or retinaldehyde in modulating the behavior of cultured cells.

7. There are examples of cell lines that both support and refute the proposition that intracellular retinoid-binding proteins are necessary to mediate the action of retinoids on cells.

It is clear that much more information is needed regarding the primary cellular targets of retinoid action before the basis of these diverse results can be analyzed.

C. Effects on Preneoplastic Progression of Cells

An important conceptual advance in our understanding of the mechanism of action of the retinoids in the suppression of carcinogenesis has come from the observations that retinoids can delay the preneoplastic progression of carcinogen or radiation-treated cells in culture (for reviews, see Sporn and Newton, 1979; Bertram *et al.*, 1982; Bertram, 1983; see also Chapters 5, Vol. 1, and 14, this volume). Harisiadis *et al.* (1978) reported that the number of transformants resulting from γ-radiation treatment of mouse fibroblast C3H/10T1/2 cells cultured in the presence of 7×10^{-6} *M* motretinid (F3) was reduced to 30% of control cultures without retinoid. Merriman and Bertram (1979) using the same cell line showed that the effects of retinoids on the suppression of transformation could be clearly separated from the "initiation" event; thus addition of retinoids to the culture as late as 3 weeks after treatment of the cells with 3-methylcholanthrene still decreased transformation by 80%. The inhibition of transformation by the retinoids was dose-dependent in the range of 10^{-8}–10^{-5} *M*, and the system has been used for bioassay of various synthetic retinoids (Bertram, 1980; see also Chapter 5). Retinol, retinyl acetate, or retinaldehyde were equally effective, though retinoic acid was without activity in 10T1/2 cells. The inhibition of transformation was fully reversible; retinoid-suppressed cells transferred to retinoid-free media for 3–5 weeks formed foci of transformed cells. Mordan *et al.* (1982) have reported the isolation of a clone of carcinogen-treated 10T1/2 cells that possesses the characteristics of initiated cells. Cultured in the presence of retinyl acetate, these cells express the phenotype characteristic of the untransformed parent cells. However, on removal of the retinoid, the cells become transformed with a latency period of about 3 weeks. The availability of a homogeneous population of initiated cells should prove invaluable in investigations of the early events of neoplastic transformation and of the role of the retinoids as chemopreventive agents.

D. Cell Cycle-Specific Locus of Retinoid Action

The observations of several investigators that the growth inhibitory effects of retinoids were not apparent for 24–48 h after exposure of the cells, suggested that a certain phase of the cell cycle might be the target for retinoid-mediated inhibition of growth. If this were the case, it would be expected that asynchronous cultures should require approximately one complete cell cycle before all cells had passed through the retinoid-sensitive stage of the cycle and thus become inhibited. Several studies have been directed at this question. Experiments with mitotically synchronous cultures of Chinese hamster ovary (CHO) cells have shown that retinol (10^{-4} *M*) can block the progression of cells midway through the G_1 phase of the cycle (Haddox and Russell, 1979; Haddox *et al.*, 1979) prior to the G_1-specific expression of ODC normally observed in a wide variety of cell types (for a review, see Russell and Haddox, 1981). Caution must be used in generalizing from these data, however, because the concentrations of retinol (10^{-4} *M*) used in the study are pharmacological (though reportedly not cytotoxic to these cells), and because of the somewhat unusual observation that retinaldehyde at more physiological concentrations (10^{-6} *M*) actually enhanced ODC activity and cell cycle progression (Haddox *et al.*, 1979; see Section V,A). Nonetheless, Dion and Gifford (1980) using synchronized cultures of a different cell line (HeLa) and a different retinoid (retinoic acid, 10^{-5} *M*) also observed a block of the cell cycle in the G_1 phase and showed that retinoic acid had to be added during or before late S phase if cells were to be blocked in the subsequent G_1 phase. Since normal cells arrest in the G_1 (or hypothetical G_0) phase of the cycle, they suggested that the retinoid-dependent inhibition of growth of these neoplastic cells might arise via a restoration of the normal growth control mechanisms operative in nonneoplastic cells.

Lotan *et al.* (1981) have extended these studies to mouse S91 melanoma cells. Using flow microfluorometric analysis, they again demonstrated that retinoic acid (10^{-5} *M*) could increase the proportion of cells in the G_1 phase of the cycle and slow the rate of entry into S phase. In further studies, these researchers have investigated the effects of retinoic acid on protein synthesis in the S91 cells (Lotan *et al.*, 1982a). Using a cell-free protein synthesis system, they found that retinoic acid treatment brought about a reduction in both total RNA and poly(A)$^+$RNA of the cell, as well as in the percentage of ribosomes in polysomes, and suggested that the resultant decrease in protein synthesis might be the cause for the growth inhibition (see also Section VII,C).

In opposition to these studies are the experiments of Schroder *et al.* (1982) on the inhibition of growth of murine Swiss 3T3 cells by retinoic acid (10^{-6} *M*). Using flow cytometric analysis of retinoic acid-treated cultures, they localized the block to the S phase or G_2 phase of the cycle. Which of these observations will prove to be correct is unclear at this time. All the studies have employed relatively high concentrations of retinoid, and all have utilized established cell lines, two neoplastic and two nonneoplastic. It will be important to test the validity of these observations on primary cultures of cells using more physiological concentrations of retinoids.

E. Modulation of Effects of Mitogens by Retinoids

The development of *in vitro* techniques for culture of eukaryotic cells has clearly demonstrated the dependence of cell proliferation on cell membrane receptor-mediated interactions with macromolecular polypeptide growth factors in serum (for reviews, see Gospodarowicz and Moran, 1976; Bradshaw and Rubin, 1980). And although a few cell lines have been adapted to grow in serum-free defined media, serum is still, for most cell types, a necessary constituent of the culture media. Therefore, it must be considered that modulation of cell proliferation by retinoids might result in certain instances from an alteration by the retinoids of the response pattern of the cell to particular hormones or growth factors present in the serum.

Before discussing in detail the effects of retinoids on the mitogenic response of cells to various stimuli, it should be pointed out that in addition to cell-specific responses, results obtained in mitogenesis experiments are extremely dependent on experimental design. The length of time of exposure of the cells to retinoids (Schroder *et al.*, 1982), the time of exposure to the labeled nucleotide, whether nucleotide uptake, nucleotide incorporation into DNA, or an increase in cell number is measured, as well as the method of achieving quiescence of the cell cultures, are all important variables to be considered. Unfortunately, the lack of a uniform experimental approach makes comparisons of results of different investigators quite difficult and at times confusing.

1. *Effects on Epidermal Growth Factor*

Epidermal growth factor (EGF) is a polypeptide of M_r 6045 that, acting through specific cell-surface receptors, is able to stimulate the growth of many cells, both epithelial and mesenchymal (for reviews, see Carpenter and Cohen, 1979; Das, 1982). In addition, it has been suggested that EGF might play an essential role in embryogenesis, differentiation, and even neoplastic transformation. Both EGF (Hirata *et al.*, 1980) and retinoids (De Ruyter *et al.*, 1979; De Leenheer *et al.*, 1982) are normal constituents of human plasma. Therefore, the observations that retinoids have specific effects on both the regulation of cellular receptors for EGF and the control of the mitogenic response of cells to EGF may have relevance to the physiological modulation of EGF activity.

a. RETINOIDS INCREASE THE NUMBER OF EGF RECEPTORS OF CERTAIN CELL LINES. Jetten (1980, 1981, 1982) has investigated the effect of retinoic acid treatment on the number of EGF receptors of several cell types, including mouse fibroblast 3T3, 3T6, and 10T1/2 cells, mouse epidermal cell lines, and normal rat kidney (NRK) cells (Jetten and Goldfarb, 1983), with greatest emphasis on the mouse fibroblast 3T6 cell line. Treatment of cells for 3 days with retinoic acid (10^{-8}–10^{-5} *M*) brought about a dose-dependent increase in the number of EGF receptors of 3T6 cells as well as of several other fibroblastic and epithelial cell lines (Jetten, 1981); receptor affinity was unaltered. The ability of EGF to "down-regulate" the receptor level was unimpaired in the retinoid-treated cells, but cell surface EGF receptors were replenished at a faster rate than in control cells. Although the EGF receptor level of many cell lines was unchanged by retinoid treatment, in 3T6 cells a strong correlation could be shown between the ability of different retinoids to increase the number of EGF receptors and to inhibit the growth of the cells. The activity of the retinoids on the 3T6 cells also correlated with their ability, in a different system, to induce differentiation of embryonal carcinoma cells (Jetten, 1981). These data suggest that these diverse effects of retinoids share at some point a common mechanism of action (for a more thorough development of this concept, see Chapter 5, Vol. 1) and that the particular response observed is a function of the cell type.

Retinoic acid-induced differentiation of embryonal carcinoma stem cells is also accompanied by an increase in the number of cellular EGF receptors. The undifferentiated stem cells have only low levels of EGF receptors, but the differentiated cell types show high EGF binding (Rees *et al.*, 1979; Jetten, 1981). Using the embryonal carcinoma stem cell line PCC4 aza lR, Jetten (1981) demonstrated that the appearance of EGF receptors could be used as a marker of the retinoic acid-induced differentiation of the cells. Whether expression of EGF receptors plays a role in the differentiation process or is a marker of the differentiated phenotype is not known. Two other observations that may also be relevant to relationships between retinoids, EGF, and embryogenesis are the findings, on the one hand, that the growth (Yoneda and Pratt, 1981) and the differentiation (Hassell and Pratt, 1977) of cells derived from embryonic palate are highly sensitive to EGF and, on the other hand, that retinoic acid treatment of embryos in culture often results in abnormalities of palate formation (Kalter and Warkany, 1961; Shenefelt, 1972; Hassell *et al.*, 1977; see also Section II).

Finally, it should be mentioned that the EGF receptor is a glycoprotein (Carpenter and Cohen, 1979; Das, 1982) and that it has been suggested that phosphorylated metabolites of retinoids may be involved in specific glycosylation reactions (De Luca *et al.*, 1979; see also Sections VI and VII). Experiments using the antibiotic tunicamycin, which inhibits protein glycosylation, showed a dramatic reduction of the retinoic acid-dependent increase in EGF receptor numbers in 3T6 cells (Jetten, 1981). Thus investigations aimed at the mechanism of EGF receptor synthesis may provide insight into the mechanism of retinoid-dependent glycoprotein synthesis.

b. RETINOIDS CAN MODULATE THE MITOGENICITY OF EGF. Although both 3T6 cells (Jetten 1980, 1981, 1982) and rat hepatoma cells (Kaneko, 1981) respond to retinoic acid treatment with an increase in the number of EGF receptors, retinoic acid inhibits the growth of each of these cell lines. More specifically, in the 3T6 cell line Jetten (1982) demonstrated that retinoic acid (10^{-9}–10^{-7} *M*) reduced the mitogenic response of the cells to EGF; retinoic acid also reduced the mitogenic effect of insulin on these cells, even though it did not alter the number of insulin receptors (Jetten, 1980). In contrast, in both primary cultures of normal adult human skin fibroblasts (Harper and Savage, 1980) and in the rat NRK cell line where retinoic acid induced a three- to sixfold increase in the number of EGF receptors (Jetten and Goldfarb, 1983), retinoic acid (10^{-7} *M*) potentiated the mitogenicity of EGF. This apparent paradox is not unique; previous investigations have demonstrated the absence of a correlation between EGF receptor numbers and the mitogenic response of cells to EGF (Aharonov *et al.*, 1978). In fact, a model for the observations in the 3T6 cell line can be found in the A-431 human epidermoid carcinoma cell line, where is it found that despite a 10-fold higher than average number of EGF receptors, EGF inhibits the growth of the cells in a reversible manner (Barnes, 1982).

c. RELATIONSHIPS BETWEEN RETINOIDS, TRANSFORMING GROWTH FACTORS (EGF-RELATED MITOGENS), AND TRANSFORMATION. Both a loss of epidermal growth factor receptors (Todaro *et al.*, 1976) and a loss of EGF requirement for growth (Cherington *et al.*, 1979) have been associated in certain instances with malignant transformation. In addition, the tumor promoters 12-*O*-tetradecanoylphorbol-13-acetate (TPA) (Lee and Weinstein, 1978) and teleocidin (Imai *et al.*, 1980) have been shown to reduce the binding of EGF to cells. The ability of retinoic acid to increase the number of EGF receptors is therefore consistent with its ability to reverse certain aspects of the neoplastic phenotype. However, retinoids are ineffective antagonists of tumor promoter-dependent decreases in EGF binding. Both TPA (Jetten, 1981) and teleocidin (Kaneko, 1981) suppressed the retinoic acid-dependent enhancement of glycoprotein and EGF receptor synthesis. Also, in a Kirsten sarcoma virus-transformed cell line, retinoic acid treatment could not replete the diminished EGF receptor binding that accompanied transformation, although it did inhibit the anchorage-independent growth of the cells (Mukherjee *et al.*, 1982).

Since retinoids are known to modulate the behavior of neoplastic cells and to suppress carcinogenesis, investigators have studied the effects of retinoids on the activity of polypeptide mediators of transformation called transforming growth factors (TGFs). The TGF family is related to EGF both chemically and in terms of its biological activity (for a review, see Roberts *et al.*, 1983). One subset of the TGF family, called type α TGF (TGF-α), is structurally related to EGF (Marquardt *et al.*, 1983) and can compete with EGF for binding to membrane receptors (De Larco and Todaro, 1978; Todaro *et al.*, 1980). Phenotypic transformation of nonneoplastic indicator cells (usually measured as the ability of cells such as rat NRK cells to grow in soft agar medium) requires the combined presence of both TGF-α or EGF and another member of the TGF family called TGF-β (Anzano *et al.*, 1982; Roberts *et al.*, 1983). The association of TGF-β with platelets has been demonstrated, and TGF-β is therefore a constituent of serum (Childs *et al.*, 1982; Assoian *et al.*, 1983).

Most studies of the effects of retinoids on the activity of TGFs have utilized a preparation called sarcoma growth factor (SGF), which is derived from the conditioned medium of Moloney murine sarcoma virus-transformed 3T3 cells (De Larco and Todaro, 1978). SGF has been shown to consist of two components, TGF-α and TGF-β (Anzano *et al.*, 1983). In the earliest experiments, Todaro *et al.* (1978) showed that retinyl acetate (10^{-8} *M*) suppressed the SGF-dependent mitogenesis of rat NRK cells. It was also shown that pretreatment with retinoids could suppress the anchorage-independent growth of NRK cells induced in response to treatment with SGF; however, retinoids could not suppress the anchorage-independent growth of the virally transformed 3T3 cells from which SGF was derived. Conflicting with these results, Jetten (1983) has shown that retinoic acid (10^{-8}–10^{-6} *M*) significantly enhanced SGF-induced colony formation of NRK cells in soft agar, and Jetten and Goldfarb (1983) have shown that retinoic acid can also enhance the formation of small colonies of NRK cells induced by EGF alone (probably acting with the TGF-β component of serum). In support of Jetten's observations, recent experiments in the authors' laboratory have shown that retinoic acid markedly enhances soft agar colony formation of NRK cells in the presence of purified TGF-β and EGF. This enhancement shows a strong correlation with the increase in the number of EGF receptors induced by retinoic acid treatment of the NRK cells (A. Roberts *et al.*, unpublished data). In contrast to these effects on NRK cells, Jetten (1981, 1982) has shown that retinoic acid suppresses the mitogenic effect of SGF on 3T6 cells in monolayer culture, even though it increases the number of EGF receptors. In mouse epidermal JB-6 cells, another cell line in which retinoic acid induces an increase in the number of EGF receptors (Jetten, 1981), retinoic acid treatment enhanced the ability of EGF to promote anchorage independent growth (Colburn *et al.*, 1981). Colony formation of these same cells induced by the tumor promoter TPA was blocked by retinoids.

In summary, not only the anticarcinogenic properties of retinoids but also the effects of retinoids on embryogenesis, differentiation, and proliferation may, in some as yet unknown way, relate to modulation of the cellular response to EGF or in some cases to TGFs. The mechanism could involve regulation of synthesis of specific glycoprotein receptor molecules or could involve genetic control of the response of cells to these growth factors. Other possible mechanisms might also be responsible for the observed effects (as discussed later in this chapter).

2. *Effects on Tumor Promoters*

The tumor promoters as a class exhibit mitogenic activity and as already discussed, the ability of retinoids to oppose or to enhance that effect depends on the particular cell culture system employed and on experimental design. Kaneko (1981) reported that retinoic acid (10^{-6} *M*) inhibited the weak mitogenic activity of the tumor promoter teleocidin on rat hepatoma cells and Jetten and De Luca (1982) and Colburn *et al.* (1981) demonstrated that retinoic acid could suppress the TPA-dependent colony formation of mouse fibroblast BALB/c 3T3 A31-1-BP-2 cells and mouse epidermal JB6 cells, respectively. However, other experiments showed that retinoic acid was not able to overcome the TPA-induced effects in BALB/c 3T6 cells (Jetten and De Luca, 1982) and that it actually enhanced TPA-dependent colony formation in soft agar of NRK cells (Jetten, 1983).

Similar discrepancies are apparent in studies of the mitogenic effects of retinoids and tumor promoters on quiescent, untransformed mouse 3T3 cells. For example, it was shown that retinoic acid potentiated the mitogenicity of TPA in Swiss 3T3 fibroblasts (Dicker and Rozengurt, 1979; Schroder *et al.*, 1980). In fact, Schroder *et al.* (1980) demonstrated that only a 4-h pretreatment of the cells with retinoic acid (10^{-5} *M*) was necessary to stimulate significantly DNA synthesis in response to TPA. Analogous to the "competence/progression" scheme outlined for the response of BALB/c 3T3 cells to platelet-derived growth factor and to plasma mitogens (Pledger *et al.*, 1977; Stiles *et al.*, 1979), these researchers demonstrated that retinoic acid treatment (10^{-6}–10^{-5} *M*) of the Swiss 3T3 cells for 4 h rendered them "competent" to respond to "progression factors" of human plasma or to TPA (Schroder *et al.*, 1980).

Schroder *et al.* (1982) have shown that treatment of Swiss 3T3 cells with retinoids for longer than 24 h results in an inhibition of growth, thus contrasting with their previous experiments demonstrating potentiating effects of short-term retinoic acid treatment on DNA synthesis. To explain these discrepancies, they have proposed that potentiation and inhibition represent two distinct response phases of cells that depend on the length of retinoid treatment, and that each of these responses may result from expansion of cellular acid-soluble nucleotide pools (Rapaport *et al.*, 1982a; see also Section V,B). Certainly it is clear that the picture is complicated not only by cell-specific differences in response patterns but also by design of mitogenesis experiments.

3. *Effects on Hormones*

In an extension of the experiments previously discussed, it was shown that the mitogenicity of insulin was inhibited by retinoid treatment of 3T6 cells (Jetten, 1982) but increased by retinoid treatment of 3T3 cells (Dicker and Rozengurt, 1979) and cultures of epidermis from newborn mice (Sporn *et al.*, 1973).

Retinoids modulate the gonadotropin response of testicular cells cultured in defined media (Mather, 1981). The growth of cells of Sertoli origin (TM4) is markedly stimulated by follicle-stimulating hormone, and this stimulation is entirely eliminated in the presence of 10^{-7} *M* retinoic acid. Primary cultures of porcine Leydig cells undergo a significant decrease in the number of luteinizing hormone receptors when grown in the presence of 5×10^{-8} *M* retinoic acid, though human chorionic gonadotropin-stimulated testosterone production by these cells is actually increased in the presence of retinoic acid.

In a different system, mouse mammary glands in organ culture in chemically defined medium in the presence of insulin respond to the hormone prolactin by exhibiting end-bud differentiation. The addition of either 10^{-6} *M* retinoic acid or *N*-(4-hydroxyphenyl)retinamide (E16) to the cultures inhibited the prolactin-induced differentiation (Mehta *et al.*, 1983). Each of these retinoids, also significantly inhibited prolactin-induced mammary DNA synthesis in a concentration-dependent manner. These results are consistent with the abilities of these retinoids to inhibit chemically induced mammary carcinogenesis in the rat *in vivo*.

V. EFFECTS OF RETINOIDS ON THE ACTIVITY AND SYNTHESIS OF CELLULAR ENZYMES AND EFFECTORS

In this section we examine the effects of retinoids on enzymes and effectors that appear to be part of the response pattern of cells to retinoid treatment. These include: (A) ornithine decarboxylase and transglutaminase, two enzymes thought to play an essential role as cells go through the cell cycle; (B) cyclic AMP and the cyclic AMP-dependent protein kinases, central elements in the stimulus–response coupling of the action of several hormones on cells; (C) plasminogen activator and (D) collagenase, two proteolytic enzymes probably removed from the primary effects of the retinoids; and (E) prostaglandins, which have effects on many diverse systems including the cyclic AMP system. If there exists a fundamental mechanism of action of retinoids in control of cell growth or differentiation, that mechanism should include elements common to most or all affected cells. With the possible exception of the cyclic AMP-dependent protein kinases, the evidence accumulated thus far on the diversity of the response patterns of cells with respect to these particular enzymes and effectors most likely rules out an obligatory role for them in the mechanism of action.

In surveying the literature of retinoid effects on these enzyme or effector systems, the prominent position occupied by studies of retinoid effects on neoplastic cells becomes apparent. In addition, the well-established antipromoter activity of retinoids in phorbol ester-mediated skin carcinogenesis experiments (for a review, see Verma and Boutwell, 1980) has fueled numerous investigations of the cellular effects of retinoids as compared to those of the phorbol ester tumor promoters (especially TPA). Unfortunately, many investigators have used this approach to come to the conclusion that if in certain instances retinoids and tumor promoters elicit identical effects on cells, then the retinoids must also have tumor-promoting activity (this approach is especially evident in Section V,C and E). In this regard, Berenblum (1978) has pointed out that many effects of phorbol esters may not be relevant to tumor promotion. In addition, cellular receptors for the tumor-promoting phorbol esters have been identified in a wide variety of species and cell types (Driedger and Blumberg, 1980). Some reports suggest that the phorbol ester receptor or a component of the receptor complex is protein kinase C (a specific calcium- and phospholipid-dependent protein kinase), and that the putative physiological analog to these lipophilic plant products might be an unsaturated diacylglycerol (Castagna *et al.*, 1982; Niedel *et al.*, 1983). These data now substantiate the caveat of Berenblum and make it clear that the stigma of tumor promotion should not be associated with all of the broad spectrum of cellular events that characterize the action of tumor promoters on cells.

A. Ornithine Decarboxylase and Transglutaminase

In studies directed toward the mechanism of retinoid-dependent regulation of cellular proliferation and differentiation, it has been found that retinoids can regulate the activity and/or the synthesis of two key enzymes, ornithine decarboxylase (ODC) and glutaminyl-peptide γ-glutamyltransferase (transglutaminase, TG). The first of these two enzymes (ODC) catalyzes the decarboxylation of ornithine to putrescine, the rate-limiting step in polyamine biosynthesis. There is now substantial evidence, in both neoplastic and nonneoplastic cells, that ODC activity is specifically induced in the G_1 phase of the cell cycle and that induction of this short-lived enzyme (half-life 10–40 min) might be obligatory for progression through the cell cycle (for a review, see Russell and Haddox, 1981). The second enzyme (TG) belongs to a family of Ca^{2+}-dependent enzymes that catalyze the cross-linking of primary amines to various proteins and polypeptides (reviewed in Williams-Ashman and Canellakis, 1980). It has been implicated in the formation of cross-linked cornified envelopes by terminally differentiating keratinocytes (Rice and Green, 1978; Green and Watt, 1982; Yuspa *et al.*, 1981, 1982) and possibly even as a regulator of ODC activity by virtue of its ability to catalyze the cross-linking of putrescine to ODC (Folk *et al.*, 1980; Williams-Ashman and Canellakis, 1980; Russell, 1981; Scott and Russell, 1982). The latter reaction has been demonstrated to occur in cell-free systems *in vitro,* but its physiological significance, including a possible function of the ODC–putrescine conjugate as a regulator of RNA polymerase I activity (Russell, 1981), is unknown.

1. Direct Effects of Retinoids on Ornithine Decarboxylase and Transglutaminase Activity

It is difficult to generalize from the data available on the reported effects of retinoids on these two related enzyme activities. Results obtained in different cell culture systems are at variance, especially because in certain systems retinoids promote proliferation whereas in others they inhibit it. Moreover, different retinoids have been employed in many of the studies, and the time points chosen for analysis vary from a few hours to several days after the onset of retinoid treatment. For example, Scott and Russell (1982), show a temporal relationship between the peaks of G_1-specific ODC and TG activity in synchronized cultures of both CHO cells and murine Cloudman S91 melanoma cells. Retinoid treatment (10^{-4} *M* retinol in the case of the CHO cells; 10^{-4} *M* retinoic acid for the S91 cells), which inhibits the growth of each of these cell lines, resulted in an increase in TG activity and a decrease in ODC activity (Scott *et al.*, 1982; Haddox and Russell, 1979; Haddox *et al.*, 1979; Russell and Haddox, 1981). However, implications of a direct effect of retinoids on TG activity are confused by the finding that α-difluoromethylornithine (DFMO), an irreversible inhibitor of ODC, increased TG activity in much the same manner as retinoid treatment (Scott *et al.*, 1982). In murine C-1300 neuroblastoma cells and rat glioma cells, a correlation could be shown between the inhibition of growth and the inhibition of ODC activity resulting from treatment with retinoic acid (10^{-5} *M*) or retinol (10^{-4} *M*), respectively (Chapman, 1980). In the latter studies, growth inhibition was significantly enhanced by treatment of the cells with a combination of retinoid and DFMO. The high concentrations of retinoids employed in all of these studies (10^{-5}–10^{-4} *M*) and the risk of generalizing an observation made from cells that have adapted to these concentrations, suggest caution in interpretation of these results. In direct conflict with these reports, are the data of Lotan *et al.* (1981) and Mukherjee *et al.* (1982) that no detectable changes in ODC concentrations were associated with the retinoic acid-dependent (10^{-5} *M*) inhibition of cellular proliferation in S91 melanoma cells and Kirsten sarcoma virus-transformed BALB/c 3T3 cells, respectively. It should be pointed out that these researchers as well as Chapman (1980) chose relatively long observation times (8 h to 5 days) and unsynchronized cell populations, whereas the conclusions of Russell and co-workers were derived from multiple sampling times during the G_1 phase (0–8 h) in synchronized cells.

An experiment by Luk *et al.* (1982) addressed the question of separating the effects of retinoids on ODC and proliferation from those on differentiation. The system chosen was the HL-60 human promyelocytic leukemia cell line that can be induced by retinoids to undergo terminal granulocytic differentiation (Breitman *et al.*, 1980b; see also Section IV,B,2). The proliferation and increased ODC activity observed upon subculturing the HL-60 stem cells could be blocked by addition of the ODC inhibitor DFMO. Simultaneous treatment of the cells with DFMO and retinoic acid (10^{-7} *M*) blocked almost completely the retinoic acid-induced increase in ODC activity but failed to interfere with the retinoic acid-promoted differentiation of the cells into mature granulocytes. An analogous effect was observed on the TPA-induced differentiation of the HL-60 cells into monocytes; again, blocking the TPA-induced ODC activity by DFMO treatment interfered with the proliferation but not the differentiation of the cells. These results suggest separate mechanisms of cellular control of proliferation and differentiation and that the effects of retinoids on these cells may be targeted to control of the differentiation process.

2. *Effects of Retinoids on TPA-Induced Ornithine Decarboxylase and Transglutaminase Activity*

Although these direct effects of retinoids on ODC activity appear to be somewhat variable, retinoids consistently have been shown to block the induction of ODC associated with tumor promotion by phorbol esters (for a review, see Verma and Boutwell, 1980). In experiments using primary cultures of epidermal basal cells (Yuspa *et al.*, 1980, 1981; Lichti *et al.*, 1981), retinoic acid (10^{-9}–10^{-6} *M*) could suppress both the TPA-induced ODC activity and the resultant induction of differentiation (Yuspa *et al.*, 1980). In the JB6 mouse epidermal cell line, which responds to TPA treatment with an irreversible induction of the transformed phenotype, retinoic acid (10^{-8}–10^{-5} *M*) inhibited both the TPA-induced colony growth in soft agar and the associated ODC induction (Colburn *et al.*, 1981). In a very preliminary study, retinoic acid (10^{-8} *M*) was also shown to block the TPA-induced ODC activity in NRK cells (Paranjpe *et al.*, 1980). Bolmer and Wolf (1982b) have demonstrated a dose-dependent antagonism by retinoic acid of TPA-dependent ODC induction in Swiss 3T3 cells; these effects were lost in enucleated cells, but antagonistic effects on a cell-surface protein fibronectin persisted. Similar effects have been observed in an *in vivo* skin carcinogenesis system where the ability of retinoids to block the TPA-induced increase in ODC activity correlated with their ability to block tumor promotion and papilloma formation (Verma and Boutwell, 1977; Verma *et al.*, 1978; Boutwell and Verma, 1981; see also Chapters 5 and 14). In each of these experimental systems, maximum inhibition was observed when the retinoic acid was added just prior to or simultaneously with the phorbol ester. Later addition of the retinoid dramatically reduced the extent of the inhibition.

In contrast to these results, increases in ODC activity brought about by treatment of mouse skin with the complete carcinogen 7,12-dimethylbenz[*a*]anthracene (Verma *et al.*, 1980) or of cultured epidermal cells with germicidal UV light (Lichti *et al.*, 1981) were not suppressed by retinoic acid. In the skin carcinogenesis system, this correlated with a lack of effect of retinoic acid on carcinogen-induced tumor formation. These experiments suggest a specific interaction of tumor-promoting phorbol esters (TPA) and retinoids, possibly on a common target. While this antagonism could possibly be a cell-surface phenomenon, it does not occur by direct interaction of retinoids with the specific phorbol ester receptors described by Driedger and Blumberg (1980).

Unlike the described antagonistic effects of TPA and retinoids on ODC induction, both of these agents induce TG activity in cultured epidermal cells (Yuspa *et al.*, 1980, 1981, 1982). In the case of the phorbol esters, the increased TG activity is associated with induction of differentiation and cornified envelope production. Paradoxically, synthesis of cornified envelopes is blocked in retinoic acid-treated cells (Yuspa *et al.*, 1982), suggesting that retinoic acid might in some way alter TG activity or possibly modify the substrate for the reaction.

3. *Conclusions*

Despite significant variations in experimental design, the available evidence suggests that, in most cells, treatment with retinoids brings about a reduction in ODC synthesis and an increase in TG synthesis. Transformed cells are generally less differentiated and have been reported to have lower TG activity (Birckbichler *et al.*, 1977; Birckbichler and Patterson, 1978) and higher ODC activity (Bachrach, 1976) than their normal counterparts. Thus the effects of retinoids on each of these activities are in the direction of suppressing the transformed phenotype.

B. Cyclic AMP and Cyclic AMP-Dependent Protein Kinases

More and more, it is being suggested that there might be one universal regulatory system underlying both the response of cells to external stimuli (Rasmussen, 1981) and the commitment of a cell to a schedule of differentiation or proliferation (Berridge, 1976). This regulatory system includes the processes that regulate intracellular cyclic nucleotide levels, on the one hand, and those that control intracellular calcium ion concentrations, on the other. Evidence is only beginning to accumulate concerning the role of the retinoids in the activation of this regulatory system. Yet, the possibility exists that this role will be of fundamental importance to the mechanism of action of retinoids on cells, and that other characteristic markers of retinoid effects on cells, some of which are discussed in this section, will be found to be secondary effects.

Before discussing effects of the retinoids on the cyclic nucleotide/calcium-dependent regulatory systems, it is necessary to identify the various components of these systems. The cyclic nucleotide system contains many elements: (1) the substrate ATP (or GTP); (2) the synthetic and degradative enzymes adenylate or guanylate cyclase and phosphodiesterases, respectively; (3) the receptor proteins, the regulatory subunits of the protein kinases; and (4) the response elements of the cell, the catalytic subunits of the protein kinases. It is the latter that translate the "message" into a change in cell function by catalyzing the phosphorylation, and thereby changing the activity, of specific enzyme or protein substrates. In this system, the kinase (catalytic subunit) is inactive when it is bound to the regulatory subunit; interaction of cyclic AMP (cAMP) with the regulatory subunit changes its conformation causing it to dissociate from the catalytic subunit, which in its free form is an active protein kinase. In the case of the calcium-dependent regulatory system, the intramembraneous events are still largely unknown (for a detailed discussion of both systems, see Rasmussen, 1981). However, in analogy to the cyclic nucleotide system, a number of intracellular receptor proteins have been identified, the best known of which is calmodulin (Klee *et al.*, 1980; Schulman *et al.*, 1980). Unlike the single class of response elements in the cyclic nucleotide system (Nimmo and Cohen, 1977), calcium-directed response elements are numerous but include calcium (calmodulin) dependent protein kinases that in certain instances have been shown to act on the same protein substrates as the cAMP-dependent protein kinases (Rasmussen, 1981). To complicate our interpretations of the effects of retinoids on these intracellular regulatory systems are the well-documented observations that the relationships of the two regulatory systems can differ from cell type to cell type, covering the entire spectrum of interactions ranging from synergistic interactions to antagonistic interactions (Berridge, 1976; Rasmussen, 1981).

To the extent that it is possible, we attempt to simplify the picture by discussing separately each of the cellular systems in which evidence exists for retinoid-mediated effects on protein kinases and protein phosphorylation. In studies of the biochemical effects of retinoids we are again faced with the central role occupied by their specific effects on neoplastic cells. The cellular systems employed for these studies have been either murine embryonal carcinoma cells (F9) or human promyelocytic leukemia cells (HL-60), each of which has been demonstrated to undergo terminal differentiation to nonneoplastic cell types following retinoid treatment (see Section IV,B,1 and 2), or murine melanoma cells in which retinoids have a marked antiproliferative effect (Section IV,B,3).

1. *Effects in F9 Murine Embryonal Carcinoma Cells*

Exposure of retinoic acid-treated F9 cells to dibutyryl cAMP results in the appearance of a stable new phenotype resembling parietal endoderm (Strickland *et al.*, 1980; Kuff and Fewell, 1980; Rizzino and Crowley, 1980). The order of addition of the retinoic acid and the cAMP analog is critical; treatment of cells with dibutyryl cAMP alone had no effect, and treatment of cells with dibutyryl cAMP followed by retinoic acid yielded a phenotype similar to that resulting from treatment with retinoic acid alone (Strickland *et al.*, 1980). Thus it can be concluded that treatment with retinoic acid has induced changes in the F9 cells that alter their response to cAMP. Since cAMP-dependent protein kinases are believed to be the only intracellular receptors for cAMP, it seemed reasonable to assume that retinoic acid might be modulating cellular levels of these proteins. Indeed, both cytosolic and plasma membrane-associated cAMP-dependent protein kinases are increased by retinoic acid treatment of F9 cells (Plet *et al.*, 1982). However, the picture is complex. Two major types of cAMP-dependent protein kinases have been identified and are referred to as types I and II. These two forms differ in the nature of their regulatory subunit; the catalytic subunits are identical (Nimmo and Cohen, 1977). Plet *et al.* (1982) have shown that retinoic acid treatment elevates both type I and II kinases in the cytosol but preferentially elevates type II in the plasma membrane. It has been suggested that such changes might localize the kinase near specific protein substrates. These effects are observed between 3 and 15 h after treatment of the cells with retinoids and remain elevated throughout 5 days of treatment. No effect was observed on cAMP-dependent protein kinases associated with the nuclear fraction.

In contrast to the rapid increase in cAMP-dependent protein kinase activity induced by retinoic acid treatment, induction of calcium- and phospholipid-dependent protein kinase activity required several days and was maximal only after 11 days of retinoic acid treatment (Kraft and Anderson, 1983). This latter kinase is the protein kinase C identified as the cellular "receptor" for TPA (Castagna *et al.*, 1982; Niedel *et al.*, 1983). A similar length of time was required for changes in hormone responsiveness of the membrane adenylate cyclase system (Evain *et al.*, 1981). Whereas calcitonin is the most potent stimulator of cAMP production in the undifferentiated F9 stem cells, calcitonin responsiveness is lost in the differentiated endodermal cells formed in response to retinoic acid treatment. The adenylate cyclase of the endodermal cells is preferentially activated by parathyroid hormone. The time course of these changes suggests that both alterations in hormone responsiveness of adenylate cyclase and changes in the cytosolic calcium- and phospholipid-dependent protein kinase activity may be the results of differentiation to a new cell phenotype. On the other hand, the very early changes observed in cAMP-dependent protein kinase activity suggest that they may be essential for the retinoic acid-mediated effects on differentiation of these cells.

2. *Effects in HL-60 Human Promyelocytic Cells*

Like the priming of cAMP-responsiveness induced by retinoic acid treatment of F9 cells, treatment of HL-60 cells with 10^{-8} *M* retinoic acid for 20 h primed the cells to respond to the cAMP-inducing agents prostaglandin E_2 or cholera toxin (Olsson *et al.*, 1982). Cells treated in this manner differentiated to mature granulocytes, but the reverse treatment sequence was ineffective. These experiments again suggest effects of retinoic acid on protein kinase activity, though concentrations of cycloheximide which blocked 85% of protein synthesis actually enhanced the priming effect. Direct support for this hypothesis comes from Durham *et al.* (1983) who showed a two- to threefold rise in cAMP-dependent protein kinase activity following treatment of HL-60 cells with retinoic acid. In this case, retinoic acid specifically increased type I activity in the cytosol and both types I and II in the particulate fraction. A cAMP-independent protamine kinase (Durham *et al.*, 1982) was also induced in these cells. Accompanying these changes in kinase activities were changes in the phosphorylation patterns of cellular proteins. Unfortunately, these experiments did not include a time course of enzyme induction; kinase activities were measured after 5 days of retinoic acid treatment. However, the short priming period of retinoic acid treatment reported by Olsson *et al.* (1982) suggests that these are early effects of retinoic acid treatment, much as was seen in the F9 cells.

3. *Effects in Murine Melanoma Cells*

It has been shown that cAMP, like retinoic acid, can inhibit the growth and stimulate melanin production by murine melanoma cells (Wong and Pawelek, 1973). However, treatment of either B16 melanoma cells (Lotan *et al.*, 1978; Ludwig *et al.*, 1980) or S91 melanoma cells (Lotan and Lotan, 1981) with levels of retinoic acid that result in an inhibition of growth has no effect on intracellular cAMP levels. On the other hand, in agreement with observations in the F9 and HL-60 cell lines in which retinoic acid induces differentiation, retinoic acid treatment of B16 cells did result in a dose-dependent increase in total cellular cAMP-dependent protein kinase activity (Ludwig *et al.*, 1980). The ED_{50} for both growth inhibition and induction of kinase activity in the B16 cells was approximately 10^{-7} *M* retinoic acid. No information was presented regarding either the type or the intracellular distribution of the kinase activity. Retinoic acid treatment of a protein kinase-deficient variant of B16 resulted in neither inhibition of growth nor increased protein kinase activity (Ludwig *et al.*, 1980). At least 24 h were required before effects of retinoic acid on either growth or kinase activity could be detected.

4. *Summary*

Much more information must be obtained before we can begin to assess whether retinoids might exert their effects on cellular differentiation and proliferation through this important calcium–cAMP intracellular regulatory system. The results obtained thus far on the retinoid-dependent induction of cAMP-dependent protein kinase activity of teratocarcinoma, promyelocytic leukemia, or melanoma cells could explain how in certain instances retinoids might potentiate the effects of hormones or growth factors known to act through this system (Section IV,E). Also, in the discussion of effects of retinoids on keratinocytes (Section IV,A) there are indications that retinoids might have some effects on the intracellular translocation of calcium, and in Swiss 3T3 cells, retinoids have been shown to enhance the mitogenic effects of calcium (Schroder *et al.*, 1980). Future investigations should test the generality of these effects.

C. Plasminogen Activator

Plasminogen activator is a cellular serine protease that functions to activate serum plasminogen to the fibrinolytic enzyme plasmin. Experiments demonstrating increases in secretion of this enzyme accompanying transformation of mesenchymal cells (Unkeless *et al.*, 1973; Ossowski *et al.*, 1973) and treatment of cells with tumor promoters (Wigler and Weinstein, 1976) suggested that its appearance might be a useful marker for tumorigenicity. However, this correlation has been proved not to exist for epithelial cells (Montesano *et al.*, 1977). Moreover, a role for the enzyme in tissue remodeling processes has been suggested (Beers *et al.*, 1975), and it has been shown to be a specific marker for differentiation in early embryogenesis (Strickland *et al.*, 1976). Its application to the assessing of retinoid-dependent differentiation of teratocarcinoma stem cells to primitive endoderm has already been discussed (see Section IV,B,1).

As in other systems, much of the impetus behind research directed at the effects of retinoids on the synthesis or secretion of plasminogen activator has been derived from the desire to compare the cellular effects of retinoid treatment with those following treatment with transforming agents or tumor promoters. Thus Wilson and Reich (1978) and Miskin *et al.* (1978) demonstrated that low concentrations of retinoic acid, like transforming viruses or phorbol ester tumor promoters, could stimulate plasminogen activator synthesis in chick embryo fibroblasts and chick embryo muscle cells, respectively. Both the induction of this enzyme (which could be detected as early as 20–40 min after addition of retinoic acid to the cultures) and its ''deinduction'' following removal of the retinoid, required mRNA synthesis. In chick embryo fibroblasts, cAMP inhibited plasminogen activator induction following sarcoma virus infection or treatment with TPA, but did not significantly affect enzyme induction by retinoids, suggesting that separate pathways of induction might be involved, at least in part (Wilson and Reich, 1978). In primary cultures of rabbit synovial fibroblasts, treatment with 10^{-6} *M* retinoic acid increased the TPA-dependent release of plasminogen activator, though 10^{-7} *M* retinoic acid was ineffective (Brinckerhoff *et al.*, 1982). Retinoic acid by itself had no effect on secretion of the enzyme. In these same cells, however, 10^{-6} *M* retinoic acid antagonized the effect of TPA on collagenase production (see Section V,E).

Though experiments in particular cell lines often form the basis of mechanistic studies, it is necessary to utilize as many cell lines as possible to test the generality of a proposed mechanism. Toward that end, Wilson and Dowdle (1980) have tested the effects of retinoids on plasminogen activator secretion of

43 different human cultured cells, both neoplastic and nonneoplastic, of both mesenchymal and epithelial origin, and Lotan *et al.* (1982b) have tested the effects of retinoids on plasminogen activator synthesis in 17 different cultured human and murine tumor cell lines. Their results conclusively show that retinoids have no consistent effect on plasminogen activator synthesis of cells, and that there is no correlation between the effects of retinoids on plasminogen activator synthesis and on the anchorage-dependent or anchorage-independent growth of tumor cells. The experiments of Wilson and Dowdle (1980) did show that on the average, the basal level of plasminogen activator secretion was higher for the epithelial cells tested than for cells of mesenchymal origin; however, no statistical differences existed between neoplastic or nonneoplastic cells within each group. Treatment for 72 h with $10^{-6}\,M$ retinoic acid increased plasminogen activator synthesis by *both* neoplastic and nonneoplastic mesenchymal cells 10- to 100-fold, whereas it had no effect on plasminogen activator synthesis by epithelial cells. Moreover, the studies of Lotan *et al.* (1982b) make it clear that the increased plasminogen activator synthesis accompanying retinoic acid treatment of certain cells does not correlate with the effect of retinoic acid on another marker of tumorigenicity, anchorage-independent growth (Kahn and Shin, 1979; Cifone and Fidler, 1980). For example, $10^{-5}\,M$ retinoic acid increased the levels of cell-associated plasminogen activator four- to fivefold in the human osteosarcoma cell line Hs781, at the same time completely inhibiting the growth of these cells in soft agar and inhibiting the growth of the cells in monolayer by 65%.

An interesting observation, possibly related to the mechanism of action of retinoids, is that Ca^{2+} could be shown to induce plasminogen activator synthesis in Swiss 3T3 cells (Chou *et al.*, 1977). Simultaneous treatment of these cells with Ca^{2+} and retinoic acid ($10^{-5}\,M$) enhanced both the Ca^{2+}-dependent induction of plasminogen activator synthesis and the mitogenic effect of Ca^{2+} on these cells (Schroder *et al.*, 1980).

D. Collagenase

By several criteria, rheumatoid arthritis bears some resemblance to neoplastic disease: (1) the fundamental lesion (the pannus) is both proliferative and invasive (Harris, 1976); (2) rheumatoid synovial cells have been shown to survive and to organize into a pannus-like structure following implantation into athymic nude mice (Brinckerhoff and Harris, 1981a); and (3) two of the cellular mediators of the disease, collagenase and prostaglandin E_2, can be induced in normal synovial fibroblasts by treatment with the tumor-promoting phorbol esters (Brinckerhoff *et al.*, 1979). Retinoids have been shown to inhibit both collagenase production in cultures of adherent rheumatoid synovial cells (Brinckerhoff *et al.*, 1980) and the TPA-induced or polyethylene glycol-induced synthesis of collagenase by normal rabbit synovial fibroblasts (Brinckerhoff and Harris, 1981b; Brinckerhoff *et al.*, 1982). These effects are cell-specific; neither retinoids nor TPA had any effects on collagenase production by primary cultures of articular chondrocytes (Brinckerhoff *et al.*, 1982; Trechsel *et al.*, 1982). Collagenase synthesis by TPA-treated synovial fibroblasts could also be inhibited by corticosteroids, and it was shown that treatment with both retinoic acid (10^{-10} *M*) and prednisolone (10^{-10} *M*) inhibited collagenase synthesis by 50%, whereas each drug alone at 10^{-10} *M* had no effect (Brinckerhoff and Harris, 1981b). In contrast to these effects on collagenase synthesis, retinoic acid was ineffective in reducing the levels of TPA-induced prostaglandin E_2 synthesis (see Section V,E) and acted synergistically with TPA to increase plasminogen activator synthesis (see Section V,C) by the synovial cells (Brinckerhoff *et al.*, 1982).

E. Prostaglandins

Prostaglandins are not stored to any appreciable extent in cells or tissues; therefore, their production is a reflection of changes in levels of the fatty acid substrates (for a review, see Levine, 1979). Two principal steps in their synthesis involve (1) the action of phospholipase on cellular lipids to release the precursor polyunsaturated fatty acids that are the substrates for prostaglandin synthesis, and (2) the action of cyclooxygenase on these fatty acids to yield endoperoxides, from which the various prostaglandins, prostacyclins, and thromboxanes are then made. As already mentioned, the realization that prostaglandins might play a role in control of cell proliferation and in neoplasia (Jaffe, 1974) and the observation that phorbol esters promote prostaglandin secretion by a variety of cell types (Levine and Ohuchi, 1978; Mufson *et al.*, 1979; Brinckerhoff *et al.*, 1982) stimulated an interest in studying the effects of retinoid treatment of cells on prostaglandin synthesis. The results are both inconsistent and inconclusive.

Mufson *et al.* (1979) reported that retinoic acid (10^{-6}–10^{-5} *M*) inhibited both arachidonic acid release and prostaglandin E_2 release induced in chick embryo fibroblasts by treatment with TPA. The opposite results were reported by Takenaga (1981), who showed that retinoic acid (ED_{50} 10^{-6} *M*) enhanced prostaglandin E_2 release by murine myeloid leukemia M1 cells, and that this effect could be inhibited by treatment with TPA. In contrast to these two systems where TPA and retinoids have an antagonistic relationship, Levine and Ohuchi (1978) reported that *both* TPA and retinoids (10^{-5} *M*) stimulated deacylation of lipids and prostaglandin E_2 synthesis in canine kidney cells (MDCK), and that the effects of the TMMP analog of retinoic acid (F2) but not of 13-*cis*-retinoic acid were actually synergistic with TPA. Both Levine (1980), using MDCK cells, and Brinckerhoff *et al.* (1982), using primary cultures of rabbit synovial fibroblasts, showed that one particular retinoid, *N*-(4-hydroxyphenyl)retinamide (E16) (10^{-8}–10^{-6} *M*), inhibited the TPA-stimulated release of prostaglandin

E_2. In the MDCK cells, treatment with *N*-(4-hydroxyphenyl)retinamide did not inhibit arachidonic acid release, suggesting that the retinoid was possibly acting as a cyclooxygenase inhibitor. Whether this enzyme inhibition is related to the retinoid character of the molecule or to a particular spatial configuration of the phenolic end group in the molecule is unknown; certainly none of the bioassay systems described in Chapter 5, Vol. 1, point to a different role or mechanism of action for this retinoid as compared to retinoic acid.

VI. RETINOID-DEPENDENT CHANGES IN GLYCOCONJUGATE BIOSYNTHESIS AND RELATED EFFECTS

Retinoids have been shown to have profound influence on the biosynthesis of a diverse class of macromolecules called glycoproteins, which, in the broadest sense, consist of carbohydrate moieties covalently linked to a polypeptide backbone (for reviews, see L. M. De Luca, 1977, 1978; Wolf, 1977, 1983; Lotan, 1980). Members of this family of glycoproteins include enzymes and hormones, secretory proteins ranging from the mucins to the immunoglobulins, cell membrane components including certain hormone and growth factor receptors, and connective tissue proteins and proteoglycans (Kornfeld and Kornfeld, 1976, 1980). In addition to this functional diversification, these glycoproteins can be classified into groups based on the chemical bond between the sugar and the amino acid involved in the carbohydrate–protein linkage. There are both O-glycosidically linked glycopeptides, involving the hydroxyl groups of serine, threonine, or hydroxylysine, and N-glycosidically linked glycopeptides, involving the amide nitrogen of asparagine. As is discussed later (Section VII,A), the biosynthesis of N-glycosidically linked oligosaccharide chains proceeds via lipid–oligosaccharide intermediates, possibly involving the direct participation of the retinoids as glycosyl donors (for reviews, see Waechter and Lennarz, 1976; Parodi and Leloir, 1979; Struck and Lennarz, 1980). Finally, there is also great diversification in the sugars found in these glycoproteins; they include galactose, mannose, glucose, *N*-acetylgalactosamine, *N*-acetylglucosamine, sialic acids, fucose, and xylose. In addition, uronic acids and sulfated amino sugars are commonly found in the proteoglycans.

It is increasingly clear that retinoids can affect the biosynthesis of all types of glycoconjugates including glycoproteins, glycolipids, and proteoglycans. The effects of retinoids have been investigated both in whole animals and *in vitro* in organ culture, cell culture, and subcellular microsomal and membrane preparations. Sugars that have routinely been used in these studies, usually in radiolabeled form, include glucosamine, galactose, mannose, and fucose; sulfate incorporation has been used in many studies of proteoglycan synthesis. Although specific target glycoconjugates have been identified in particular experimental systems, it is clear from the available literature that glycosylation of a broad and diverse set of molecules is affected by retinoid status. This suggests involvement of retinoids at some point common to the biosynthesis of the oligosaccharide portion of many different glycoconjugates. De Luca and Wolf and co-workers have provided evidence for a scheme involving direct participation of retinoid phosphate derivatives in glycosyl transfer reactions (for reviews, see L. M. DeLuca 1977, 1978, 1982; Wolf, 1977, 1983). This intriguing mechanism is discussed in detail in Section VII,A. Others have suggested that retinoids might affect the activity of specific glycosyltransferases (Plotkin and Wolf, 1980; Creek and Morré, 1980; Moskal *et al.*, 1980; Lotan *et al.*, 1983a) or possibly the biosynthesis of the lipid glycosyl carrier dolichol (Rosso *et al.*, 1981). Glycoconjugates have been shown to play central roles in cell-surface changes related to regulation of cell growth, differentiation, migration, and even transformation (for reviews, see Hakomori, 1975; Hynes, 1976; Atkinson and Hakimi, 1980; Schachter and Roseman, 1980). These findings suggest that an understanding of the biochemical mechanism of retinoid control of glycoconjugate synthesis will substantially increase our understanding of the mechanism of retinoid control of the cellular processes relating to proliferation and differentiation.

A. Glycoproteins

The early studies of Wolbach and Howe (1925) demonstrating the prevalence of a keratinizing metaplasia of epithelial structures of the vitamin A-deficient rat, and evidence for mucus metaplasia of these same tissues in vitamin A excess (Moore, 1957) provided the first hints of retinoid control of the glycoprotein profile of affected tissues. Many years later, the work principally of De Luca, Wolf, and their collaborators has firmly established the biochemistry of retinoid-mediated changes in glycoprotein synthesis. The earlier biochemical studies have been thoroughly reviewed by L. M. De Luca (1977) and Wolf (1977) and are not discussed in detail here. To summarize briefly, *in vivo* studies of glycoprotein synthesis in intestinal mucosa (De Luca *et al.*, 1969, 1970a, 1971; Olson *et al.*, 1981), in liver (De Luca *et al.*, 1975), and rat liver lysosomal membranes (Adhikari and Vakil, 1980), as well as *in vitro* studies in cultured rat corneal epithelia (Kim and Wolf, 1974; Kiorpes *et al.*, 1979), cultured hamster (Bonanni *et al.*, 1973) and rat tracheal epithelia (Bonanni and De Luca, 1974), and cultured mouse epidermal cells (De Luca and Yuspa, 1974; Adamo *et al.*, 1979b) all showed retinoid-dependent incorporation of radiolabeled sugars into glycoconjugates. In every case, incorporation was significantly reduced in vitamin A deficiency and could be restored after retinoid repletion. In epidermal cell culture, an effect of retinyl acetate on glycoprotein synthesis could be detected after only 2 h of treatment (Adamo *et al.*, 1979b), suggesting that these changes might constitute a primary response of the cell to retinoid treatment.

Other *in vitro* experiments utilizing a wide variety of cell lines, both epithelial and fibroblastic, both nonneoplastic and neoplastic, have both supported and amplified these earlier observations. To simplify the presentation of the complex effects of retinoids on glycoprotein synthesis, various experimental systems have been listed in Table I, A. Only general features common to these investigations are discussed here.

1. *General Observations*

Analysis of the collective effects of retinoids on glycoprotein synthesis has demonstrated that (1) retinoid treatment does not uniformly affect the biosynthesis of all glycoconjugates of an animal or even of a particular cell; (2) the sugar composition of retinoid-dependent glycoproteins is not constant; (3) the incorporation of different sugars into glycoproteins of a single tissue is differentially affected; and (4) the degree of glycosylation of a particular protein can be affected to a different extent than synthesis of the protein core. Thus, regarding the first point, the synthesis of ovalbumin in chick oviduct was not reduced even in severe vitamin A deficiency (Sneider and Wolf, 1976), whereas synthesis and glycosylation of the rat serum glycoprotein α_1-macroglobulin (Kiorpes *et al.*, 1976; Kiorpes and Anderson, 1981) and of the various tissue glycoproteins previously discussed (reviewed in L. M. De Luca, 1977) was affected. Also, in S91 murine melanoma cells treated with retinoic acid, sugar incorporation into a specific 160-K glycoprotein was increased fourfold, while overall incorporation of sugars into protein was reduced 20–40% (Lotan *et al.*, 1983a). To illustrate the second point, the sugar compositions of specific retinoid-dependent glycopeptides isolated from rat intestine (L. M. De Luca, 1977), from rat trachea (Bonanni and De Luca, 1974), and from hamster trachea (Bonanni *et al.*, 1973) have been reported; no consistent trends are apparent. Experiments on sugar incorporation into glycoproteins of mouse epidermal cells in culture most clearly support the third point; fucose incorporation was unaffected, while incorporation of glucosamine, mannose, and galactose was increased 3-fold, 4-fold, and 11-fold, respectively, in retinyl acetate-treated cultures as compared to controls (Adamo *et al.*, 1979b). Finally, regarding the last point, Kiorpes and Anderson (1981) have shown that the initial change in rat serum α_1-macroglobulin in vitamin A deficiency is a decrease in the extent of glycosylation of the protein and not in its synthesis. Similarly, Lotan *et al.* (1983a) have shown that the incorporation of [^{3}H]glucosamine or [^{3}H]mannose into a specific 160-K glycoprotein of murine S91 melanoma cells is increased 4-fold following retinoic acid treatment, whereas the incorporation of either [^{35}S]methionine or [^{3}H]leucine into that same protein remains unchanged. One possible explanation for such

TABLE I

Retinoid-Dependent Alterations in Glycoconjugate Biosynthesis

Cells or tissue	Radiolabeled precursor	Retinoid[a]	Reference
A. Glycoproteins			
1. Nonneoplastic cells			
Rat tracheal organ culture	Glucosamine	Retinyl acetate	Clark and Marchok (1979)
Mouse epidermal JB6 cells	Mannose	Retinoic acid	Dion *et al.* (1981)
Mouse epidermal cells (primary)	Fucose, galactose	Retinyl acetate	Adamo *et al.* (1979b)
	Glucosamine, mannose		
Rat corneal epithelium	Glucosamine	Retinoic acid[b]	Hassell and Newsome (1981)
Rat conjunctival epithelium			Hassell *et al.* (1980)
Mouse JB6 cells	Mannose	Retinoic acid	Dion *et al.* (1981)
Mouse 3T3 fibroblasts	Galactose	Retinol	Patt *et al.* (1978)
Mouse 3T6 fibroblasts	None[c]	Retinoic acid	Jetten *et al.* (1979b)
Hamster NIL2K fibroblasts	Galactose	Retinol	Patt *et al.* (1978)
2. Neoplastic cells			
Hamster NILpy fibroblasts	Galactose	Retinol	Patt *et al.* (1978)
Mouse 3T12 fibroblasts	Mannose, fucose	Retinoic acid	Sasak *et al.* (1980)
Rat tracheal cell lines	Glucosamine	Retinoic acid	Marchok *et al.* (1981)
Mouse S91 melanoma	Glucosamine, mannose	Retinoic acid	Lotan *et al.* (1983a)
Human HeLa	Glucosamine	Retinoic acid	Lotan *et al.* (1980a)

(*continued*)

TABLE I *Continued*

Cells or tissue	Radiolabeled precursor	Retinoid[a]	Reference
B. Glycolipids			
Hamster NIL2K, NILpy fibroblasts	Galactose	Retinol	Patt *et al.* (1978)
Mouse 3T3 fibroblasts	Galactose	Retinol	Patt *et al.* (1978)
Mouse 10T1/2 fibroblasts	None	Retinyl acetate	Bertram *et al.* (1982)
Mouse S91 melanoma	Galactose	Retinoic acid	Lotan *et al.* (1983a)
Mouse 3T12 fibroblasts	None	Retinoic acid	Morré *et al.* (1981)
Hepatomas	None	Retinoic acid	Morré *et al.* (1981)
Human epithelial carcinoma	None	Retinoic acid	Moskal *et al.* (1980)
C. Proteoglycans			
1. Nonneoplastic cells			
Pig ear epidermis	Sulfate, fucose	Retinoic acid	King and Tabiowo (1981)
Human intestinal cell line	Sulfate	Retinoic acid	Shapiro and Poon (1979)
Mouse dermal cells (primary)	Sulfate	Retinyl acetate	Shapiro and Poon (1978)
Mouse epidermal cells (primary)	Sulfate	Retinoic acid	Shapiro and Mott (1981)
Rat costal cartilage chondrocytes	Sulfate	TMMP retinoic acid	Shapiro and Poon (1976)
Chick sternal chondrocytes	Glucosamine, sulfate	Retinyl palmitate	Vasan and Lash (1975)
Mouse limb mesenchymal cells	Glucosamine, mannose	Retinoic acid	Pennypacker *et al.* (1978)
Human/rabbit corneal fibroblasts	Glucosamine	Retinoic acid	Dahl and Axelsson (1980)
Mouse 3T3/3T6 fibroblasts	Sulfate	Retinoic acid	Jetten *et al.* (1979b)
2. Neoplastic cells			
Mouse B16F10 melanoma	Glucosamine, sulfate	Retinyl acetate	Maniglia and Sartorelli (1981)

[a] In all systems glucoconjugate synthesis was shown to be retinoid-dependent, as evidenced by a change in glycoprotein pattern on either addition or removal of retinoids. Reports prior to 1977 have been reviewed (L. M. De Luca, 1977) and are not included.

[b] Corneas were removed from rats that were treated with retinoids *in vivo;* tissues were then labeled by incubation *in vitro.*

[c] Cell-surface proteins were iodinated using the lactoperoxidase technique.

data would be that the carbohydrate moieties of the glycoprotein turn over at a faster rate than the protein core. In support of this, Kreisel *et al.* (1980) have purified a 110-K glycoprotein from plasma membranes of rat liver and have determined that the half-lives of the carbohydrates L-fucose, *N*-acetylneuraminic acid, and galactose were 12.5, 33, and 20 h, respectively, whereas the half-lives of methionine and arginine of the protein were 70 and 78 h, respectively.

2. *Cell Surface Glycoproteins*

Several methods have been employed as a diagnostic test for the cell surface location of glycoproteins. These treatments are generally of two types. The first consists of brief treatment of metabolically labeled (usually with mannose or glucosamine) intact cells with nonpenetrating enzymes such as neuraminidase, or various combinations of proteolytic enzymes such as trypsin, papain, and pronase. A change in the electrophoretic mobility of a glycoprotein following such treatment is interpreted as indicating a cell surface location of that glycoprotein. The second treatment consists of direct surface labeling of exposed tyrosyl or histidyl residues of glycoproteins by lactoperoxidase-catalyzed iodination or related methods. These methods have been applied to analysis of retinoid-dependent changes in cell surface glycoproteins of mouse fibroblast 3T3 or 3T6 cells (Jetten *et al.*, 1979b) or spontaneously transformed 3T12 cells (Sasak *et al.*, 1980), of human HeLa cells (Lotan *et al.*, 1980a), and of murine S91 melanoma cells (Lotan *et al.*, 1983a). In each of these cases, retinoid treatment induced increased synthesis of cell surface glycoproteins. In general, these changes were detectable within 12–24 h after retinoid treatment and preceded any effects of retinoids on the growth of the cells. Retinoid effects on specific identifiable cell surface glycoproteins are discussed in greater detail below.

a. FIBRONECTIN. Fibronectin is a large (250 K) cell-surface glycoprotein, also called "large external transformation-sensitive" protein (LETS). This protein is thought to play a role in attachment of fibroblasts via an association with collagens and proteoglycans of the extracellular matrix (reviewed by Hynes, 1976; Yamada and Olden, 1978). It has also been observed that transformed cells frequently synthesize reduced amounts of fibronectin relative to their non-transformed counterparts (Hynes, 1976), although there are notable exceptions to this trend (Kahn and Shin, 1979).

Several investigations have implicated retinoids in the control of fibronectin synthesis. *In vivo* studies by Bolmer and Wolf (1982a) showed increased fibronectin synthesis in mouse skin papillomas treated with retinoic acid. Mouse limb bud mesenchymal cells *in vitro* synthesize a 240-K glycoprotein that is lost on differentiation of these cells to chondrocytes; retinoic acid (10^{-5} *M*) inhibits chondrogenesis and stabilizes synthesis of the 240-K glycoprotein by the mesenchymal cells, increasing their adhesiveness (Lewis *et al.*, 1978; Hassell *et al.*,

1978b). However, it is impossible to assess whether retinoic acid is affecting glycoprotein synthesis directly or whether the effects are a consequence of the inhibition of differentiation. These same reservations apply to the ability of retinoic acid to induce fibronectin synthesis accompanying differentiation of embryonal carcinoma stem cells (Jetten *et al.*, 1979a). In nondifferentiating fibroblast cell lines, retinoic acid increased fibronectin synthesis in murine 3T3 and 3T6 cells (Jetten *et al.*, 1979b), whereas retinol decreased fibronectin synthesis by hamster NIL2K cells (Patt *et al.*, 1978). Retinoid treatment had no effect on fibronectin of virally transformed counterparts to either the murine or hamster fibroblast cells.

In yet other studies, Hassell *et al.* (1979) found that although retinoic acid did not directly affect fibronectin synthesis in chick sternal chondrocytes, it did seem to increase the ability of these cells to bind fibronectin tightly to the cell surface. This finding is interesting in light of the reported retinoid-induced changes of cell surface glycolipids (Section VI,B); certain complex gangliosides have been implicated as putative cell surface receptors for fibronectin (Kleinman *et al.*, 1979). Alternatively, retinoids could be regulating the synthesis of collagens (Section VI,A,2,6) and proteoglycans (Section VI,C), which have an affinity for fibronectin. Bolmer and Wolf (1982b) have addressed the question of the mechanism of retinoid effects on binding of fibronectin. They have shown that retinoic acid antagonized the effects of TPA on loss of cell-associated fibronectin in Swiss 3T3 cells. These effects persisted in enucleated cells, suggesting that they are mediated by posttranscriptional or possibly even direct cell-surface events.

b. COLLAGEN. Intimately associated with the ability of a cell to bind fibronectin are effects on cellular collagen synthesis. The observations have been made that a rapid loss of both fibronectin (Blumberg *et al.*, 1976) and of procollagen (Delclos and Blumberg, 1979) follows either viral transformation or TPA treatment of chick embryo fibroblasts. A similar loss of procollagen has been shown to accompany TPA treatment of mouse epidermal JB-6 cells (Dion *et al.*, 1982); retinoic acid treatment of the cells increases synthesis of this 180-K glycoprotein (Dion *et al.*, 1981). Therefore, in the JB-6 cells, TPA and retinoic acid have antagonistic effects on transformation (Colburn *et al.*, 1981), on adhesion of the cells to the substratum, and on procollagen synthesis (Dion *et al.*, 1981, 1982).

It has been suggested that other retinoid-induced 180-K glycoproteins might also be collagens. Thus, though the evidence is only circumstantial, retinoid enhancement of 180-K glycoprotein synthesis has been observed in hamster fibroblasts (Patt *et al.*, 1978) and in HeLa cells (Lotan *et al.*, 1980a). In addition, Strickland *et al.* (1980) have identified as type IV collagen a retinoid-induced glycoprotein that is formed on differentiation of F9 murine teratocarcinoma stem cells to parietal endoderm cells. Again, in this latter system, kinetic studies must be undertaken to answer the question whether the procollagen synthesis is a direct effect of the retinoid or a consequence of the retinoid-induced differentiation.

An opposite effect of retinoids on collagen synthesis has been observed in cultures of rabbit articular chondrocytes. Trechsel *et al.* (1982) showed that retinol (10^{-7}–10^{-5} *M*) produced a concentration-dependent reduction in net collagen synthesis. However, synthesis of α_2-collagen chains was increased by retinol treatment, suggesting a switch from type II to type I collagen.

3. Glycoprotein Membrane Receptors

As discussed earlier in this chapter, retinoids can modulate the levels of at least two membrane receptors known to be glycoproteins. Thus retinoic acid treatment of several cell types brought about increases in the number of EGF receptors (Jetten, 1980, 1981, 1982; Kaneko, 1981; Jetten and Goldfarb, 1983), and this increase could be abolished in the presence of the glycosylation inhibitor tunicamycin (Jetten, 1981). The opposite effect of retinoic acid has been demonstrated on another glycoprotein receptor, luteinizing hormone receptor. Mather (1981) showed that retinoic acid treatment of primary cultures of porcine Leydig cells significantly reduced the number of luteinizing hormone receptors compared to control cultures grown in retinoid-free defined medium. Also, Steinberg and Sgoutas (1981) showed that Leydig cell-enriched preparations from vitamin A-deficient rats maintained on retinoic acid had approximately one-half the number of gonadotropin receptors for luteinizing hormone compared to cells prepared from vitamin A-normal rats. These results could be interpreted either as a retinoic acid-induced decrease in receptor glycosylation or, in the case of the latter study, as a retinol-specific glycosylation of the receptors. Karl and Griswold (1980) have shown that retinol appears to act synergistically with several hormones to prolong the ability of Sertoli cells to secrete androgen-binding protein. Quite possibly these results relate to the inability of retinoic acid to maintain normal testicular function in rats. Because the gonadotropins themselves are glycoproteins, retinoid-dependent glycosylation of these ligands might be important as well. In this regard, it has been shown that alteration of the glycosylation of both human luteinizing hormone and human chorionic gonadotropin can alter their receptor binding and biological activity (Catt *et al.*, 1974).

B. Glycolipids

Membrane glycolipids, and in particular the gangliosides, are also thought to play an important though poorly understood role in growth, differentiation, and transformation (for a review, see Schachter and Roseman, 1980). Although there are notable exceptions to the pattern (Schachter and Roseman, 1980), there is a trend toward incomplete glycosylation of ganglioside components of transformed cells, arising in some cases from decreases in specific glycosyltransferases (for a review, see Hakomori, 1975). The gangliosides have thus been the focus of several studies of retinoid-dependent effects (see Table I,B).

Most interesting of these studies are those employing the murine C3H/10T1/2 fibroblast cell line, in which retinoids have been shown to have a strong anti-promoter effect and can suppress either radiation or carcinogen-induced transformation (Harisiadis *et al.*, 1978; Merriman and Bertram, 1979; see also Section IV,C). Neoplastic transformation of 10T1/2 cells is accompanied by a trend toward underglycosylation and simplification of glycosylation patterns of membrane glycoconjugates (Langenbach and Kennedy, 1978). In contrast, retinyl acetate treatment of the untransformed 10T1/2 cells leads to a more complex glycosylation of gangliosides; gangliosides containing a single sialic acid residue (GM3, GM2, GM1) are relatively unchanged, whereas those containing linked sialic acid residues (GD3, GD1b, and GT1) are increased (Bertram *et al.*, 1982). These results might have implications for the increased adhesion of 10T1/2 cells accompanying retinyl acetate treatment (Bertram, 1980; see Section VI,D) because it has been suggested that di- and trisialogangliosides might function as membrane receptor sites for fibronectin (Kleinman *et al.*, 1979). In human epithelioid carcinoma cells (KB), the sialyltransferase responsible for GM3 synthesis was increased 14-fold by retinoic acid treatment (Moskal *et al.*, 1980). Thus in both of these systems, retinoid treatment opposes the trend accompanying transformation of cells and accentuates the normal phenotype.

The results of other studies are more difficult to interpret. In hamster fibroblasts (NIL2K and their transformed counterparts NILpyT) GM3 synthesis was increased following treatment with retinol (Patt *et al.*, 1978), whereas in murine S91 melanoma cells, GM3 levels were decreased following retinoic acid treatment (Lotan *et al.*, 1983a). In transformed 3T12 mouse fibroblasts and S91 cells, GM1 levels were increased, and in 3T12 cells GD1b levels were decreased (Morré *et al.*, 1981). The cell-dependent variations in these changes and the absence of any obvious conceptual link to the biological consequences of retinoid treatment make it impossible to evaluate the importance of these changes at this time.

C. Proteoglycans and Glycosaminoglycans

Proteoglycans, the carbohydrate components of which are called glycosaminoglycans (GAG), are classified separately from other glycoproteins and contain various uronic acids and sulfated amino sugars in addition to the other commonly occurring sugars (for reviews, see Kraemer, 1979; Rodén, 1980). The principal classes of glycosaminoglycans are hyaluronic acid, chondroitin sulfates, keratan sulfate, dermatan sulfate, and heparan sulfate. In addition to being a major component of connective tissue matrix, proteoglycans are associated with the cell membrane of nearly all eukaryotic cells and, like the other glycoconjugates already discussed, have been implicated in control of cell growth, differentiation, and transformation, presumably by participating in cell-to-cell and cell-to-substratum interactions (for a review, see Atkinson and Hakimi, 1980). Just as retinoids have been shown to influence glycoprotein and glycolipid biosynthesis, investigations have shown that they can alter GAG and therefore proteoglycan biosynthesis as well. All but one of these studies (Pennypacker *et al.*, 1978), listed in Table I,C, have focused on changes in the GAG component of the proteoglycan; tissues are digested proteolytically prior to analysis. Some of these results have been reviewed (Lotan, 1980).

Although no prevalent pattern of retinoid regulation of GAG synthesis has emerged from these studies, several general observations deserve mention: (1) The effect of retinoids on GAG synthesis is cell-specific; synthesis is increased in certain cell types and decreased in others. Thus synthesis was decreased by retinoid treatment of corneal fibroblasts (Dahl and Axelsson, 1980), primitive mesenchyme (Pennypacker *et al.*, 1978), and chondrocytes (Vasan and Lash, 1975; Shapiro and Poon, 1976); synthesis was increased in epidermal and dermal cells (Shapiro and Poon, 1978; King and Tabiowo, 1981) and in mouse fibroblast cell lines (Jetten *et al.*, 1979b). (2) GAG levels are affected differentially in the medium, the cell matrix, and inside the cell. The best demonstration of this is provided by the very detailed studies of Shapiro and Poon (1976, 1978, 1979) and Shapiro and Mott (1981). Using three different retinoids at eight different concentrations, they have shown, for example, that retinoid treatment of dermal cells results in increased GAG levels in the cellular and matrix compartments and decreased levels in the medium; in epidermal cells GAG levels were increased in all three compartments. (3) Different GAG groups are affected differentially by retinoid treatment. Thus, for example, the studies of Shapiro and Poon (1976) on the effects of retinoic acid on cell-matrix GAG synthesis in rat chondrocytes showed a dose-dependent increase in heparan sulfate levels, a decrease in chondroitin 4-sulfate levels, and no effect on the levels of chondroitin 6-sulfate. (4) Not only the synthesis of the individual GAGs, but also the degree of sulfation and the type of sulfation can be affected. Thus in intestinal cells, retinoic acid treatment changes the ratio of *N*-sulfate to *O*-sulfate in heparan sulfate without a major change in total sulfate incorporation (Shapiro and Mott, 1981). (5) Finally, changes in GAG levels can be detected within 4–8 h of treatment, prior to effects of retinoids on growth or adhesion (King and Tabiowo, 1981), and the direction of the changes in GAG levels may vary with the time of treatment (Maniglia and Sartorelli, 1981).

Obviously these effects of retinoid treatment on GAG synthesis are extremely complex. A result indicating no change in overall incorporation of sulfate into GAGs may overlook compensating changes in individual GAGs, and even in instances where no changes occur in sulfate incorporation into a particular GAG,

Shapiro and Mott (1981) have shown that retinoids may be altering the ratio of sulfamido to ester sulfate groups. Whether retinoid-dependent regulation of GAG synthesis shares a common mechanism with the regulation of glycoprotein and glycolipid synthesis is unknown. Regardless, it is reasonable to assume that some of the observed retinoid-dependent changes in cell behavior are a consequence of changes in proteoglycan synthesis.

D. Effects on Cellular Adhesion

From the foregoing it is clear that retinoid treatment of cells often has profound influence on the synthesis of and composition of cell surface glycoconjugates and might therefore be expected to have an effect on cell-to-cell and cell-to-substratum interactions. In culture, these changes can be detected as changes in resistance to trypsinization. As listed in Table II, cell types that respond to retinoid treatment with increased cell-to-substratum adhesiveness include epidermal cells (Yuspa and Harris, 1974; Christophers and Wolff, 1975; Dion *et al.*, 1981) and several different lines of mouse fibroblasts including 3T3, 3T6, 3T12, and C3H/10T1/2 (Jetten *et al.*, 1979b; Adamo *et al.*, 1979a; Bertram, 1980). Other cells such as transformed 3T3SV mouse fibroblasts (Jetten *et al.*, 1979b) or human fibroblasts (Lacroix *et al.*, 1981) showed no changes in adhesion accompanying treatment with retinoids. At the other extreme of the spectrum are cells that responded to retinoid treatment with a decrease in cell-to-substratum adhesion; these include several transformed cell lines derived from trachea (Marchok *et al.*, 1981; Klann and Marchok, 1982) or from bladder (Tchao and Leighton, 1979) and an intestinal epithelial cell line (Shapiro and Poon, 1979). Although no correlation can be drawn among effects on growth, adhesion, and glycoprotein synthesis in epithelial cells, it is clear that in fibroblast cell lines, whether tumorigenic or nontumorigenic, a strong correlation exists among retinoid-dependent decreases in growth, increases in cell-to-substratum adhesion, and increases in glycoconjugate synthesis (Table II). Maximum effects on fibroblast adhesion required 3–4 days of treatment with retinoids and were rapidly reversible; they could, therefore, be a consequence of the retinoid-induced changes in glycoconjugate synthesis.

The 3T6 (Jetten *et al.*, 1979b), 3T12 (Adamo *et al.*, 1979a), and 10T1/2 (Bertram, 1980) murine fibroblast cell lines have been used for studies of structure–activity relationships of various retinoids (see also Chapter 5, Section IV,F, Vol. 1). In 3T6 cells, a good correlation exists between the ability of a particular retinoid to inhibit the growth of the cells and to increase adhesion. Retinoic acid and its analogs had the highest activity in both assays (Jetten *et al.*, 1979b). In the 3T12 cells, however, retinoids had no consistent effect on cell growth (Adamo *et al.*, 1979a); the order of activity in increasing cellular adhesion was retinol and 5,6-epoxyretinol > retinoic acid and its 5,6-epoxy derivative (D6) > 13-*cis*-

TABLE II

Retinoid-Dependent Changes in Growth Rate, Adhesion, and Glycoprotein Synthesis

Cells	Retinoid	Changes[a] in			References
		Growth	Adhesion	Glycoprotein synthesis	
1. Epithelial cells					
a. Nontumorigenic					
Guinea pig epidermal cells	Retinoic acid	↑	↑	ND	Christophers and Wolff (1975)
Human intestinal epithelial cells	Retinoic acid	↓	↓	↓ (GAG)	Shapiro and Poon (1979)
Mouse JB6 epidermal cells	Retinoic acid	ND	↑	↑	Dion *et al.* (1981)
b. Tumorigenic					
Tracheal epithelial cells	Retinyl acetate	↓	↓	↑	Marchok *et al.* (1981)
Tracheal epithelial cells	Retinoic acid	↑	NC	ND	Klann and Marchok (1982)
Bladder	Retinyl palmitate	↑	↓	ND	Tchao and Leighton (1979)
2. Fibroblasts					
a. Nontumorigenic					
Mouse 3T3 cells	Retinoic acid	↓	↑	↑ [d]	Jetten *et al.* (1979b)
Mouse 3T6 cells	Retinoic acid[b]	↓	↑	↑ [d]	Jetten *et al.* (1979b)
Mouse 10T1/2 cells	Retinyl acetate[b]	↓ [c]	↑	↑ [e]	Bertram (1980), Bertram *et al.* (1982), and Mordan and Bertram (1983)
Human skin cells	Retinoic acid	↓	NC	ND	Lacroix *et al.* (1981)
b. Tumorigenic					
Mouse 3T12 cells	Retinol[b]	↓ [f]	↑	↑	Adamo *et al.* (1979a) Sasak *et al.* (1980)
Mouse 3T3SV cells	Retinoic acid	NC	NC	NC	Jetten *et al.* (1979b)

[a] ND, not determined; NC, no change; ↑, increased; ↓, decreased.
[b] Structure–activity relationships of several retinoids were investigated. The most active retinoid is listed here.
[c] Exponential growth rate was not affected by retinoids, but saturation density was decreased.
[d] An increase in both fibronectin and GAG synthesis was found.
[e] Changes were noted on glycolipid synthesis; effects on synthesis of other glycoconjugates were not measured.
[f] Variable reductions in saturation density were observed.

retinoic acid and the TMMP analog of retinoic acid. Ester or amide derivatives of retinoic acid were inactive in the adhesion assay. Although the latter is easily explained by metabolic effects, the order of activity of the other retinoids in promoting adhesion in the 3T12 cells correlates poorly with their activity in other biological assays (see Chapter 5). In the 10T1/2 cells, retinoids can be assayed for their ability to suppress transformation induced by 3-methylcholanthrene (Merriman and Bertram, 1979; Bertram *et al.*, 1981) and to increase adhesion (Bertram, 1980). Although retinyl acetate coordinately affected each of these parameters, retinylidene dimedone and *N*-(4-hydroxyphenyl)retinamide, which were each highly active in suppression of transformation of 10T1/2 cells and in other biological assays (see Chapter 5), either had no effect or actually decreased the adhesion of the cells, respectively. Retinoic acid, itself, is without activity on 10T1/2 cells.

Regarding the effects of retinoids on glycoconjugate synthesis in these fibroblast cell lines, it is interesting to note that all three cell lines responded with changes which could be interpreted to result directly in increased adhesion. In 3T6 cells, fibronectin synthesis was increased (Jetten *et al.*, 1979b); in 3T12 cells, there was a shift toward an increase in higher molecular weight complex glycopeptides at the cell surface (Sasak *et al.*, 1980); and in 10T1/2 cells, there was an increased synthesis of complex gangliosides that are thought to function as fibronectin receptors (Bertram *et al.*, 1982). However, correct assignment of such a cause and effect relationship requires careful scrutiny; at the minimum, complete analysis of the effects of some of the synthetic retinoids mentioned above on glycoconjugate synthesis must be obtained. The availability of such analogs is a powerful tool for analysis of biochemical mechanisms and should not be overlooked.

E. Effects on Membrane Microviscosity

Retinoids are lipid-soluble compounds that readily dissolve in cellular membranes. Furthermore, retinoid-induced changes in membrane glycosylation may also alter the mobility of macromolecules within the membrane. Therefore, it might be expected that there would be retinoid-dependent changes in membrane microviscosity. Indeed, in mouse fibroblast 3T6 cells that respond to retinoic acid treatment with increased cell-to-substratum adhesiveness and increased glycosylation (Jetten *et al.*, 1979b), retinoic acid in the concentration range 10^{-7}–10^{-5} M induced a dose-dependent decrease in membrane viscosity (Jetten *et al.*, 1981). However, the nature of the change was nonspecific as evidenced by the observations that biologically inactive as well as active retinoids affected a change in microviscosity. Also, retinoid-induced changes in membrane microviscosity of different fibroblastic cell lines did not correlate with the effects of retinoids on the growth or adhesion of the cells.

Interestingly, changes in membrane microviscosity that accompany differentiation of embryonal carcinoma cells in response to treatment with low concentrations of retinoid (10^{-7} *M*) do correlate with the biological activity of the retinoids in inducing differentiation (Jetten *et al.*, 1982). At higher retinoid concentrations, nonspecific effects are again observed. Therefore, it seems likely that these changes in microviscosity result from the differentiation of the stem cells to cells having different surface properties and not from a primary effect of the retinoids on the cell surface.

VII. MECHANISM OF ACTION OF THE RETINOIDS

In attempting to define mechanisms of action of the retinoids, the ultimate goal is the identification of universal retinoid-dependent biochemical events that underlie the diverse patterns of phenotypic expression of cells in response to retinoid treatment (pleiotypic response). These patterns of response include changes in membrane glycoconjugate synthesis, changes in cellular adhesion, changes in metabolite patterns and enzyme levels, and at a more general level, changes in cellular proliferation and differentiation. The diversity of these cell-specific responses has been the source of much confusion. However, to reiterate the concept introduced at the beginning of this chapter, it should be considered that the information for these varied responses is not in the retinoid itself but in the context of the particular cellular system in which it is acting. Applying this concept, it is reasonable to assume that many of the varied effects of retinoids may share a common mechanism.

Before proceeding to discuss these mechanisms, it should be said that there is substantial support among investigators in the retinoid field for the consideration of separate roles for retinol, on the one hand, and for retinoic acid, on the other hand (see Wolf, 1983; Zile and Cullum, 1983). The unique role of retinaldehyde in the visual process (Chapter 10) as well as characterization of two distinct, separately regulated, intracellular binding proteins, one for retinol and one for retinoic acid (Chapter 9), provides evidence for such a model. Others feel that retinoic acid, or a closely related derivative, may represent the "active form" of the vitamin in all functions related to growth, differentiation, and transformation (see Chapter 11). Support for this concept comes from the ability of retinoic acid to fulfill most of the *in vivo* functions of vitamin A despite its inability to be reduced to retinol (Dowling and Wald, 1960), as well as from the significantly greater activity of retinoic acid, as compared to retinol, in most *in vitro* bioassay systems (Chapter 5, Vol. 1). Superimposed on these differences of opinion with respect to the "active" biochemical form of the retinoids, is the controversy whether one common mechanism will underlie all effects, or whether two mechanisms, possibly one operating at an epigenetic level and another at the level of gene expression, might be jointly operative (Wolf, 1983; Zile and Cullum, 1983). For the moment, these questions must remain unanswered. However, the simplest approach that can be derived from analysis of structure–function relationships of hundreds of different retinoids, including the new chemistry of the retinoidal benzoic acid derivatives, in diverse assays of growth, differentiation, and suppression of carcinogenesis, is that one common mechanism will be found (see Chapter 5, Vol. 1).

We now review the two dominant theories of retinoid mechanism, the proposed cofactor role of retinoids in glycosyl transfer and the proposed steroid model for retinoid control of gene expression. In addition, we discuss retinoid control of gene expression per se (independent of specific binding proteins) since some involvement at the level of the genome, either direct or indirect, is a *sine qua non* of altered phenotypic expression.

A. Cofactor Role of Retinoids in Glycosyl Transfer Reactions

A little over 10 years ago it became clear, largely through the work of Leloir and co-workers, that the polyisoprenol dolichol functioned as a lipid carrier for assembly of an oligosaccharide–lipid intermediate in the biosynthesis of N-glycosidically linked oligosaccharide chains of certain glycoproteins (Behrens and Leloir, 1970; see also reviews by Waechter and Lennarz, 1976; Parodi and Leloir, 1979; Struck and Lennarz, 1980). At about the same time, preliminary evidence was obtained for the participation of retinol in the formation of a mannolipid (De Luca *et al.*, 1970b) that was later conclusively established to be β-mannosylphosphorylretinol (A13) (Rosso *et al.*, 1975, 1976). Since that time, there has been extensive development, largely in the laboratories of De Luca and Wolf, of the concept of posttranscriptional control by retinoids of glycoprotein synthesis through the formation of glycosylphosphoryl retinoid intermediates (for reviews, see L. M. De Luca, 1977, 1978, 1982; De Luca *et al.*,1979; Wolf, 1977, 1983; Wolf *et al.*, 1979). However, in contrast to the dolichol-dependent pathway for N-linked oligosaccharide synthesis, for which many of the intermediates have been characterized (Fig. 1), firm evidence for the steps involved in the retinoid-dependent pathway is still lacking. It is well established that retinoids have profound effects on glycoconjugate synthesis both *in vivo* and *in vitro* (see Section VI). Furthermore, it has been amply demonstrated that retinol can be phosphorylated to an intermediate, retinyl phosphate (A12) (RetP), which can then accept mannose from GDP-mannose to form β-mannosylretinyl phosphate (MRP), analogous to the intermediate β-mannosyldolichyl phosphate (MDP). However, some ambiguity still exists with respect to the existence of separate and specific enzymes for the formation of either MRP or MDP, and many recent investigations have been directed at this issue. Finally, very little information is presently available concerning description of both the specific

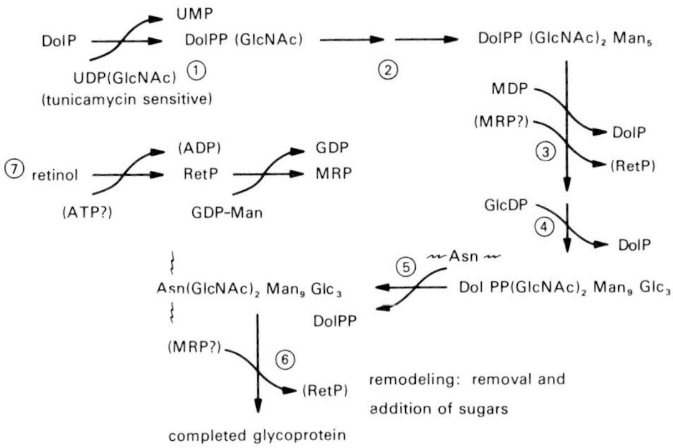

Fig. 1. Possible role of retinol in the biosynthesis of N-glycosidically linked oligosaccharide chains. Synthesis of the dolichol–oligosaccharide intermediate begins with synthesis of the primer *N*-acetylglucosaminylpyrophosphoryldolichol (reaction 1) and proceeds, possibly by direct donation from GDP-Man, to the Man_5 intermediate (reaction 2). The chain is then built up to the Man_9Glc_3 intermediate by the addition of dolichol-linked mannosyl (reaction 3) and glucosyl residues (reaction 4), whereupon the lipid–oligosaccharide is transferred to an asparagine residue of the acceptor protein (reaction 5). Remodeling to the final "complex" or polymannose glycoprotein then occurs (reaction 6). (Adapted from Parodi and Leloir, 1979; Struck and Lennarz, 1980.) The phosphorylation of retinol to RetP and the formation of a mannosyl adduct (MRP) analogous to MDP have been demonstrated (reaction 7). It has been postulated that MRP can participate in donation of mannosyl residues to protein, either in a manner analogous to MDP by donating residues to the growing oligosaccharide precursor (reaction 3) or in a novel mechanism involving direct donation to glycoprotein (reaction 6). Abbreviations used in this figure are: DolP, dolichyl phosphate; RetP, retinyl phosphate; UDP, uridine diphosphate; GDP, quanosine diphosphate; MDP, mannosyl dolichyl phosphate; MRP, mannosyl retinyl phosphate; Man, mannose; GlcNAc, *N*-acetylglucosamine; Asn, asparagine.

steps involved in the retinoid-mediated transfer of glycosyl residues to protein, and the role of retinoic acid and related retinoids in this glycosylation process. Some of the more important aspects of these points are discussed below. In light of the many reviews on the subject, no attempt is made to cite all relevant contributions. Rather, emphasis is on the most current literature in the field.

1. Retinoid Effects on Glycoconjugate Synthesis in Vivo and in Vitro

The ability of retinoids to affect the synthesis of all classes of glycoconjugates has been discussed at length in the preceding section. However, it is particularly noteworthy to this discussion of the role of retinoids as donors of mannosyl residues to point out that certain of the retinoid-sensitive glycoconjugates such as the intestinal mucins (L. M. De Luca, 1977; Olson *et al.*, 1981), the collagens (Section VI,A,2,b), and the glycosaminoglycans (Section VI,C) do not contain mannose. In addition, it was shown that retinoids induced a change in the ratio of N- to O-sulfation of heparan sulfate of intestinal cells, rather than a change in the basic glycosylation pattern (Shapiro and Mott, 1981). At the least, these results suggest that some alternative mechanism of retinoid control of glycoconjugate synthesis must be operative in addition to the proposed role of retinoids in the donation of mannosyl residues. One such possibility, already discussed in Section VI would be control of certain specific glycosyltransferases by the retinoids. Several reports document the effects of retinoids on the activity of galactosyltransferase (Plotkin and Wolf, 1980; Creek and Morré, 1980) and sialyltransferase (Moskal *et al.*, 1980).

2. *Biosynthesis of Retinyl Phosphate and Retinyl Phosphate Mannose*

Although neither the particular kinase nor the phosphate donor involved in the synthesis of RetP have been identified, experiments utilizing chemically synthesized RetP have shown that it can act as a direct acceptor for mannose from GDP-mannose (L. M. De Luca, 1977; De Luca *et al.*, 1977, 1979, 1982; Sato *et al.*, 1978; see also reaction 3, Fig. 1 and Chapter 11). Both RetP and MRP have been identified from epidermal cells cultured in the presence of retinol (Adamo *et al.*, 1979b) and in a variety of membrane preparations from rat or hamster liver (Sato *et al.*, 1978; Shidoji and De Luca, 1981; De Luca *et al.*, 1982), from lactating bovine mammary tissue (Prakash *et al.*, 1981), and from human platelets (S. De Luca, 1977). The development of new methods of separation of RetP and MRP by high-pressure liquid chromatography (HPLC) (Bhat *et al.*, 1980; Kurokawa and De Luca, 1982) has facilitated identification of both RetP and MRP in some studies.

Studies by Shidoji and De Luca (1981) have shown that the formation of MRP from either endogenous or exogenous retinol or RetP is greatly facilitated by the addition of bovine serum albumin to the incubation mixture (2–5 mg/ml), presumably by functioning to increase solubilization of RetP in the aqueous incubation medium. The amount of MRP formed from RetP in the presence of bovine serum albumin was increased almost 100-fold compared to the same reaction in the presence of detergent (0.5% Triton X-100). Synthesis of MDP, on the other hand, requires detergent.

Experiments using different synthetic retinoid phosphates demonstrated that *all* retinoid phosphate esters can function as acceptors of mannosyl residues from GDP mannose, including the biologically inactive perhydromonoeneretinyl phosphate. In contrast, the mixed anhydride formed from retinoic acid (retinoyl phosphate) was inactive (De Luca *et al.*, 1977). Interestingly, some studies have shown that MRP but not the perhydromonoeneretinol analog can function as a mannosyl donor to endogenous protein (Shidoji *et al.*, 1982).

3. *Comparison of Dolichol and Retinol as Glycosyl Carriers*

a. PHYSICAL PROPERTIES. Dolichol is not a single compound but a family of polyisoprenols of chain length C_{80}–C_{100} (16–20 isoprenoid units) terminating with an α-saturated unit; retinol is a tetraisoprenoid derivative containing five conjugated double bonds. The phosphate ester of dolichol is chemically more stable than the allylic phosphate ester of retinol. It has been suggested that these differences in reactivity may form the basis for specificity of the proposed mannosyl carrier function of retinol (De Luca, 1982).

b. SPECIFICITY FOR GLYCOSYL RESIDUES. Dolichol is known to form derivatives of *N*-acetylglucosamine, mannose, glucose, and xylose, by reaction of dolichol phosphate (DolP) with the respective sugar nucleotides (Struck and Lennarz, 1980). Retinol, on the other hand, appears to be specific for formation of a mannosyl derivative, although some evidence exists for a carrier function for galactose in mouse mastocytoma (Peterson *et al.*, 1976; L. M. De Luca, 1977) and in cultured mouse epidermal cells (Adamo *et al.*, 1979b). In the latter experiments it was concluded, however, that the galactolipid formed was not a derivative of RetP. Moreover, experiments in rat liver membranes have not supported the role of retinol as a galactosyl donor (De Luca *et al.*, 1977; Shidoji and De Luca, 1981; Lehle *et al.*, 1983).

c. EVIDENCE FOR A SPECIFIC MANNOSYLTRANSFERASE IN THE FORMATION OF MRP. Much controversy exists concerning the issue of whether MRP is formed by the action of a specific mannosyltransferase or whether RetP merely serves as a poor analog for DolP. The evidence will be fully resolved only when the respective transferases have been purified to homogeneity; meanwhile, circumstantial evidence is beginning to accumulate in support of separate and specific enzymes.

In nonmammalian systems, Lehle *et al.* (1983) using *Saccharomyces cerevisiae* obtained evidence for one mannosyltransferase that can utilize either RetP or DolP as substrate. However, the relevance of these results to mammalian systems remains questionable, especially until the occurrence of retinoids can be demonstrated in yeast. In the mold *Aspergillus niger* van Tieghem, all polyprenyl phosphates *except* RetP were substrates for the transfer of mannose from GDP-mannose; this was interpreted to indicate enzyme selectivity for RetP (Letoublon *et al.*, 1982).

In rat liver membranes, Kato *et al.* (1980) obtained indirect evidence for a lack of specificity of a mannosyltransferase that can utilize DolP as a substrate: phenyl phosphate can serve as a mannose acceptor and can inhibit the acceptor activity of endogenous DolP. But Shidoji and De Luca (1981) have shown that phenyl phosphate in 10-fold excess does *not* inhibit MRP synthesis from exogenous RetP in the presence of bovine serum albumin. In other experiments, Shidoji *et al.* (1981) provided additional evidence for the existence of two separate enzymes: one a high-affinity enzyme responsible for the synthesis of MDP with a K_m for GDP-mannose of 1.7 μM, and the other a low-affinity enzyme responsible for the synthesis of MRP with a K_m for GDP-mannose of 12.5 μM. These enzymes also respond differentially to solubilization with Triton X-100.

d. FORMATION OF MRP OR MDP FROM ENDOGENOUS SUBSTRATES. The relative amounts of endogenous RetP or DolP available for mannosylation by GDP-mannose appear to be highly sensitive to assay conditions, particularly to the concentration of sugar nucleotide (Shidoji *et al.*, 1981) and to the use of detergent or bovine serum albumin (Shidoji and De Luca, 1981). For example, Prakash *et al.* (1981) found a 30:1 ratio of MDP to MRP formed from endogenous acceptors in a membrane preparation of lactating bovine mammary gland in the absence of detergents; the ratio could be changed to 1:5.5 by addition of 10^{-4} *M* RetP. On the other hand, De Luca *et al.* (1982) utilizing bovine serum albumin in the assay, found very different ratios of 3:1 or 1.5:1 of MDP to MRP formed from endogenous DolP or RetP in rat or hamster liver microsomes, respectively. In vitamin A-deficient hamster microsomes this ratio was changed to approximately 12:1.

4. *Mannosyl Transfer from MRP to Endogenous Proteins*

Whereas a clear picture has emerged for the function of dolichol in the synthesis of N-linked oligosaccharides (Parodi and Leloir, 1979; Struck and Lennarz, 1980), the mode of action of MRP is not understood. As shown in Fig. 1, it has been suggested that MRP might function in a manner analogous to MDP in donation of mannosyl residues to the growing lipid–oligosaccharide intermediate (reaction 3), or that MRP might function in a manner distinct from MDP and donate mannosyl residues directly to protein during the remodeling that occurs after transfer of the core oligosaccaride–lipid intermediate to protein (reaction 6).

Possibly in support of the first hypothesis, Rosso *et al.* (1981) have shown that retinoid-deficient rats accumulate an oligosaccharide–lipid of the composition $Man_5(GlcNAc)_2PPDol$ (reaction 2, Fig. 1), and that this shortened intermediate is rapidly replaced by the normal intermediate $Glc_3Man_9(GlcNAc)_2PPDol$ upon treatment of the rats with retinyl acetate (reactions 3 and 4, Fig. 1). One interpretation of these data is that MRP might function in the synthesis of the oligosaccharide–lipid intermediate or that either retinoids or MRP, specifically, affect the activity of the MDP-dependent mannosyltransferase. In further support of this mechanism, De Luca *et al.* (1982) have demonstrated an accumulation of both DolP and DolPP(GlcNAc)$_2$ in vitamin A-deficient rat liver membranes, and Lehle *et al.* (1983) have shown that MRP can participate directly in elongation of the oligosaccharide–lipid intermediate in yeast. To evoke specificity in this mechanism, the existence of distinct mannosyltransferases for MRP and MDP must be postulated; preliminary evidence of Quill and Wolf (1981) does not support such a distinction.

In contrast to these results, other investigations have shown that RetP is inactive in stimulating mannose incorporation into lipid–oligosaccharide intermediates but rather donates mannose directly to protein (reaction 6, Fig. 1) (Frou-Coutaz *et al.*, 1979; Sasak and De Luca, 1980; Shidoji *et al.*, 1982). In the face of these conflicting results, it is clear that the role of retinol in transfer of glycosyl residues to protein is still unsolved. Thus, it cannot yet be said whether retinol functions in a novel distinct pathway, or rather in a manner analogous to dolichol. Nor is it yet clear, if the latter is so, whether there are different enzymes that specifically require retinol as the lipid "donor" or whether the participation of retinol reflects a lack of specificity on the part of the enzymes of the dolichol-dependent pathway? Firm unequivocal evidence in several different responsive cell types is needed to resolve these issues.

5. *Role of Retinoic Acid and its Derivatives in Mannosyl Transfer*

Ultimately any formulation of a mechanism of action of the retinoids must be able to include retinoic acid and its derivatives. It is well-known that retinoic acid can substitute for retinol in supporting all the *in vivo* functions of the vitamin in growth and differentiation (with the exception of reproduction), and that retinoic acid, with few exceptions, is 100–1000 times as active as retinol in most cell-culture systems (see Section IV). Most of the various diverse biological assay systems discussed in Chapter 5, Vol. 1, show retinoic acid and its derivatives to be the most active members of the retinoid family. Furthermore, several studies have shown that retinoic acid can stimulate the incorporation of mannose into both glycoproteins and mannolipids in a variety of tissues and cells, both *in vivo* and *in vitro* (L. M. De Luca, 1977; Sato *et al.*, 1978; Sasak *et al.*, 1980). Since it has been shown that the mixed anhydride retinoyl phosphate cannot serve as an acceptor of mannose from GDP-mannose (De Luca *et al.*, 1977; Sato *et al.*, 1978), an alternative intermediate must be found if this mechanism is to remain viable.

Bhat and De Luca (1981) addressed this question by studying the effect of radioactively labeled retinoic acid on mannolipid formation in transformed mouse 3T12 cells. Using both thin-layer chromatography and HPLC, they were able to demonstrate the formation of a retinoic acid-derived mannolipid (MXP) with properties closely related to those of MRP. However, positive identification of MXP is still lacking. Although their results exclude the reduction of retinoic acid to retinol, they nonetheless do suggest reduction of retinoic acid to a "retinol-like" derivative. Reduction seems highly unlikely, as a general phenomenon, in light of the classic experiments of Dowling and Wald (1960) demonstrating that the health of rats maintained on a vitamin A-free diet supplemented with retinoic acid was unimpaired, but that the animals became blind, owing to the inability of retinoic acid to be reduced to retinaldehyde or to retinol. Conceptually, the possibility that retinol might be oxidized to retinoic acid and then again reduced to a retinol-like metabolite seems unlikely. Rather, since the metabolism of retinol in mouse 3T12 cells has already been demonstrated to be unusual [it is reduced to anhydroretinol (A8) (Bhat *et al.*, 1979)], it is essential that formation of the metabolite MXP be demonstrated in other cell types in which retinoic acid is known to have an effect on glycoprotein synthesis.

It has been demonstrated that a new series of retinoids that may be viewed as retinoidal benzoic acid derivatives (G1–G18) (Loeliger *et al.*, 1980) are even more potent than retinoic acid in several test systems that measure control of both growth and differentiation *in vivo* and *in vitro* (Strickland *et al.*, 1983; Chapter 5, Vol. 1). If such effects are to be explained by a mechanism of glycosyl transfer, then it will be necessary to demonstrate that these retinoidal benzoic acid derivatives, which bear no structural resemblance to dolichol, can form mannosyl retinoid phosphate intermediates. More importantly, evidence must be provided for the participation of any such derivatives (and of the MXP derivative of retinoic acid) in donation of mannosyl residues to endogenous protein acceptors. The previously discussed example of perhydromonoeneretinyl phosphate has, in fact, provided evidence for a broader specificity in the mannosyltransferase responsible for the formation of the monoglycosylphosphoryl retinoid derivatives than for transfer of mannose to endogenous protein acceptors (Shidoji *et al.*, 1982).

6. *Summary*

The mechanism of direct participation of retinoids in the transfer of mannose to glycoconjugates is appealing in light of the broad implications of changes in cell surface glyconjugates in control of cell growth, differentiation, and transformation (see Section VI). However, firm conclusive evidence must still be provided to demonstrate a role for retinoids distinct from that of dolichol. Furthermore, a satisfactory metabolic scheme for the direct participation of retinoic acid and its derivatives in sugar transfer reactions does not yet exist. Future experiments must be addressed to these two important questions.

B. Steroid Model—Participation of Cellular Retinoid-Binding Proteins

In the mid-1970s, two intracellular retinoid-binding proteins were described; one was specific for retinol (Bashor *et al.*, 1973) and one was specific for retinoic acid (Sani and Hill, 1974; Ong and Chytil, 1975). These have been the subject of several reviews (Chytil and Ong, 1978, 1979; Ross and Goodman, 1979) and are treated in depth in Chapter 9 of this volume. For the purposes of this discussion of mechanism of action of retinoids, it should be reiterated that the intracellular binding proteins for retinol (CRBP) and for retinoic acid (CRABP) appear to be under independent regulation. Moreover, the concentration of these binding proteins appears to be under developmental control and also appears, in many instances, to change in response to neoplastic transformation (see Chapter 9). A natural extension of these observations was the development of a theory for the involvement of CRBP and CRABP in the mechanism of action of retinoids (Chytil and Ong, 1978).

The working hypothesis for the steroid model is that the retinoids, being small hydrophobic molecules like the steroid hormones, and thought to alter genomic expression like the steroids, might act in a way similar to the accepted model for steroid hormone action (O'Malley and Means, 1974). By analogy, retinoids would bind to a specific intracellular receptor (in this case CRBP or CRABP) and be translocated to the nucleus as the receptor–ligand complex that would bind to chromatin and lead to altered genomic expression. Indeed, data have been obtained to support this hypothesis, and they point to possible differences in the mode of action of retinol and retinoic acid. Takase *et al.* (1979) have made the observation that purified CRBP complexed with retinol delivers retinol to specific binding sites on isolated rat liver nuclei. The binding could be competed with excess holo-CRBP or the CRBP–retinol complex but not with excess retinol or with excess CRABP–retinoic acid complex. The number of nuclear binding sites was found to be higher in nuclei isolated from livers of vitamin A-deficient rats than from livers of control rats. The number of binding sites in the vitamin A-deficient nuclei could be restored to normal levels 2 h after administration of retinyl acetate. In an extension of this work, Liau *et al.* (1981) have shown that the nuclear binding sites for retinol delivered as the CRBP–retinol complex reside on the chromatin and not on the nuclear envelope. They further conclude from their evidence that CRBP dissociates from the chromatin after it has presumably delivered the retinol to specific sites on the chromatin. This latter observation is different from the steroid model in which the steroid–receptor complex can be detected bound to chromatin. However, in apparent agreement with the steroid model, Wiggert *et al.* (1977) have provided evidence for a translocation of the retinoic acid–CRABP complex to the nucleus of retinoblastoma cells and Sani and Donovan (1979) have demonstrated its appearance in the nuclei of Lewis lung tumor and chick embryo skin.

Thus, although the details must be worked out more thoroughly, the framework exists for the participation of the intracellular retinoid-binding proteins CRBP and CRABP in a steroid hormone-like mechanism of control of gene expression. The question that must be asked is whether these binding proteins can be shown to be *necessary* for retinoid action (see also Chapter 9). In experiments with cultured cells, there is evidence both for and against this theory. In support of the concept are the experiments of Jetten *et al.* (1979b) and Schindler *et al.* (1981) that show a convincing relationship between the existence of CRABP and the action of retinoic acid on suppressing the growth of mouse fibroblasts or promoting the differentiation of mouse teratocarcinoma stem cells, respectively. The data of Matthaei *et al.* (1983) that human teratocarcinoma cell lines having high levels of CRABP nonetheless fail to differentiate in response to retinoic acid, demonstrate that the presence of CRABP is not in itself a sufficient criterion to predict the response pattern of the cells. A similar situation exists in mutant S91 melanoma cells that are resistant to the growth inhibitory effects of retinoic acid; the levels of CRABP found in these mutant cells do not differ significantly from control levels (Lotan *et al.*, 1983b). In other studies, Lotan *et al.* (1980c) were unable to develop a correlation between CRABP levels and the growth effects of retinoic acid in 11 different cell lines. In addition, there are the observations that neither CRBP nor CRABP have been detected in myeloid leukemia cells (Breitman *et al.*, 1981; Douer and Koeffler, 1982c) or in the mouse fibroblast 10T1/2 cell line (Libby and Bertram, 1982), despite the fact that these systems are exquisitely sensitive to retinoid effects on differentiation and suppression of carcinogenesis, respectively. It can be argued that the methods of detection of CRBP or CRABP are not sufficiently sensitive; however, at the least, these data indicate a lack of a quantitative relationship between the levels of binding protein and the sensitivity of the cells to retinoids. Another argument might be that the mode of action of the retinoids is different in certain cell systems such as the HL-60 cells or 10T1/2 cells. However, the strong correlation between structure–activity relationships of retinoids in these cells as compared to many other diverse assay systems, suggests that this is not so (see Chapter 5, Vol. 1). At the time of this writing, then, it is not possible to say with confidence that the intracellular retinoid-binding proteins play an obligatory role in the mechanism of action of the retinoids. However, with the very rapid progress being made in the area of steroid hormone interaction with genetic elements, it is likely that definitive answers will be forthcoming from future investigations.

C. General Mechanism for Regulation of Gene Expression

The molecular hypothesis of retinoid action that is compatible with the broadest range of experimental data on retinoid control of cell growth, differentiation, and transformation is to suggest that retinoids modify genomic expression. This, of course is not a new idea. Almost 15 years have passed since it was first demonstrated that retinoids might be controlling RNA synthesis (Zachman, 1967; Johnson *et al.*, 1969; Zile and DeLuca, 1970); however, these early studies were subject to criticism of experimental design. Thus, it is only lately that we have been able to address experimentally the questions of (1) *which* genes are controlled by the retinoids, and (2) *how* are these genes controlled by the retinoids? Obvious questions relating to the second point are whether retinoids are controlling gene expression directly or indirectly, and whether control is at the level of gene transcription itself, the processing of primary gene transcripts, or regulation of translation of the processed transcript. The extraordinary developments that have occurred in the past few years in the area of molecular biology (e.g., involving recombinant DNA and molecular cloning techniques) will undoubtedly soon have impact on the retinoid field and will begin to provide answers to some of these questions concerning mechanism of action. Some aspects of retinoid control of genomic expression, including effects on protein synthesis, on oncogene expression and on RNA processing are presented below.

1. Retinoid Control of Protein Synthesis

The contention that retinoids modulate synthesis of specific cellular proteins has been amply illustrated in this chapter. Retinoids have been shown to influence synthesis of cellular receptors (Section IV,E), of enzymes and effectors (Section V), of glyconjugates and cytoskeletal elements (Section VI), and of keratins (Section IV,A,1,a). Taken together, it is clear that retinoids can both activate and suppress genomic expression. Two experiments, one *in vivo* and one *in vitro,* are beginning to provide information about these processes. Omori and Chytil (1982) have examined retinoid-dependent changes in cytoplasmic poly(A)$^+$RNA in rat testis, intestinal mucosa, and liver by molecular hybridization and *in vitro* translation, and Lotan *et al.* (1982a) using a cell-free *in vitro* translation system, have investigated the effect of retinoic acid on the protein synthesis machinery of mouse S91 melanoma cells.

In the studies of Omori and Chytil (1982), it was found that the levels of poly(A)$^+$RNA of all tissues were decreased in retinoid-deficient rats as compared to controls. Both homologous and heterologous hydridization experiments revealed qualitative changes in the poly(A)$^+$RNA population; specifically, there appeared to be a reduction in the amount of fast-annealing RNA in preparations of retinoid-deficient testis as compared to controls. *In vitro* translation of the isolated poly(A)$^+$RNA showed qualitative differences in the protein products. Administration of retinyl acetate or retinoic acid to the deficient animals produced a decrease in poly(A)$^+$RNA accompanied by changes in the translatable protein pattern that were detectable 1 h after oral dosing. Qualitative changes distinguished the testicular protein patterns observed following administration of retinyl acetate or retinoic acid. At least in the testis where different effects of retinol and retinoic acid on reproductive capacity have long been observed, these results point to both common and separate modes of control of protein synthesis by these two retinoids. Although these experiments are complicated by mixed cell populations and lack of characterization of the affected gene products, they represent an interesting attempt to study the molecular biology of retinoid action.

The experiments of Lotan *et al.* (1982a) showed decreases in the poly(A)$^+$RNA population of S91 melanoma cells cultured for 5 days in the presence of 10^{-5} *M* retinoic acid as compared to controls. Such treatment results in nearly 90% inhibition of the growth of the cells. Protein synthesis by these cells was depressed relative to control cultures, as evidenced not only by the decreased poly(A)$^+$RNA levels but also by decreases in the proportions of ribosomes in polysomes and decreases in polysome-associated mRNA. In contrast to the effects of Omori and Chytil (1982) on a differentiating epithelial tissue, Lotan *et al.* (1982a) found little qualitative differences in the protein patterns of treated or control cells. Further experiments should help answer the question of whether the underlying mechanisms of retinoid control of growth and differentiation are distinct or identical.

2. *Retinoid Control of Oncogene Expression*

Explosive developments in the understanding of normal cellular development and of carcinogenic transformation have accompanied the emergence of the "oncogene story." These evolutionarily conserved genes, which at first were thought to be specific for cellular transformation, have now been shown to be expressed in normal cells and differentially in the development of the embryo (Mintz and Fleischman, 1981; Müller *et al.*, 1982) and in the regenerating liver (Goyette *et al.*, 1983). The elucidation of possible links between retinoids and oncogenes will undoubtedly be intensively pursued in the near future. A hint at a possible relationship is provided by the experiments of Westin *et al.* (1982) who used specific molecular probes to show that expression of the *c-myc* but not of the *c-abl* cellular oncogenes of HL-60 human myeloid leukemia cells is suppressed following the induction of differentiation of the cells by retinoic acid. However, similar effects on gene expression accompanied differentiation of the cells following treatment with dimethyl sulfoxide. The important question of the kinetics of the suppression, which would shed light on the cause–effect relationship of *c-myc* expression and differentiation, was not addressed, though other evidence does suggest that *c-myc* plays a role in the arrest of differentiation particular to these cells (Favera *et al.*, 1982; Collins and Groudine, 1982).

3. *Retinoid Control of RNA Processing*

One specific example of *how* retinoids might control gene expression is provided by investigations of the susceptibility of murine teratocarcinoma stem cells to transformation by papovaviruses. It had been noted that infection of F9 stem cells by SV40 virus produced unspliced, nonfunctional viral RNA transcripts (Segal *et al.*, 1979), whereas infection of the differentiated endoderm cells resulting from treatment of the F9 cells with retinoic acid led to the production of functional viral(T)antigens (Segal and Khoury, 1979). This suggested a defect in the RNA splicing mechanism of the stem cells. Linnenbach *et al.* (1980) then infected a thymidine kinase-deficient mutant F9 cell line with a recombinant plasmid genome consisting of the pBR322 genome linked to a herpes simplex thymidine kinase gene and a single copy of the SV40 genome. The thymidine kinase gene was functionally expressed in both the stem cell line and the differentiated cells, but functional SV40(T)antigens were expressed only after the induction of differentiation by retinoic acid. The questions must be asked whether these effects on expression of the viral genome are relevant to control of expression of normal cellular genes and what temporal relationship they have to the differentiation event. In the broadest interpretation, these results suggest that posttranscriptional control of RNA processing may be one of the mechanisms of retinoid-mediated regulation of protein synthesis, and experiments of the sort just described using defined genetic elements under selected control, might be one way to approach this problem.

D. Other Possible Mechanisms of Action

Still other mechanisms of action might be considered. Rapaport *et al.* (1982a,b) have shown that treatment of Swiss mouse 3T3 cells with retinoic acid significantly expands total cellular ATP pools in a dose-dependent manner, and that these changes alter functional compartmentalization of nuclear ATP pools (Schroder *et al.*, 1983). It is proposed that these changes in nucleotide pools are directly responsible for the inhibition of DNA replication observed in these cells and that a similar mechanism might apply to other cells whose growth rate is modulated by retinoids (Schroder *et al.*, 1983).

The described effects of retinoids on the cAMP-dependent protein kinases (Section V,B) or on transport or compartmentalization of Ca^{2+} are yet other possibilities, though it is perhaps difficult to understand the specificity of mechanisms such as these. For future investigations, the use of defined cell-culture systems will provide a specific experimental approach. However, the generality of the results must then be tested in other *in vitro* systems as well as *in vivo*.

VIII. CONCLUSIONS

The single theme that continually recurs in all studies of the cellular mechanism of action of retinoids is that they modify cell differentiation and cell proliferation. However extensive the data that we have just summarized on the effects of retinoids on individual cellular enzymes, effectors, structural proteins, or glycoconjugates, they fail to provide a satisfactory explanation of the molecular mechanism of action of retinoids because they fail to consider the central role of the genes of the cell in control of differentiation and proliferation. Ultimately, it would appear that the problem of the molecular mechanism of action of retinoids will have to converge on one of the central problems in all fields of biology, namely the mechanism of the control of gene expression during cell differentiation. The question of how to approach this problem experimentally is very difficult. It would appear that studies of abnormal cell function may provide important information regarding the normal function of retinoids in control of cell differentiation; the extensive use of both preneoplastic and neoplastic cells to investigate the cellular mechanism of action of retinoids reflects this approach. The description of oncogenes as normal cellular genes, whose function may be aberrantly expressed (either in amount or in time sequence) in the cancer cell, may now provide some new experimental approaches to this problem (Bishop, 1982). It is clear that particular oncogenic retroviruses cause malignant transformation of cells by means of inappropriate expression of specific genes, which are none other than cellular genes that are associated with normal cellular function, and in particular with normal embryological development. In this context, study of aberrant function of such genes in neoplastic cells is only a means to understand the normal function of these genes in nonneoplastic cells. Thus, it has been stressed (Mintz and Fleischman, 1981) that "the malregulated expression of developmentally relevant cellular genes . . . appears to underlie all cancers; and that the genes at issue are any of the numerous 'banal' or ordinary genes involved in cell growth and differentiation, rather than special or exotic cancer genes." The striking effects of retinoids on suppression of malignant transformation of cells suggest that this type of experimental analysis may ultimately yield important molecular information on the role of retinoids in normal cell differentiation. Similarly, the *selective* teratogenic effects of retinoids on normal embryonic development and study of the specific retinoid requirements of the embryo for appropriate differentiation and development may also provide systems in which the activation or inactivation of specific genes by retinoids can be investigated with modern techniques of genetic analysis. A satisfactory explanation of the cellular and molecular mechanism of action of retinoids will probably not be available until we have an adequate understanding of their role in control of gene expression. New methods to pursue such studies are now available, and they should give important new information about an old problem.

REFERENCES

Abb, J., and Deinhardt, F. (1981). *Oncology* **38,** 346–350.

Abe, E., Miyaura, C., Sakagami, H., Takeda, M., Konno, K., Yamazaki, T., Yoshiki, S., and Suda, T. (1981). *Proc. Natl. Acad. Sci. U.S.A.* **78,** 4990–4994.

Adamo, S., De Luca, L. M., Akalovsky, I., and Bhat, P. V. (1979a). *JNCI, J. Natl. Cancer Inst.* **62,** 1473–1478.

Adamo, S., De Luca, L. M., Silverman-Jones, C. S., and Yuspa, S. H. (1979b). *J. Biol. Chem.* **254,** 3279–3287.

Adhikari, H. R., and Vakil, V. K. (1980). *Biochim. Biophys. Acta* **633,** 465–478.

Aharonov, A., Pruss, R., and Herschman, H. (1978). *J. Biol. Chem.* **253,** 3970–3977.

Anzano, M. A., Roberts, A. B., Meyers, C. A., Komoriya, A., Lamb, L. C., Smith, J. M., and Sporn, M. B. (1982). *Cancer Res.* **42,** 4776–4778.

Anzano, M. A., Roberts, A. B., Smith, J. M., Sporn, M. B., and De Larco, J. E. (1983) *Proc. Natl. Acad. Sci. U.S.A.* **80,** 6264–6268.

Assoian, R. K., Komoriya, A., Meyers, C. A., Miller, D. M., and Sporn, M. B. (1983). *J. Biol. Chem.* **258,** 7155–7160.

Atkinson, P. H., and Hakimi, J. (1980). *In* "The Biochemistry of Glycoproteins and Proteoglycans" (W. J. Lennarz, ed.), pp. 191–239. Plenum, New York.

Bachrach, U. (1976). *Ital. J. Biochem.* **25,** 77–93.

Barnes, D. (1982). *J. Cell. Biol.* **93,** 1–4.

Bashor, M. M., Toft, D. O., and Chytil, F. (1973). *Proc. Natl. Acad. Sci. U.S.A.* **70,** 3483–3487.

Beers, N. H., Strickland, S., and Reich, E. (1975). *Cell* **6,** 387–394.

Behrens, N. H., and Leloir, L. F. (1970). *Proc. Natl. Acad. Sci. U.S.A.* **66,** 153–159.

Berenblum, I. (1978). *In* "Mechanisms of Tumor Promotion and Cocarcinogenesis" (T. J. Slaga, A. Sivak, and R. K. Boutwell, eds.), pp. 1–10. Raven, New York.

Berridge, M. J. (1976). *In* "SEB Symposium 30, Calcium in Biological Systems" (C. J. Duncan, ed.), pp. 219–231. Cambridge Univ. Press, London and New York.

Bertram, J. S. (1980). *Cancer Res.* **40,** 3141–3146.

Bertram, J. S. (1983). *Cancer Surveys* **2,** 243–262.

Bertram, J. S., Mordan, L. J., Blair, S. J., and Hui, S. (1981). *Ann. N. Y. Acad. Sci.* **359,** 218–236.

Bertram, J. A., Mordan, L. J., Domanska-Janik, K., and Bernaki, R. J. (1982). *In* "Molecular Interrelations of Nutrition and Cancer" (M. S. Arnott, J. van Eys, and Y. M. Wang, eds.), pp. 315–335. Raven, New York.

Bhat, P. V., and De Luca, L. M. (1981). *Ann. N. Y. Acad. Sci.* **359,** 135–149.

Bhat, P. V., De Luca, L. M., Adamo, S., Akalovsky, I., Silverman-Jones, C. S., and Peck, G. L. (1979). *J. Lipid Res.* **20,** 357–362.

Bhat, P. V., De Luca, L. M., and Wind, M. L. (1980). *Anal. Biochem.* **102,** 243–248.

Birckbichler, P. J., Orr, G. R., Conway, E., and Patterson, M. K. (1977). *Cancer Res.* **37,** 1340–1344.

Birckbichler, P. J., and Patterson, M. K. (1978). *Ann. N. Y. Acad. Sci.* **312,** 354–365.

Bishop, J. M. (1982). *Adv. Cancer Res.* **37,** 1–32.

Blumberg, P. M., Driedger, P. E., and Rossow, P. W. (1976). *Nature (London)* **264,** 445–447.

Bolmer, S. D., and Wolf, G. (1982a). *Cancer Res.* **42,** 4465–4472.

Bolmer, S. D., and Wolf, G. (1982b). *Proc. Natl. Acad. Sci. U.S.A.* **79,** 6541–6545.

Bonanni, F., and De Luca, L. M. (1974). *Biochim. Biophys. Acta* **343,** 632–637.

Bonanni, F., Levinson, S. S., Wolf, G., and De Luca, L. M. (1973). *Biochim. Biophys. Acta* **297,** 441–451.

Boutwell, R. K., and Verma, A. K. (1981). *Ann. N. Y. Acad. Sci.* **359,** 275–280.

Bradshaw, R. A., and Rubin, J. S. (1980). *J. Supramol. Struct.* **14,** 183–199.

Breitman, T. R., Collins, S. J., and Keene, B. R. (1980a). *Exp. Cell Res.* **126,** 494–498.

Breitman, T. R., Selonick, S. E., and Collins, S. J. (1980b). *Proc. Natl. Acad. Sci. U.S.A.* **77,** 2936–2940.

Breitman, T. R., Collins, S. J., and Keene, B. R. (1981). *Blood* **57,** 1000–1004.

Brinckerhoff, C. E., and Harris, E. D., Jr. (1981a). *Am. J. Pathol.* **103,** 411–419.

Brinckerhoff, C. E., and Harris, E. D., Jr. (1981b). *Biochim. Biophy. Acta* **677,** 424–432.

Brinckerhoff, C. E., McMillin, R. M., Fahey, J. V., and Harris, E. D., Jr. (1979). *Arthritis Rheum.* **22,** 1109–1116.

Brinckerhoff, C. E., McMillin, R. M., Dayer, J.-M., and Harris, E. D., Jr. (1980). *N. Engl. J. Med.* **303,** 432–436.

Brinckerhoff, C. E., Nagase, H., Nagel, J. E., and Harris, E. D. (1982). *J. Am. Acad. Dermatol.* **6,** 591–602.

Brinster R. L. (1974). *J. Exp. Med.* **140,** 1049–1056.

Camisa, C., Eisenstat, B., Ragaz, A., and Weissmann, G. (1982). *J. Am. Acad. Dermatol.* **6,** 620–629.

Carpenter, G., and Cohen, S. (1979). *Annu. Rev. Biochem.* **48,** 193–216.

Castagna, M., Takai, Y., Kaibuchi, K., Sano, K., Kikkawa, U., and Nishizuka, Y. (1982). *J. Biol. Chem.* **257,** 7847–8751.

Catt, K. J., Tsuruhara, T., Mendelson, C., Ketelslegers, J. M., and Dufau, M. L. (1974). *Curr. Top. Mol. Endocrinol.* **1,** 1–30.

Chapman, S. K. (1980). *Life Sci.* **26,** 1359–1366.

Cherington, P. V., Smith, B., and Pardee, A. B. (1979). *Proc. Natl. Acad. Sci. U.S.A.* **76,** 3937–3941.

Childs, C. B., Proper, J. A., Tucker, R. F., and Moses, H. L. (1982). *Proc. Natl. Acad. Sci. U.S.A.* **79,** 5312–5316.

Chopra, C. P., and Flaxman, B. A. (1975). *J. Invest. Dermatol.* **64,** 19–22.

Chou, I.-N., Roblin, R. O., and Black, P. H. (1977). *J. Biol. Chem.* **252,** 6256–6259.

Christophers, E. (1974). *J. Invest. Dermatol.* **63,** 450–455.

Christophers, E., and Wolff, H. H. (1975). *Nature (London)* **256,** 209–210.

Chytil, F., and Ong, D. E. (1978). *Vitam. Horm. (N. Y.)* **36,** 1–32.

Chytil, F., and Ong, D. E. (1979). *Fed. Proc., Fed. Am. Soc. Exp. Biol.* **38,** 2510–2514.

Cifone, M. A., and Fidler, I. J. (1980). *Proc. Natl. Acad. Sci.* **77,** 1039–1043.

Clark, J. N., and Marchok, A. C. (1979). *Differentiation* **14,** 175–183.

Clarkson, B. D. (1972). *Cancer (Philadelphia)* **30,** 1572–1582.

Colburn, N. H., Ozanne, S., Lichti, U., Ben, T., Yuspa, S., Wendel, E., Jardini, E., and Abruzzo, G. (1981). *Ann. N. Y. Acad. Sci.* **359,** 251–259.

Collins, S., and Groudine, M. (1982). *Nature (London)* **298,** 679–681.

Collins, S. J., Ruscetti, F. W., Gallagher, R. E., and Gallo, R. C. (1978). *Proc. Natl. Acad. Sci. U.S.A.* **75,** 2458–2462.

Collins, S. J., Ruscetti, F. W., Gallagher, R. E., and Gallo, R. C. (1979). *J. Exp. Med.* **149,** 969–974.

Creek, K. E., and Morré, D. J. (1980). *Biochem. Biophys. Res. Commun.* **95,** 175–178.

Dahl, I. M. S., and Axelsson, I. (1980). *Exp. Eye Res.* **31,** 443–450.

Das, M. (1982). *Int. Rev. Cytol.* **78,** 233–256.

Delclos, K. B., and Blumberg, P. M. (1979). *Cancer Res.* **39,** 1667–1672.

De Larco, J. E., and Todaro, G. J. (1978). *Proc. Natl. Acad. Sci. U.S.A.* **75,** 4001–4005.

De Leenheer, A. P., Lambert, W. E., and Claeys, I. (1982). *J. Lipid Res.* **23,** 1362–1367.

De Luca, L. M. (1977). *Vitam. Horm. (N. Y.)* **35,** 1–57.

De Luca, L. M. (1978). *In* "Handbook of Lipid Research" (H. F. DeLuca, ed.), Vol. 2, pp. 1–67. Plenum, New York.

De Luca, L. M. (1982). *J. Am. Acad. Dermatol.* **6,** 611–619.

De Luca, L. M., and Yuspa, S. H. (1974). *Exp. Cell Res.* **86,** 106–110.

De Luca, L. M., Little, E. P., and Wolf, G. (1969). *J. Biol. Chem.* **244,** 701–708.

De Luca, L. M., Schumacher, M., and Wolf, G. (1970a). *J. Biol. Chem.* **245,** 4551–4558.

De Luca, L., Rosso, G., and Wolf, G. (1970b). *Biochem. Biophys. Res. Commun.* **41,** 615–620.

De Luca, L. M., Schumacher, M., and Nelson, D. P. (1971). *J. Biol. Chem.* **246,** 5762–5765.

De Luca, L. M., Silverman-Jones, C. S., and Barr, R. M. (1975). *Biochim. Biophys. Acta* **409,** 342–359.

De Luca, L. M., Frot-Coutaz, J. P., Silverman-Jones, C. S., and Roller, P. R. (1977). *J. Biol. Chem.* **252,** 2575–2579.
De Luca, L. M., Bhat, P. V., Sasak, W., and Adamo, S. (1979). *Fed. Proc., Fed. Am. Soc. Exp. Biol.* **38,** 2535–2539.
De Luca, L. M., Brugh, M. R., Silverman-Jones, C. S., and Shidoji, Y. (1982). *Biochem. J.* **208,** 159–170.
De Luca, S. (1977). *Biochim. Biophys. Acta* **498,** 341–348.
De Ruyter, M. G., Lambert, W. E., and De Leenheer, A. P. (1979). *Anal. Biochem.* **98,** 402–409.
De Ruyter, T. H. (1934). *Acta Brevia Neerl. Physiol. Pharmaeol., Microbiol.* **4,** 122.
Dicker, P., and Rozengurt, E. (1979). *Biochem. Biophys. Res. Commun.* **91,** 1203–1210.
Dion, L. D., and Gifford, G. E. (1980). *Proc. Soc. Exp. Biol. Med.* **163,** 510–514.
Dion, L. D., Blalock, J. E., and Gifford, G. E. (1977). *JNCI, J. Natl. Cancer Inst.* **58,** 795–800.
Dion, L. D., Blalock, J. E., and Gifford, G. E. (1978). *Exp. Cell Res.* **117,** 15–22.
Dion, L. D., De Luca, L. M., and Colburn, N. H. (1981). *Carcinogenesis* **2,** 951–958.
Dion, L. D., Bear, J., Bateman, J., De Luca, L. M., and Colburn, N. H. (1982). *JNCI, J. Natl. Cancer Inst.* **69,** 1147–1154.
Douer, D., and Koeffler, H. P. (1982a). *J. Clin. Invest.* **69,** 1039–1041.
Douer, D., and Koeffler, H. P. (1982b). *Exp. Cell Res.* **138,** 193–198.
Douer, D., and Koeffler, H. P. (1982c). *J. Clin. Invest.* **69,** 277–283.
Dowling, J. E., and Wald, G. (1960). *Proc. Natl. Acad. Sci. U.S.A.* **46,** 587–608.
Driedger, P. E., and Blumberg, P. M. (1980). *Proc. Natl. Acad. Sci. U.S.A.* **77,** 567–571.
Durham, J. P., Butcher, F. R., and Fontana, J. A. (1982). *Biochem. Biophys. Res. Commun.* **108,** 840–845.
Durham, J. P., Emler, C. A., McClung, J. K., Butcher, F. R., and Fontana, J. A. (1983). *In* "Modulation and Mediation of Cancer by Vitamins" (F. L. Meyskens and K. Prasad, eds.), pp. 194–203. Karger, Basel.
Evain, D., Binet, E., and Anderson, W. B. (1981). *J. Cell. Physiol.* **109,** 453–459.
Favera, R. D., Wong-Staal, F., and Gallo, R. C. (1982). *Nature (London)* **299,** 61–63.
Fell, H. B., and Mellanby, E. (1953). *J. Physiol. (London)* **119,** 470–488.
Findlay, G. M., and McKenzie, R. D. (1922). *J. Pathol. Bacteriol.* **25,** 402–403.
Folk, J. E., Park, M. H., Chung, S. I., Schrode, J., Lester, E. P., and Cooper, H. L. (1980). *J. Biol. Chem.* **255,** 3695–3700.
Frank, M. (1935). *Med. Klin. (Munich)* **31,** 486–488.
Frolik, C. A., Tavela, T. E., Peck, G., and Sporn, M. (1978). *Anal. Biochem.* **86,** 734–750.
Frot-Coutaz, J., Letoublon, R., and Got, R. (1979). *FEBS Lett.* **107,** 375–378.
Fuchs, E., and Green, H. (1980). *Cell* **19,** 1033–1042.
Fuchs, E., and Green, H. (1981). *Cell* **25,** 617–625.
Garg, L. G., and Brown, J. C. (1983). *Differentiation* (in press).
Gootwine, E., Webb, C. G., and Sachs, L. (1982). *Nature (London)* **299,** 63–65.
Gospodarowicz, D., and Moran, J. S. (1976). *Annu. Rev. Biochem.* **45,** 531–558.
Goyette, M., Petropoulos, C. J., Shank, P. R., and Faust, N. (1983). *Science* **219,** 510–512.
Green, H. (1977). *Cell* **11,** 405–416.
Green, H., and Watt, F. M. (1982). *Mol. Cell. Biol.* **2,** 1115–1117.
Haddox, M. K., and Russell, D. H. (1979). *Cancer Res.* **39,** 2476–2480.
Haddox, M. K., Scott, K. F. F., and Russell, D. H. (1979). *Cancer Res.* **39,** 4930–4938.
Hakomori, S. (1975). *Biochim. Biophys. Acta* **417,** 55–89.
Harisiadis, L., Miller, R. C., Hall, E. J., and Borek, C. (1978). *Nature (London)* **274,** 486–487.
Harper, R. A., and Savage, C. R. (1980). *Endocrinology (Baltimore)* **107,** 2113–2114.
Harris, E. D., Jr. (1976). *Arthritis Rheum.* **19,** 68–72.

Hassell, J. R., and Newsome, D. A. (1981). *Ann. N. Y. Acad. Sci.* **359,** 368–365.
Hassell, J. R., and Pratt, R. M. (1977). *Exp. Cell Res.* **106,** 55–62.
Hassell, J. R., Greenberg, J. H., and Johnston, M. C. (1977). *J. Embryol. Exp. Morphol.* **39,** 267–271.
Hassell, J. R., Pennypacker, J. P., and Lewis, C. A. (1978a). *Exp. Cell Res.* **112,** 409–417.
Hassell, J. R., Pennypacker, J. P., Yamada, K. M., and Pratt, R. M. (1978b). *Ann. N. Y. Acad. Sci.* **312,** 406–409.
Hassell, J. R., Pennypacker, J. P., Kleinman, H. K., Pratt, R. M., and Yamada, K. M. (1979). *Cell* **17,** 821–826.
Hassell, J. R., Newsome, D. A., and De Luca, L. M. (1980). *Invest. Ophthal.* **19,** 642–647.
Hennings, H., Michael, D., Cheng, C., Steinert, P., Holbrook, K., Yuspa, S. H. (1980). *Cell* **19,** 245–254.
Hirata, Y., Moore, G. W., Bertagna, C., and Orth, D. N. (1980). *J. Clin. Endocrinol. Metab.* **50,** 440–444.
Hoal, E., Wilson, L., and Dowdle, E. B. (1982). *Cancer Res.* **42,** 5191–5195.
Hodges, R. E., Sauberlich, H. E., Canham, J. E., Wallace, D. L., Rucker, R. B., Mejia, L. A., and Mohanram, M. (1978). *Am. J. Clin. Nutr.* **31,** 876–885.
Hogan, B. L. M., Taylor, A., and Adamson, E. (1981). *Nature (London)* **291,** 235–237.
Honma, Y., Takenaga, K., Kasukabe, T., and Hozumi, M. (1980). *Biochem. Biophys. Res. Commun.* **95,** 507–512.
Howe, C. C., and Solter, D. (1981). *Dev. Biol.* **84,** 239–243.
Hynes, R. O. (1976). *Biochim. Biophys. Acta* **458,** 73–107.
Imai, Y., Kaneko, Y., Matsuzaki, F., Endo, Y., and Oda, T. (1980). *Biochem. Biophys. Res. Commun.* **97,** 926–931.
Jaffe, B. M. (1974). *Prostaglandins* **6,** 453–461.
Jetten, A. M. (1980). *Nature (London)* **284,** 626–629.
Jetten, A. M. (1981). *Ann. N. Y. Acad. Sci.* **359,** 200–217.
Jetten, A. M. (1982). *J. Cell. Physiol.* **110,** 235–240.
Jetten, A. M. (1983). *Cancer Res.* **43,** 68–72.
Jetten, A. M., and De Luca, L. M. (1982). *In* "Carcinogenesis" (E. Hecker, N. E. Fusenig, W. Kunz, F. Marks, and H. W. Thielmann, eds.), Vol. 7, pp. 513–518. Raven, New York.
Jetten, A. M., and Goldfarb, R. H. (1983). *Cancer Res.* **43,** 2094–2099.
Jetten, A. M., and Jetten, M. E. R. (1979). *Nature (London)* **278,** 180–182.
Jetten, A. M., Jetten, M. E. R., and Sherman, M. I. (1979a). *Exp. Cell Res.* **124,** 381–391.
Jetten, A. M., Jetten, M. E. R., Shapiro, S. S., and Poon, J. P. (1979b). *Exp. Cell Res.* **119,** 289–299.
Jetten, A. M., Meeks, R. G., and De Luca, L. M. (1981). *Ann. N. Y. Acad. Sci.* **359,** 398–400.
Jetten, A. M., De Luca, L. M., and Meeks, R. G. (1982). *Exp. Cell Res.* **138,** 494–498.
Johnson, B. C., Kennedy, M., and Chiba, N. (1969). *Am. J. Clin. Nutr.* **22,** 1048–1058.
Kahn, P., and Shin, S. (1979). *J. Cell Biol.* **82,** 1–16.
Kalter, H., and Warkany, J. (1961). *Am. J. Pathol.* **38,** 1–21.
Kaneko, Y. (1981). *Int. J. Cancer* **27,** 841–846.
Karl, A. F., and Griswold, M. C. (1980). *Biochem. J.* **186,** 1001–1003.
Kato, S., Tsuji, M., Nakanishi, Y., and Suzuki, S. (1980). *J. Biochem. (Tokyo)* **87,** 929–939.
Kim, Y. L., and Wolf, G. (1974). *J. Nutr.* **104,** 710–718.
King, I. A., and Tabiowo, A. (1981). *Biochem. J.* **194,** 341–350.
Kiorpes, T. C., and Anderson, R. S. (1981). *Ann. N. Y. Acad. Sci.* **359,** 401–404.
Kiorpes, T. C., Molica, S. J., and Wolf, G. (1976). *J. Nutr.* **106,** 1659–1667.
Kiorpes, T. C., Kim, Y.-C. L., and Wolf, G. (1979). *Exp. Eye Res.* **28,** 23–25.
Klann, R. C., and Marchok, A. C. (1982). *Cell Tissue Kinet.* **15,** 473–482.

Klee, C. B., Crouch, T. H., and Richman, P. G. (1980). *Annu. Rev. Biochem.* **49,** 489–516.
Kleinman, H. K., Martin, G. R. Fishman, P. H. (1979). *Proc. Natl. Acad. Sci. U.S.A.* **76,** 3367–3371.
Kleinsmith, L. J., and Pierce, G. B., Jr. (1964). *Cancer Res.* **24,** 1544–1551.
Kochhar, D. M. (1967). *Acta Pathol. Microbiol. Scand.* **70,** 398–404.
Kochhar, D. M. (1968). *Teratology* **1,** 299–310.
Kochhar, D. M. (1973). *Teratology* **7,** 289–298.
Kornfeld, R., and Kornfeld, S. (1976). *Annu. Rev. Biochem.* **45,** 217–237.
Kornfeld, R., and Kornfeld, S. (1980). *In* "The Biochemistry of Glycoproteins and Proteoglycans" (W. J. Lennarz, ed.), pp. 1–33. Plenum, New York.
Kraemer, P. M. (1979). *In* "Surfaces of Normal and Malignant Cells" (R. O. Hynes, ed.), pp. 149–198. Wiley, New York.
Kraft, A. S., and Anderson, W. B. (1983). *J. Biol. Chem.* **258,** 9178–9183.
Kriesel, W., Volk, B. A., Buchsel, R., and Reutter, W. (1980). *Ann. N. Y. Acad. Sci.* **359,** 401–404.
Kuff, E. L., and Fewell, J. W. (1980). *Dev. Biol.* **77,** 103–115.
Kurokawa, T., and De Luca, L. M. (1982). *Anal. Biochem.* **119,** 428–432.
Kwasigroch, T. E., and Kochhar, D. M. (1975). *Exp. Cell Res.* **95,** 269–278.
Lacroix, A., and Lippman, M. E. (1980). *J. Clin. Invest.* **65,** 586–591.
Lacroix, A., Anderson, G. D., and Lippman, M. E. (1981). *Ann. N. Y. Acad. Sci.* **359,** 405–406.
Langenbach, R., and Kennedy, S. (1978). *Exp. Cell Res.* **112,** 361–372.
Lasnitzki, I. (1955). *Br. J. Cancer* **9,** 434–441.
Lee, L., and Weinstein, I. B. (1978). *Science* **202,** 313–315.
Lehle, L., Haselbeck, A., and Tanner, W. (1983). *Biochim. Biophys. Acta* **757,** 77–84.
Letoublon, R., Mayet, B., Fort-Coutaz, J., and Got, R. (1982). *J. Lipid. Res.* **23,** 1053–1057.
Levine, L. (1979). *Cold Spring Harbor Conf. Cell Proliferation* **6,** 623–640.
Levine, L. (1980). *Prostaglandins Med.* **4,** 285–296.
Levine, L., and Ohuchi, K. (1978). *Nature (London)* **276,** 274–275.
Lewis, C. A., Pratt, R. M., Pennypacker, J. P., and Hassell, J. R. (1978). *Dev. Biol.* **64,** 31–47.
Liau, G., Ong, D. E., and Chytil, F. (1981). *J. Cell Biol.* **91,** 63–68.
Libby, P. R., and Bertram, J. S. (1982). *Carcinogenesis* **3,** 481–484.
Lichti, U., Patterson, E., Hennings, H., and Yuspa, S. H. (1981). *Cancer Res.* **41,** 49–54.
Linder, S., Krondahl, U., Sennerstam, R., and Ringertz, N. R. (1981). *Exp. Cell Res.* **132,** 453–460.
Linnenbach, A., Huebner, K., and Croce, C. M. (1980). *Proc. Natl. Acad. Sci. U.S.A.* **77,** 4875–4879.
Loeliger, P., Bollag, W., and Mayer, H. (1980). *Eur. J. Med. Chem.* **15,** 9–15.
Lotan, R. (1979). *Cancer Res.* **39,** 1014–1019.
Lotan, R. (1980). *Biochim. Biophys. Acta* **605,** 33–91.
Lotan, R., and Lotan, D. (1980). *Cancer Res.* **40,** 3345–3350.
Lotan, R., and Lotan, D. (1981). *J. Cell. Physiol.* **106,** 179–189.
Lotan, R., and Nicolson, G. L. (1977). *JNCI, J. Natl. Cancer Inst.* **59,** 1717–1722.
Lotan, R., and Nicolson, G. L. (1979). *Cancer Res.* **39,** 4767–4771.
Lotan, R., Giotta, G., Nork, E., and Nicolson, G. L. (1978). *JNCI, J. Natl. Cancer Inst.* **60,** 1035–1041.
Lotan, R., Kramer, R. H., Neumann, G., Lotan, D., and Nicolson, G. L. (1980a). *Exp. Cell Res.* **130,** 401–414.
Lotan, R., Neumann, G., and Lotan, D. (1980b). *Cancer Res.* **40,** 1097–1102.
Lotan, R., Ong, D. E., and Chytil, F. (1980c). *JNCI, J. Natl. Cancer Inst.* **64,** 1259–1262.
Lotan, R., Neumann, G., and Lotan, D. (1981). *Ann. N. Y. Acad. Sci.* **359,** 150–170.

Lotan, R., Fischer, I., Meromsky, L., and Moldave, K. (1982a). *J. Cell Physiol.* **113,** 47–55.
Lotan, R., Lotan, D., and Kadouri, A. (1982b). *Exp. Cell Res.* **141,** 79–86.
Lotan, R., Neumann, G., and Deutsch, V. (1983a). *Cancer Res.* **43,** 303–312.
Lotan, R., Stolarsky, T., and Lotan, D. (1983b). *Cancer Res.* **43,** 2868–2875.
Lotan, R., Thein, R., and Lotan, D. (1983c). *In* "The Modulation and Mediation of Cancer by Vitamins" (F. L. Meyskens and K. Prasad, eds.), pp. 211–222. Karger, Basel.
Lotem, J., and Sachs, L. (1982). *Proc. Natl. Acad. Sci. U.S.A.* **79,** 4347–4351.
Ludwig, K. W., Lowey, B., and Niles, R. M. (1980). *J. Biol. Chem.* **255,** 5999–6002.
Luk, G. K., Civin, C. I., Weissman, R. M., and Baylin, S. B. (1982). *Science* **216,** 75–77.
Maniglia, C. A., and Sartorelli, A. C. (1981). *Ann. N. Y. Acad. Sci.* **359,** 322–330.
Marchok, A. C., Rhoton, J. C., and Nettesheim, P. (1978). *Cancer Res.* **38,** 2030–2037.
Marchok, A. C., Clark, J. N., and Klein-Szanto, A. (1981). *JNCI, J. Natl. Cancer Inst.* **66,** 1165–1174.
Marquardt, H., Hunkapiller, M. W., Hood, L. E., Twardzik, D. R., De Larco, J. E., Stephenson, J. R., and Todaro, G. J. (1983). *Proc. Natl. Acad. Sci. U.S.A.* **80,** 4684–4688.
Mather, J. P. (1981). *Ann. N. Y. Acad. Sci.* **359,** 412–413.
Matthaei, K. I., Andrews, P. W., and Bronson, D. L. (1983). *Exp. Cell Res.* **143,** 471–474.
Mehta, R. G., Cerny, W. L., and Moon, R. C. (1983). *Carcinogenesis* **4,** 23–26.
Merriman, R. L., and Bertram, J. S. (1979). *Cancer Res.* **39,** 1661–1666.
Meyskens, F. L., and Fuller, B. B. (1980). *Cancer Res.* **40,** 2194–2196.
Meyskens, F. L., and Salmon, S. E. (1979). *Cancer Res.* **39,** 4055–4057.
Mintz, B., and Fleischman, R. A. (1981). *Adv. Cancer Res.* **34,** 211–276.
Mintz, B., and Illmensee, K. (1975). *Proc. Natl. Acad. Sci. U.S.A.* **72,** 3585–3589.
Miskin, R., Easton, T. G., and Reich, E. (1978). *Cell* **15,** 1301–1312.
Miyaura, C., Abe, E., Kuribayaski, T., Tanaka, H., Konno, K., Nishii, Y., and Suda, T. (1981). *Biochem. Biophy. Res. Commun.* **102,** 937–943.
Montesano, R., Drevon, C., Kuroki, T., Saint Vincent, L., Handleman, S., Sanford, K. K., DeFeo, D., and Weinstein, I. B. (1977). *JNCI, J. Natl. Cancer Inst.* **59,** 1651–1656.
Moore, T. (1957). "Vitamin A," pp. 645. Elsevier, New York.
Mordan, L. J., and Bertram, L. M. (1983). *Cancer Res.* **43,** 567–571.
Mordan, L. J., Bergin, L. M., Budnick, J. E. L., Meegan, R. R., and Bertram, J. S. (1982). *Carcinogenesis* **3,** 279–285.
Morré, D. J., Creek, K. E., Morré, D. M., and Richardson, C. L. (1981). *Ann. N. Y. Acad. Sci.* **359,** 367–382.
Morriss, G. M. (1976). *In* "New Approaches to the Evaluation of Abnormal Embryonic Development" (D. Neubert and H. Merker, eds.), p. 678. Thieme, Stuttgart.
Morriss, G. M., and Steele, G. E. (1977). *Teratology* **15,** 109–119.
Moskal, J. R., Lockney, M. W., Marvel, G. C., Mason, P. A., Sweeley, C. C., Warren, S. J., and Trosko, J. E. (1980). *In* "Cell Surface Glycolipids" (C. C. Sweeley, ed.), Vol. 128, pp. 241–260. Am. Chem. Soc., Washington, D. C.
Mufson, R. A., DeFeo, D., and Weinstein, I. B. (1979). *Mol. Pharmacol.* **16,** 569–578.
Mukherjee, B. B., Mobry, P. M., and Pena, S. D. J. (1982). *Exp. Cell Res.* **138,** 95–107.
Müller, R., Slamon, D. J., Tremblay, J. M., Cline, M. J., and Verma, I. M. (1982). *Nature (London)* **299,** 640–644.
Murray, T., and Russell, T. R. (1980). *J. Supramol. Struct.* **14,** 255–266.
New, D. A. T. (1966). "The Culture of Vertebrate Embryos," pp. 245. Academic Press, New York.
New, D. A. T., Coppola, P. T., and Cockroft, D. L. (1976). *J. Embryol. Exp. Morphol.* **36,** 133–144.
Niedel, J. E., Kuhn, L. J., and Vandenbark, G. R. (1983). *Proc. Natl. Acad. Sci. U.S.A.* **80,** 36–40.
Nimmo, H. G., and Cohen, P. (1977). *Adv. Cyclic Nucleotide Res.* **8,** 146–266.

Oegema, T. R., and Parzych, S. M. (1981). *JNCI, J. Natl. Cancer Inst.* **67,** 99–106.
Olson, J. A., Rojanapo, W., and Lamb, A. J. (1981). *Ann. N. Y. Acad. Sci.* **359,** 181–191.
Olsson, I. L. and Breitman, T. R. (1982). *Cancer Res.* **42,** 3924–3927.
Olsson, I. L., Breitman, T. R., and Gallo, R. C. (1982). *Cancer Res.* **42,** 3928–3933.
O'Malley, B. W., and Means, A. L. (1974). *Science* **183,** 610–620.
Omori, M., and Chytil, F. (1982). *J. Biol. Chem.* **257,** 14370–14374.
Ong, D. E., and Chytil, F. (1975). *J. Biol. Chem.* **250,** 6113–6117.
Ossowski, L., Unkeless, J. C., Tobia, A., Quigley, J. P., Rifkin, D. B., and Reich, E. (1973). *J. Exp. Med.* **137,** 112–126.
Paranjpe, M., De Larco, J. E., and Todaro, G. (1980). *Biochem. Biophys. Res. Commun.* **94,** 586–591.
Parodi, A. J., and Leloir, L. F. (1979). *Biochim. Biophys. Acta* **559,** 1–37.
Patt, L. M., Itaya, K., and Hakomori, S.-I. (1978). *Nature (London)* **273,** 279–381.
Pennypacker, J. P., Lewis, C. A., and Hassell, J. R. (1978). *Arch. Biochem. Biophys.* **186,** 351–358.
Peterson, P. A., Rask, L., Helting, T., Ostberg, L., and Fernstedt, Y. (1976). *J. Biol. Chem.* **251,** 4986–4995.
Pledger, W. J., Stiles, C. D., Antoniades, H. N., and Scher, C. D. (1977). *Proc. Natl. Acad. Sci. U.S.A.* **74,** 4481–4485.
Plet, A., Evain, D., and Anderson, W. B. (1982). *J. Biol. Chem.* **257,** 889–893.
Plewig, G., and Wagner, A. (1981). *Arch. Dermatol. Res.* **270,** 89–94.
Plotkin, G. M., and Wolf, G. (1980). *Biochim. Biophys. Acta* **615,** 94–102.
Prakash, C., Cheng, T., and Vijay, I. K. (1981). *Biochem. Int.* **2,** 153–160.
Quill, H., and Wolf, G. (1981). *Ann. N. Y. Acad. Sci.* **359,** 331–344.
Rapaport, E., Schroder, E. W., and Black, P. H. (1982a). *J. Cell. Physiol.* **110,** 318–322.
Rapaport, E., Schroder, E. W., Kabcenell, A. K., and Black, P. H. (1982b). *Cancer Res.* **42,** 4918–4920.
Rasmussen, H. (1981). "Calcium and Cyclic AMP as Synarchic Messengers," pp. 370. J. Wiley, New York.
Rees, A., Adamson, E., and Graham, C. (1979). *Nature (London)* **281,** 309–311.
Rice, R. H., and Green, H. (1978). *J. Cell. Biol.* **76,** 705–711.
Rizzino, A., and Crowley, C. (1980). *Proc. Natl. Acad. Sci. U.S.A.* **77,** 457–461.
Rhodes, J., and Oliver, S. (1980). *Immunology* **40,** 467–472.
Roberts, A. B., Frolik, C. A., Anzano, M. A., and Sporn, M. B. (1983). *Fed. Proc., Fed. Am. Soc. Exp. Biol.* **42,** 2621–2626.
Rodén, L. (1980). *In* "The Biochemistry of Glycoproteins and Proteoglycans" (W. J. Lennarz, ed.), pp. 267–371. Plenum, New York.
Ross, A. C., and Goodman, D. S. (1979). *Fed. Proc., Fed. Am. Soc. Exp. Biol.* **38,** 2515–2518.
Rosso, G. C., De Luca, L. M., Warren, C. D., and Wolf, G. (1975). *J. Lipid Res.* **16,** 235–243.
Rosso, G. C., Masushige, S., Warren, C. D., Kiorpes, T. C., and Wolf, G. (1976). *J. Biol. Chem.* **251,** 6465–6467.
Rosso, G. C., Bendrick, C. J., and Wolf, G. (1981). *J. Biol. Chem.* **256,** 8341–8347.
Roth, J. (1981). *Diabetes Care* **4,** 27–32.
Rothblat, G. H., Arborgast, L. Y., Ovellett, L., and Howard, B. V. (1976). *In Vitro* **12,** 554–557.
Russell, D. H. (1981). *Med. Biol.* **59,** 286–295.
Russell, D. H., and Haddox, M. K. (1981). *Ann. N. Y. Acad. Sci.* **359,** 281–297.
Sachs, L. (1978). *Nature (London)* **274,** 535–539.
Sachs, L. (1980). *Proc. Natl. Acad. Sci. U.S.A.* **77,** 6152–6159.
Sani, B. P., and Donovan, M. K. (1979). *Cancer Res.* **39,** 2492–2496.
Sani, B. P., and Hill, D. L. (1974). *Biochem. Biochem. Res. Commun.* **61,** 1276–1282.

Sasak, W., and De Luca, L. M. (1980). *FEBS Lett.* **114,** 313–318.
Sasak, W., De Luca, L. M., Dion, L. D., and Silverman-Jones, C. S. (1980). *Cancer Res.* **40,** 1944–1949.
Sato, M., De Luca, L. M., and Muto, Y. (1978). *J. Nutr. Sci. Vitaminol.* **24,** 9–23.
Sato, M., Hiragun, A., and Mitsui, H. (1980). *Biochem. Biophys. Res. Commun.* **95,** 1839–1845.
Schachter, H., and Roseman, S. (1980). *In* "The Biochemistry of Glycoproteins and Proteoglycans" (W. J. Lennarz, ed.), pp. 127–160. Plenum, New York.
Schindler, J., Matthaei, K. I., and Sherman, M. I. (1981). *Proc. Natl. Acad. Sci. U.S.A.* **78,** 1077–1080.
Schroder, E. W., Chou, I.-N., and Black, P. H. (1980). *Cancer Res.* **40,** 3089–3094.
Schroder, E. W., Rapaport, E., Kabcenell, A. K., and Black, D. H. (1982). *Proc. Natl. Acad. Sci. U.S.A.* **79,** 1549–1552.
Schroder, E. W., Rapaport, E., and Black, P. H. (1983). *Cancer Surveys* **2,** 223–241.
Schulman, H., Huttner, W. B., and Greengard, P. (1980). *In* "Calcium and Cell Function. I. Calmodulin" (W. Y. Cheung, ed.), pp. 217–252. Academic Press, New York.
Scott, K. F. F., and Russell, D. H. (1982). *J. Cell. Physiol.* **111,** 111–116.
Scott, K. F. F., Meyskens, F. L., Jr., and Russell, D. H. (1982). *Proc. Natl. Acad. Sci. U.S.A.* **79,** 4093–4097.
Segal, S., and Khoury, G. (1979). *Proc. Natl. Acad. Sci. U.S.A.* **76,** 5611–5615.
Segal, S. S., Levine, A. J., and Khoury, G. (1979). *Nature (London)* **280,** 335–337.
Shapiro, S. S., and Mott, D. J. (1981). *Ann. N. Y. Acad. Sci.* **359,** 306–321.
Shapiro, S. S., and Poon, J. P. (1976). *Arch. Biochem. Biophys.* **174,** 74–81.
Shapiro, S. S., and Poon, J. P. (1978). *Connect. Tissue Res.* **6,** 101–108.
Shapiro, S. S., and Poon, J. P. (1979). *Exp. Cell Res.* **119,** 349–357.
Shenefelt, R. E. (1972). *Teratology* **5,** 103–118.
Sherman, M. I., Matthaei, K. I., and Schindler, J. (1981). *Ann. N. Y. Acad. Sci.* **359,** 192–199.
Shidoji, Y., and De Luca, L. M. (1981). *Biochem. J.* **200,** 529–538.
Shidoji, Y., Sasak, W., Silverman-Jones, C. S., and De Luca, L. M. (1981). *Ann. N. Y. Acad. Sci.* **359,** 345–357.
Shidoji, Y., Silverman-Jones, C. S., and De Luca, L. M. (1982). *Biochem. J.* **208,** 865–868.
Sidell, N. (1982). *JNCI, J. Natl. Cancer Inst.* **68,** 589–593.
Sneider, W. D., and Wolf, G. (1976). *J. Nutr.* **106,** 1515–1525.
Speers, W. C. (1982). *Cancer Res.* **42,** 1843–1849.
Sporn, M. B., and Newton, D. L. (1979). *Fed. Proc., Fed. Am. Soc. Exp. Biol.* **38,** 2528–2534.
Sporn, M. B., and Roberts, A. B. (1983). *Cancer Res.* **43,** 3034–3040.
Sporn, M. B., Dunlop, N. M., and Yuspa, S. H. (1973). *Science* **182,** 722–723.
Sporn, M. B., Dunlop, N. M., Newton, D. L., and Henderson, W. R. (1976). *Nature (London)* **263,** 110–113.
Sporn, M. B., Newton, D. L., Roberts, A. B., De Larco, J. E., and Todaro, G. J. (1981). *In* "Molecular Actions and Targets for Cancer Chemotherapeutic Agents" (A. Sartorelli, J. Bertino, and J. Lazo, eds.), pp. 541–554. Academic Press, New York.
Steinberg, K. K., and Sgoutas, D. S. (1981). *Proc. Soc. Exp. Biol. Med.* **167,** 110–116.
Stiles, C. D., Capone, G. T., Sher, C. D., Antoniades, H. N., Van Wyk, J. J., and Pledger, W. J. (1979). *Proc. Natl. Acad. Sci. U.S.A.* **76,** 1279–1283.
Strickland, S., and Mahdavi, M. (1978). *Cell* **15,** 393–403.
Strickland, S., and Sawey, M. J. (1980). *Dev. Biol.* **78,** 76–85.
Strickland S., Reich, E., and Sherman, M. E. (1976). *Cell* **9,** 231–240.
Strickland, S., Smith, K. K., and Marotti, K. R. (1980). *Cell* **21,** 347–355.
Strickland, S., Breitman, T. R., Frickel, F., Hädicke, E., Nürrenbach, A., and Sporn, M. B. (1983). *Cancer Res.* **43,** 5268–5272.

Struck, D. K., and Lennarz, W. J. (1980). *In* "The Biochemistry of Glycoproteins and Proteoglycans" (W. J. Lennarz, ed.), pp. 35–81. Plenum, New York.

Sun, T.-T., and Green, H. (1978). *Nature (London)* **269,** 489–493.

Takase, S., Ong, D. E., and Chytil, F. (1979). *Proc. Natl. Acad. Sci. U.S.A.* **76,** 2204–2208.

Takenaga, K. (1981). *Gann* **72,** 488–497.

Takenaga, K., Hozumi, M., and Sakagami, Y. (1980). *Cancer Res.* **40,** 914–919.

Takenaga, K., Honma, Y., Okabe-Kado, J., and Hozumi, M. (1981). *Cancer Res.* **41,** 1948–1953.

Takigawa, M., Ishida, H., Takano, T., and Suzuki, F. (1980). *Proc. Natl. Acad. Sci. U.S.A.* **77,** 1481–1485.

Takigawa, M., Takano, T., and Suzuki, F. (1982). *Mol. Cell. Biochem.* **42,** 145–153.

Tchao, R. (1980). *In Vitro* **16,** 407–414.

Tchao, R., and Leighton, J. (1979). *Invest. Urol.* **16,** 476–482.

Thein, R., and Lotan, R. (1982). *Cancer Res.* **42,** 4771–4775.

Thompson, J. N., Howell, J. M., Pitt, G. A., and McLaughlin, C. I. (1969). *Br. J. Nutr.* **23,** 471–490.

Todaro, G. J., De Larco, J. E., and Cohen, S. (1976). *Nature (London)* **264,** 26–31.

Todaro, G. J., De Larco, J. E., and Sporn, M. B. (1978). *Nature (London)* **276,** 272–274.

Todaro, G. J., Fryling, C., and De Larco, J. E. (1980). *Proc. Natl. Acad. Sci. U.S.A.* **77,** 5258–5262.

Toyoshima, K., and Leighton, J. (1975). *Cancer Res.* **35,** 1873–1879.

Trechsel, U., Dew, G., Murphy, G., and Reynolds, J. J. (1982). *Biochim. Biophys. Acta* **720,** 364–370.

Trown, P. W., Buck, M. J., and Hansen, R. (1976). *Cancer Treat. Rep.* **60,** 1647–1653.

Tsao, D., Morita, A., Bella, A., Jr., Luu, P., and Kim, Y. S. (1982). *Cancer Res.* **42,** 1052–1058.

Unkeless, J. C., Tobia, A., Ossowski, L., Quigley, J. P., Rifkin, D. B., and Reich, E. (1973). *J. Exp. Med.* **137,** 85–111.

Vasan, N. S., and Lash, J. W. (1975). *Calcif. Tissue Res.* **19,** 99–107.

Verma, A. K., and Boutwell, R. K. (1977). *Cancer Res.* **37,** 2196–2201.

Verma, A. K., and Boutwell, R. K. (1980). *In* "Polyamines in Biomedical Research" (J. M. Gaugas, ed.), pp. 185–201. Wiley, New York.

Verma, A. K., Rice, H. M., Shapas, B. G., and Boutwell, R. K. (1978). *Cancer Res.* **38,** 793–801.

Verma, A. K., Conrad, E. A., and Boutwell, R. K. (1980). *Carcinogenesis* **1,** 607–611.

Waechter, C. J., and Lennarz, W. J. (1976). *Annu. Rev. Biochem.* **45,** 95–112.

Weiner, N. (1980). *In* "The Pharmacological Basis of Therapeutics" (A. G. Gilman, L. S. Goodman, and A. Gilman, eds.), 6th ed., pp. 138–175. Macmillan, New York.

Westin, E. H., Wong-Staal, F., Gelmann, E. P., Favera, R. D., Papas, T. S. Lautenberger, J. A., Eva, A., Reddy, E. P., Tronick, S. R., Aaronson, S. A., and Gallo, R. C. (1982). *Proc. Natl. Acad. Sci. U.S.A.* **79,** 2490–2494.

Wiggert, B., Russell, P., Lewis, M., and Chader, G. (1977). *Biochem. Biophys. Res. Commun.* **79,** 218–225.

Wigler, M., and Weinstein, I. B. (1976). *Nature (London)* **259,** 232–233.

Wilkinson, D. I. (1978). *Arch. Determatol. Res.* **263,** 75–81.

Wilkinson, D. I., and Orenburg, E. K. (1980). *Arch. Dermatol. Res.* **267,** 25–31.

Williams-Ashman, H. G., and Canellakis, Z. N. (1980). *Physiol. Chem. Phys.* **12,** 457–472.

Wilson, E. L., and Dowdle, E. B. (1980). *Cancer Res.* **40,** 4817–4820.

Wilson, E. L., and Reich, E. (1978). *Cell* **15,** 385–392.

Witz, G., Goldstein, B. D., Amoruso, M., Stone, D. S., and Troll, W. (1980). *Biochem. Biophys. Res. Commun.* **97,** 883–888.

Wolbach, S. B., and Howe, P. R. (1925). *J. Exp. Med.* **42,** 753–777.

Wolf, G. (1977). *Nutr. Rev.* **35,** 97–99.

Wolf, G. (1980). *In* "Human Nutrition" (R. B. Alfin-Slater and D. Kritchevsky, eds.), Vol. 3B, pp. 97–203. Plenum, New York.

Wolf, G. (1983). *Physiol. Rev.* (in press).

Wolf, G., Kiorpes, T. C., Masushige, S., Schreiber, J. B., Smith, M. J., and Anderson, R. S. (1979). *Fed. Proc., Fed. Am. Soc. Exp. Biol.* **38,** 2540–2543.

Wong, G., and Pawelek, J. (1973). *Nature (London) New Biol.* **241,** 213–215.

Yaar, M., Stanley, J. R., and Katz, S. I. (1981). *J. Invest. Dermatol.* **76,** 363–366.

Yamada, K. M., and Olden, K. (1978). *Nature (London)* **275,** 179–184.

Yoneda, T., and Pratt, R. M. (1981). *Science* **213,** 563–565.

Yuspa, S. H., and Harris, C. C. (1974). *Exp. Cell Res.* **86,** 95–105.

Yuspa, S. H., Elgjo, K., Morse, M. A., and Wiebel, F. J. (1977). *Chem.-Biol. Interact.* **16,** 251–264.

Yuspa, S. H., Ben, T., Hennings, H., and Lichti, U. (1980). *Biochem. Biophys. Res. Commun.* **97,** 700–708.

Yuspa, S. J., Lichti, U., Ben, T., and Hennings, H. (1981). *Ann. N. Y. Acad. Sci.* **359,** 260–274.

Yuspa, S. H., Ben, T., and Steinert, P. (1982). *J. Biol. Chem.* **257,** 9906–9908.

Zachman, R. D. (1967). *Life Sci.* **6,** 2207–2213.

Zile, M. H., and Cullum, M. E. (1983). *Proc. Soc. Exp. Biol. Med.* **172,** 139–152.

Zile, M., and DeLuca, H. F. (1970). *Arch. Biochem. Biophys.* **140,** 210–214.

13

Preclinical and Clinical Toxicology of Selected Retinoids

JEROME J. KAMM
Department of Toxicology and Pathology
Roche Research Center
Hoffmann-La Roche Inc.
Nutley, New Jersey

KATHLEEN O. ASHENFELTER
Department of Medical Research
Roche Research Center
Hoffmann-La Roche Inc.
Nutley, New Jersey

CARL W. EHMANN*
Department of Medical Research
Roche Research Center
Hoffmann-La Roche Inc.
Nutley, New Jersey

*Present address: Research and Development, Bristol-Myers Company, Buffalo, New York 14213.

THE RETINOIDS, VOL. 2

ISBN 0-12-658102-9

I. PRECLINICAL TOXICOLOGY OF SELECTED RETINOIDS

A. Introduction

The toxicity of relatively high single doses or of relatively low multiple doses of retinol in laboratory animals has been described extensively in the scientific literature. Although the earlier studies were carried out with a variety of "vitamin A" concentrates including fish oils and polar bear liver, most, if not all, findings have been confirmed subsequently by studies using highly purified, crystalline forms of the vitamin. The early literature on hypervitaminosis A in laboratory animals has been reviewed (Nieman and Obbink, 1954; Hayes and Hegsted, 1973) and is not described in detail in this chapter. In brief, it has been shown that single large doses of vitamin A can be lethal and that chronic intoxication has adverse effects on many tissues and organ systems. The most characteristic effects of chronic hypervitaminosis A in laboratory animals are weight loss, erythema, hair loss, internal hemorrhage, and fractures. Many adverse effects of chronic hypervitaminosis A are reversible after treatment is stopped; body weight increases, erythema subsides, hair grows again, and fractures heal. Hypervitaminosis A has been associated with teratogenicity in a wide variety of laboratory animals (Schardein, 1976; Geelen, 1979).

Interest in the toxicity of the retinoids has resulted from their increased use in the treatment of dermatological disorders (Peck, 1980; Pardini, 1981; Elias and Williams, 1981; Windhorst, 1982; Chapter 16, this volume) and neoplasms (Sporn, 1977; Sporn and Newton, 1981; Sporn, 1982; Bollag, 1979; Lotan, 1980; Newberne and Rogers, 1981; Chapter 14, this volume). Isotretinoin (13-*cis*-retinoic acid, Ro 4-3780, Accutane) has become available in the United States for use in the treatment of severe recalcitrant cystic acne; in Europe, etretinate (Ro 10-9359, Tegison) is used for treatment of psoriasis. The toxicity of these retinoids has been exhaustively studied in laboratory animals as part of the regulatory approval process.

The first part of this chapter is devoted to the results of toxicologic, mutagenic, teratogenic, and carcinogenic studies with isotretinoin and etretinate; when appropriate, a comparison with results for retinol and tretinoin (all-*trans*-retinoic acid) is presented. The second part of this chapter is devoted to a description of clinical or human toxicity.

B. Acute Toxicity

Acute toxicity, elicted by relatively large single doses of a substance, is usually expressed as an LD_{50}, the amount of material required to kill 50% of a population of animals. An LD_{50}, often expressed in terms of mg/kg, is calculated from a dose–response curve and varies with a host of factors, some of which are route of administration, species, strain, sex, age, nutritional status, and environmental conditions. This variability should be kept in mind when comparing data generated in different laboratories under conditions that are often inadequately defined (Table I). The reader is referred to the work of Beyer (1978) for an excellent brief discussion of acute toxicity studies and their use in safety assessment.

Single large doses of retinoids are only moderately toxic. Table I presents some LD_{50} values for vitamin A alcohol, tretinoin, isotretinoin, and etretinate given orally or intraperitoneally to rats, mice, and rabbits. For comparison, the approximate LD_{50} for botulinus toxin is 0.00001 mg/kg and for ethyl alcohol 10,000 mg/kg (Klaassen and Doull, 1980).

The data presented in Table I indicate that, in most instances, differences between the retinoids with respect to their acute oral toxicity, within or between species, are small. Acute intraperitoneal toxicity of retinoids is generally greater than acute oral toxicity, and differences between species are usually relatively minor.

Nieman and Obbink (1954) have described the symptoms noted in rats and mice following the administration of a single large dose of retinol. In the animals that succumb, death is preceded by convulsions and paralysis; in those that survive, intoxication is characterized by malaise, decreased motor activity, stu-

TABLE I

Acute Toxicity of Selected Retinoids in Laboratory Animals

Retinoid	Species	Route of administration	LD_{50} (mg/kg)	Reference
Retinol	Mouse	Oral	2,570	Kamm (1982)
	Mouse	Oral	1,510	Kamm (1982)
Tretinoin	Mouse	Oral	4,000	Herold *et al.* (1975)
	Mouse	Oral	1,100	Hixson and Denine (1978)
	Mouse	Intraperitoneal	350–500	Kretzschmar and Leuschner (1975)
	Mouse	Intraperitoneal	31	Hixson and Denine (1978)
	Rat	Oral	2,000	Herold *et al.* (1975)
	Rat	Intraperitoneal	350–500	Kretzschmar and Leuschner (1975)
Isotretinoin	Mouse	Oral	3,389	Kamm (1982)
	Mouse	Oral	26,000	Hixson and Denine (1978)
	Mouse	Intraperitoneal	904	Kamm (1982)
	Mouse	Intraperitoneal	138	Hixson and Denine (1978)
	Rat	Oral	>4,000	Kamm (1982)
	Rabbit	Oral	~1,960	Kamm (1982)
Etretinate	Mouse	Oral	>4,000	Teelmann (1981)
	Mouse	Intraperitoneal	>4,000	Teelmann (1981)
	Rat	Oral	>4,000	Teelmann (1981)
	Rat	Intraperitoneal	>1,000	Teelmann (1981)

por, muscular weakness, and in some instances "changes in the pelt." It is interesting that animals that survive, recover without apparent residual adverse effects.

Macapinlac and Olson (1981) reported that young monkeys given single lethal doses of retinyl acetate become progressively weak, have difficulty breathing, lapse into coma, lose simple reflexes, and then die; in some instances death is preceded by convulsions. Symptoms preceding death following single large doses of tretinoin, isotretinoin, or etretinate have not been described in detail.

An interesting observation for some retinoids is that toxicity, as defined by their LD_{50}, is cumulative. Thus Teelmann (1981) has reported that the single-dose oral LD_{50} for etretinate in rats and mice is greater than 4000 mg/kg. In contrast, in animals treated with repeated daily doses of etretinate for 10 days, the LD_{50} was about 50 mg/kg/day. Similar cumulative toxicity has been described for tretinoin (Kretzschmer and Leuschner, 1975).

C. Subchronic Toxicity

Subchronic toxicity studies in laboratory animals are generally done for periods of time that range from 2 weeks to 3 months. In practice, doses are selected that range from low levels with no observed toxic effects to levels sufficiently high to produce well-defined toxic effects. The ultimate goal is to estimate doses that will be safe in the clinic. Toward this end, subchronic toxicity studies should establish a "no-observed-effect" dose, hopefully greater than the dose intended for clinical use, and should characterize the early signs of toxicity so as to alert the clinician to adverse side effects that potentially might be encountered in the clinic.

1. *Retinol*

Nieman and Obbink (1954) have provided an extensive review of the subchronic toxicity of vitamin A preparations and of retinol that appeared in the literature up to the early 1950s. The studies described were carried out in a variety of animal species but primarily in the rat; dosages ranged from 3 to 180 mg retinol equivalents/day and the duration of treatment varied from a few days to several weeks.

Hypervitaminosis A was produced in laboratory animals by doses as low as 3 mg retinol equivalents/day. The time to first appearance of signs was dependent on the route of administration, the species under study, the age of the animal, the duration of treatment, the specific toxic manifestation in question, and the size of the dose. As summarized by Nieman and Obbink (1954) hypervitaminosis A is characterized as follows.

a. GENERAL SIGNS OF TOXICITY. Hypervitaminosis A has been associated with anorexia, weight loss, emaciation, anemia, cachexia, and finally death.

Spontaneous fracture(s), often observed as hind-leg paralysis, are frequently observed in rodents; young rats are more susceptible to fractures than are mature rats. When treatment is stopped, the fractures heal and do not recur. Extensive subcutaneous and/or intramuscular hemorrhages commonly occur and are often manifestations of hypothrombinemia. Internal hemorrhage is frequently observed. Inflammation of nasal passages, gut, and conjunctivae are sometimes noted.

b. EFFECTS ON THE SKIN. Hair loss, localized hyperemia and/or erythemia and thickened epithelium have been reported to be characteristics associated with hypervitaminosis A. Topical administration of retinol is irritating and may lead to peeling skin, reduced keratinization, and hyperplasia of the dermal papillae and blood vessels.

c. EFFECTS ON INTERNAL ORGANS. Macroscopic and microscopic changes have been described in animals that have died of hypervitaminosis A. Most significant is fatty infiltration of the liver. Fatty changes in pulp cells of the spleen and hemosiderosis of the spleen have also been reported as have fatty changes in the heart and kidney. Glomerulonephritis and necrotizing nephrosis have been described in the kidney. Testicular hypertrophy has been noted in adult rats and degenerative testicular changes have been observed in weanling rats, but not in adults, following prolonged treatment. In addition to these changes, Nichele and Carlino (1953a,b) reported extensive foci of degenerative myocardial fibers in rats treated with 3 or 6 mg retinol equivalents/kg/day for 3 months; these changes were clinically associated with EKG changes.

d. EFFECTS ON BLOOD. Hypochromic anemia with hyperplastic bone marrow has been reported. Decreased hemoglobin concentration and transient increases in total circulating lipid and serum cholesterol have also been observed.

e. EFFECTS ON THE SKELETAL SYSTEM. The toxic manifestation of hypervitaminosis A in laboratory animals that has been investigated most is the effect on the skeletal system. In the earliest description of hypervitaminosis A in the rat by Takahashi *et al.* (1925), one of the signs described was a limping gait observed in rats and mice after about 1 week of feeding concentrates of vitamin A; the daily doses were about 10,000 times the amount required to sustain normal growth. This observation was subsequently confirmed for vitamin A concentrates by others (Collazo and Rodriguez, 1933a,b; Bomskov and Seemann, 1933; Davies and Moore, 1934).

Moore and Wang (1945), using pure crystalline retinyl acetate, established that the skeletal changes were indeed a direct result of excessive ingestion of nonlethal, toxic doses of the vitamin. These workers fed 8.5–13.6 mg retinol equivalents/day (as retinyl acetate) to rats as a dietary admixture; after about 1 week, they noted a limping gait in some animals and eventually all animals were affected. X-Ray examination of the rats showed fractures. Bone-ash determinations suggested normal mineralization, and the authors concluded that the toxicity was associated with the formation or destruction of bone matrix. The effect of age on the toxicity of pure retinyl acetate was also confirmed by Moore and Wang (1945). Thus, young rats suffered fractures and adult rats suffered hemorrhages.

The histopathological changes in bone elicited by retinol have been summarized by Nieman and Obbink (1954). Characteristically, there is increased activity of osteoclasts (Strauss, 1934) or alternatively, decreased activity of osteoblasts while osteoclastic activity remains unchanged (Irving, 1949). In either case, longitudinal bone growth is more than circumferential bone growth, resulting in a thin fragile cortex that ultimately breaks. On the basis of microscopic studies, Wolbach and Bessey (1942) suggested that periosteal bone remodeling is accelerated by osteoclasis.

The bone toxicity of hypervitaminosis A has also been demonstrated in dogs (Maddock *et al.*, 1949), cats (Seawright *et al.*, 1967; Clark *et al.*, 1970), calves (Grey *et al.*, 1965), and hogs (Wolke *et al.*, 1968; Dobson, 1969). Toxicity is not restricted to the skeletal system. Irving (1949) has reported reduced formation of dentine and atrophy of lingual odontoblasts, and Rodahl (1949) has described degeneration of pulp and odontoblasts accompanied by amorphous calcification of dentine. The effect of excess retinol on bone *in vitro* has been studied by Fell and Mellanby (1952).

2. *Tretinoin*

Kretzschmar and Leuschner (1975) studied the toxicity of tretinoin administered orally to rats and dogs for 13 weeks at doses of 5 and 50 mg/kg/day. In rats, the low dose produced hair loss, dermal and mucosal alterations (not defined), inhibition of spermatogenesis, and weight loss. At 50 mg/kg/day, serum transaminase and serum alkaline phosphatase activities were elevated and total serum protein declined; about 20% of the rats died. In dogs, similar signs of toxicity were seen at 5 mg/kg/day; 50 mg/kg/day was lethal to about 50% of the dogs after about 11 weeks of treatment. In mice, doses of ~150–250 mg/kg/day for 5 days caused alopecia, weight loss, and undefined skin and mucous membrane lesions.

In another study (Herold *et al.*, 1975), the only adverse effect reported in rats given tretinoin orally at a dose of 0.4 or 10 mg/kg/day for 90 days was elevation of serum alkaline phosphatase. Although no attempt was made to identify the isozymes, the elevated values of alkaline phosphatase were attributed to "osteoblastic hyperactivity." Evidence for bone remodeling was seen only at the higher dose (Cahn *et al.*, 1975).

Zbinden (1975a) reported the results of a 12-week study in rats given tretinoin by dietary admixture at doses of approximately 0.02, 0.1, and 0.5 mg/kg/day. There were no adverse effects on food consumption, growth, or behavior. Hematological parameters remained within normal limits. Although serum parameters such as blood urea nitrogen (BUN), creatinine, and serum glutamic pyruvic transaminase (SGPT) were normal, alkaline phosphatase was moderately increased in female rats at all doses and in male rats at the two higher doses. Liver weight was increased at the two higher doses, and in the high-dose rats, there was evidence for increased basophilia and perinuclear rarefaction of the hepatocyte protoplasm. There were no other pathological changes; skin, cartilage, and bone appeared normal.

Relatively high doses of tretinoin may be applied to the skin without causing systemic toxicity. However, as has been reviewed by Zbinden (1975a), marked cutaneous changes follow the topical application of tretinoin. Dermal toxicity studies with tretinoin in rats and rabbits have been described by Herold *et al.* (1975). In rats, there was no evidence for systemic toxicity following 7 weeks of daily application of 0.075 or 0.15 mg of tretinoin per animal, applied as either a solution or as a cream. Epidermal changes, described as acanthosis and hyperkeratosis, were reported. In rats treated with a cream formulation of tretinoin, there was slight edema of the upper skin layers. Rabbits tolerated daily topical applications of 0.35 or 1.5 mg tretinoin/day/rabbit for 6 weeks with no evidence of systemic toxicity. Histological changes reported in the skin included epidermal hyperplasia, acanthosis, dyskeratosis, and parakeratosis; in some rabbits epidermal erosion and psoriasis-like changes were observed. Epidermal hyperplasia and acanthosis, accompanied in some animals by hyperkeratosis or psoriasis-like lesions, were also reported in nude rats treated topically with a 0.1% solution of tretinoin (Cahn *et al.*, 1975). The changes were much more pronounced and occurred in all nude rats treated with a cream formulation containing 0.05% tretinoin.

3. *Etretinate*

The subchronic toxicity of etretinate has been summarized by Teelmann (1981). Two-week intravenous toxicity studies were carried out in rats and dogs at doses of 0.5 and 2 mg/kg/day. Oral toxicity studies were performed at doses up to 20 mg/kg/day for 4 weeks or 13 weeks in rats and at doses up to 30 mg/kg/day in dogs for 13 weeks.

The rat was the most sensitive of the species studied. Signs of etretinate toxicity were noted initially as a dose-dependent decrease in body weight gain accompanied at the higher doses (10–15 mg/kg/day) by a marked deterioration of general body condition. In contrast, dogs tolerated doses as high as 15 mg/kg/day for 13 weeks without apparent untoward effects. Circumscribed reddening and erythema of the skin were dose-dependent and appeared in both rats and dogs. Signs appeared after 8–10 days of treatment but gradually disappeared after 2–3 weeks, even when treatment was continued. Histologically, slight basal cell hyperplasia was noted. Those areas of skin that were erythematous frequently showed hair loss and, in the dog, hyperpigmentation was frequently observed.

Hematologically, decreased red blood cell counts and decreased hemoglobin concentration were observed in both rats and dogs. The effects were slight to moderate, dose-dependent and persisted throughout the trials. Compensatory, slight-to-moderate increases in reticulocyte counts and, in rats, compensatory extramedullary hematopoesis in the spleen and liver were observed.

Clinical chemistry changes were confined mainly to dose-dependent increases in serum alkaline phosphatase. The effect was marked in rats but slight in dogs. Elevations in serum triglycerides were also observed in rats.

In rats, the most striking manifestation of etretinate toxicity was fractures. The occurrence of fractures was noted as early as 10–14 days at doses of 10–15 mg/kg/day. Fractures were preceded by "intensified remodelling of the bones with deformations, hemorrhages, increased periosteal osteoclastic activity and thickening of the periosteum;" there was a marked reduction in bone diameter and increased porosity, especially of the long bones as a result of "an augmented endosteal osteoblastic activity" (Teelmann, 1981). Interestingly, similar effects of etretinate were not observed in adult dogs, but occasional periosteal hemorrhages and painful legs and joints were sometimes observed. Young dogs treated with etretinate at a dose of 5 mg/kg/day for 90 days did not have bone changes.

The signs of etretinate toxicity, body-weight deficits, skin alterations, hematological changes, clinical chemistry changes, and bone alterations, were reversible following cessation of treatment. Some hematological and biochemical alterations attributed to treatment with high doses of etretinate have also been reported for tretinoin and isotretinoin (Hixson and Denine, 1978; Hixson *et al.*, 1979; Gerber and Erdman, 1979a) and for retinol (Studer *et al.*, 1962).

4. *Isotretinoin*

In contrast to retinol, tretinoin, and etretinate, little has been published on the subchronic toxicity of isotretinoin, with the notable exception of two comparative toxicity studies reported by Hixson and co-workers (1978, 1979). A qualitative description of studies that have been done has been presented by Kamm (1982). These studies, which were carried out by the oral route in rats and dogs, suggest that isotretinoin toxicity is qualitatively very similar to the toxicity already described for retinol, tretinoin, and etretinate. Dose-related effects include decreased food consumption, decreased body-weight gain, erythema, alopecia, mucosal changes (i.e., hemorrhage and/or inflammation), long-bone fracture (in rats only), apparent joint pain (in dogs only), elevated serum alkaline phosphatase activity, elevated serum triglycerides (in rats only), and some elevations in serum transaminase activity. Gross anatomic findings at autopsy included fractures (in rats), increased liver weights, pale and/or mottled livers and decreased testicular weight. Histologic changes were generally confined to the bone in rats and to decreased spermatogenesis in dogs and were generally noted in animals treated with doses of isotretinoin in excess of 40–50 mg/kg/day.

Hixson and co-workers have carried out direct comparative studies of the toxicity of tretinoin and isotretinoin in mice (Hixson and Denine, 1978) and in rats (Hixson *et al.*, 1979). For the subchronic toxicity studies in mice, treatment was for 21 days at doses of tretinoin of 3–60 mg/kg/day intraperitoneally and 10–100 mg/kg/day orally; for isotretinoin the doses were 30–300 mg/kg/day intraperitoneally and 60–400 mg/kg/day orally. A general conclusion of this study was that, dose for dose, tretinoin was more toxic than isotretinoin. The authors noted that isotretinoin produced the same number and incidence of fractures at doses three to five times that of tretinoin administered by the same route. It is of interest that fractures were not invariably accompanied by elevations in serum alkaline phosphatase. Dose-related decreases in red cell counts were observed with isotretinoin but not with tretinoin. Hixson and Denine (1978) concluded on the basis of differences in onset, severity, and frequency that tretinoin was more toxic than isotretinoin in the Swiss mouse.

In a later study, Hixson *et al.* (1979) compared the oral toxicity of tretinoin to isotretinoin administered to Sprague–Dawley rats for 12 weeks. Doses of 0, 0.5, 2, and 5 mg/kg/day for tretinoin and 0, 4, 15, and 40 mg/kg/day for isotretinoin were administered to groups of rats (5/sex). Fracture was not a prominent feature of toxicity in this study; one rat in the high-dose tretinoin group and one rat in the high-dose isotretinoin group had fractures that were detected radiologically. Despite the absence of a significant incidence of fractures, increases in plasma alkaline phosphatase were observed in rats treated with either retinoid. The increases were dependent on dose and duration of treatment. Hematology and clinical chemistry values were, for the most part, within normal limits for rats treated with either retinoid.

Decreases in plasma albumin were observed in female rats treated with tretinoin or isotretinoin. The decreases were related to both duration of treatment and dose. No gross or microscopic lesions were observed in the livers or kidneys of affected rats that could account for the decreases in plasma albumin. Effects on plasma albumin concentration in male rats were less clear, although some statistically significant reductions were observed. In the absence of evidence for hepatic or renal lesions, the authors (Hixson *et al.*, 1979) have suggested that the decreased plasma albumin concentration may reflect an effect of the retinoids on protein synthesis.

Decreases in hemoglobin concentration were observed in rats treated with 15 or 40 mg isotretinoin/kg/day as early as after 4 weeks of treatment and in rats treated with 4 mg/kg/day after 12 weeks of treatment; female rats appeared to be more sensitive than male rats to this effect. In contrast, tretinoin had little or no effect on hemoglobin concentration even after 12 weeks of treatment at 5 mg/kg/day (Hixson *et al.*, 1979).

As was concluded in the comparative study in mice (Hixson and Denine,

1978), Hixson *et al.* (1979) concluded that in rats, tretinoin was more toxic than isotretinoin. Thus, doses of isotretinoin three to eight times higher than doses of tretinoin were required to produce similar adverse effects on hematologic or clinical chemistry values. (Hixson *et al.*, 1979) suggested that changes in plasma albumin concentration, alkaline phosphatase activity and hemoglobin concentration might be useful early indicators of clinical toxicity in humans because, in animal studies, their onset preceded the appearance of lesions in tissues or organs.

5. *Retinamides*

Many retinamides are less toxic than the corresponding free carboxylic acids or retinyl esters. Thus, the studies of Moon *et al.* (1979) and Hixson and Denine (1979) showed the relatively low toxicity of 4-hydroxyphenylretinamide and ethylretinamide, compared either to retinoic acid or retinyl acetate.

Comparative toxicity studies have been reported for tretinoin, isotretinoin, etretinate, and two retinamides, *N*-(4-ethoxycarbopheny)retinamide and *N*-(4-hydroxycarbophenyl)retinamide (Du *et al.*, 1982). Two-week oral toxicity studies in mice at doses of 150 mg/kg/day of tretinoin and 175 mg/kg/day of etretinate resulted in weight loss, hair loss, and fractures; these signs of toxicity were not seen with 150 mg/kg/day of isotretinoin, 600 mg/kg/day of *N*-(4-hydroxycarbophenyl)retinamide or 860 mg/kg/day of *N*-(4-ethoxycarbophenyl)-retinamide. In another experiment, 60 and 300 mg/kg/day of tretinoin and *N*-(4-ethoxycarbophenyl)retinamide, respectively, were administered for 3 months; tretinoin proved to be more toxic than the retinamide.

The comparative toxicity of retinoids and retinamides have also been studied by Sani and Meeks (1982). On the basis of clinical signs of toxicity, lethality, and histopathological findings in rats and mice, the toxicity of the retinoids and retinamides were ranked in the order: tretinoin > 2-hydroxypropylretinamide > isotretinoin > 2-hydroxyethylretinamide > ethylretinamide > 3-hydroxypropyl-retinamide > 4-hydroxyphenylretinamide. Treatment with the retinoids (doses not stated) caused an anemia characterized by erythrocytopenia and decreases in hemoglobin concentration, packed-cell volume, and hematocrit. Dose-dependent increases in serum alkaline phosphatase activity and decreases in serum albumin were also observed.

6. *General Comments*

Although a major adverse effect of the retinoids is an increase in serum alkaline phosphatase, an association of this effect with fractures has not been unequivocally established. There have been no studies demonstrating that the increase in the activity of the serum enzyme is due to an increase in bone isozyme and there has been no clear-cut association between the incidence of fractures and serum alkaline phosphatase activity.

Many studies with retinol and other retinoids have demonstrated effects on lipid metabolism. Indeed, Strauss *et al.* (1978a,b) have suggested that isotretinoin may be effective in acne because of its ability to inhibit lipid metabolism in human sebaceous glands and to block sebum production in the skin (Strauss *et al.*, 1980). Sebaceous gland atrophy in the hamster flank organ following treatment with isotretinoin, but not with etretinate, has been reported by Gomez (1981).

As has been described in many subacute toxicity studies, treatment with retinoids is associated with an elevation in serum triglycerides. Because increased serum triglyceride concentrations may be associated with an increased risk for the development of coronary heart disease (Carlson and Böttiger, 1972), the effects of retinoids on lipid metabolism in laboratory animals have been intensively investigated.

Gerber and Erdman (1979a, 1980) studied the effect of treatment of rats with tretinoin and isotretinoin on serum triglycerides. Although both retinoids induced hypertriglyceridemia, tretinoin was more potent than isotretinoin; neither retinoid had a consistent effect on serum cholesterol. These workers postulated that the elevated triglyceride levels were associated with a very low-density lipoprotein-linked hypertriglyceridemia. In other studies, Gerber and Erdman (1979b) showed that in the rat tretinoin causes hypertriglyceridemia but not hypercholesterolemia.

Effects of retinol on serum triglycerides have also been reported. Hypertriglyceridemia has been demonstrated in rats fed 33 mg retinol/day (Setty and Misra, 1975; Ahuja and Misra, 1975) or 550–1100 mg retinyl palmitate/day (Bayerle *et al.*, 1973); in the latter case, the effect on serum lipids was reversed following discontinuation of treatment. On the other hand, Singh and Singh (1978) reported that relatively low doses of retinyl palmitate (about 185 mg/100 g body wt) did not affect serum triglyceride levels.

D. Chronic Toxicity and Carcinogenicity

Relatively few chronic toxicity studies with retinoids in laboratory animals have been reported. These studies have been carried out mainly in the laboratories of Hoffmann-La Roche and are unpublished except for some brief descriptive reports on retinyl palmitate (Randall, 1980), isotretinoin (Kamm, 1982), and etretinate (Teelmann, 1981). Table II presents the protocol details for chronic toxicity studies in rats and dogs with retinyl palmitate, isotretinoin, and etretinate.

1. Chronic Toxicity of Retinol in Rats and Dogs

The high dose of retinyl palmitate selected for studies in rats (27.5 mg/kg/day) and dogs (13.8 mg/kg/day) was chosen on the basis that it was tolerated by humans and was about 250 times higher than the human recommended daily allowance for vitamin A palmitate (approximately 100 IU or 0.06 mg/kg/day vitamin A palmitate). No adverse effects were observed in rats or dogs treated for

TABLE II

Protocols for Chronic Toxicity and Carcinogenicity Studies

Variable	Retinyl palmitate[a]		Isotretinoin[b]		Etretinate[c]	
Species	Dog	Rat	Dog	Rat	Dog	Rat
Strain	Mongrel	Sprague–Dawley	Beagle	Fischer-344	Beagle	Sprague–Dawley
Duration of treatment	10 months	10 months	55 weeks	104 weeks	52 weeks	104 weeks
Route of administration	Oral (capsule)	Oral (intubation)	Oral (capsule)	Oral (dietary admix)	Oral (capsule)	Oral (dietary admix)
Dose (mg/kg/day)[d]	0,0.6,2.8,13.8	0,5.5,13.8,27.5	0,3,20,120(60)[e]	0,1(2),[f] 1(8),[f] 1(32)[f]	0,2.5,5.0,15.0	0,0.5,1.0,3.0
Animals/sex/group	6 (sex unspecified) controls; 3 (sex unspecified) per treatment group	30 (male only) controls; 10 (males only) per treatment group	9/sex/group	135/sex for the control group; 80/sex/treatment group	4/sex/group	100/sex for the control group; 50/sex/treatment group
Interim sacrifices	None	None	30 weeks	39 and 78 weeks	None	78 weeks
General behavior and clinical signs of toxicity	Recorded daily; body weights recorded weekly	Frequency of observation not noted; body weight recorded weekly	Recorded daily; body weight recorded weekly	Mortality and moribundity observed daily; other signs of toxicity and body weight recorded weekly	Recorded daily; body weight recorded weekly	Recorded daily; body weight recorded weekly
Hematology	Measured after 8, 18, 28, and 38 weeks	Measured after 18 and 35 weeks	Measured at ~ 3-month intervals	Measured at ~ 3-month intervals	Measured at ~ 3-month intervals	Measured at ~ 3-month intervals
Blood chemistries	Not done	Not done	Measured at ~ 3-month intervals	Measured at ~ 3-month intervals	Measured at ~ 3-month intervals	Measured at ~ 3-month intervals

Ophthalmological examination	Not done	Not done	Done at ~ 3-month intervals	Done at ~ 3-month intervals	Done after 6 and 12 months	Done after 12 and 18 months
Neurological examination	Not done	Not done	Done at ~ 3-month intervals	Done at ~ monthly intervals	Not done	Not done
Urinalysis	Not done	Not done	Done at ~ 3-month intervals	Done at ~ 3-month intervals	Done after 6 and 12 months	Done after 3, 12, and 18 months
Electrocardiography	Not done	Not done	Not done	Not done	Done after 6 and 12 months	Not done
Gross autopsy	Not done	Rats that died or were sacrificed for humane reasons	All dogs	All rats	All dogs	All rats
Histological examination	Not done	Not done	All dogs—28 organs and tissues	All rats—~ 30 organs and tissues	All dogs—~ 35 organs and tissues	Major organs and tissues from 10/sex/group and organs with gross lesions from rats sacrificed for humane reasons

[a] Randall (1980) and data on file, Hoffmann-LaRoche, Nutley, New Jersey.

[b] Kamm (1982) and data on file, Hoffmann-LaRoche, Nutley, New Jersey.

[c] Teelmann (1981) and data on file, Hoffmann-LaRoche, Nutley, New Jersey.

[d] Dosing was 5 days/week for the studies with retinyl palmitate and 7 days/week for studies with isotretinoin and etretinate.

[e] Dose was decreased to 60 mg/kg/day after 27 doses of 120 mg/kg/day because of signs of severe toxicity.

[f] All treated groups were given 1.0 mg/kg/day for the first 13 weeks of the study; thereafter the doses were 2, 8, and 32 mg/kg/day for the low-, mid-, and high-dose groups, respectively.

10 months with vitamin A palmitate. Body growth and hematological parameters were within normal limits in both species (Randall, 1980).

2. *Chronic Toxicity of Isotretinoin in Dogs*

Dogs treated with the high dose of isotretinoin (120 mg/kg/day) developed severe signs of toxicity characterized by weight loss and general debilitation, and dosing was stopped after about 4 weeks of treatment. During a period of 8 weeks of no treatment, there appeared to be an almost complete reversal of the signs of toxicity. Treatment was restarted but at a reduced dose level of 60 mg/kg/day. Severe signs of toxicity again developed and after 50 days of treatment, dosing was again stopped. Once again, signs of toxicity disappeared and after 6 weeks of no treatment dosing was restarted at the 60 mg/kg/day level. Thereafter, the dogs were maintained on a cycle of 2 weeks of no treatment followed by 6 weeks of treatment with 60 mg/kg/day until the end of the study, 55 weeks from the time treatment was started. Dogs in the lower-dose groups (3 and 20 mg/kg/day) underwent uninterrupted treatment with isotretinoin for 55 weeks (Kamm, 1982; Hoffmann-La Roche, unpublished).

Severe toxicity observed in dogs treated with either 120 or 60 mg/kg/day included weight loss, skin lesions, bloody feces, various ophthalmological changes (epiphora, corneal opacities, and vascularity), slight decreases in hemoglobin and hematocrit, increased serum alkaline phosphatase activity, and a suggestion of albuminuria. In general, most clinical signs of toxicity, including some instances of corneal opacity, disappeared or diminished when treatment with isotretinoin was stopped, only to reappear when treatment was restarted. There were no fractures noted in this study. However, there were clinical observations of transient lameness characterized by abnormal stance and/or apparent sore foot pads.

Pathological changes observed in dogs treated with the high dose of isotretinoin (120 or 6 mg/kg/day) included cardiac lesions such as degeneration of elastic fibers and fibrosis and/or focal calcification of the left atrium and/or the aorta, focal lesions in the gastrointestinal tract, increased liver weight (with a few instances of some mild fatty changes), testicular atrophy with evidence for spermatogenic arrest, and edema and/or erythrophagocytosis in lymph nodes. Amyloidosis was noted in a single dog.

The lower dosages of isotretinoin (3 and 20 mg/kg/day) were much better tolerated by dogs. Although many clinical signs of toxicity seen at higher dosages were seen at 20 mg/kg/day, there was a decreased frequency of occurrence and a longer time to first appearance than was noted at the higher doses of isotretinoin. The low dose of 3 mg/kg/day of isotretinoin produced no clinical signs of toxicity in dogs, even after treatment for 55 weeks.

Testicular atrophy with spermatogenic arrest and increased liver weight with some fatty infiltration were observed in dogs treated with 20 mg/kg/day of isotretinoin for 55 weeks but not in dogs treated for the same length of time at the lower dose of 3 mg/kg/day. The only change observed in all dogs treated with isotretinoin was the edema and/or erythrophagocytosis in lymph nodes.

3. *Chronic Toxicity of Etretinate in Dogs*

The high dose of etretinate in the chronic dog toxicity study was considerably below that used in the study with isotretinoin because preliminary studies had suggested that a dose in excess of 15 mg/kg/day would lead to excessive mortality.

Dogs treated with 15 mg/kg/day of etretinate had diffuse generalized alopecia; hair loss was first observed after about 15–20 weeks of treatment and progressed until the end of the study at 52 weeks, Occasional ataxia and stiffness of gait were frequent clinical signs. Food consumption and body-weight gain were slightly compromised. Hematologically, an increased sedimentation rate was noted after 26 weeks of treatment, and slight decreases in hemoglobin content and percent hematocrit were seen after 52 weeks of treatment. Pathologically, decreased testes, prostate gland, and kidney weights were observed; slight, localized testicular atrophy was the only microscopic change associated with the decreased organ weights. Except for some hyperkeratosis of the skin in a single dog, there were no pathological findings in skin or bone joints of the dogs treated with 15 mg/kg/day of etretinate for 1 year.

As was demonstrated for isotretinoin, the chronic toxicity of etretinate in the dog was dose-related with respect to the number of dogs affected, the severity of the effect, and the time to onset of occurrence. Finally, Teelmann (1981) has pointed out that the toxicity of etretinate in terms of reduced weight gain, hair loss, decreased hemoglobin concentration, and spermatogenic depression were reversible upon cessation of treatment.

4. *Chronic Toxicity and Carcinogenicity of Isotretinoin and Etretinate in Rats*

For both isotretinoin and etretinate, the chronic toxicity studies in rats were designed to yield information on chronic toxicity and carcinogenicity of the retinoids. In general, data generated after the first 18 months of treatment provided information on chronic toxicity, and data generated after 24 months of treatment provided information on carcinogenicity.

A combined chronic toxicity–carcinogenicity study of isotretinoin was carried out in Fischer-344 rats (Kamm, 1982). The retinoid was administered as a dietary admixture, 7 days a week, for 2 years. An initial dose of 1 mg/kg/day of isotretinoin was administered to all retinoid-treated groups so that the young, rapidly growing rats would be exposed to isotretinoin but not at doses high enough to cause fractures. A high incidence of bone fractures would have compromised the assessment of other toxicity, including carcinogenicity. The low dose of 1 mg/kg/day of isotretinoin was administered for the first 13 weeks of the study; thereafter, doses of 2, 8, and 32 mg/kg/day were administered for the remaining 19 months of the experiment.

Dose-related findings in Fischer-344 rats consisted of increased mortality (noted primarily in male rats after about 100 weeks of treatment) and decreased food intake and body-weight gain. Clinical evidence (i.e., altered gait or awkward posture) for fracture of long bones was noted in the high-dose group. Decreases in hemoglobin concentration and hematocrit were observed only in the 8- and 32-mg/kg/day dose groups; these changes generally appeared after about 25 weeks of treatment and persisted for the duration of the study. Increased alkaline phosphatase activity was observed for all dose groups; elevations were first noted after 39 weeks of treatment. Except for the low-dose group, serum triglycerides were elevated when first measured after 78 weeks of treatment. Increased weights of the liver and kidneys were noted in all treatment groups and for both sexes; adrenal weights in treated male rats were also increased. The increases in organ weights were minimal for rats treated for 2 years with 2 mg/kg/day of isotretinoin.

Nonneoplastic microscopic changes associated with treatment with isotretinoin consisted of focal fibrosis and focal chronic inflammation in the heart, adrenal medullary hyperplasia, arteritis, arterial calcification, focal tissue calcification, and focal osteolysis of the bone. The severity and distribution of these lesions were dose-related.

The incidence of histologically confirmed tumors in Fischer-344 rats treated for 2 years with 2, 8, or 32 mg/kg/day of isotretinoin was no greater than the incidence in control rats except for pheochromocytoma. An increased incidence of the adrenal medullary tumor was noted in male and female rats treated with 32 mg/kg/day of isotretinoin and in male rats treated with 8 mg/kg/day; pheochromocytoma occurred in rats treated with 2 mg/kg/day of isotretinoin with about the same incidence as in control rats. It is of interest that the incidence of leukemia in female rats treated with the high dose of isotretinoin and the incidences of liver adenomas and liver angiomas in male rats treated with 8 or 32 mg/kg/day of the retinoid were decreased relative to control rats.

The finding of a decrease in the incidence of certain tumors in rats treated with isotretinoin is not surprising in view of the demonstrated antineoplastic activity of the retinoids (Sporn, 1982). Of more concern is the relevance of the increase in the incidence of pheochromocytoma to the use of isotretinoin in the treatment of dermatological disorders in humans. Pheochromocytoma has been shown to be the most common adrenal tumor in aging Fischer-344 rats (Goodman *et al.*, 1979); in contrast, it is ordinarily a very rare tumor in humans (Symington, 1969). Thus, the relatively high spontaneous incidence of pheochromocytoma in the aging Fischer-344 rat (about 8% in males and about 3% in females; Goodman *et al.*, 1979) suggests that this strain of rat may be a poor model of the human population in which the incidence is about 0.005% (Symington, 1969). Although a mechanism for the isotretinoin-induced increase in pheochromocytoma in the Fischer-344 rat has not been established, it is possible that the phenomenon may be limited to species and strains that have a relatively high spontaneous incidence of the tumor. The occurrence of pheochromocytoma in rats and factors that may be related to changes in the spontaneous incidence of the tumor have been reviewed by Cheng (1980).

The administration of etretinate by dietary admixture at doses of 0.5, 1, and 3 mg/kg/day for 18 months was generally well tolerated by Sprague-Dawley rats of both sexes (Teelmann, 1981). Adverse effects were, for the most part, confined to the high-dose group and included a reduction in body-weight gain, slight decreases in hemoglobin concentration and hematocrit, and elevations in serum alkaline phosphatase activity. After about 6 months of treatment, approximately half of the rats in the high-dose group had long-bone fractures. Thus, with respect to fractures, etretinate is more toxic than isotretinoin; relatively few fractures were observed with a dose of 32 mg/kg/day of isotretinoin, but fractures were relatively common with 3 mg/kg/day of etretinate.

There was no increase in the spontaneous incidence of any tumors in rats treated for 2 years at doses of etretinate up to 3 mg/kg/day. On the contrary, an antineoplastic effect of etretinate was observed. At doses of 1 and 3 mg/kg/day, the total number of rats bearing multiple tumors (type not specified) was decreased and at the dose of 3 mg/kg/day the number of mammary tumors in female rats was decreased by etretinate.

5. *Effects of Retinoids on Chemically Induced Carcinogenesis and Photocarcinogenesis*

The antineoplastic activity of the retinoids has been alluded to earlier in this chapter and is discussed in greater detail in Chapter 14 of this volume. It is of interest, however, that there have been studies demonstrating that some retinoids both enhance and inhibit chemical carcinogenesis and/or photocarcinogenesis.

With respect to chemically induced carcinogenesis, studies reported in the late 1960s showed that 9,10-dimethyl-1,2-benzanthracene (DMBA)-induced carcinogenesis in the hamster cheek pouch was enhanced when a 10% retinyl palmitate lotion was applied prior to (Levij *et al.*, 1969), together with (Levij and Polliack, 1968), or following (Polliack and Levij, 1969) application of the DMBA. Tretinoin also has been shown to play a role in DMBA-induced keratoacanthoma carcinogenesis. Thus, Prutkin (1968) has demonstrated that keratoacanthomas induced on rabbit ears by DMBA are converted to mucus-secreting tumors following topical application of a 3% tretinoin lotion; the tumors reverted to type when tretinoin treatment was stopped.

In 1977 Epstein reported that the topical application of a 0.3% solution of tretinoin accelerated the appearance of skin tumors in hairless mice exposed to ultraviolet radiation from a medium-pressure mercury arc lamp (Epstein, 1977). This finding was subsequently confirmed by Forbes *et al.* (1979), who demonstrated that 1 or 10 μg of tretinoin per day, applied immediately following exposure to ultraviolet radiation from a xenon arc lamp filtered to simulate natural sunlight, enhanced tumor yield and decreased the time to first appearance of tumors in hairless mice. The effect of tretinoin in photocarcinogenesis is complex and dose-related. Thus, the topical application of relatively dilute solutions of tretinoin (0.005% and 0.025%) had no effect on ultraviolet light-induced carcinogenesis, but a higher concentration (0.05%) significantly inhibited photocarcinogenesis in hairless mice (Epstein and Grekin, 1981a); as already indicated, a 0.3% solution of tretinoin stimulated photocarcinogenesis (Epstein, 1977).

The effects of orally administered retinoids on photocarcinogenesis has also been studied. Epstein (1981) has reported preliminary results of a study with orally administered isotretinoin (10 mg/kg/day) in which no increase in tumor incidence was observed in hairless mice irradiated under the same conditions used in his studies with tretinoin.

Kligman and Kligman (1981) reported that tretinoin (0.05% or 0.001%) applied topically together with ultraviolet irradiation or following the appearance of UV-induced tumors did not enhance photocarcinogenesis with respect to latency, tumor yield, or tumor progression in lightly pigmented hairless mice. When treatment with topical or oral tretinoin was initiated about 18 weeks prior to UV exposure and then continued during the subsequent 20-week UV exposure period, there was an inhibition of tumorigenesis. Oral treatment with etretinate (0.5 or 1.5 mg/kg/day) together with exposure to simulated sunlight had no effect on photocarcinogenicity in hairless mice (Hartmann and Teelmann, 1981).

Finally, Israili *et al.* (1982) reported that neither tretinoin (20 mg/kg/day), isotretinoin (80 mg/kg/day), nor retinol (23 mg/kg/day) had any effect on the incidence of squamous cell carcinoma in hairless mice exposed to UV light generated by a solar simulator.

Although the mechanisms are not completely understood, it is clear that, under appropriate conditions, retinoids can inhibit or potentiate chemical carcinogenesis and photocarcinogenesis. It is likely, though not yet proved, that the effects of the retinoids on carcinogenesis may be related to their known effects on DNA metabolism (Epstein and Grekin, 1981b), immunocompetence (Lotan and Dennert, 1979), and/or cell differentiation (Fell and Mellanby, 1952, 1953). For a more complete discussion of the biochemistry and cellular biology of the retinoids, see Chapter 12 of this volume.

E. Genotoxicity

Genetic toxicology, or more simply genotoxicity, is a special category of toxicology that deals specifically with toxic effects on DNA. Genotoxicity is assuming increased importance in the overall evaluation of the toxicity of substances to humans. The reader is referred to a chapter by Brusick (1982) for an excellent review of the subject.

Perhaps the most widely used assay for mutagenic potential is the Ames test (Ames *et al.*, 1973, 1975; Ames, 1979). In this test the compound under investigation is studied for its ability to bring about a reverse mutation in bacteria. The bacterial tester strains are selected to detect two types of mutation: base-pair substitution (TA 1535, TA 100) and frameshift (TA 1537, TA 1538, TA 98). Generally, the assay is carried out in the presence and absence of an enzyme preparation (S-9) isolated from rodent liver in order to study the mutagenicity of potential metabolites.

Retinol, tretinoin, isotretinoin, and etretinate have all been shown to be nonmutagenic or, in the case of etretinate, weakly mutagenic in the Ames test (Kamm, 1982; Hummler and Schüpbach, 1981). An unpublished study done in the laboratories of Hoffmann-La Roche has shown that isotretinoin may be weakly mutagenic in tester strain TA 100 when the assay was carried out in the presence but not in the absence of activating enzyme.

The potential for genotoxicity may also be evaluated in the mouse micronucleus test and in a host-mediated assay in the mouse. The former procedure is a measure of the clastogenic activity of the compound in erythrocyte stem cells (Heddle, 1973; Jenssen *et al.*, 1974), and the latter is similar to the *in vitro* Ames test, except that exposure of the tester organisms occurs in an animal that has been dosed with the compound to be tested. Etretinate is negative in both tests (Hummler and Schüpbach, 1981).

In the sister chromatid exchange (SCE) assay, the test substance is evaluated for its ability to induce SCEs in Chinese hamster ovary cells; this assay frequently incorporates an activating enzyme system as in the Ames test (Brusick, 1982). Retinol (Huang *et al.*, 1982) and isotretinoin (Hoffmann-La Roche, unpublished) have been found to be negative in this assay. The frequency of occurrence of SCEs in human lymphocytes from subjects treated with etretinate was shown to be no different than before treatment was started (Hoffmann-La Roche, unpublished).

F. Reproductive Toxicology and Teratology

The effects of retinoids on reproduction have been studied in a series of experiments designed to evaluate each step beginning with fertility and general reproductive performance, continuing through the gestation period, and concluding with parturition and postnatal development of the offspring. The dosing schedules and the parameters generally assessed in these reproduction studies are presented in Table III. For a description of the methods used, the reader is referred to Manson *et al.* (1982). Kamm (1982) has summarized the results of experiments with retinol, tretinoin, isotretinoin, and etretinate; Hummler and

TABLE III

Protocols for Studies in Reproductive Toxicology and Teratology

Study type	Dosing schedule	Parameters evaluated
Segment I Fertility and reproduction Usually done in rats	Males Dosing is initiated 60–80 days prior to mating and continued through successful mating. Females Dosing is initiated 14 days prior to mating. For one-half the animals, dosing is continued until sacrifice at day 13 of pregnancy; for the other half of the group, dosing is continued through weaning.	Gonadal function, estrus cycles, mating behavior, conception rates, embryotoxicity, fetotoxicity, teratogenesis, parturition, natal toxicity, and postnatal growth of pups
Segment II Teratology Generally done in two species, usually rats and rabbits	Males Untreated Females Dosing on days 6–15 (rats and mice) or 6–18 (rabbits) of gestation	Fetal growth and development and external, visceral, and skeletal anomalies
Segment III Peri- and postnatal development Usually done in rats	Males Untreated Females Dosing is initiated beginning with the last trimester of gestation (day 15 of pregnancy in the rat) and continues through lactation and weaning	Late fetal development, labor, delivery, lactation, and neonatal viability and growth

Schüpbach (1981) have described studies with etretinate in somewhat greater detail.

1. Effects on Fertility and Reproduction

Retinol has long been recognized as being essential to reproduction, and the effects of hypovitaminosis A on the reproductive process have been described (Moore, 1957). Although hypervitaminosis A has not been systematically studied for effects on fertility and reproductive performance, there are scattered reports suggesting that high intake of retinol is associated with adverse effects on several reproductive functions. These include reports of decreased sperm motility and decreased sperm survival in rabbits (Dubiel *et al.*, 1975) and testicular changes in rats (Weslaw *et al.*, 1938; Maddock *et al.*, 1953; Lamano-Carvalho *et al.*, 1978a). As with other toxic effects of hypervitaminosis A, testicular changes appear to be reversible when treatment is stopped (Lamano-Carvallo *et al.*, 1978b). In female rats, Gellert (1977) reported that chronic treatment with 2.8 mg of retinyl palmitate administered three times per week for 9 months resulted in an inhibition of cycling ovulatory activity. The significance of these changes with respect to reproductive behavior and ability to conceive and bear young was not established.

Effects of tretinoin, isotretinoin, and etretinate on fertility and reproductive performance of the rat have been studied in detail according to the segment I protocol described in Table III (Kamm, 1982; Hummler and Schüpbach, 1981). No adverse effects on gonadal function, fertility, conception rate, gestation, parturition, or neonatal viability were observed in rats treated with tretinoin, isotretinoin, or etretinate at doses up to 2.0, 32.0, and 2.5 mg/kg/day, respectively. At a dose of 5.0 mg/kg/day, neonatal survival was somewhat decreased in pups born of rat dams treated with tretinoin; treatment with the same dose of etretinate resulted in an increase in the rate of resorption, so that the number of viable pups in the litter was decreased and their body-weight gain and survival during lactation were reduced.

2. Teratogenesis

Teratogenicity associated with hypervitaminosis A has been reviewed briefly by Miller and Hayes (1982) and in greater detail by Geelen (1979). Since the first demonstration that high doses of retinol were teratogenic in the pregnant rat (Cohlan, 1953), the vitamin has been shown to produce more than 70 types of anomalies, affecting almost every organ or tissue system, in mice, hamsters, guinea pigs, rabbits, dogs, pigs, and monkeys (Shenefelt, 1972). Abnormalities of the face, ears, eyes, and nervous system are the most commonly encountered terata (Miller and Hayes, 1982). It is of interest that teratological effects are produced in rat fetuses at doses of retinol that do not cause overt toxicity in the dam (Morriss and Steele, 1974).

The teratogenicity of tretinoin has been reviewed by Zbinden (1975b). Minimal teratogenic doses were about 4 mg/kg in mice, about 0.4 mg/kg in Long-Evans rats, greater than 1.7 mg/kg in Wistar rats, about 5 mg/kg in the hamster, and about 26 mg/kg in the rhesus monkey. In addition to causing a variety of skeletal and soft tissue abnormalities, these doses were also frequently embryotoxic. Kretschmer and Leuschner (1975) reported that the minimum teratogenic dose of tretinoin in rats treated from day 6 to day 15 of pregnancy was 1–2 mg/kg/day; in rabbits treated from day 6 to day 18 of pregnancy, the minimal teratogenic dose was between 2 and 10 mg/kg/day. Observed malformations in both species included cleft palates, stunted tails, and, in rats, exencephaly and deformed extremities.

Isotretinoin was teratogenic in the rat at a dose of 150 mg/kg/day but not at 50 mg/kg/day; in the rabbit, teratogenesis could be demonstrated at a dose of 10 mg/kg/day (Kamm, 1982). As with the other retinoids studied, the malformations produced involved the skeletal system, the nervous system, and other soft tissues. Doses of 150 mg/kg/day and 10 mg/kg/day were frequently embryotoxic in the rat and rabbit, respectively; 10 mg/kg/day of isotretinoin was an abortifacient dose in the rabbit.

Etretinate-induced teratogenesis and embryotoxicity have been described by Hummler and Schüpbach (1981). Minimal effective doses were 2, 2, and 4 mg/kg/day in the rat, rabbit, and mouse, respectively. Embryotoxicity was characterized by increased resorption rates, increased incidences of stillbirths, and increased pup mortality during lactation. Teratological findings included anophthalmia, exencephaly, cleft palate, renal aplasia, and skeletal anomalies.

3. *Effects on Peri- and Postnatal Development*

Adverse effects on neonatal rats have been demonstrated when tretinoin, isotretinoin, or etretinate were administered beginning with the last trimester of pregnancy and continuing through lactation until weaning. Doses of tretinoin in excess of 2 mg/kg/day resulted in a decreased survival of the newborn during the first 23 days following birth (Kamm, 1982). Higher doses of etretinate were tolerated without adverse effects, but at a dose of 8 mg/kg/day (highest dose tested) there was an increased number of stillbirths and a marked decrease in neonatal survival (Hummler and Schüpbach, 1981). As with other aspects of reproductive toxicology, pregnant rats tolerated higher doses of isotretinoin without adverse effects on their offspring. Thus, neonatal toxicity was not noted until the dose to the pregnant rat was increased to 32 mg/kg/day (Kamm, 1982). Even then the decreased neonatal viability was considered to be secondary to maternal toxicity characterized by decreased food consumption and bone fracture.

Effects of hypervitaminosis A during the last trimester of pregnancy have not been studied in the same type of experiment as was the case for the other retinoids discussed (see Table III). However, minor brain defects, growth disturbances, and behavioral abnormalities have been observed during postnatal life following *in utero* exposure to retinol, even in the absence of growth defects at birth (Miller and Hayes, 1982).

G. Miscellaneous Studies

The resurgence of interest in the use of retinoids as therapeutic agents may be traced in large part to the development of new synthetic retinoids that retain efficacy but are generally less toxic than retinol. However, as is evident from the preceding discussion, the newer retinoids are not completely free of potentially serious side effects.

The search for more effective and safer retinoids is a continuing process but is time-consuming and very expensive. In an attempt to systematize the search for therapeutically useful retinoids having minimal undesirable side effects, Bollag and co-workers (Bollag, 1981; Bollag and Matter, 1981) have developed a screening procedure based on a mouse papilloma model and clinical signs of toxicity. These workers have defined a therapeutic index as the ratio between the lowest dose causing "hypervitaminosis A" (weight loss, desquamation of the skin, hair loss, and fractures) and the dose producing a 50% regression of papillomas. In this screen, the therapeutic index for tretinoin, isotretinoin, and etretinate is 0.2, 0.5, and 2.0, respectively (Bollag, 1981). The use of this screen has led to the discovery of retinoidal benzoic acid derivatives (arotinoids), which may prove to be a useful new class of therapeutic agents potentially free of undesirable side effects (Bollag, 1982; Loeliger *et al.*, 1980).

H. Summary

The toxicological characteristics of substances intended to be administered to humans, or which might be added to the food chain or which may pollute the environment, are studied in animals in order to assist in predicting safety in humans. It should be understood that this goal is not always achieved. Frequently, adverse effects occur in humans that were not first identified in animal studies, and toxicities observed in laboratory animals do not always have their counterpart in humans. The reasons for predictive failures are varied and it is not appropriate to discuss them here; the interested reader can find many excellent reviews of the subject (Balazs, 1976; Rall, 1979; Garattini, 1982). However, as will be demonstrated in the remainder of this chapter, almost all of the adverse side effects described thus far in laboratory animals have been reported to occur in humans.

II. Clinical Toxicology of Selected Retinoids

A. Retinol Toxicity in Humans

For nearly four decades the scientific literature has reported that the excessive intake of retinol or hypervitaminosis A can have harmful effects in animals and man. The International Vitamin A Consultative Group has summraized all reported cases of hypervitaminosis A from 1850 to 1979 (Bauernfeind, 1980); 579 cases in approximately 200 literature reports were described. There were great differences in the daily doses given and in the frequency and duration of administration associated with the observation of toxicity, indicating wide variability among individuals in their tolerance to retinol. Hypervitaminosis A can be divided into two categories: acute toxicity following a single massive ingestion and chronic toxicity resulting from the ingestion of lower amounts over several months to years.

1. Acute Toxicity

For centuries, Eskimos and arctic travelers who ingested polar bear or seal liver experienced acute and severe illness that was later found to be a result of acute hypervitaminosis A. Rodahl and Moore (1943) found that polar bear livers were frequently rich in retinol, often containing 5.4–8.1 mg retinol equivalents/g liver.

Acute hypervitaminosis A is now rarely seen in adults but is still fairly common in infants due to accidental ingestion or to overdosing by misguided parents. Acute symptoms usually occur within a few hours following consumption; the most prominent effects are on the central nervous system followed by effects on the gastrointestinal tract (Koerner and Voellm, 1975).

Koerner and Voellm (1975) have described the toxic effects of acute hypervitaminosis A; these are listed in decreasing order of occurrence in Table IV

TABLE IV

Clinical Findings in Patients with Acute Hypervitaminosis A in Decreasing Order of Occurrence[a]

Increased cerebrospinal fluid pressure
children—pseudotumor cerebri, hyperostosis
adults—headache (occipital)
Anorexia, vomiting, nausea
Scaling of skin, dry mucous membranes, cheilitis, hair loss
Fatigue, somnolence, vertigo
Edema
Tenderness of long bones
Hepatomegaly and spleenomegaly

[a] Adapted from Koerner and Voellm (1975).

Laboratory abnormalities include elevated plasma levels of retinol, calcium, alkaline phosphatase, serum glutamic oxaloacetic transaminase (SGOT), and serum glutamic pyruvic transaminase (SGPT).

As late as 1979, acute retinol poisoning was reported in fishermen who had consumed 20–300 g of fried halibut liver (Nater and Doeglas, 1970). Approximately 5 h after the meal, symptoms included headache, dizziness, nausea, vomiting, and immobility. The following day, redness, and desquamation of the face and trunk, and later of the palms and soles, were observed. Recovery occurred within a few weeks.

Acute reactions have also been seen in patients treated for dermatologic conditions. Frey and Schock (1952) reported that psoriatic adults given high oral doses of retinyl acetate experienced symptoms of acute toxicity. While doses of 136 mg retinol equivalents (as retinyl acetate) were well tolerated, doses of 340 mg resulted in dryness, scaling, and lip fissuring, and doses of 680–1360 mg resulted in redness, scaling, nausea, and dizziness.

2. *Chronic Toxicity*

Chronic hypervitaminosis A, recognized as a clinical entity in the mid-twentieth century, is more commonly observed and is a more subtle disease than the acute form. Many more cases of chronic hypervitaminosis A have been documented in children than in adults.

The first human case of chronic hypervitaminosis A was described by Josephs (1944). A 3-year-old boy who had been given halibut liver oil (72 mg retinol equivalents/day for 3 months) presented with an enlarged liver and spleen and abnormal skeletal development.

The most prominent signs of chronic hypervitaminosis A are cutaneous, including dryness of the mucous membranes. Laboratory abnormalities include: radiologically detected bone changes in children, increased serum alkaline phosphatase, increased serum calcium, increased cerebrospinal fluid pressure, and disturbed blood clotting. Adverse effects associated with chronic hypervitaminosis A, in decreasing order of occurrence, are presented in Table V.

Toomey and Morissette (1947) reported that the major abnormality following chronic hypervitaminosis A was thickening of the posterior and medial aspect of both ulnas. Skeletal abnormalities have been variable in chronic hypervitaminosis A patients. Periosteal calcification of long bones has been observed in children and sometimes in adults. The partial skeleton of the most complete *Homo erectus* skeleton so far known, found in Kenya, showed pathological changes consistent with chronic hypervitaminosis A and attributable to the high dietary intake of animal liver (Walker *et al.*, 1982). Skeletal manifestations included a superiosteal diaphyseal deposit of coarse-woven bone containing enlarged, subspherical, and randomly placed lacunae.

Pease (1962) reported seven pediatric cases in which most symptoms abated

TABLE V

Clinical Findings in Patients with Chronic Hypervitaminosis A in Decreasing Order of Occurrence[a]

Skin scaling, erythema, pruritus, disturbed hair growth
Dry mucous membranes, cheilitis, gingivitis
Pain, tenderness of bones
Occipital headache
Papillary edema
Anorexia, weight loss
Edema
Fatigue, lassitude, somnolence
Hemorrhage, epistaxis

[a] Adapted from Koerner and Voellm (1975).

after excess retinol intake was discontinued; bony changes were the longest lasting. Growth of the long bones was retarded and occasional premature epiphyseal closure occurred.

Babb and Kieraldo (1978) reported a case of cirrhosis and portal hypertension attributable to hypervitaminosis A in a 72-year-old male who had been taking 12 mg/day of retinol for 6–7 years. Meunter *et al.* (1971) reviewed 17 cases of chronic vitamin A toxicity in patients receiving 12.3–180 mg retinol equivalents/day for several months up to 9 years. Two patients showed marked liver function impairment at $2\frac{1}{2}$ and 12 years, respectively, after discontinuing retinol. Marked elevations of plasma lipid levels were also noted. Deuel (1957) reported that hypervitaminosis A is accompanied by hyperlipidemia involving most of the plasma lipid fractions.

Farris and Erdman (1982) described a case of chronic retinol intoxication following long-term, low-level intake. A 16-year-old male experienced headache, nausea, papilledema, and elevated SGOT following ingestion of 15 mg retinol equivalents/day for $2\frac{1}{2}$ years; this daily dose is only ten times the recommended daily dose. Unlike most cases of chronic hypervitaminosis A, symptoms failed to improve within several days of cessation of intake but persisted for several weeks. Silverman and Lecks (1982) reported that a severely atopic boy, suffering from protein–calorie deficiency and receiving doses of only 6 mg/day for 1 year, was hospitalized for stiff neck, fever, and hepatomegaly. The symptoms cleared in 2 months following discontinuation of retinol.

A case of malformations of the fetal central nervous system following hypervitaminosis A in early pregnancy was reported by Stange *et al.* (1978). The mother had been treated with 31.5 mg retinol equivalents/day during gestation days 19–40, when she realized she was pregnant. Following a normal pregnancy, labor was induced 10 days after term; the baby was born microcephalic and died after 18 min. The child had multiple malformations of the central nervous system and very small adrenal glands. The malformations were considered to have been related to the high doses of retinol given to the mother during her pregnancy. Two other cases had been reported previously indicating a possible relationship between hypervitaminosis A and malformations (Pilotti and Scorta, 1965; Bernhardt and Dorsey, 1974). In these instances, urinary tract malformations were observed in infants born of women who had taken daily doses of 7.5–12.0 mg retinol equivalents during the first trimester of pregnancy.

The mode of transport of retinol may be an important determinant in the development of hypervitaminosis A. Retinol is transported in blood bound to specific carrier proteins, retinol-binding proteins (RBP). According to Goodman (1981), the data suggest that retinol toxicity occurs when the capacity of RBP to transport retinol is exceeded. Hypervitaminosis A occurs when the intake of retinol exceeds the liver's capacity to remove the vitamin from the blood stream and transport it to storage sites (Smith and Goodman, 1976). The toxicity in hypervitaminosis A is then a consequence of relatively high circulating levels of free (i.e., unbound) retinol. For a more complete discussion of retinoid binding proteins, see Chapters 8 and 9 of this volume.

B. Tretinoin Toxicity in Humans

Topical tretinoin was a major advance in dermatologic therapeutics, finding application in a wide variety of dermatoses including acne. Skin irritation is the most common and frequently limiting side effect and reversible hypopigmentation may occur as they do with other irritants. Treatment of some diseases may be impeded because of poor penetration through the thickened stratum corneum.

In a study by Peachy and Conner (1971), 22 patients were treated for acne with topical tretinoin. Almost all patients experienced unacceptable degrees of erythema and peeling. The skin irritation associated with the treatment presented a major barrier to successful treatment. When the side effects were sufficiently diminished by decreasing the intensity of therapy, there was no improvement in the acne. According to Christianson *et al.* (1974), 80% of patients treated with tretinoin experience side effects of erythema and/or desquamation. Contact dermatitis, severe irritation, and reversible hypopigmentation were side effects experienced by patients with acne vulgaris (Handajo, 1977).

Kligman *et al.* (1969) report that within 48 h of onset of topical tretinoin therapy, the skin becomes red and scaly. The irritancy does not diminish with prolonged use, and the skin burns and stings after each application. Once topical tretinoin treatment is discontinued, the skin recovers within a few days. These investigators also noted that in approximately 20% of the patients, new pustules and papules appeared during the initial weeks of therapy. These inflammatory "explosions" of preexisting comedones healed but left no visible scars.

In a study by Frost and Weinstein (1969) irritation, most commonly as erythema and sometimes pruritus, occurred in 4 of 35 patients treated with a 0.2% preparation of tretinoin and in 6 of 21 patients treated with a 0.3% preparation of tretinoin. The intensity of erythema in 60 patients with actinic keratoses, followed by Bollag and Ott (1975), was shown to be dependent on the concentration of the topical preparation of tretinoin and duration of therapy. Although tretinoin has had its widest use as a topical agent, studies have also been carried out using oral therapy. The side effects of oral tretinoin therapy are similar to effects noted with vitamin A. In a study by Stuttgen (1975), 30 patients given oral tretinoin at doses of 50–100 mg/day for 4 weeks showed toxicities similar to those for vitamin A: headache, other central nervous system problems, gastrointestinal effects, and mucocutaneous problems (Table VI). Serious hepatotoxic reactions have not been reported with oral tretinoin presumably because, unlike vitamin A, it is not stored in the liver.

Koch (1978) conducted a study comparing various retinoids in the treatment of precancerous oral lesions. Fifty percent of the patients on oral tretinoin exhibited general symptoms of hypervitaminosis A such as dizziness and headache; in 11% of the patients, the severity of the side effects necessitated the interruption of therapy. In patients with leukoplakia, Koch and Schettler (1973) found a strikingly larger incidence of side effects, including intolerable headaches, dizziness, and nausea among patients under 50 years of age, particularly when dosage was rapidly increased.

TABLE VI

Percent Incidence of Adverse Effects in Thirty Patients Given Oral Tretinoin[a,b]

Clinical finding	Incidence (%)
Cheilitis	80
Headache	80
Xerosis/inflammation	50
Lethargy, fatigue	43
Anorexia	30
Dry mucous membranes	30
Nausea, vomiting	30
Visual disturbances	30
Pruritus	23
Epistaxis	17
Increased bulb pressure (eye)	11
Hair thinning	10
Psychologic changes	10

[a] Adapted from Stuttgen (1975).

[b] The daily dose was 50–100 mg/day for 4 weeks.

TABLE VII

Percent Incidence of Adverse Effects in Patients Given Oral Isotretinoin[a,b]

Clinical finding	Incidence[c] (%)	Laboratory finding	Incidence[c] (%)
Cheilitis	90	Elevated sedimentation rates	50
Conjunctivitis, eye irritation	50	Elevated triglycerides	25
Dry mouth	30	Decreased red cell parameters	10
Xerosis, desquamation (especially face)	30	Decreased white cell counts	10
		Increased platelet counts	10
Epistaxis	25	White cells in urine	10
Pruritus	25	Altered liver function tests	10
Nausea, vomiting, other gastrointestinal disorders	20		
Bone/joint/muscle symptoms	15		
Hair thinning	10		
Headache	10		
Lethargy, fatigue	10		
Palmoplantar desquamation	5		

[a] Adapted from Windhorst and Nigra (1982).
[b] The mean daily dose was 109 mg/day and the mean duration of treatment was 150 days.
[c] 523 patients were studied.

C. Isotretinoin Toxicity in Humans

The development of biologically active, less-toxic synthetic derivatives of retinoic acid has lead to a major advance in dermatologic therapeutics. Isotretinoin (13-*cis*-retinoic acid, Accutane) is the first of these synthetic retinoids to be marketed for the treatment of severe, recalcitrant cystic acne in the United States.

Acute side effects of isotretinoin have mainly been limited to the skin and mucous membranes. The side effects described for hypervitaminosis A are, in general, greatly reduced in frequency and severity in patients treated with oral isotretinoin, except for cheilitis and conjunctivitis, which occur with greater frequency and/or severity.

Windhorst and Nigra (1982) have summarized the clinical results in 523 patients treated with isotretinoin for various disorders of keratinization and cystic acne. Cutaneous and mucous membrane symptoms were by far the most prominent side effects (Table VII). Mucosal effects of isotretinoin have most often included cheilitis, dry nose, conjunctivitis, sore mouth, excessive thirst, and epistaxis. Integumentary effects have included desquamation, pruritus, erythema, skin fragility, finger-tip peeling, and, less frequently, hair loss (Farrell *et al.*, 1980; Gilgor, 1980; Jones *et al.*, 1980; Peck, 1981). Other less-frequent side effects that have been reported included inflammation of urethral meatus, severe earaches, gastrointestinal bleeding, nausea, headaches, appetite changes, and bone/joint pain (Table VII).

In thc study by Windhorst and Nigra (1982), there was a 50% incidence of conjunctivitis or eye irritation. Thirty-four patients were given complete ophthalmologic examinations which, in some patients, included electroretenography and dark adaptation studies; in all cases the exams were reported as normal (Blackman *et al.*, 1979). All these patients had disorders of keratinization and were receiving high doses of isotretinoin. The superficial eye irritations were found to be secondary to blepharokeratoconjunctivitis that responded to antibiotic ophthalmic ointment. Corneal opacities have also been reported in five patients while receiving chronic isotretinoin treatment, but these opacities regressed following discontinuation of the drug.

Of the 523 patients treated with isotretinoin, a 16% incidence of bone, joint, and muscular symptoms was reported (Windhorst and Nigra, 1982). Findings have shown that after more than 2 years of treatment with high doses of isotretinoin in patients with Darier's disease and other disorders of keratinization, skeletal abnormalities have been observed. Hyperostosis with osteophyte and bony bridge formation has occurred in some adults and premature calcification of the epiphyses has been reported in two children. The relationship of these radiologic changes to more than 2 years of treatment with isotretinoin remains unclear and is under further investigation.

The incidence of side effects observed during clinical studies with isotretinoin may be summarized as follows: cheilitis was seen in more than 90% of patients and conjunctivitis was noted in 38% of patients. In cystic acne patients, dry skin, pruritus, epistaxis, and dry nose and mouth were demonstrated in up to 80% of patients. In all patients treated with isotretinoin, 16% had musculoskeletal pain; $< 10\%$ had thinning of the hair; $\sim 5\%$ had desquamation of the palms and soles, nonspecific urogenital findings, gastrointestinal symptoms (e.g., transient nausea), fatigue, headache, or photosensitivity. To date in patients treated for cystic acne, all of the foregoing adverse reactions have been demonstrated to be reversible following discontinuation of drug.

Of the more than 500 patients followed in clinical studies conducted in the United States, only 13 were dropped from study because of laboratory abnormalities. Twelve of these patients had elevations of their serum triglycerides, and one patient had elevated liver enzymes (this patient also had a history of alcohol consumption). Seven of the twelve patients with elevated triglycerides were able to restart isotretinoin therapy after dietary manipulation and adjustment of the dose of isotretinoin.

Katz *et al.* (1980) first reported elevations of triglycerides and reduced level of high-density lipoproteins (HDL) in seven of ten patients with disorders of keratinization treated with isotretinoin; however, normal serum cholesterol levels were maintained. Elevations of serum triglycerides were seen at doses > 1.0 mg/

kg/day of isotretinoin. Serum lipoprotein electrophoresis values obtained while triglycerides were elevated showed type IV patterns with elevations of very low-density lipoproteins (VLDL) and no chylomicrons present except in one patient. This effect was found to be dose-related and reversible.

Many of the patients who have received isotretinoin have had congenital skin diseases. One sibship reported by Dicken and Connolly (1980) consisted of a brother and two sisters with Darier's disease, both of whom were obese; their laboratory data suggested preexisting diabetes. The brother developed eruptive xanthomas when his triglyceride level was 4768 mg/dl; the xanthomas cleared within 1 month of discontinuing the drug. The clinical significance of hypertriglyceridemia remains unclear since elevations are usually modest and are not accompanied by significant hypercholesterolemia. It is possible that hypertriglyceridemia is only encountered in patients predisposed to the condition, such as those with obesity, diabetes mellitus, high alcohol intake, and/or a positive family history of hypertriglyceridemia.

Peck *et al.* (1978) studied the effects of isotretinoin in 81 patients with various dermatologic disorders and reported elevated erythrocyte sedimentation rate in 13 patients, slight increases in SGPT in three patients, increased SGOT in one patient, and increased alkaline phosphatase in two patients. In a 50-patient study conducted by Yoder and Peck (1977), the only abnormal laboratory test was elevated erythrocyte sedimentation rate noted in seven patients. Drug-related hypercalcemia was seen in 3 of 42 patients treated with isotretinoin for various malignancies (Cassidy *et al.*, 1981).

Investigations on sperm conducted by Schill *et al.* (1981) showed that long-term treatment with isotretinoin at therapeutic doses had no adverse effects on male reproductive functions. In another study (Scheiber and Plewig, 1978) of 27 patients with lichen planus mucosae, one patient had decreased potency but the relationship to treatment was not clear.

Although Plewig *et al.* (1980) reported that the incidence of side effects involving drying of skin and mucous membranes were independent of dosage, several other studies have demonstrated that most side effects of isotretinoin are dose-related in severity and reversible on discontinuation of therapy (Gilgor *et al.*, 1980; Frank, 1981; Plewig, 1981). Jones *et al.* (1980) reported that doses of 0.5 mg/kg/day clearly result in a higher incidence of side effects and abnormal laboratory tests than did doses of 0.1 mg/kg/day.

A potential side effect of isotretinoin is its teratogenicity which has been demonstrated in laboratory animals at doses greater than 50 mg/kg/day in rats and greater than 10 mg/kg/day in rabbits (Kamm, 1982). There are no adequate and well-controlled studies in pregnant women but major fetal abnormalities related to isotretinoin administration have been reported. When isotretinoin is used in a therapeutic situation, females of child-bearing potential are required to use effective contraceptive methods. Because the elimination half-life of isotretinoin following chronic therapy is 12–20 h, patients are advised to avoid pregnancy for a minimum of one full menstrual cycle after therapy is discontinued.

D. Etretinate Toxicity in Humans

Etretinate, a synthetic aromatic retinoid, has been used as an investigational drug in many thousands of patients throughout the world for various dermatologic disorders; it is commercially available by prescription in several European countries for treatment of psoriasis and some disorders of keratinization. As with isotretinoin, the side effects manifested during therapy with etretinate are predominantly mucocutaneous.

In a 97-patient double-blind, placebo-controlled, cross-over study conducted by Lassus (1980), side-effects of uninvolved skin and mucous membranes were largely dose-dependent. Three patients discontinued the study because of alopecia and one because of paronychia. The most significant side effects reported were erythema and scaling, which were observed at doses of 50 and 100 mg/day of etretinate. Thinning of skin, rhagades, and ulceration of skin were observed only at 100 mg/day; pruritus occurred at both doses but at a greater frequency at the higher dose. Mucocutaneous side effects including dryness of lip and nasal and oral mucosa were more frequent at doses of 100 mg/day. No laboratory abnormalities attributable to drug treatment were reported.

Mahrle *et al.* (1982) treated 124 patients with keratinizing dermatoses with 25–100 mg/day etretinate. The most striking clinical adverse experiences reported were cheilitis (70%), dryness of mucous membranes (27%), and hair loss (27%). Other side effects included palm/sole exfoliation, paronychia, pruritus, shivering, pyoderma, onychorrhexis, rhagades, and headache. Symptoms occurring in < 3% of the patients included hair thinning, erythema, gingival bleeding, impotence, arthralgia, sweating, burning sensation, dizziness, nausea, hoarseness, anorexia, and urethritis. Treatment was discontinued in nine patients owing to dermatological effects or abnormal clinical chemistry.

Goerz and Orfanos (1978) treated 203 psoriatics with 75 mg/day etretinate for 3–4 weeks, followed by 50 mg/day for up to 6 months, followed by a 25–50 mg/day maintenance dose. Table VIII shows the number of patients experiencing side effects attributable to etretinate therapy in this study. Side effects such as hair loss, paronchia, and elevation of serum transminases caused discontinuation of treatment in 14% of the cases.

Erosions and dryness of mucous membranes were observed in 8 of 24 patients with severe psoriasis given etretinate at doses of 50–100 mg/day (Fredrikison and Petterssen, 1978). A trend toward a dose-related severity of side effects was observed for all the reported effects, but the differences were not statistically significant. Ruest and Rufli (1979) also observed dose-dependent adverse reactions at doses of 0.5–1 mg/kg/day. Beside these commonly observed side ef-

TABLE VIII

Incidence of Clinical and Laboratory Adverse Effects in Patients Given Oral Etretinate

Clinical findings (N = 203)[a]	Incidence (%)	Laboratory findings[b]	Incidence (%)
Cheilitis	78	Elevated bilirubin	32 (N = 188)
Palm/sole exfoliation	31	Elevated SGPT	25 (N = 235)
Hair loss	29	Elevated alkaline phosphatase	22 (N = 226)
Dry mucosa	21	Elevated SGOT	21 (N = 235)
Pruritus	15	Increased prothrombin time	20 (N = 148)
Skin thinning	10		
Paronychia	8		
Sweating	6		
Epistaxia	4		
Shivering	4		
Headache	2		
Irritation of urethral meatus	1		
Vomiting	0.5		

[a] Adapted from Goerz and Orfanos (1978). The initial dose was 75 mg/day for 3–4 weeks; the dose was then decreased to 50 mg/day for 6 months, followed by a daily maintainance dose of 25–50 mg.

[b] Adopted from Mahrle *et al.* (1982). The dose was 25–100 mg/day administered for up to 30 months.

fects, retinoid dermatitis in the form of follicular papules and/or vesicles was observed in 9 of the 23 patients described in this study.

In a multicenter study conducted by Mahrle *et al.* (1979), hair loss occurred in 27 patients, with maximum loss occurring between weeks 5 and 8 after a total dose of 1.9 g of etretinate had been given. The hair loss was reversible by decreasing dose or discontinuing drug. Seventy percent of the effluvium was found to be teologenic, and in 22% of cases teologen and dystrophic anagen roots were found.

Laboratory abnormalities associated with etretinate treatment are similar to those reported for isotretinoin with only transient elevations of transminases and serum triglycerides being those most commonly observed. Psoriatic patients, treated up to 4 weeks with 50–100 mg/day followed by a 25–50 mg/day maintenance dose, had increases in SGOT, SGPT, alkaline phosphatase, bilirubin, and prothrombin time (Mahrle *et al.*, 1982) (Table VIII).

Ott and Bollag (1975) noted prolonged bleeding time in 6 of 24 patients treated with etretinate; bleeding time returned to normal with dosage reduction. All other hematological and chemistry investigations showed no deviation from normal.

One case of centrilobular necrosis of the liver was reported by Thune and Mork (1980) in a 54-year-old woman with nonbullous congenital ichthyosiform erythroderma treated for 3 months with 50–75 mg/day etretinate. Besides mucocutaneous side effects, nausea, and hair loss, the patient noticed dark urine. In addition, SGOT, SGPT, and alkaline phosphatase were abnormally high; bilirubin and lactate dehydrogenase were also elevated. These tests began to normalize after 1 month after therapy ceased, but SGOT and SGPT were still elevated 6 months after stopping treatment with etretinate. Although several contributing factors (alcohol intake, etc) were present in this patient, etretinate could not be ruled out as a contributing factor.

Only one of five patients with known liver damage showed deterioration of hepatic function possibly associated with etretinate treatment (Ott, 1977). In a long-term tolerance study of etretinate by Ott (1981), 107 patients were treated for 1–6 years with no disturbance of liver function, as measured by normal serum bilirubin, SGOT, SGPT, and alkaline phosphatase. In 15 of 20 patients with severe psoriasis prospectively followed by Glazer *et al.* (1982), no change in liver biopsy class after 6 months of etretinate therapy was noted. Three patients showed progressive fatty metamorphosis and two showed liver cell necrosis and progressive fibrosis. The increase in fatty infiltration could be explained by sampling variations, obesity, alcohol ingestion, or retinoid toxicity. Based on these data, etretinate does not appear to produce significant hepatotoxicity.

Ellis *et al.* (1982) investigated changes in serum triglycerides, cholesterol, and high-density lipoprotein (HDL) in 21 patients treated with etretinate for 16 weeks. During treatment, 77% of the patients had elevations in serum triglycerides and 25% had elevated serum cholesterol levels (outside the normal range). After 8 weeks off treatment, triglyceride values returned to approximately baseline but 6 of 19 patients had cholesterol values 20% or greater above baseline values. Mean HDL–cholesterol did not change significantly during the study. Five of 25 patients followed by Gollnick (1981) had elevated serum triglycerides. In each case, the patient had one or more risk factors (e.g., obesity, alcohol abuse, lipid metabolism problems) prior to starting etretinate therapy. The long-term importance of these observations is yet to be determined.

Etretinate has been shown to be teratogenic in animals at doses approximately in the human therapeutic range (Kamm, 1982). Furthermore, etretinate has been found to have a prolonged elimination half-life after chronic therapy (Paravicini, 1981). Several instances of abnormalities have been observed in fetuses from women who received etretinate in the first trimester of pregnancy (Windhorst and Nigra, 1982), although some women have been reported to have normal children (Cordero *et al.*, 1981). Since the elimination half-life of etretinate exceeds 80 days, extremely long-term contraception is required after discontinuing the drug. Although etretinate may not be a drug of choice for women of child-bearing potential, results of spermatological studies by Schill *et al.* (1981) indicate that etretinate has no adverse effect on sperm count, sperm morphology, or sperm motility.

TABLE IX

Comparison of Four Oral Retinoids: Frequency of Adverse Reactions[a]

Reaction	Retinol (%)	Tretinoin (%)[b]	Isotretinoin (%)[c]	Etretinate (%)[c]
Cheilitis	37	80	100	100
Pruritus	34	23	12	12
Thinning of hair	37	1	16	39
Dryness of mucous membranes	—	30	50	69
Epistaxis, petechiae	—	17	35	46
Headache	40	80	6	rare
Visual disturbances	37	30	—	—
Conjunctivitis	—	—	77	77
Bone/joint/muscle complaints	56	—	12	20

[a] Adapted from Windhorst and Peck (1982).
[b] Stuttgen (1975).
[c] Peck *et al.* (1981).

E. Summary of Retinoid Toxicity in Humans

As discussed in another chapter of this book, the specificity of various retinoids with respect to their biological effects and their use in different disease conditions can be demonstrated.

The frequency and severity of adverse reactions are also specific for the various retinoids (Table IX). For example, cheilitis, facial dermatitis, and conjunctivitis are frequently occurring side effects of isotretinoin therapy; whereas telogen effluvium, palmar–plantar desquamation, skin fragility, paronychia, arthralgias, and pruritus are more commonly observed in patients receiving etretinate.

These differences allow for targeting of retinoids to given diseases and/or patients. The efficacy of retinoids are well established and it is anticipated that a major thrust of future retinoid research will be toward developing newer retinoids with a narrower therapeutic spectrum and a lower frequency of adverse side effects.

REFERENCES

Ahuja, H. C., and Misra, U. K. (1975). *Agric. Biol. Chem.* **39,** 637–644.
Ames, B. N. (1979). *Science* **204,** 587–593.
Ames, B. N., Durston, W. E., Yamasaki, E., and Lee, F. D. (1973). *Proc. Nat. Acad. Sci. U.S.A.* **70,** 2281.
Ames, B. N., McCann, J., and Yamasaki, E. (1975). *Mutat. Res.* **31,** 347–364.

Babb, R. R., and Kieraldo, J. K. (1978). *West. J. Med.* **128,** 244–246.
Balazs, T. (1976). *Adv. Mod. Toxicol.* **1** 141–153.
Bauernfeind, J. C. (1980). "The Safe Use of Vitamin A. Report of the International Vitamin A Consultative Group." Nutrition Foundation, Washington, D. C.
Bayerle, H., Meister, P., and Wolff, H. (1973). *Arzneim. Forsch.* **23,** 1082–1085.
Bernhardt, I. B., and Dorsey, D. J. (1974). *Obstet. Gynecol. (N. Y.)* **43,** 750–755.
Beyer, K. H., Jr. (1978). "Discovery, Development, and Delivery of New Drugs." Spectrum Publ., New York.
Blackman, H. J., Peck, G. L., Olsen, T. G., and Bergsma, D. R. (1979). *Ophthalmology (Rochester, Minn.)* **86,** 753–759.
Bollag, W. (1979). *Cancer Chemother. Pharmacol.* **3,** 207–215.
Bollag, W. (1981). *In* "Retinoids: Advances in Basic Research and Therapy" (C. E. Orfanos, O. Braun-Falco, E. M. Farber, C. Grupper, M. K. Polano, and R. Schuppli, eds.), pp. 5–11. Springer-Verlag, Berlin and New York.
Bollag, W. (1982). *In* "Psoriasis" (E. M. Farber, A. J. Cox, L. Nall, and P. H. Jacobs, eds.), pp. 175–183. Grune & Stratton, New York.
Bollag, W., and Matter, A. (1981). *Ann. N. Y. Acad. Sci.* **359,** 9–23.
Bollag, W., and Ott, F. (1975). *Acta Derm. Venerol., Suppl. 74,* No. 55, pp. 163–166.
Bomskov, C., and Seemann, G. (1933). *Z. Gesamte Exp. Med.* **89,** 771–779.
Brusick, D. (1982). *In* "Principles and Methods of Toxicology" (A. W. Hayes, ed.), pp. 223–272. Raven, New York.
Cahn, J., Bordier, P., Herold, M., and Meunier, M. T. (1975). *Acta Derm. Venerol., Suppl. 74,* No. 55, pp. 33–35.
Carlson, L. A., and Böttiger, L. E. (1972). *Lancet* **i,** 865–868.
Cassidy, J., Lippman, M., Lacroix, A., Peck, G., and Ozols, R. (1981). *Proc. Soc. Clin. Oncol.* **22,** 441.
Cheng, L. (1980). *J. Environ. Pathol. Toxicol.* **4,** 219–228.
Christiansen, J. V., Gadborg, E., Ludvigsen, K., Meici, C. H. K., Norholm, A., Osmundsen, P. E., Pedersen, D., Rasmussen, K. A., Reiter, H., Reymann, F., Rooman, N., Sylvest, B., Unna, P., Wehnert, R., Aastrup, B., Andersen, B., and Holm, P. (1974). *Dermatologica* **149,** 121–128.
Clark, L., Seawright, A. A., and Gartner, R. J. W. (1970). *J. Comp. Pathol.* **80,** 113–121.
Cohlan, S. Q. (1953). *Science* **117,** 535–536.
Collazo, J. A., and Rodriquez, J. S. (1933a). *Klin. Wochenschr.* **12,** 1732–1734.
Collazo, J. A., and Rodriguez, J. S. (1933b). *Klin Wochenschr.* **12,** 1768–1771.
Cordero, A. A., Allevato, M. A., and Donatti, L. (1981). *In* "Retinoids: Advances in Basic Research and Therapy" (C. E. Orfanos, O. Braun-Falco, E. M. Farber, C. Grupper, M. K. Polano, and R. Schuppli, eds.), p. 501. Springer-Verlag, Berlin and New York.
Davies, A. W., and Moore, T. (1934). *Biochem. J.* **28,** 288–295.
Deuel, H. J., Jr. (1957). "The Lipids: Their Chemistry and Biochemistry." Wiley (Interscience), New York.
Dicken, C., and Connolly, S. (1980). *Arch. Dermatol.* **116,** 951–952.
Dobson, K. J. (1969). *Aust. Vet. J.* **45,** 570–573.
Du, C., Su, X., and Yu, Y. (1982). *Yaoxue Xuebao* **17***(5),* 331–337. [See *Chem. Abst.* **97,** 103953V (1982).]
Dubiel, A., Monkiewicz, J., and Graczyk, S. (1975). *Med. Weter.* **31,** 439–442.
Elias, P. M., and Williams, M. L. (1981). *Arch. Dermatol.* **117,** 160–180.
Ellis, C. N., Swanson, N. A., Grekin, R. C., Goldstein, N. G., Bassett, D. R., Anderson, T. F., and Voorhees, J. J. (1982). *Arch. Dermatol.* **118,** 559–562.
Epstein, J. H. (1977). *Aust. J. Dermatol.* **18,** 57–61.
Epstein, J. H. (1981). *J. Invest. Dermatol.* **76,** 144–146.

Epstein, J. H., and Grekin, D. A. (1981a). *J. Invest. Dermatol.* **76,** 178–180.
Epstein, J. H., and Grekin, D. A. (1981b). *Photochem. Photobiol.* **33,** 265–266.
Farrell, L., Strauss, J., and Stranieri, A. (1980). *J. Am. Acad. Dermatol.* **3,** 602–611.
Farris, W. A., and Erdman, J. W. (1982). *JAMA, J. Am. Med. Assoc.* **247,** 1317–1318.
Fell, H. B., and Mellanby, E. (1952). *J. Physiol. (London)* **116,** 320–349.
Fell, H. B., and Mellanby, E. (1953). *J. Physiol. (London)* **119,** 470–488.
Forbes, P. D., Urbach, F., and Davies, R. E. (1979). *Cancer Lett. (Shannon, Irel.)* **7,** 86–90.
Frank, S. P. (1981). *Mod. Med. (Minneapolis)* **49***(6),* 31–37, 37–38, 43.
Fredrikison, T., and Petterssen, U. (1978). *Dermatologica* **157,** 238–244.
Frey, J. R., and Schock, M. A. (1952). *Dermatologica* **104,** 80–86.
Frost, P., and Weinstein, G. (1969). *JAMA, J. Am. Med. Assoc.* **207,** 1863–1868.
Garattini, S. (1982). *In* "Animals in Toxicological Research" (I. Bartosek, G. Uitain, and E. Pacet, eds.), pp. 201–208. Raven, New York.
Geelen, J. A. (1979). *Crit. Rev. Toxicol.* **6***(4),* 351–375.
Gellert, R. J. (1977). *J. Reprod. Fertil.* **50,** 223–229.
Gerber, L. E., and Erdman, J. W., Jr. (1979a). *Fed. Proc., Fed. Am. Soc. Exp. Biol.* **38,** 761.
Gerber, L. E., and Erdman, J. W., Jr. (1979b). *J. Nutr.* **109,** 580–589.
Gerber, L. E., and Erdman, J. W., Jr. (1980). *J. Nutr.* **110,** 343–351.
Gilgor, R., Chiaramonti, A., Goldsmith, L., and Lazarus, G. (1980). *Cutis* **25,** 380–381, 385.
Glazer, S. O., Roenigk, H. H., Yokoo, H., and Sparberg, M. (1982). *J. Am. Acad. Dermatol.* **6,** 683–687.
Goerz, G., and Orfanos, C. E. (1978). *Dermatologica, Suppl. 1,* **157,** 38–44.
Gollnick, H. (1981). *In* "Retinoids: Advances in Basic Research and Therapy" (C. E. Orfanos, O. Braun-Falco, E. M. Farber, C. Grupper, M. K. Polano, and R. Schuppli, eds.), pp. 503–505. Springer-Verlag, Berlin and New York.
Gomez, E. C. (1981). *J. Invest. Dermatol.* **76,** 68–69.
Goodman, D. S. (1981). *In* "Retinoids: Advances in Basic Research and Therapy" (C. E. Orfanos, O. Braun-Falco, R. M. Farber, C. Grupper, M. K. Polano, and R. Schuppli, eds.), pp. 31–39. Springer-Verlag, Berlin and New York.
Goodman, D. G., Ward, J. M., Squire, R. A., Chu, K. C., and Linhart, M. S. (1979). *Toxicol. Appl. Pharmacol.* **48,** 237–248.
Grey, R. M., Nielsen, S. W., Rousseau, J. E., Jr., Calhoun, M. C., and Eaton, H. D. (1965). *Pathol. Vet.* **2,** 446–467.
Handojo, I. (1977). *Mod. Med. Asia* **13,** 5–6, 11–12.
Hartmann, H. R., and Teelmann, K. (1981). *In* "Retinoids: Advances in Basic Research and Therapy" (C. E. Orfanos, O. Braun-Falco, E. M. Farber, C. Grupper, M. K. Polano, and R. Schuppli, eds.), pp. 447–451. Springer-Verlag, Berlin and New York.
Hayes, K. C., and Hegsted, D. M. (1973). "Toxicants Occurring Naturally in Foods," 2nd ed., Natl. Acad. Sci., Washington, D. C.
Heddle, J. (1973). *Mutat. Res.* **18,** 187–190.
Herold, M., Cahn, J., and Gomont, P. (1975). *Acta Dermatol. Venerol., Suppl. 74,* No. 55, pp. 29–32.
Hixson, E. J., and Denine, E. P. (1978). *Toxicol. Appl. Pharmacol.* **44,** 29–40.
Hixson, E. J., and Denine, E. P. (1979). *JNCI, J. Natl. Cancer Inst.* **63,** 1359–1364.
Hixson, E. J., Burdeshaw, J. A., Denine, E. P., and Harrison, S. D., Jr. (1979). *Toxicol. Appl. Pharmacol.* **47,** 359–365.
Huang, C. C., Hsueh, J. L., Chen, H. H., and Batt, T. R. (1982). *Carcinogenesis* **3***(1),* 1–5.
Hummler, H., and Schüpbach, M. E. (1981). *In* "Retinoids: Advances in Basic Research and Therapy" (C. E. Orfanos, O. Braun-Falco, E. M. Farber, C. Grupper, M. K. Polano, and R. Schuppli, eds.), pp. 49–59. Springer-Verlag, Berlin and New York.

Irving, J. T. (1949). *J. Physiol. (London)* **108,** 92–101.
Israili, Z., Razdan, R., and Willis, I. (1982). *Proc. Am. Assoc. Cancer Res.* **23,** 204.
Jenssen, D., Ramel, C., and Gothe, R. (1974). *Mutat. Res.* **26,** 553–555.
Jones, H., Blanc, D., and Currliffe, W. (1980). *Lancet* **2,** 1048–1049.
Josephs, H. W. (1944). *Am. J. Dis. Child.* **67,** 33–34.
Kamm, J. J. (1982). *J. Am. Acad. Dermatol.* **6** *(No. 4, Part 2),* 652–659.
Katz, R. A., Jorgensen, H., and Nigra, T. P. (1980). *Arch. Dermatol.* **116,** 1369–1372.
Klaassen, C. D., and Doull, J. (1980). *In* "Casarett and Doull's Toxicology: The Basic Science of Poisons" (J. Doull, C. D. Klaassen, and M. O. Amdur, eds.), 2nd ed., pp. 11–27. Macmillan, New York.
Kligman, A., Fulton, J., and Plewig, G. (1969). *Arch. Dermatol.* **99,** 469–476.
Kligman, L. H., and Kligman, A. M. (1981). *In* "Retinoids: Advances in Basic Research and Therapy" (C. E. Orfanos, O. Braun-Falco, E. M. Farber, C. Grupper, M. K. Polano, and R. Schuppli, eds.), pp. 411–415. Springer-Verlag, Berlin and New York.
Koch, H. F. (1978). *J. Maxillofacial Surg.* **6,** 59–63.
Koch, H., and Schettler, D. (1973). *Dtsch. Zahnaerztl. Z.* **28,** 623–627.
Koerner, W. F., and Voellm, J. (1975). *Int. J. Vitam. Nutr. Res.* **45,** 363–372.
Kretzschmar, R., and Leuschner, F. (1975). *Acta Derm. Venerol., Suppl. 74,* No. 55, pp. 25–28.
Lamano-Carvalho, T. L., Lopes, R. A., Azoubel, R., and Ferreira, A. L. (1978a). *Int. J. Vitam. Nutr. Res.* **48,** 307–315.
Lamano-Carvalho, T. L., Lopes, R. A., Azoubel, R., and Ferreira, A. L. (1978b). *Int. J. Vitam. Nutr. Res.* **48,** 316–324.
Lassus, A. (1980). *Br. J. Dermatol.* **102,** 195–202.
Levij, I. S., and Polliack, A. (1968). *Cancer (Philadelphia)* **23,** 300–306.
Levij, I. S., Rwomushano, J. W., and Polliack, A. (1969). *J. Invest. Dermatol.* **53,** 228–231.
Loeliger, P., Bollag, W., and Mayer, H. (1980). *Eur. J. Med. Chem.* **15,** 9–15.
Lotan, R. (1980). *Biochim. Biophys. Acta* **605,** 33–91.
Lotan, R., and Dennert, G. (1979). *Cancer Res.* **39,** 55–58.
Macapinlac, M. P., and Olson, J. A. (1981). *Int. J. Vitam. Nutr. Res.* **51,** 331–341.
Maddock, C. L., Wolbach, S. B., and Maddock, S. (1949). *J. Nutr.* **39,** 117–137.
Maddock, C. L., Cohen, J., and Wolbach, S. B. (1953). *Arch. Pathol.* **56,** 333–340.
Mahrle, G., Orfanos, C., Ippen, H., and Hofbauer, M. (1979). *Dtsch. Med. Wochenschr.* **104,** 473–477.
Mahrle, G., Meyer-Hamme, S., and Ippen, H. (1982). *Arch. Dermatol.* **118,** 97–100.
Manson, J. M., Zenick, H., and Costlow, R. D. (1982). *In* "Principles and Methods of Toxicology" (A. W. Hayes, ed.), pp. 141–184. Raven, New York.
Meunter, M. D., Perry, H. O., and Ludwig, J. (1971). *Am. J. Med.* **50,** 129–136.
Miller, D. R., and Hayes, K. C. (1982). *In* "Nutritional Toxicology" (J. N. Hathcock, ed.), Vol. I, pp. 81–133. Academic Press, New York.
Moon, R. C., Thompson, H. J., Becci, P. J., Grubbs, C. J., Gander, R. J., Newton, D. L., Smith, J. M., Phillips, S. L., Henderson, W. R., Mullen, L. T., Brown, C. C., and Sporn, M. B. (1979). *Cancer Res.* **39,** 1339–1346.
Moore, T. (1957). "Vitamin A." Elsevier, New York.
Moore, T., and Wang, Y. L. (1945). *Biochem. J.* **39,** 222–228.
Morriss, G. M., and Steele, C. (1974). *J. Embryol. Exp. Morphol.* **32,** 505–514.
Nater, J. P., and Doegtas, H. M. G. (1970). *Acta Derm. Venerol.* **50,** 109–113.
Newberne, P. M., and Rogers, A. E. (1981). *In* "Nutrition and Cancer: Eitiology and Treatment" (G. R. Newell and N. M. Ellison, eds.), pp. 217–232. Raven, New York.
Nichele, G., and Carlino, L. (1953a). *Pediatr. Int.* **3,** 401–434.
Nichele, G., and Carlino, L. (1953b). *Atti. Soc. Ital. Cardiol.* **14,** 231–233.

Nieman, C., and Obbink, H. J. K. (1954). *Vitam. Horm. (N. Y.)* **12,** 69–99.
Ott, F. (1977). *Schweiz, Med. Wochenschr.* **107,** 144–147.
Ott, F. (1981). *In* "Retinoids: Advances in Basic Research and Therapy" (C. E. Orfanos, O. Braun-Falco, E. M. Farber, C. Grupper, M. K. Polano, and R. Schuppli, eds.), pp. 355–357. Springer-Verlag, Berlin and New York.
Ott, F., and Bollag, W. (1975). *Schweiz, Med. Wochenschr.* **105,** 439–441.
Paravicini, U. (1981). *In* "Retinoids: Advances in Basic Research and Therapy" (C. E. Orfanos, O. Braun-Falco, E. M. Farber, C. Grupper, M. K. Polano, and R. Schuppli, eds.), pp. 13–20. Springer-Verlag, Berlin and New York.
Pardini, D. A. (1981). *West. J. Med.* **134,** 47–48.
Peachey, R., and Connor, B. (1971). *Br. J. Dermatol.* **85,** 209–213.
Pease, G. M. (1962). *JAMA, J. Am. Med. Assoc.* **182,** 980–985.
Peck, G. L. (1980). *Arch. Dermatol.* **116,** 283–284.
Peck, G. L. (1981). *In* "Progress in Diseases of the Skin" (R. Fleischmajor, ed.), Vol. 1, pp. 227–269. Grune and Stratton, New York.
Peck, G. L., Yoder, F., Olsen, T., Pandya, M., and Butkus, D. (1978). *Dermatologica, Suppl. 1,* No. 157, pp. 11–12.
Peck, G. L., Gross, E. G., and Butkus, D. (1981). *In* "Retinoids: Advances in Basic Research and Therapy" (C. E. Orfanos, O. Braun-Falco, E. M. Farber, C. Grupper, M. K. Polano, and R. Schuppli, eds.), pp. 279–286. Springer-Verlag, Berlin and New York.
Pilotti, G., and Scorta, A. (1965). *Minerva Ginecol.* **17,** 1103–1108.
Plewig, G. (1981). *J. Invest. Dermatol.* **76,** 417.
Plewig, G., Wagner, A., and Braun-Falco, O. (1980). *Klin. Ergeb. Munich Med. Wochenschr.* **122,** 1287–1293.
Polliack, A., and Levij, I. S. (1969). *Cancer Res.* **29,** 327–332.
Prutkin, L. (1968). *Cancer Res.* **28,** 1021–1030.
Randall, L. O. (1980). *In* "Evaluation of the Health Aspects of Vitamin A, Vitamin A Acetate, and Vitamin A Palmitate as Food Ingredients," p. 18. Fed. Am. Soc. Exp. Biol., Bethesda, Maryland.
Rall, D. P. (1979). *Environ. Health Perspec.* **32,** 297–300.
Rodahl, K. (1949). *Nature (London)* **164,** 531.
Rodahl, K., and Moore, T. (1943). *Biochem. J.* **37,** 166–168.
Ruest, D., and Rufli, T. (1979). *Schweiz, Med. Wochenschr.* **109,** 1921–1925.
Sani, B. P., and Meeks, R. G. (1982). *Proc. Am. Assoc. Cancer Res.* **23,** 89.
Schardein, J. L. (1976). "Drugs as Teratogens." CRC Press, Boca Raton, Florida.
Scheiber, W., and Plewig, G. (1978). *Dermatologica* **157,** 171–180.
Schill, W. B., Wagner, A., Nikolow, J., and Plewig, G. (1981). *In* "Retinoids: Advances in Basic Research and Therapy" (C. E. Orfanos, O. Braun-Falco, E. M. Farber, C. Grupper, M. K. Polano, and R. Schuppli, eds.), pp. 389–395. Springer-Verlag, Berlin and New York.
Seawright, A. A., English, P. B., and Gartner, R. J. W. (1967). *J. Comp. Pathol.* **77,** 29–39.
Setty, O. H., and Misra, U. K. (1975). *Int. J. Vitam. Nutr. Res.* **45,** 107–112.
Shenefelt, R. E. (1972). *Am. J. Pathol.* **66,** 589–592.
Silverman, S. H., and Lecks, H. I. (1982). *Clin. Pediat.* **21,** 172–175.
Singh, M., and Singh, V. N. (1978). *Am. J. Physiol.* **234,** E511–E514.
Smith, F. R., and Goodman, D. S. (1976). *N. Engl. J. Med.* **294,** 805–808.
Sporn, M. B. (1977). *Nutr. Rev.* **35,** 65–69.
Sporn, M. B. (1982). *Psoriasis, Proc. 3rd Int. Symp.* pp. 191–196.
Sporn, M. B., and Newton, D. L. (1981). *In* "Inhibition of Tumor Induction and Development" (M. S. Zedeck and M. Lipkin, eds.), pp. 71–100. Plenum, New York.
Stange, L., Carlström, K., and Eriksson, M. (1978). *Acta Obstet. Gynecol. Scand.* **57,** 289–291.

Strauss, J. S., Peck, G. L., Olsen, T. G., Downing, D. T., and Windhorst, D. B. (1978a). *J. Invest. Dermatol.* **70,** 228.

Strauss, J. S., Peck, G. L., Olsen, T. G., Downing, D. T., and Windhorst, D. B. (1978b). *Clin. Res.* **26,** 577a.

Strauss, J. S., Stranieri, A. M., Farrell, L. N., and Downing, D. T. (1980). *J. Invest. Dermatol.* **74,** 66–67.

Strauss, K. S. (1934). *Beitr. Pathol. Anat. Allg. Pathol.* **94,** 345–352.

Studer, A., Zbinden, G., and Uehlinger, E. (1962). *In* "Handbuch der allgemeiner Pathologic" (F. Buchner, E. Letterr, and F. Roulet, eds.), Vol. 11, Part 1, pp. 936–976. Springer–Verlag, Berlin and New York.

Stuttgen, G. (1975). *Acta Derm. Venerol, Suppl. 74,* No. 55, pp. 174–179.

Symington, T. (1969). "Functional Pathology of the Human Adrenal Gland." Livingstone, Edinburgh.

Takahashi, L., Nakamiya, Z., Kawakimi, K., and Kitasato, T. (1925). *Sci. Pap. Inst. Phys. Chem. Research (Jpn.)* **3,** 81–148.

Teelmann, K. (1981). *In* "Retinoids: Advances in Basic Research and Therapy" (C. E. Orfanos, O. Braun-Falco, E. M. Farber, C. Grupper, M. K. Polano, and R. Schuppli, eds.), pp. 41–47. Springer-Verlag, Berlin and New York.

Thune, P., and Mork, N. (1980). *Dermatologia* **160,** 405–408.

Toomey, J. A., and Morissette, R. A. (1947). *Am. J. Dis. Child.* **73,** 473.

Walker, A., Zimmerman, M. R., and Leaky, R. E. G. (1982). *Nature (London)* **296,** 248–250.

Weslaw, W., Wronski, B., Wroblewski, A., and Wroblewski, B. (1938). *Klin. Wochenschr.* **17,** 777–781.

Windhorst, D. B. (1982). *Psoriasis, Proc. 3rd Int. Symp.* pp. 185–189.

Windhorst, D. B., and Nigra, T. (1982). *J. Am. Acad. Dermatol.* **6,** 675–682.

Windhorst, D. B., and Peck, G. (1982). *In* "Update: Dermatology in General Medicine" (T. Fitzpatrick, A. Z. Eisen, K. Wolff, I. Freedberg, and K. F. Austen, eds.), pp. 226–237. McGraw-Hill, New York.

Wolbach, S. B., and Bessey, O. A. (1942). *Physiol. Rev.* **22,** 233–289.

Wolke, R. E., Nielsen, S. W., and Rousseau, J. E., Jr. (1968). *Am. J. Vet. Res.* **29,** 1009–1024.

Yoder, F., and Peck, G. (1977). *Clin. Res.* **25,** 288A.

Zbinden, G. (1975a). *Acta Dermatol. Venerol., Suppl. 74,* No. 55, pp. 21–24.

Zbinden, G. (1975b). *Acta Dermatol. Venerol., Suppl. 74,* No. 55, pp. 36–40.

14

Retinoids and Cancer

RICHARD C. MOON
Life Sciences Division
IIT Research Institute
Chicago, Illinois

LORETTA M. ITRI
Department of Medical Oncology
Roche Research Center
Hoffmann-La Roche Inc.
Nutley, New Jersey

I. INTRODUCTION

The importance of retinoids in the field of cancer studies has been discussed at length in these two volumes, particularly in Chapters 5 and 12. The ability of retinoids to modify the differentiation and proliferation of both preneoplastic and neoplastic cells is one of the fundamental properties of this class of substances and has been successfully used as the basis for several biological assays for retinoids, as documented in detail in Chapter 5. The use of preneoplastic and neoplastic cells to study the cellular and molecular mechanism of action of

THE RETINOIDS, VOL. 2

ISBN 0-12-658102-9

retinoids provides a central theme for Chapter 12. Studies of neoplastic cell function have provided important information regarding the normal function of retinoids in control of cell differentiation, and the retinoids have already been used successfully to study the nature and regulation of oncogene function. The present chapter deals with the topic of the use of retinoids to prevent or treat cancer in animals and man, and the reader is referred to Chapters 5 (Vol. 1) and 12 (this volume) for mechanistic details, of which we provide only the briefest summary here.

Although Wolbach and Howe (1925) noted that there were histological resemblances between the epithelia of vitamin A-deficient organs and neoplastic tissues, they were not able to provide any mechanistic analysis of this important observation. Significant advances in understanding the mechanistic relationship between retinoids and cancer did not occur until *in vitro* systems were used as experimental tools. Most notable was the introduction of organ culture methodology, particularly the studies of Lasnitzki (1955), who showed that the premalignant phenotype of mouse prostate glands that had been treated with the carcinogen 3-methylcholanthrene could be altered by retinoids. In prostatic organ cultures, she showed that retinoid treatment caused disappearance of atypical epithelial cells that had been induced by the carcinogen, with replacement by cells with more normal morphology. The introduction of cell culture techniques allowed further analysis of such preneoplastic phenomena, and in 1979 Merriman and Bertram showed that retinoids can act directly on nonneoplastic mouse fibroblastic cells to suppress the process of malignant transformation induced by a chemical carcinogen such as methylcholanthrene. In these experiments it was clearly shown that retinoids were effective in suppressing transformation, even when they were applied to cell cultures a full week after the original exposure to carcinogen. It was thus shown that one of the roles of retinoids in suppressing the development of neoplastic cells was to alter events during the promotion or progression phase of carcinogenesis, rather than to modify the process of initiation. This finding is of fundamental importance in analyzing the animal and human studies that are discussed at length in the balance of this chapter since it clearly indicates that in order to prevent cancer, retinoids need *not* be present when carcinogenesis is initiated.

Indeed, it has even been shown that there are some fully neoplastic, transformed cells that can be induced to undergo terminal differentiation in response to treatment with retinoids, with loss of their neoplastic properties. The most striking examples of this phenomenon are the induction of terminal differentiation in murine F9 teratocarcinoma cells (Strickland and Mahdavi, 1978; Strickland, 1981) and human promyelocytic leukemia cells (Breitman *et al.*, 1980, 1981, 1983). However, this does not appear to be a general effect of retinoids, and there are unfortunately only a limited number of instances in which such profound effects of retinoids on differentiation of invasive tumor cells have been shown.

The balance of this chapter surveys the extensive data that have been obtained on the use of retinoids in prevention or treatment of cancer in experimental animals and then considers the application of these results to clinical studies. Such studies have only begun, and it is apparent that at present there is little in the way of definitive clinical data, in contrast to the striking clinical successes in the use of retinoids in dermatology, as documented in Chapter 16. However, the data that have been obtained in animals are of sufficient magnitude to make it apparent that it is only a matter of time before the retinoids will be important agents in clinical oncology. There will be many problems related to selection of specific retinoids, appropriate dosage scheduling, use in combination with other pharmacological agents, monitoring of toxic side effects, and the like. In spite of these problems, it is already clear that retinoids have unique ability to prevent cancer in experimental systems, and the challenge now is to find suitable methods for their practical application in clinical oncology.

II. CHEMOPREVENTION OF CANCER IN EXPERIMENTAL ANIMALS

A. Chemoprevention of Skin Cancer

Most of the experimental evidence regarding chemoprevention of skin cancer by retinoids has been derived from studies utilizing the mouse skin carcinogenesis model. The process of induction of cancer in mouse skin involves at least two well-defined stages, initiation and promotion (Berenblum and Shubik, 1947; Boutwell, 1964, 1974). Initiation is accomplished by the application of one (Boutwell, 1974) or two (Bollag, 1972a,b) subcarcinogenic doses of polycyclic aromatic hydrocarbons, such as benzo[*a*]pyrene or 7,12-dimethylbenz[*a*]-anthracene (DMBA). If the skin is treated only with the carcinogen, few, if any, tumors will appear during the life span of the animal. However, visible tumors do develop when the initiated skin is subjected to repeated applications of a second noncarcinogenic chemical known as a promoter. Croton oil and a potent constituent of croton oil 12-*O*-tetradecanoyl-phorbol-13-acetate (TPA) have been the most frequently used promoters for such studies. Application of tumor promoters alone does not result in the formation of skin tumors. It is only following initiation that the application of these compounds to the skin results in the appearance of tumors. This two-stage model of skin carcinogenesis has been used extensively to study the chemopreventive effects of retinoids, and it has also proved useful in determining the therapeutic effect of the retinoids on established skin papillomas and carcinomas.

Davies (1967) was the first to show that papilloma formation in mice could be modified by retinoids. Following a single topical application of 10 μg DMBA, Davies observed a decrease in the rate of appearance and an accelerated rate of regression of papillomas in rhino mice fed a defined diet supplemented with 100 IU "vitamin A" per gram diet, as compared to that of animals fed a similar diet but deficient in the vitamin. Although papilloma incidence was similar in both groups of mice, the number of papillomas per animal was substantially reduced in the animals receiving the vitamin A-supplemented diet. These initial observations have been greatly extended in a series of reports by Bollag (1971a,b, 1972a,b, 1974, 1975a,b). In these studies, papillomas and carcinomas were induced in mice by the topical application of two 150-μg doses of DMBA administered 1 week apart, followed by twice weekly applications of 0.5 mg croton oil beginning 30 days after the second carcinogen dose. These initiation/promotion experiments led to the development of papillomas 3–8 months following the original application of carcinogen, whereas carcinomas usually did not appear until 5–12 months after DMBA administration. Using this model, Bollag (1972a,b) found that the intragastric administration of all-*trans*-retinoic acid (D1) (200 mg/kg) once every 2 weeks during the promotion phase resulted in a reduction of papilloma and carcinoma multiplicity as well as a decrease in carcinoma incidence. In addition, papilloma growth was retarded and papilloma regression was enhanced. These results led Bollag to conclude that retinoic acid had a dual effect in skin carcinogenesis: a prophylactic or preventive effect (delayed appearance and reduced incidence), as well as a therapeutic effect, manifested by the retarded growth and enhanced regression of the papillomas. In experiments in which retinoid administration was delayed until several well-established papillomas were present, the administration of retinoic acid or retinyl palmitate (A4) resulted in regression of the established papillomas (Bollag, 1971b).

Although these early studies with all-*trans*-retinoic acid and retinyl palmitate demonstrate efficacy of natural retinoids in the inhibition of carcinogenesis, the use of retinoic acid or retinyl esters in cancer chemoprevention is severely limited by their toxicity when administered in pharmacological doses. Thus it has been necessary to develop synthetic retinoids possessing pharmacological properties significantly different from those of the natural retinoids, so that toxicity is reduced while the preventive activity is retained. In an effort to measure both retinoid toxicity and tumor inhibition, Bollag developed two test systems for screening these activities in mice: the hypervitaminosis A and papilloma regression assays. The hypervitaminosis A assay determines the lowest daily intraperitoneal dose of a retinoid that causes a defined degree of hypervitaminosis within a 14-day period. The papilloma regression assay measures the dose of a retinoid that results in a 50% regression of papillomas during a 2-week period of intraperitoneal administration. The ratio between the hypervitaminosis assay dose and the papilloma regression ED_{50} dose is defined by Bollag as the therapeutic index; the higher the therapeutic index, the greater the potential biological effectiveness of the retinoid. [In earlier publications by Bollag (1974), the inverse relationship was termed "therapeutic ratio".] During the next decade, Bollag (1974, 1975a,b) and his colleagues (Mayer *et al.*, 1978) as well as Trown *et al.* (1980) and Pawson *et al.* (1979, 1982) established a therapeutic index for a large number of synthetic retinoids. Although the therapeutic index has proved of some value in studies of structure–function relationships, it has provided little insight into the cancer chemopreventive potential of retinoids in tumor systems other than the skin (Moon and Mehta, 1982; McCormick and Moon, 1982a). The therapeutic index has, however, provided a means of selecting retinoids for further studies of cancer inhibition. One such compound, etretinate (F1) has been studied extensively by Bollag (1974, 1975b) and has been shown to be highly effective in inhibiting the development of papillomas and carcinomas, as well as causing regression of established papillomas (Bollag, 1975b).

In 1971 Shamberger showed the natural retinoids to be effective chemopreventive agents for mouse skin when applied topically, and this work led to further studies by Boutwell and his associates, who have investigated the effects of topically applied natural and synthetic retinoids on skin tumor promotion in a large series of experiments. It was shown in Boutwell's laboratory (O'Brien *et al.*, 1975a,b; O'Brien, 1976; Verma and Boutwell, 1977) that the induction of epidermal ornithine decarboxylase (ODC) activity by tumor-promoting agents is an important and possibly obligatory step in the process of skin tumor promotion. The evidence supporting this suggestion and the correlation between induction of enzyme activity and the promotion of skin tumor formation has been reviewed by Boutwell (1977) and Verma and Boutwell (1980).

Important studies by Verma and Boutwell (1977) showed that the induction of epidermal ODC activity by the potent promoter TPA was markedly inhibited by the topical application of retinoic acid 1 h prior to the application of TPA. Other natural retinoids also inhibited TPA-induced ODC activity, and the degree to which ODC activity was inhibited correlated well with the inhibitory effect of the retinoid on skin tumor promotion. In addition to retinoic acid, Verma *et al.* (1979a) showed that the topical application of some of the synthetic retinoids effectively inhibited TPA-promoted skin papilloma formation. Those synthetic retinoids that did not inhibit papilloma formation also were ineffective in inhibiting TPA-induced ODC activity (see, for example, Table I, Chapter 5, Vol. 1). These findings lend further support to the premise that TPA-induced ODC activity may be an essential step in the mechanism of skin tumor promotion. However, retinoic acid will not prevent skin carcinogenesis caused by DMBA when this chemical is used as a complete carcinogen without a promoting agent. Thus, Verma *et al.* (1980) showed that retinoic acid not only failed to inhibit ODC induction by repeated applications of small doses of DMBA, but also failed to prevent formation of skin papillomas by the carcinogen; in some experiments, retinoic acid actually potentiated the number of papillomas per mouse.

Although the evidence for retinoid inhibition of the promotion phase of chemically induced skin carcinogenesis is substantial, the effect of retinoids on skin carcinogenesis induced by ultraviolet (UV) radiation is less clear. Epstein (1977a) initially reported that all-*trans*-retinoic acid enhanced UV-induced skin carcinogenesis. In this study, approximately 0.5 g of a 0.3% retinoic acid cream was applied to both the exposed and unexposed areas of the skin of hairless mice immediately after UV irradiation. Although UV exposure continued three times a week for 10 months, the application of retinoic acid was discontinued after 4.5 months owing to the development of erythema and desquamation, as well as a high mortality rate in those animals receiving the retinoid. Both the rate of appearance and growth of the tumors were more rapid in mice treated with retinoic acid up to 34 weeks following the initial UV exposure. Although total tumor number was not given, Epstein did indicate that tumors greater than 50 mm^3 were more prevalent in mice treated with retinoic acid during the 34-week treatment period. Thereafter, no significant difference was apparent in tumor incidence between the placebo- and retinoic acid-treated mice, and the percentage of animals bearing tumors in size > 100 mm^3 was equal between the two groups. Epstein interpreted these results to indicate that retinoic acid promoted UV-induced cancer formation. In a subsequent study, Epstein and Grekin (1981) examined the effects of lower, nonirritating concentrations of retinoic acid on UV-induced carcinogenesis. The conditions of this study were essentially the same as that of Epstein's previous study except that retinoic acid was dissolved in an alcohol–propylene glycol solution at concentrations of 0.05%, 0.025%, and 0.005%. The retinoic acid solutions or vehicle were applied to the backs of mice in volumes of 0.1 ml, three times a week for the duration of the study. The parameters for evaluating tumor development and growth were comparable to that of the previous report. Although tumor development in mice receiving topical application of the two lower concentrations of retinoic acid did not differ from that of those receiving only the retinoic acid vehicle, tumor incidence was significantly lower and tumor growth was significantly slower in mice treated with 0.05% concentration of retinoic acid.

Experiments by Forbes *et al.* (1979) in which hairless mice received daily application of 0.01% or 0.001% retinoic acid in methanol and were exposed to daily UV irradiation resulted in marked enhancement of tumor formation. Although these concentrations of retinoic acid resulted in slight scaling and flaking of the skin (indicative of some toxicity), gross evidence of erythema or irritation was absent. The magnitude of the response resulting from applications of these essentially nonirritating concentrations of retinoic acid left little doubt that under these conditions, retinoic acid enhanced UV tumorigenesis; both tumor incidence and total number of tumors in animals treated with retinoic acid were substantially greater at all observation periods when compared to that of the respective controls. Additional studies by Forbes (1981) showed that retinoic acid applied either concurrently with UV irradiation or applied after a course of UV treatment can enhance the development of tumors. Furthermore, retinoic acid administered in the diet produces a similar but a somewhat reduced effect, although Epstein (1981) cited an experiment in which orally administered 13-*cis*-retinoic acid had essentially no effect on tumor formation up to 34 weeks after the initiation of UV exposure.

In summary, the evidence accumulated thus far indicates that retinoic acid, as well as other natural and synthetic retinoids, inhibits two-stage phorbol ester-promoted skin carcinogenesis. In contrast, under certain conditions retinoic acid may enhance tumor formation induced with UV irradiation. Whether this effect of retinoic acid on photocarcinogenesis is limited to this particular retinoid, or whether it is representative of the effects of other natural and synthetic retinoids must await further investigation. Although the biochemical events in photocarcinogenesis are not well defined, UV irradiation has been shown to induce epidermal ODC activity (Verma *et al.*, 1979b); however, the effect of retinoic acid on UV induction of ODC activity differs markedly from that of the retinoid on the induction of the enzyme by TPA (Lichti *et al.*, 1981). Although the results of studies of retinoic acid on photocarcinogenesis are ambiguous, it must be borne in mind that β-carotene, which is metabolized to retinaldehyde and retinol, effectively inhibits UV-induced skin carcinogenesis when given intraperitoneally (Epstein, 1977b) or in the diet (Mathews-Roth, 1982). Moreover, in another experimental model, the dietary administration of retinyl acetate was found to inhibit the induction of rat skin tumors by monoenergetic electrons (Burns *et al.*, 1982).

B. Chemoprevention of Lung Cancer

Studies of the inhibition of chemically induced tumorigenesis of the respiratory tract by retinoids have been equivocal and in some cases, contradictory. Saffiotti and co-workers (1967) were the first to describe an inhibitory effect of retinoids on respiratory carcinogenesis. Tumors were induced in hamsters by the intratracheal instillation of benzo[*a*]pyrene adsorbed onto ferric oxide particles (3 mg benzo[*a*]pyrene plus 3 mg ferric oxide suspended in 0.2 ml saline), once a week for 10 weeks. Some animals received, in addition to the carcinogen, twice weekly intragastric instillations of 5 mg retinyl palmitate in 0.1 ml corn oil beginning 7 days after the last carcinogen dose and continuing for the life span of the animals. As can be seen from Table I, treatment with retinyl palmitate markedly reduced both the incidence and number of respiratory tumors that developed; more significantly, only one squamous tumor developed in animals receiving the benzo[*a*]pyrene and retinyl palmitate, whereas 13 squamous tumors

TABLE I

Tumor Incidence of Respiratory Tract of Hamsters Treated with Benzo[a]pyrene (BP) and Retinyl Palmitate

		Animals with						Total number		
		Respiratory tumors (all types)		Squamous changes		Squamous tumors				
Treatment	Effective number of animals at risk[a]	Number	%	Number	%	Number	%	Respiratory tumors (all types)	Squamous tumors	Squamous metaplasias
BP only	53	17	32	18	34	11	21	20	13	13
BP + retinyl palmitate	46	5	11	2	4	1	2	6	1	1

[a] Survivors at time of first respiratory tumor (33 weeks), less animals lost through cannibalism. For details, see Saffiotti *et al.* (1967).

arose in animals receiving only the carcinogen. Thus, these experiments dramatically demonstrated that supplemental retinyl palmitate was capable of modulating the carcinogenic response of the respiratory tract.

Other investigators have not been as successful in the use of retinoids to prevent the development of respiratory tumors induced by the Saffiotti technique. Smith *et al.* (1975a), using 12 weekly intratracheal instillations of benzo[*a*]pyrene:ferric oxide, treated hamsters with twice weekly intragastric instillations of retinyl acetate (A3) at total weekly dosages of 100 μg, 1600 μg, and 3300 μg beginning 1 week after the cessation of the intratracheal instillations. At 24 weeks following the initiation of the retinoid treatment, the high dose of retinyl acetate was reduced to 2400 μg per week because of toxicity. These studies failed to demonstrate a protective effect of retinyl acetate on respiratory tract carcinogenesis, and furthermore, a significantly greater tumor incidence was reported for animals treated with the high dose of retinyl acetate. A second report by Smith *et al.* (1975b) sheds some light on the results of their previous study. The experimental conditions of the second study were the same as that of the first except that the animals received a semisynthetic diet and were housed either conventionally or under laminar air flow conditions. The results of the second study indicated that there was no statistically significant difference in the incidence of respiratory tract tumors in animals treated with the various doses of retinyl acetate irrespective of whether the hamsters were housed under conventional or laminar air flow conditions. However, the animals kept in laminar air flow units did develop tumors more slowly than did the hamsters housed conventionally. In addition, the incidence of respiratory tract infections was significantly less in hamsters housed under laminar air flow conditions than that of animals housed conventionally. When the results of the two studies are considered together, it is apparent that several factors could have contributed to failure of retinyl acetate to inhibit lung cancer induction: (1) the high effective dose of the carcinogen which resulted in a 60–80% tumor incidence as compared to the 32% incidence of the Saffiotti study; (2) the toxicity of the high doses of retinyl acetate used in these studies; and (3) the increase in respiratory infections in animals housed under conventional conditions and the cocarcinogenic and enhancing effect of such infections on respiratory carcinogenesis (Schreiber *et al.*, 1972).

In another study, Port *et al.* (1975) utilized a relatively low intratracheal dose of benzo[*a*]pyrene:ferric oxide to induce an incidence of respiratory carcinoma in hamsters that approximated the incidence of lung cancer in a human population of heavy smokers. In a life-time study, 10% of 134 control hamsters developed respiratory carcinoma. When the animals were treated with 13-*cis*-retinoic acid (E6) following cessation of carcinogen dosing, the incidence of carcinoma was dramatically reduced. Only 2 of 152 hamsters receiving twice weekly intragastric instillations of 1.5 mg of 13-*cis*-retinoic acid developed respiratory carcinoma.

No carcinomas developed in 158 animals receiving 4.5 mg of 13-*cis*-retinoic acid two times a week.

Nettesheim and his colleagues have evaluated the effects of retinoids in a lung tumor model in which metaplastic lung nodules and pulmonary carcinomas were induced by the intratracheal instillation of 3-methylcholanthrene in rats. Using this model, Nettesheim *et al.* (1976) found that 42% of the rats receiving a vitamin A-deficient diet supplemented with 10 μg retinyl acetate per week developed preneoplastic lung nodules, whereas only 3% of the animals receiving 10,000 μg retinyl acetate per week developed these lesions. In another study, Nettesheim and Williams (1976) failed to achieve any greater protection against the development of lung nodules when animals fed either large doses of retinyl acetate or 13-*cis*-retinoic acid were compared to animals fed a normal diet. From these studies, they concluded that retinoid deficiency enhances the susceptibility of the respiratory epithelium to carcinogenesis induced by 3-methylcholanthrene, but that excessive levels of retinoids have little chemopreventive efficacy when compared to adequate levels. By contrast, an earlier study of Cone and Nettesheim (1973) had shown that large doses of retinyl acetate did indeed reduce the number and volume of tumors that developed in response to 3-methylcholanthrene treatment of rats fed a normal chow diet.

Studies from three different laboratories have indicated that a number of synthetic retinoids have been ineffective in inhibiting tracheal tumors after repeated exposure of respiratory epithelium to methylnitrosourea (MNU). Grubbs *et al.* (1980) found that *N*-(4-hydroxyphenyl)retinamide (E16) administered in the diet neither enhanced nor inhibited tracheal tumors induced by this carcinogen. Similar findings were reported by Yarita *et al.* (1980) for 13-*cis*-retinoic acid and for the aromatic retinoid motretinid (F3) when administered in the diet. Both tumor incidence and tumor multiplicity were essentially the same in animals receiving diets supplemented with the synthetic retionids as in those fed a placebo diet. On the other hand, Stinson *et al* (1981) have reported an increased incidence in tracheal tumorigenesis induced with MNU in animals receiving diets supplemented with 13-*cis*-retinoic acid or *N*-ethylretinamide (E11). It is of interest that the animal model used in all these experiments with MNU does not utilize an initiation–promotion design.

C. Chemoprevention of Mammary Cancer

Some of the most convincing evidence for the chemoprevention of cancer by retinoids comes from chemical carcinogenesis studies of the mammary gland. The earliest study to report an inhibition of mammary carcinogenesis was that of Moon *et al.* (1976), who found a 52% reduction in the incidence of mammary cancer in rats treated with DMBA and 2.5 mg of retinyl acetate per day, compared to that of animals receiving DMBA and a placebo diet. A previous report by Schmähl *et al.* (1972) had indicated that retinyl palmitate was ineffective in altering DMBA-induced mammary tumorigenesis. The discrepancy between these two studies was attributed to the mode of administration of the retinoids; the retinoid was administered intragastrically once a week in the study of Schmähl *et al.*, whereas in the study of Moon *et al.*, the retinoid was administered in the diet, thereby resulting in a more sustained exposure of the target cells to the retinoid. Subsequent studies by Grubbs *et al.* (1977a) not only confirmed the suppressive effects of retinyl acetate on DMBA-induced mammary carcinogenesis but also indicated that retinyl methyl ether (A11) was much more effective in preventing the appearance of mammary cancer than retinyl acetate; retinyl methyl ether, when compared to retinyl acetate, had previously been shown to yield a greater concentration of retinoid in the mammary gland and adjacent fat pad after oral dosing of equivalent amounts of these two compounds. In these studies, feeding of the retinoids was not started until 7 days after DMBA treatment, a time at which the uptake and binding of the carcinogen by mammary parenchymal cells was essentially complete (Janss *et al.*, 1972); the results thus indicate that the retinoids were inhibiting progression rather than initiation of mammary tumorigenesis. In addition to the effect of retinyl acetate on DMBA-induced mammary tumorigenesis, Moon *et al.* (1977) have also shown that retinyl acetate inhibits mammary carcinogenesis induced by the intravenous injection of the direct-acting carcinogen MNU. Both the incidence of mammary cancer as well as tumor number were diminished at all doses of the carcinogen used, and those animals receiving the low dose of carcinogen and retinyl acetate did not develop any tumors. Again, the prevention of MNU-induced cancer was not due to an inhibition of the initiation process since feeding of the retinyl acetate was not begun until after all free carcinogen had disappeared.

The temporal relationships of carcinogen and retinoid administration have been explored in other experiments. McCormick *et al.* (1980) found a significant inhibition of tumorigenesis when retinoid treatment was begun 1 week after the administration of DMBA and continued until termination of the experiment; they also noted that the administration of retinyl acetate during a restricted period (2 weeks before to 1 week after DMBA) during the early phase of mammary carcinogenesis inhibited the development of mammary tumors. The treatment with retinyl acetate during this early phase appears to impart a permanent inhibition on tumorigenesis, while treatment with retinyl acetate beginning 1–12 weeks after DMBA appears to require continued treatment with the retinoid since the mammary tumor response approaches that of the control animals after cessation of retinoid treatment. The studies of McCormick *et al.* (1980) confirmed those of Thompson *et al.* (1979), who also showed that a continuous intake of retinoid was necessary to sustain the chemopreventive effect during the progression stage of mammary carcinogenesis. Althouth continuous treatment with reti-

TABLE II

Retinoids Evaluated for Chemopreventive Activity against Mammary Cancer Induced in Rats by Methylnitrosourea

Active	Inactive
Axerophthene (A7)	13-*cis*-Retinoic acid (E6)
Retinyl acetate (A3)	Retinyl butyl ether
Retinyl methyl ether (A11)	*N*-Ethylretinamide (E11)
N-(4-Hydroxyphenyl)retinamide (E16)	*N*-(2-Hydroxyethyl)retinamide (E12)
N-(4-Hydroxyphenyl)-13-*cis*-retinamide (E21)	*N*-(4-Pivaloyloxyphenyl)retinamide
	N-(4-Propionoxyphenyl)retinamide
	Retinylidene dimedone (C6)
	Retinylidene acetylacetone (C4)
	Etretinate (F1)
	Motretinid (F3)
	TMMP[a] analog of retinyl methyl ether

[a] TMMP = 4-methoxy-2,3,6-trimethylphenyl. For details, see Moon *et al.* (1983a).

noid appears necessary to inhibit the progression phase of mammary carcinogenesis, retinoid treatment may be delayed for some time after initiation and still exert a suppressive effect (McCormick *et al.*, 1980). Thus, McCormick and Moon (1982a) showed that retinoid administration could be delayed for various periods up to 16 weeks after carcinogen administration with a significant retention of preventive activity. Delaying the administration of the retinoid for 20 weeks, however, resulted in a carcinoma multiplicity that approached that of control animals.

Using the MNU mammary tumor system, Moon *et al.* (1983a) have evaluated a number of synthetic retinoids for prevention of mammary cancer (Table II). All the active compounds listed in Table II are highly effective in reducing the incidence and number of mammary cancers. 4-Hydroxyphenylretinamide (4-HPR) (E16) is the most efficacious synthetic retinoid in mammary cancer prevention when both chemopreventive activity and toxicity are taken into consideration.

Several studies have demonstrated synergistic effects between retinoids and other modifiers of mammary carcinogenesis, with combined treatment being significantly more effective in cancer prevention than treatment with either agent alone. Chief among these is the interaction between hormonal manipulation and retinoids. Thus, both retinyl acetate and ovariectomy inhibit mammary cancer induction by MNU; however, the combination of ovariectomy plus retinyl acetate is significantly more effective in cancer prevention than is either treatment regimen alone (Moon *et al.*, 1983a,b). Similar results have been obtained with 4-HPR and ovariectomy; not only is the latency of tumor appearance greatly lengthened, but cancer incidence and tumor multiplicity are dramatically reduced in animals treated with both 4-HPR and ovariectomy (McCormick *et al.*, 1982b). A similar combination effect has been demonstrated by Welsch *et al.* (1980), who found a synergistic inhibition of MNU-induced mammary carcinogenesis by concomitant administration of retinyl acetate and 2-bromo-α-ergocryptine (CB-154), an inhibitor of pituitary prolactin secretion. Blood prolactin levels of the rats treated with retinyl acetate did not differ from that of control animals, indicating that the enhanced combination effect was not due to a further suppression of prolactin secretion. Subsequent studies using combination regimens of either retinyl acetate (DeHoog and Welsch, 1982) or 4-HPR (McCormick and Moon, 1982b) and the antiestrogen tamoxifen also resulted in a much greater chemopreventive effect than that resulting from treatment with either compound alone. The role of hormones in the genesis and modulation of experimental mammary tumorigenesis is well established (Moon, 1981), and it is now clear from the evidence cited above that the retinoids also effectively modulate mammary tumorigenesis. The experimental data suggest the existence of populations of preneoplastic or neoplastic cells with differential sensitivity to the retinoids and hormones. Thus, the retinoids might preferentially affect the development of hormone-independent tumors or those that do not regress following either ovariectomy or treatment with agents that suppress hormonal secretion and/or function.

Combination chemoprevention has also been demonstrated with retinoids and other agents that inhibit development of mammary cancer. Thompson *et al.* (1981b) were the first to show an enhanced inhibition of MNU-induced rat mammary carcinogenesis with retinyl acetate and selenium. The effect was confirmed by Ip and Ip (1981) using the DMBA-induced mammary tumor model. Although both groups of workers found that the combined effect of retinyl acetate and selenium was substantially greater than that of either treatment alone, both studies were complicated by the significant reduction in food intake and body weight gain in animals receiving these chemopreventive agents. Attempts to use combined modalities for prevention of mammary cancer are not always successful. For example, 4-HPR and MVE-2 (a maleic anhydride–divinyl ether copolymer), an immunostimulatory agent, are both effective inhibitors of mammary carcinogenesis induced in rats by MNU. However, combined administration of 4-HPR and MVE-2 was no more effective in cancer inhibition than was either agent alone (McCormick *et al.*, 1982a).

The evidence supporting the chemopreventive activity of retinoids in mammary tumor models in rats is substantial. However, only three studies describing the use of retinoids in mammary tumorigenesis in mice have been reported. In the initial study, Maiorana and Gullino (1980) found that retinyl acetate did not influence (neither inhibited nor enhanced) tumor incidence, latency, or tumor number in C_3H-A^{vy} female mice positive for the mammary tumor virus (MTV). Moon *et al.* (1983a) using C_3H mice negative for the MTV found that the number of hyperplastic alveolar nodules, (putative precancerous lesions) developing in animals receiving dietary 4-HPR was significantly less than that of control mice receiving the placebo diet, while retinyl acetate did not affect nodulgoenesis in these experiments. Welsch *et al.* (1981) have reported an enhancement of tumor development in nulliparous and multiparous mice of the GR strain fed a diet supplemented with retinyl acetate. GR mice are unique in that the tumors of these animals contain mammary tumor virus that is transmitted not only by the milk but also by the female and male gametes. They also differ from most strains of mice in that development and growth of these MTV-positive mammary tumors requires intense stimulation by mammotropic hormones, either through pregnancy or the exogeneous administration of estrogen and progesterone (Sluyser and Van Nie, 1974; Welsch *et al.*, 1979). Although the mechanism by which retinyl acetate apparently enhanced tumor development is unknown, the increase in mammary development and uterine weight in GR mice receiving the retinoid would suggest that retinyl acetate may facilitate the responsiveness of these tissues to hormone action.

The overall mechanism by which retinoids inhibit mammary carcinogenesis is presently unclear, although some insight into the process has been gained by the study of direct effects of retinoids on the mammary gland. Retinyl acetate and 4-HPR both exert an antiproliferative effect on the mammary epithelium. This is shown by the significant inhibition of ductal branching and end bud proliferation (Fig. 1A,B, and C) of the glands of rats fed the retinoids chronically in the diet for 6 months (Moon *et al.*, 1979). Retinyl acetate has also been shown to inhibit the increase in mammary gland DNA synthesis induced by chemical carcinogens (Mehta and Moon, 1980), as well as inhibiting the induction by carcinogens of terminal ductal hyperplasia, a putative precancerous lesion (McCormick *et al.*, 1981). The addition of 4-HPR or retinoic acid to organ culture of mouse mammary glands inhibits the prolactin-induced increase in DNA synthesis in such glands (Mehta *et al.*, 1983), which is reflected in a decreased glandular differentiation. Furthermore, *in vivo* studies (Moon *et al.*, 1983a) in C_3H mice also suggest an antiproliferative effect for 4-HPR since hyperplastic alveolar nodulogenesis is reduced in animals maintained on a diet supplemented with the retinoid. These effects on the mammary gland are probably not mediated via an influence on host hormonal levels since retinoid administration has little effect on either circulating prolactin levels (Welsch *et al.*, 1980) or normal ovarian function (Moon *et al.*, 1976, 1979). Moreover, the synergistic effects of retinoids and hormonal manipulation in the combination studies cited above would also appear to substantiate this view.

D. Chemoprevention of Urinary Bladder Cancer

The induction of urinary bladder cancer in rodents by chemical carcinogens is a multistage process involving initiation, promotion, and propagation (Hicks,

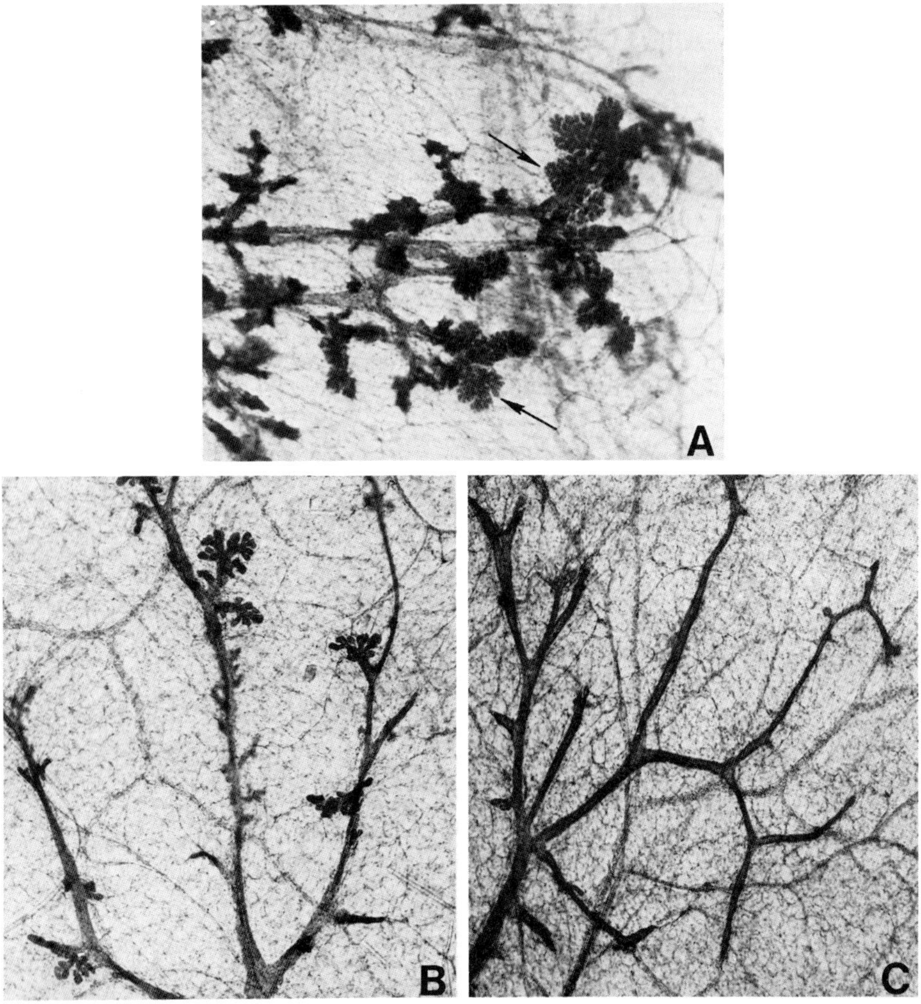

Fig. 1. Representative areas of mammary gland whole mounts from virgin female rats fed special diets from age 2 months to 8 months. (A) Control diet. Note the marked end-bud proliferation (arrows). Alum carmin, ×30. (B) Retinyl acetate (328 mg/kg of diet). A marked decrease in end-bud proliferation is noted. Compare to A. Alum carmine, ×30. (C) 4-Hydroxyphenylretinamide (782 mg/kg of diet). An almost total absence of end-bud proliferation is noted. Compare to A. Alum carmine, ×30. [Reprinted with permission from Moon *et al.* (1979).]

1980). As in other epithelial tissues, development of neoplasia of the urothelium is sensitive to modulation by retinoids. Early studies showed that vitamin A deficiency in rats resulted in squamous metaplasia of the normal transitional epithelium of the urinary bladder (Hedenberg, 1954; Hicks, 1968). Cohen *et al.* (1976) found that rats fed a vitamin A-deficient diet containing the carcinogen *N*-[4-(5-nitro-2-furyl)-2-thiazolyl]formamide (FANFT) were more highly susceptible to bladder carcinogenesis than animals receiving FANFT and the same diet supplemented with physiological levels of retinyl palmitate. Carcinogen-treated animals fed the deficient diet had a higher incidence of lesions that invaded the submucosal tissues and also had tumors of the ureters and renal pelves several months before tumors appeared in animals receiving either laboratory chow or the deficient diet supplemented with retinyl palmitate. Although it is evident from these experiments that retinoid deficiency increases the susceptibility of the bladder urothelium to chemical carcinogens, further supplementation of adequate diets with high levels of retinyl palmitate did not afford any protection against urinary bladder carcinogenesis induced with FANFT (Cohen *et al.*, 1976).

Sporn *et al.* (1977) were the first to demonstrate the efficacy of retinoids in the inhibition of bladder carcinogenesis in experimental animals. They found that 13-*cis*-retinoic acid not only inhibited the incidence but also reduced the severity of bladder neoplasms induced by the intravesical administration of MNU to female Wistar–Lewis rats. A detailed histological evaluation of this study (Squire *et al.*, 1977) indicated that all measurable histological parameters used in the evaluation of the retinoid effect demonstrated the efficacy of 13-*cis*-retinoic acid in inhibiting bladder carcinogenesis. In this study, intravesical administration of MNU resulted in a high percentage of the malignancies that were classified as squamous cell carcinomas. Since the majority of bladder cancers that develop in humans are of the transitional rather than squamous type, several studies were subsequently conducted by Moon and his colleagues (Grubbs *et al.*, 1977b; Becci *et al.*, 1978, 1979, 1981; Thompson *et al.*, 1981a; Moon *et al.*, 1982; McCormick *et al.*, 1982a) to determine the efficacy of several retinoids in an experimental system in which the tumors that develop are almost entirely transitional cell carcinomas. Thus, Grubbs *et al.* (1977b) determined the effect of 13-*cis*-retinoic acid on the induction of bladder cancer in male rats by the intragastric instillation of the bladder-specific carcinogen *N*-butyl-*N*-(4-hydroxybutyl)nitrosamine (OH-BBN). It was found that 13-*cis*-retinoic acid not only inhibited the incidence of transitional cell carcinoma but also reduced the severity as well as the number of other proliferative lesions resulting from the administration of this carcinogen. Similar findings were reported by Miyata *et al.* (1978) in rats receiving OH-BBN in drinking water and treated with dietary supplements of retinyl acetate.

In the study of Grubbs *et al.*, as in previous studies, treatment with 13-*cis*-

retinoic acid was started 1 week after the termination of the carcinogen dosing. Since urinary bladder carcinogenesis is a multistage process and since in man clinically observable bladder cancer is preceded for many years by foci of dysplasia and preneoplastic changes in the urothelium, Becci *et al.* (1979) conducted a study in rats to determine whether the administration of a retinoid could be significantly delayed following carcinogen treatment without diminishing its ability to inhibit bladder carcinogenesis. It was found that the administration of 13-*cis*-retinoic acid may be delayed up to 9 weeks following the last carcinogen exposure without loss of ability to inhibit bladder carcinogenesis. The average number and severity of transitional cell carcinomas was significantly reduced by 13-*cis*-retinoic acid when compared to that of the control animals, irrespective of the carcinogen dose. Not only did the retinoid treatment reduce the incidence, number, and severity of carcinoma, but it also reduced the average atypia score of all urothelial lesions as well as the number of animals displaying severe atypia, resulting in a concomitant increase in the number of animals showing only minimal atypia.

Following these reports utilizing the natural retinoids and 13-*cis*-retinoic acid, several other analogs have subsequently been shown to have anticarcinogenic activity in the rodent urinary bladder. Particularly notable among these compounds are the retinamides, a group of synthetic retinoids in which the terminal carboxyl group of retinoic acid is replaced by N-substituted carboxyamide groups. Compared to natural retinoids, retinamides have the desirable qualities of reduced toxicity (Hixson and Denine, 1979) while retaining much or all of the *in vitro* biological activity of the natural compounds (Newton *et al.*, 1980). In an initial study with retinamides (Thompson *et al.*, 1981a) the efficacy of *N*-ethylretinamide (E11) and *N*-(2-hydroxyethyl)retinamide (E12) against the induction of both low-grade papillary transitional cell carcinomas of the urinary bladder in rats and highly invasive urinary bladder carcinomas in mice was determined. The 2-hydroxyethylretinamide was highly effective in reducing the incidence, number,and severity of transitional cell carcinomas that developed in both male and female rats exposed to OH-BBN. Ethylretinamide had no significant effect on either the number or incidence of transitional cell carcinoma in male rats, although the severity of the lesions that developed was less in the retinoid-treated group than in those that received only the placebo diet. However, in female rats ethylretinamide was highly effective in inhibiting urinary bladder carcinogenesis and had chemopreventive activity similar to that of 2-hydroxyethylretinamide irrespective of the parameter measured. In OH-BBN-treated mice (C57BL/6 $\times$ DBA/2F_1), both ethyl- and 2-hydroxyethylretinamide significantly inhibited the induction of highly invasive transitional cell carcinomas at a dose of 1 mmol/kg of diet, although reduction of carcinoma incidence was not significant in animals receiving a low dose (0.5 mmol/kg of diet) of these two retinamides.

The first studies on chemoprevention of bladder carcinogenesis induced by OH-BBN used relatively large doses of carcinogen and limited treatment (4–6 months) with the retinoids. Subsequently, in a long-term study by Hicks *et al.* (1982), female rats were treated with lower dosages of carcinogen (300 mg, 600 mg, and 1200 mg OH-BBN), as contrasted to the studies of Thompson *et al.* (1981a) in which a total of 1800 mg of the carcinogen was used. Sequential time–kill studies by Hicks *et al.* indicated a rapid proliferative response in the urothelium during the period of repeated intragastric instillations of the carcinogen. On completion of the carcinogen treatment, these hyperplastic lesions remained static for a period of months, then increased in size and complexity and produced macroscopic, rapidly growing tumors. The duration of the latent period before the ensuing rapid tumor growth began was inversely related to the dose of OH-BBN employed, being on the order of 2–3 months for the 1200-mg, 4–6 months for the 600-mg, and 12–14 months for the 300-mg dosages of the carcinogen. Animals treated with ethylretinamide and examined 1 year after dosing with OH-BBN did not have any significant difference in bladder cancer incidence when compared to control animals. However, mean tumor volume was significantly reduced in the animals receiving the retinoid both at the 600-mg and 1200-mg dosages of OH-BBN. Although it was not possible to determine the growth rate of individual tumors, examination of bladders of animals killed sequentially indicated that tumor growth rate per se was probably not affected by the retinoid. Thus, the decrease in tumor volume found in animals treated with ethylretinamide most likely resulted from an inhibition of progression; the initiated urothelia remained quiescent for a longer period (increase in latent period) before entering into a period of rapid growth.

Moon *et al.* (1982) have evaluated the efficacy of several alkylretinamides as inhibitors of chemical carcinogenesis of the urinary bladder in C57BL/6 × DBA/2F_1 mice treated with OH-BBN. Various hydroxy-substituted propylretinamides were tested for anticarcinogenic activity, and it was shown that 2-hydroxypropylretinamide was just as active as 2-hydroxyethylretinamide in the inhibition of bladder carcinogenesis at the dosage levels administered in this study. However, moving the hydroxyl group from the 2 position to the 3 position to form 3-hydroxypropylretinamide resulted in a complete loss of cancer inhibitory activity. Similarly, 2,3-dihydroxypropylretinamide was not active in inhibiting bladder carcinogenesis.

In addition to the propylretinamides, the chemopreventive activity of six other all-*trans*-retinamides was compared with their respective 13-cis isomers (Moon *et al.*, 1982). A comparison of the all-trans and 13-cis isomers of ethyl- and 4-hydroxyphenylretinamide, respectively, indicated that 13-cis isomers were as effective as all-trans isomers in inhibiting the formation of bladder carcinomas. The same was found to be true with the all-trans and 13-cis isomers of 2-hydroxyethylretinamide, whereas only the all-trans isomer of 5-tetrazolyretinamide was active; 13-*cis*-5-tetrazolylretinamide did not reduce the development of transitional cell carcinomas in this mouse model system. Neither all-trans nor 13-cis isomers of butyl- or 4-hydroxybutylretinamide were effective in inhibiting OH-BBN-induced bladder carcinogenesis. At present, no pattern of structure–activity relationships is apparent for the ability of retinoids to prevent experimental bladder cancer, in spite of the fact that all of the retinoids that have been tested for prevention of bladder cancer possess significant *in vitro* activity in controlling epithelial cell differentiation in the hamster trachea organ culture system (Newton *et al.*, 1980). Thus, in addition to the intrinsic biological activity of retinoids, other factors such as their bioavailability or their metabolism *in vivo* play a key role in determining cancer prevention in the whole animal.

In addition to the retinamides, some of which have proved highly effective in inhibiting bladder carcinogenesis when administered after completion of carcinogen dosing, etretinate (F1) was apparently more effective in inhibiting OH-BBN-induced bladder neoplasms in rats if the anmals were treated with the retinoid prior to rather than during or after the administration of the carcinogen (Murasaki *et al.*, 1980). In line with these results, Schmahl and Habs (1978) indicated that etretinate had little effect on OH-BBN-induced bladder carcinogenesis when it was given three times a week by stomach tube and the carcinogen was administered chronically in the drinking water until death of the animal. Additional studies (Miyata *et al.*, 1980) indicate that etretinate is also effective in preventing DNA damage in bladder epithelial cells by the 3-carboxypropyl metabolite of OH-BBN. These studies not only indicate that the retinoids effectively inhibited progression of urinary bladder tumor growth but also may be capable of inhibiting initiation.

TABLE III

Retinoids Evaluated for Chemopreventive Activity against Urinary Bladder Cancer in Rats and/or Mice[a]

Active	Inactive
Retinoic acid (D1)	Retinyl palmitate (A4)
13-*cis*-Retinoic acid (E6)	*N*-(4-Carboxyphenyl)retinamide
Retinyl acetate (A3)	*N*-(4-Carboxypropyl)retinamide
N-Ethylretinamide (E11)	*N*-(3-Hydroxypropyl)retinamide
N-Ethyl-13-*cis*-retinamide (E19)	*N*-(2,3-Dihydroxypropyl)retinamide
N-(2-Hydroxyethyl)retinamide (E12)	*N*-(*n*-Butyl)retinamide (E13)
N-(2-Hydroxyethyl)-13-*cis*-retinamide (E20)	*N*-(*n*-Butyl)-13-*cis*-retinamide
N-(4-Hydroxyphenyl)retinamide (E16)	*N*-(4-Hydroxybutyl)retinamide
N-(4-Hydroxyphenyl)-13-*cis*-retinamide (E21)	*N*-(4-Hydroxybutyl)-13-*cis*-retinamide
N-(2-Hydroxypropyl)retinamide	*N*-(5-Tetrazolyl)-13-*cis*-retinamide
N-(5-Tetrazolyl)retinamide (E17)	Motretinid (F3)
Etretinate (F1)	

[a] For details, see Moon *et al.* (1983a).

In conclusion, it is clear that several retinoids effectively inhibit the development of carcinogen-induced bladder cancer in rats and mice. Table III provides a summary of these results. Furthermore, it is also apparent that several of the synthetic retinoids, particularly the retinamides, are less toxic and more effective than the natural retinoids for this purpose. Although some retinoids may be effective in preventing initiation of bladder carcinogenesis, the majority appear to act during the progression phase of the carcinogenic response. If these compounds are to be of use as chemopreventive agents for bladder cancer in man, then the ability of the retinoids to delay tumor growth in urothelium already neoplastically transformed is of critical importance. As noted by Hicks *et al.* (1982), "even a delay in tumor development of as little as 7 to 10 weeks in the rat could equate to an extra 5 or 6 years of symptom-free life for the human bladder cancer patient."

E. Chemoprevention of Cancer of the Digestive Tract and Associated Organs

The use of retinoids for the inhibition of chemical carcinogenesis of the epithelia of the digestive tract has been somewhat disappointing compared to the positive results obtained with the skin, mammary gland, and urinary bladder. For the most part, the retinoids tested, both natural and synthetic, have had little effect in modulating the carcinogenic response of these epithelia to chemical carcinogens. However, isolated experiments have suggested that, in some instances, retinoids may either inhibit or enhance tumorigenesis in a particular epithelium.

1. *Forestomach and Esophagus*

In the initial retinoid chemoprevention study of Chu and Malmgren (1965), papillomas and carcinomas were induced in the forestomach of Syrian hamsters by the polycyclic hydrocarbons 7,12-dimethylbenz[*a*]anthracene and benzo[*a*]pyrene. These workers observed that the oral administration of 100 mg retinyl palmitate per week completely prevented the occurrence of carcinoma in animals receiving carcinogens, although a few papillomas and dyskeratotic lesions were found in a few animals receiving the retinyl palmitate. A similar reduction in the incidence of forestomach papillomas was also noted by Saffiotti *et al.* (1967) and Smith *et al.* (1975a,b) in hamsters treated with benzo[*a*]pyrene–ferric oxide and retinyl esters. In addition to the effect on the forestomach, retinyl palmitate also significantly inhibited the appearance of papillomas and dyskeratotic lesions of the esophagus of animals receiving 7,12-dimethylbenz[*a*]anthracene.

The experiments of Iizuka *et al.* (1979) have also demonstrated that esophageal cancer induced in rats by *N*-methyl-*N*-amylnitrosamine could be markedly inhibited by the concomitant administration of retinyl palmitate. A subsequent study by Schmahl and Habs (1981), although not definitive, has indicated that etretinate may afford some protection against the induction of esophageal tumors by the carcinogen *N*-methyl-*N*-benzylnitrosamine. Although an increase in the latent period of induction and a decrease in the incidence of esophageal tumors was observed in animals treated with etretinate, the interpretation of these data was complicated by the marked toxicity induced by the doses of retinoid utilized in the study. Moon and McCormick (1982) subsequently found that administration of 13-*cis*-retinoic acid to rats treated with *N*-methyl-*N*-benzylnitrosamine exerted a protective effect on mortality attributable to esophageal tumors. Although the above studies do suggest a protective role for retinoids in esophageal carcinogenesis, a definitive answer will not be forthcoming until additional retinoids are evaluated more extensively for esophageal cancer.

2. *Colon*

Several studies have been conducted to determine the effect of retinoids on chemical carcinogenesis of the colon. Although several carcinogens and retinoids have been used in these colon carcinogenesis studies, the results indicate that, for the most part, retinoids have been ineffectual in inhibiting chemically induced colon tumorigenesis. In a series of studies by Newberne and his colleagues (Newberne and Rogers, 1973; Rogers *et al.*, 1973; Newberne and Suphakarn, 1977), it was found that rats deficient in vitamin A and exposed to either of the two carcinogens, aflatoxin B_1 or 1,2-dimethylhydrazine, develop a higher incidence of colon cancer than do animals receiving the vitamin-deficient diet supplemented with retinyl esters. On the other hand, Narisawa *et al.* (1976) found that vitamin A deficiency in rats significantly suppressed the development of large bowel tumors caused by rectally administered *N*-methyl-*N'*-nitro-*N*-nitrosoguanidine, However, this vitamin A-deficient suppression of colon tumorigenesis was only observed with high doses of carcinogen; with lower doses of carcinogen no effect was noted. Although supplementation of vitamin A-deficient diets with retinyl esters had little effect in inhibiting colon carcinogenesis, the study of Newberne and Suphakarn (1977) did show some efficacy of 13-*cis*-retinoic acid in the prevention of colon tumorigenesis induced with aflatoxin B_1. As with the retinyl esters, Schmahl *et al.* (1976) found that all-*trans*-retinoic acid had no effect on suprressing the induction of colon tumors induced with 1,2-dimethylhydrazine.

In a subsequent study by Ward *et al.* (1978) in which colon carcinomas were induced by the intrarectal administration of *N*-methyl-*N*-nitrosourea, 13-*cis*-retinoic acid (240 mg/kg of diet) had no effect on either the incidence, number, or location of the tumors that developed. Motretinid (F3) was also ineffective in inhibiting colon carcinogenesis in this tumor model. Additional studies by Wenk *et al.* (1981) also indicated that the synthetic retinoids *N*-ethylretinamide (E11),

N-(2-hydroxyethyl)retinamide (E12), and retinylidene dimedone (C6) were ineffective in altering the tumor response to MNU. Similar findings have been reported by Silverman *et al.* (1981) with ethylretinamide and 2-hydroxyethylretinamide. Silverman *et al.* also showed that retinyl acetate and 4-hydroxyphenylretinamide were also ineffective against MNU-induced colon carcinogenesis. Using these same retinoids in the 1,2-dimethylhydrazine-induced colon tumor model, Silverman *et al.* found that neither the retinamides nor retinyl acetate had any effect on induction of adenocarcinomas of the colon, although both retinyl acetate and 4-hydroxyphenylretinamide reduced the number of colon adenomas that developed.

3. *Liver*

A few studies of the effects of retinoids on liver carcinogenesis have been reported. Newberne and Rogers (1973) and Newberne and Suphakarn (1977) indicated that neither retinyl palmitate nor retinyl acetate affected the development of liver tumors in rats treated with aflatoxin B_1. On the other hand, Daoud and Griffin (1980) found that 90% of rats receiving laboratory chow containing the carcinogen 3′-methyl-4-dimethylaminoazobenzene developed liver tumors, whereas only 11% of the animals receiving the same diet supplemented with 13-*cis*-retinoic acid had liver tumors at the termination of the experiment. Similar findings were reported by Maiorana and Gullino (1980) for hepatomas arising spontaneously in C_3H-A^{vy} female mice given various doses of retinyl acetate; a dose–response reduction of the percent of incidence of hepatomas was evident in those animals receiving the retinyl acetate.

4. *Pancreas*

Longnecker *et al.* (1981) were the first to report an inhibition of pancreatic carcinogenesis with retinoids. They found that male rats treated with azaserine and fed a purified diet (AIN-76) supplemented with 0.025% retinyl acetate during the progression phase of carcinogenesis developed fewer pancreatic neoplasms (adenomas and carcinomas) than did rats receiving only the purified diet containing the carcinogen. Furthermore, when the cross-sectional area of atypical acinar cell nodules developing in the carcinogen-treated animals was determined, the size of these lesions was significantly smaller in the rats receiving retinyl acetate. Subsequent studies by Longnecker *et al.* (1982) showed a positive preventive effect for three of four synthetic retinoids tested; 2-hydroxyethylretinamide, 4-propionyloxyphenylretinamide and retinylidene dimedone all significantly reduced the incidence of pancreatic carcinoma. However, in a study by Birt *et al.* (1981) feeding of several retinoids (13-*cis*-retinoic acid, ethylretinamide, 2-hydroxyethylretinamide, and 4-hydroxyphenylretinamide) did not significantly influence the development of pancreatic cancer induced in hamsters by a single subcutaneous injection of *N*-nitrosobis-(2-oxopropyl)amine. It is evident that the response to retinoids in these two model systems for pancreatic carcinogenesis differs considerably. Whether this is due to differences in the pathogenesis of the tumors induced by the two different carcinogens or to species differences in metabolizing the retinoids is unknown at this time. However, the studies do warrant further investigation with other retinoids.

5. *Oral Epithelium*

Much of the controversy about enhancement of carcinogenesis by retinoids comes from the early studies of the effect of retinyl palmitate on DMBA carcinogenesis in the hamster cheek pouch. In the studies of Levij and Polliack (1968), cheek pouches of hamsters were painted three times weekly with a combination of 0.5% DMBA and 10% retinyl palmitate dissolved in liquid paraffin for up to 12 weeks. Although these workers reported a potentiating effect of retinyl palmitate on DMBA-induced carcinogenesis of the hamster cheek pouch, their data actually show little effect. The incidence and number of carcinomas that developed in the cheek pouch of hamsters treated with the carcinogen alone or with the carcinogen and retinyl palmitate were the same, irrespective of the time period at which the animals were sacrificed. The only indication of "potentiation" of DMBA-induced carcinogenesis by retinyl palmitate was the somewhat larger size of the tumors at the end of the experiment in the animals treated with the retinoid. Of critical importance is the marked histological changes that occurred in the cheek pouches of hamsters receiving repeated applications of the 10% retinyl palmitate alone. These changes are the result of a toxic, irritating effect of retinyl palmitate on the cheek pouch epithelium and account for the potentiating effect of retinyl plamitate on cheek pouch carcinogenesis. Subsequent experiments by these workers resulted in similar findings (Polliack and Levij, 1969); the toxic effect of large doses of retinyl palmitate on the hamster cheek pouch epithelium was confirmed (Polliack and Levij, 1968; Levij and Polliack, 1969). Studies by White and Al-Azzawi (1981) on the cheek pouch strongly suggest that nontoxic topical application of retinyl palmitate after the appearance of the first neoplasms induced with DMBA actually result in the regression of these lesions.

Other studies by Gilmore and Giunta (1981) in which toxic levels of 13-*cis*-retinoic acid (300 mg/kg of diet) were fed to hamsters receiving cheek pouch applications of DMBA indicated that the retinoid did not affect either induction or progression of DMBA-induced carcinomas, even though retinoid toxicity resulted in prominent metaplastic changes in the cheek pouch epithelium. On the other hand, Shklar *et al.* (1980b) found that when 13-*cis*-retinoic acid was administered in nontoxic doses (10 mg p.o. twice a week) to hamsters receiving buccal pouch applications of DMBA, a significant delay in DMBA-induced carcinogenesis was observed as indicated by the number, size, and histological characteristics of the tumors developing during various periods throughout the experiment. Similar results have been obtained by Shklar *et al.* (1980a) with 13-*cis*-retinoic acid in the inhibition of DMBA-induced carcinoma of the tongue of hamsters.

It is readily apparent from the experiments cited above that enhancement of buccal carcinogenesis by retinyl palmitate is the result of toxicity caused by excessive retinoid dosage. It is also apparent that when nontoxic doses of retinyl palmitate are used, regressive changes occur within the lesions induced by topical applications of carcinogen. The use of 13-*cis*-retinoic acid in experimental models for oral carcinogenesis would appear to warrant additional studies in this area using other synthetic retinoids that are now available.

F. Chemotherapeutic Effects of Retinoids

Many investigators have attempted to inhibit the growth of transplantable tumors or tumor cells *in vivo,* with both the natural and synthetic retinoids. Most of these attempts have been disappointing, as documented in detail by Bollag (1971a) and in a review by Hill and Grubbs (1982). A large number of retinoids have been screened against many of the commonly used transplantable tumors, and little, if any, data pointing to therapeutic usefulness have been obtained. One notable exception is the striking effects of retinoids on a transplantable rat chondrosarcoma. Heilman and Swarm (1975) originally reported that 13-*cis*-retinoic acid significantly inhibited the growth of this tumor. Subsequent studies by Trown *et al.* (1976) showed that a dose-dependent inhibition of the growth of this chondrosarcoma could be achieved with etretinate, motretinid, and their carboxylic acid analog (F2), given by oral or intraperitoneal dosing. These regressions of tumor growth occurred with doses of retinoid that did not appreciably affect body weight.

The use of retinoids in combination with cytotoxic cancer chemotherapeutic drugs to inhibit tumor growth has also been unrewarding. Most of the tests have involved the use of retinoids in combination with several chemotherapeutic agents against L1210 leukemia, a tumor system in which retinoids per se show little efficacy (Hill and Grubbs, 1982). The combination chemotherapeutic studies with retinoids have failed to reveal any consistent pattern. No one has shown a particularly effective retinoid, tumor test system, or other chemotherapeutic agent to be used in combination with retinoids. Furthermore, the retinoids selected for combination studies in a specific tumor system were, in all cases, those that were ineffective against these particular tumors when the retinoid was administered alone. It should also be noted that combination chemotherapeutic studies have not been conducted using synthetic retinoids that have been shown to possess significant anticancer activity.

III. RETINOIDS AND HUMAN CANCER

The topic of retinoids and human cancer is considered in the following areas:

A. *Epidemiology:* This section reviews studies that relate dietary intake of vitamin A or serum concentration of retinol to the incidence of cancer. These data suggest an inverse relation between cancer risk and low intake of vitamin A or low serum retinol.

B. *Clinical Cancer Studies:* This section examines the clinical use of retinoids for the treatment of established cancer. While results of these studies are quite preliminary, initial data suggest definite antitumor activity in certain disorders.

C. *Prevention Trials:* The third section reviews the use of retinoids as cancer preventives in high-risk populations.

D. *Future Directions:* Finally, other possible uses of retinoids in clinical oncology are examined.

A. Epidemiologic Evidence for the Relationship between Retinoids and Human Cancer

Epidemiologic evidence indicates that an inverse relationship exists between cancer risk and vitamin A consumption. In most epidemiologic investigations, estimates of vitamin A intake have been based on the frequency of ingestion of foods known to have a high content of β-carotene (e.g., green and yellow vegetables) or a few foodstuffs that contain preformed retinol (e.g., liver or whole milk).

β-Carotene (provitamin A) is one of several naturally occurring precursors of vitamin A. Owing to inefficient intestinal conversion, only one-sixth of the dietary intake of β-carotene is ultimately converted to retinol (Goodman, 1979). Additionally, as the amount of β-carotene in the diet increases, the conversion to retinol decreases so that systemic vitamin A toxicity is not caused even by massive doses. A variable amount of β-carotene is also absorbed unchanged from the intestine, enters into circulation, and is stored in adipose tissue.

β-Carotene is the primary source of retinol in most underdeveloped countries and is an important source of vitamin A in many of the following reports. It is therefore difficult to discern whether the observed anticancer effects are due to retinol, β-carotene itself, or some other dietary constituent. Nevertheless, increased consumption of foods rich in vitamin A content appears to be inversely related to the incidence of several types of carcinoma. In the following discussion, the term "vitamin A" will be used to include both retinoids and carotenoids since most epidermiological reports unfortunately do not distinguish between these two classes of substances.

1. Serum Retinol Levels and Cancer Incidence

Several studies have analyzed retinol levels in stored samples of blood taken some years previously for unrelated research. All participants were apparently free of cancer at the time the samples were taken. Kark *et al.* (1981) followed 3102 individuals from Evans County, Georgia, for periods up to 14 years. During this time, 129 new cases of cancer were diagnosed. Retinol levels were measured on stored sera from 85 cancer patients and from 174 controls matched for age, race, and sex. Compared to controls, persons who eventually developed cancer had significantly lower serum retinol levels at least 12 months before the diagnosis of cancer. The association was noted in all four race–sex groups and was stronger for males than females.

Wald *et al.* (1980) prospectively followed 16,000 men who had samples of serum collected and frozen. Retinol levels were analyzed in stored samples from 86 men who subsequently developed cancer and in samples from 172 controls. Low retinol levels were again associated with an increased risk for malignant disease. This association was independent of age, smoking habits, and serum cholesterol levels, and it was highest in men who developed lung cancer. The risk of cancer at any site for men with retinol levels in the lowest quintile was 2.2 times greater than the risk for men with levels in the highest quintile. Although the apparent protective effects are statistically significant, so few cancers were observed in these studies that there is substantial statistical uncertainty as to the magnitude of the effect.

Peto *et al.* (1981) examined the role of β-carotene in cancer prevention. Included in this extensive review were eight retrospective studies from several countries in which serum retinol levels of recently diagnosed cancer patients were significantly lower than in controls. These countries included India (Wahi *et al.*, 1962), East Africa (Clifford, 1972), Pakistan (Ibrahim *et al.*, 1977), Britain (Basu *et al.*, 1976; Atukorala *et al.*, 1979; Capel and Williams, 1979), and the United States (Abels *et al.*, 1941; Cohen *et al.*, 1977).

Despite these data, evidence for the inverse association between cancer and serum levels of vitamin A or β-carotene is still incomplete. In some cases it is not certain that the control populations were appropriately selected. Moreover, depression of β-carotene or retinol might have occurred secondary to already established cancers. Further studies are needed in individuals with early cancers whose disease has not already had a major impact upon their diet or general health.

2. Retinoid Intake and Cancer Incidence by Site

As noted in the previous section, there is an inverse relation between vitamin A intake and overall incidence of malignant disease. This relation has been further developed with respect to individual histologic types of cancer. The epidemiologic studies in the following section have been categorized by site, of which the most widely studied is lung. Studies of other sites are fewer in number and, with certain exceptions, generally comprise smaller numbers of subjects. The reader is referred to the original publications for the precise methodologic and statistical considerations of individual studies.

a. CARCINOMA OF THE LUNG. In 1975, Bjelke first reported epidemiologic data that suggested that vitamin A may protect against lung cancer. Using a mail survey, 8278 Norwegian men were asked to report their smoking and dietary habits. Investigators then calculated a vitamin A index based on reported consumption of vegetables and dairy products. Thirty-six cases of lung cancer were subsequently diagnosed in these men. After controlling for cigarette smoking, a lower incidence of lung cancer was noted in the patient group with high values of the vitamin A index in all age groups. Although this study was prospective, the number of lung cancer deaths reported was small; therefore, the magnitude of the association with vitamin A intake cannot be accurately established. Moreover, variation in the index among subjects mainly reflected differences in the consumption of vegetables and milk. Other sources of vitamin A (e.g., liver, cheese, butter, and pharmaceutical preparations) were not covered in the questionnaire.

In a case-control study among Chinese females in Singapore, MacLennan *et al.* (1977) found an inverse association between the consumption of leafy green vegetables (rich in vitamin A) and lung cancer. This population had a particularly high incidence of lung cancer that was not related to smoking.

Gregor *et al.* (1980) obtained estimates of dietary vitamin A intake from 100 patients with lung cancer and 173 controls. Intake of vitamin A was significantly lower in male cancer patients relative to male controls, but it was higher in female cases than female controls. However, intake from vitamin A supplements was lower in cases than controls for both sexes.

Smith and Jick (1978) compared use of vitamin A pharmacologic preparations among 800 newly diagnosed cancer patients and 3433 patients with nonmalignant conditions. Overall, there was no convincing evidence that regular consumption of these preparations protected against the development of cancer. However, a suggestion of negative association between the development of cancer and the frequent use of vitamin A preparations was noted in males.

A retrospective study of the dietary and smoking habits of 292 male patients with lung cancer and 801 control patients with nonrespiratory nonneoplastic diseases was conducted by Mettlin *et al.* (1979). An index of vitamin A intake was again computed. It was noted that lung cancer patients had lower values of vitamin A intake than controls. The reduced relative risk of lung cancer associated with vitamin A intake was most evident among men who smoked heavily. Frequency of daily milk drinking was lower among patients with lung cancer and a lower relative risk was found among the men who smoked heavily but frequently consumed carrots. In another study of 28 patients with lung cancer, Basu *et al.* (1976) reported that patients with either squamous or oat cell carcinoma had significantly lower plasma vitamin A levels than those with large cell undifferentiated carcinomas.

In a prospective study of 1954 middle-aged men followed for 19 years, the intake of dietary β-carotene was inversely related to the incidence of lung cancer (Shekelle *et al.*, 1981). The incidence of lung cancer was inversely related to carotene intake both before and after adjustment for cigarette smoking. Intake of preformed vitamin A (retinol) and of other nutrients was not significantly related to the risk of lung cancer. Neither carotene nor retinol intake was significantly related to the total risk for other types of carcinoma, although carotene intake was lower in males who developed epithelial carcinoma of the head and neck.

Hirayama (1982) conducted a prospective study of 265,118 adults who were interviewed in 1965 at the time of the Japanese census. During the initial interview, participants were grouped according to smoking habits, consumption of green and yellow vegetables, and socioeconomic status. Persons smoking more than 50 cigarettes per day had a risk of lung cancer ten times higher than nonsmokers. Persons who smoked and consumed green–yellow vegetables on a daily basis had age-adjusted death rates for lung cancer of 37.6, whereas nonsmokers had rates of 68.4. Consumers of green–yellow vegetables showed a lower risk of lung cancer in both smokers and nonsmokers. Additionally, among ex-smokers, there was a greater reduction of cancer risk after cessation of smoking for daily consumers of green–yellow vegetables, suggesting an effect on the recovery process.

b. CARCINOMA OF THE ORAL CAVITY AND OROPHARYNX. Two studies have reported mean plasma levels of vitamin A and β-carotene in patients with carcinoma of the oral cavity and oropharynx compared to a control population (Ibrahim *et al.*, 1977; Chaudhy *et al.*, 1980). Both studies showed decreased levels of β-carotene in the cancer population. Moreover, a progressive decrease in plasma vitamin A was also observed as patients developed progressive disease.

c. CARCINOMA OF THE LARYNX. Three hundred and seventy-four males with carcinoma of the larynx were interviewed and compared with 381 control patients with nonneoplastic diseases (Graham *et al.*, 1981). Men who ingested low amounts of vitamin A had approximately twice the cancer risk relative to men who consumed large amounts, even after the population was controlled for ethanol intake and smoking. A similar relation was found for vitamin C consumption. Both statistical observations were highly significant and a dose–response relationship was noted.

d. CARCINOMA OF THE ESOPHAGUS. In 1961 150 males with squamous cell carcinoma of the esophagus were compared with 150 control patients with cancers of other sites (Wynder and Bross). Consumption of milk and vegetables was found to be lower in patients with esophageal cancer. A recent retrospective review of the dietary habits of 147 patients with cancer of the esophagus was compared to those of 264 men of similar age with noncancer diagnoses (Mettlin *et al.*, 1981). Even after controlling for possible associations with smoking or ethanol consumption, an overall higher frequency of vegetable and fruit consumption was associated with a significantly lower cancer risk. Moreover, a dose–response relation was also observed.

In the Caspian littoral of Iran (a region with particularly high incidence of esophageal cancer), 38 villages were chosen by random sampling and were evaluated for environmental characteristics. An analysis of food intake indicated a higher cancer risk for areas with a low consumption of fruit and green vegetables. Areas of low risk generally had higher intakes of vitamins A and C (Hormozdiari *et al.*, 1975; Joint Iran-International Agency for Research on Cancer Study Group, 1977). A subsequent case-control study of this region has confirmed the earlier findings. Moreover, the risk of esophageal cancer was strongly associated with low socioeconomic status and a low consumption of fresh fruit and vegetables (Cook-Mozaffari, 1979; Cook-Mozaffari *et al.*, 1979).

e. CARCINOMA OF THE STOMACH. An early case-control study relating the role of diet and the incidence of cancer of the stomach was conducted by Hirayama (1967) in Japan. The results indicated an inverse association between daily milk consumption and stomach cancer. A later prospective cohort study by the same author (Hirayama, 1977) of greater than 26,000 Japanese citizens revealed a similar inverse relationship between the incidence of gastric cancer and milk consumption. Additionally, nonsmokers who consumed green and yellow vegetables were also found to be at a lower risk.

Graham *et al.* (1972) studied 228 patients with gastric cancer who were matched with control patients by age, country of birth, background nationality, and absence of neoplastic or gastrointestinal disease. When subjects were analyzed for intake of groups of food, no relationships were noted for fried food preparation or consumption, meat or fish consumption, or alcohol ingestion. However, a substantial difference was associated with the ingestion of various raw vegetables, and control patients were noted to have eaten larger numbers of raw vegetables than the cancer patients.

A similar association of decreased gastric cancer risk and increased consumption of raw vegetables was noted by Haenszel *et al.* (1972). In a study of 220 Japanese stomach cancer patients and 440 hospital patient controls in Hawaii, increased risk was noted for consumers of pickled vegetables and dry or salted fish, while low risk was suggested for consumers of raw vegetables.

f. CARCINOMA OF THE COLON/RECTUM. Paymaster *et al.* (1968), in an epidemiologic study of gastrointestinal neoplasias in India, noted that fewer cases of colon and rectum cancers were found in populations consuming a strict vegetarian diet. A case-control dietary study of 198 patients with colon cancer and two matched control groups was conducted by Modan *et al.* (1975). Comparison of these groups revealed a significantly lower consumption of fiber and certain vegetables among the cancer patients. In a large, ongoing case-control study of over 40,000 persons in Minnesota and Norway, Bjelke (1978) has observed a decreased consumption of several vegetables in colorectal cancer patients. This same group also had a lower index of vitamin A intake that correlated with vegetable consumption.

g. CARCINOMA OF THE BLADDER. Mettlin and Graham (1979) obtained retrospective dietary information from 569 patients with bladder cancer and 1025 age-matched controls. Using a computed index of vitamin A intake, the authors found an increased risk for patients with a lower index. A similar elevated risk was associated with infrequent milk and carrot consumption. In a small study of ten patients with bladder cancer, both serum retinol and β-carotene levels were similarly lower than those found in controls (Mahmoud and Robinson, 1982).

h. CARCINOMA OF THE PROSTATE. Schuman *et al.* (1982) conducted a study of 358 patients with prostatic carcinoma and 715 matched controls. Dietary information was obtained by nurse interviews and dietary questionnaires. In addition to other findings, the authors noted a strong inverse relationship between decreased consumption of certain vegetables and liver among the prostatic cancer cases. A prospective epidemiologic study of 122,000 Japanese men was conducted by Hirayama (1979). In this 10 year follow-up study of men aged 40 years or older, a significantly lower age-standardized death rate for prostate cancer was noted for men who consumed green and yellow vegetables on a daily basis.

i. SUMMARY. Epidemiologic studies have several major advantages: (1) they are based on the direct observation of human populations (which avoids the problems of extrapolating results from animal studies); (2) they can test the degree of environmental exposure (i.e., dose–response correlations); and (3) the results of such studies may permit intervention in the form of elimination or reduction of risk factors before the fundamental mechanisms of the disease are completely understood.

Unfortunately, certain difficulties are inherent in these trials. Because research has relied primarily on observational data (such as recalled dietary information), imprecise measurement is virtually guaranteed. Investigators must frequently depend on data sources with a high potential for error (e.g., variability of cancer reporting in different countries). In studies of diet and cancer, the long latency period between manifestation of illness and risk exposure has serious implications. Either the investigator must assume that recent intake reflects past food-consumption patterns, or that the participant can accurately recall events long past, as in retrospective case-control studies. Alternatively, in prospective cohort studies, the researcher collects current data and must wait many years for the disease to appear. Moreover, epidemiologic studies are very costly because of the large populations and the lengthy follow-up periods required to achieve a statistically significant result. Nevertheless, the results of epidemiologic studies are often assumed to represent a conservative estimate of the true risk for cancer. The epidemiologic data presented, when considered in conjunction with laboratory results, provide compelling evidence for a protective effect of foods rich in β-carotene or vitamin A.

B. Clinical Cancer Studies

Retinol has been used as a treatment for malignant tumors since the 1930s and was considered effective palliative therapy when applied topically or given orally. However, since the majority of cases received other therapy, which included surgery, radiation, or hormones, it is difficult to assess efficacy accurately. Upon review, it appears that objective responses were indeed rare in the early trials. The reader is referred to the publication of Bollag (1970) for an excellent review of the early literature. The use of retinoids for the treatment of premalignant lesions is discussed in the next section of this chapter (Section III,C). The following discussion is limited to studies in which retinoids were administered systemically.

1. Retinol and Tretinoin

Some trials have utilized retinol in combination with chemotherapeutic or immunomodulatory drugs. Mickshe *et al.* (1977) treated a small group of patients with advanced squamous carcinoma of the lung with a combination of radiation and retinol or isotretinoin. Some stimulation of immune parameters was reported, although no increase in response rate or survival duration was noted. Komiyama *et al.* (1978) studied 48 patients with carcinoma of the esophagus or head and neck. Treatment consisted of a combination of 5-fluorouracil, irradiation, and vitamin A (50,000 IU i.m.). The authors reported an improved response rate for patients with esophageal carcinoma and possibly improved survival in patients with laryngeal cancer.

Thatcher *et al.* (1980) combined weekly intramuscular injections of retinol with adriamycin, bleomycin, 5-fluorouracil, and methotrexate for the treatment of 25 patients with advanced squamous cell carcinoma of the head and neck. No effect on response rate or duration by the addition of vitamin A can be determined; however, the authors state that the addition of retinol may have ameliorated drug-induced mucositis.

Israel and Aguilera (1980) treated seven patients with advanced squamous carcinomas with orally administered retinol (150,000–300,000 IU daily) and reported several clinical remissions. Kucera (1980) performed a randomized trial of irradiation with or without oral retinyl palmitate (1.5 million IU/day) in 42 patients with advanced cervical cancer. No significant difference in relapse rate was noted between the groups. However, patients who received vitamin A and radiation developed increased lymphocyte responsiveness to PHA (phytohemagglutinin) stimulation.

An adjuvant study of patients with stage I (Clark's levels III–V) or II melanoma is ongoing. Patients are stratified according to age, sex, location of primary tumor, and stage and are randomized to receive either bacillus Calmette-Guerin (BCG) alone or BCG in combination with vitamin A for 18 months. Early results indicate that both drugs are well tolerated and suggest a more favorable outcome for patients receiving retinol with BCG (Meyskens *et al.*, 1981).

All-*trans*-retinoic acid (tretinoin) has been used in only a small number of cancer trials. Smith *et al.* (1974) conducted a phase I study of retinoic acid in 11 patients with advanced cancer, using doses which ranged from 50 to 1000 mg/day. The most common toxic effects were angular cheilitis and pruritic rash. Two patients with brain metastases developed severe headaches and dizziness. There was no evidence of liver, kidney, or bone marrow toxicity in any patient. No therapeutic effects were seen in any patient with measurable tumor. Some activity has been reported in the treatment of basal cell carcinoma (Bollag and Ott, 1971; Ryssel *et al.*, 1971); however, doses high enough to achieve response were accompanied by unacceptable side effects that included inflammation or desquamation of the skin and mucous membranes, bone tenderness, elevation of serum alkaline phosphatase levels, hypercalcemia, headaches, and hyperirritability ("hypervitaminosis-A syndrome").

2. *Etretinate and Isotretinoin*

a. TOXICITY STUDIES. Most recent trials in advanced cancer patients have utilized either etretinate or isotretinoin because of a more favorable therapeutic index. Isotretinoin and etretinate have a similar spectrum of toxicity. Etretinate may cause more palmoplantar peeling and alopecia, whereas isotretinoin is associated with a greater frequency of facial dermatitis (Windhorst and Peck, 1983). In humans, etretinate has a biological half-life of approximately 3 months, whereas isotretinoin is almost entirely eliminated within 5 days. In the United States, most cancer trials have utilized isotretinoin, whereas European studies have primarily employed etretinate. To date, no convincing evidence exists to suggest superiority of either compound for the prophylaxis or treatment of human cancer. Toxic reactions associated with low doses of retinoids are discussed in Chapter 13. Most clinical studies of retinoids in patients with cancer utilize higher doses in the range of 2–3 mg/kg/day (100 mg/m^2 of isotretinoin or 50 mg/day (total dose) of etretinate. However, most adjuvant studies and all prevention trials employ lower doses.

Gold *et al.* (1983b) performed a phase I trial of 13-*cis*-retinoic acid in cancer patients at doses that ranged from 40 to 120 mg/m. The most common toxic effects were facial hyperkeratosis and cheilosis. These effects were seen in > 90% of patients taking doses ≥ 80 mg/m. Hepatotoxicity was dose limiting at 120 mg/m^2 and was manifested by hyperbilirubinemia and increased serum glutamic-oxaloacetic transaminase levels. At these high doses, elevation in liver enzymes is seen in approximately 25% of patients; however, the effect is readily reversible following discontinuation of the drug. In another study of advanced cancer patients that utilized doses of 3 mg/kg/day, no abnormalities of hepatic function were observed; however, headache and mild emotional changes were reported in 25% of patients (Meyskens *et al.*, 1982).

Elevation of serum triglycerides (> 25% of baseline) with concomitant decreases in high-density lipoproteins occurs in approximately 25% of patients treated with higher doses of isotretinoin. Although alterations in serum lipids are not of major concern in patients with advanced cancer, most protocols specify that patients have normal fasting serum cholesterol and triglycerides and normal liver function. Less common effects (including joint pains, epistaxis, conjunctivitis, mucositis, urethritis, nausea, and vomiting) occur in approximately 10–15% of patients. These quickly resolve with discontinuation of therapy.

Importantly, retinoids are devoid of myelosuppressive effects and, in certain instances, actually stimulate myelopoesis. This finding is of special relevance in cancer therapy. Since many drugs used for cancer treatment are myelosuppressive, retinoids may eventually be introduced into combination cancer therapy regimens. Animal reproduction studies indicate that both etretinate and isotretinoin are teratogenic. Therefore, caution must be exercised in the treatment of female patients of reproductive age. With respect to semen, no abnormalities of ejaculate volume, sperm count, motility, or morphology have been demonstrated (Schill *et al.*, 1981).

b. EFFICACY STUDIES. Etretinate has undergone only limited testing for the treatment of malignant disorders. Beretti *et al.* (1981) treated 21 patients with multiple superficial basal cell carcinomas arising from arsenical or solar exposure or basal cell nevus syndrome. Patients were treated for 4–5 months. The initial dose was 1 mg/kg/day, which was reduced after the first 2 months to 0.5 mg/kg/day. Complete clearing occurred in two patients (10%); improvement was also noted in eight additional patients (38%). Relapses occurred on dose reduction or withdrawal of etretinate. The authors concluded that maintenance therapy was necessary for the treatment of this disorder. However, dose and duration of treatment remain to be determined. Other investigators have reported similar findings (Neering and Zonnenberg, 1981; Zonnenberg *et al.*, 1980).

Souteyrand *et al.* (1981) treated 10 patients with mycosis fungoides using etretinate alone or in combination with PUVA (psoralen + long wavelength ultraviolet light). As sole treatment, etretinate did not cause significant clinical improvement; however, when used in combination with PUVA, rapid improvements with complete clearing of lesions occurred in several cases. Similar responses have also been reported by Zachariae *et al.* (1982).

Weber *et al.* (1982) treated 46 patients with squamous lung carcinoma in a randomized trial of bleomycin and CCNU (cyclohexyl chlorethyl nitrosourea) with or without etretinate. No statistical difference in response rate or duration of remission was noted between the two groups. Other clinical trials that utilize etretinate as adjuvant therapy after primay surgical or radiation treatment are ongoing in patients with carcinomas of the bladder, esophagus, and lung. Data from these European studies are very preliminary.

Isotretinoin has also had only limited clinical testing as an anticancer agent; however, several clinical trials have been initiated. In two studies (1978, 1979), Peck *et al.* reported complete clinical regression of several lesions in patients with multiple basal cell carcinomas. Larger tumors (> 10 mm in diameter) responded poorly to treatment. High doses of approximately 4.5 mg/kg/day were used in this trial. Meyskens *et al.* (1982) have also reported clinical activity of isotretinoin in basal cell carcinoma, using doses of 2–3 mg/kg/day. In the same study, 6 of 24 patients (25%) with epithelial carcinomas of the head and neck or lung had partial or mixed regression of disease. One of 12 patients with malignant melanoma achieved a partial response of 4-months duration. No significant benefit was observed in 66 patients with hematologic or nonepithelial malignancies. This group has also reported therapeutic efficacy in a small number of patients with laryngeal papillomatoses (personal communication). A randomized trial is planned to verify these findings.

Cassidy *et al.* (1982) studied 18 patients with advanced breast carcinoma refractory to standard therapies. Patients were treated with starting doses of 0.5 mg/kg/day and escalated to 8 mg/kg/day. No objective responses were observed. Gold *et al.* (1983a) conducted a trial of isotretinoin in patients with advanced germ cell tumors who had failed prior chemotherapy. In the 10 patients currently evaluable, no objective responses have been observed. However, two patients experienced disease stabilization. Autopsy findings in one of these patients revealed that most of the tumor represented mature teratoma, whereas the original pathology was pure embryonal cell carcinoma.

Presant and Bearman (1981) reported preliminary findings of a randomized trial that used isotretinoin either alone or in combination with busulfan for the treatment of patients with chronic phase chronic myelocytic leukemia. No differences have been noted in the time to development of blastic transformation in either group. The addition of the retinoid did not increase the acute toxicity of busulfan. Saccomanno *et al.* (1982) administered isotretinoin in 26 patients who exhibited varying degrees of cytologic atypia in exfoliated cells obtained from sputum. Subsequent samples revealed a variety of degenerative changes that suggested some cytopathic effect of the retinoid.

In a phase I trial of isotretinoin in 19 patients with myelodysplastic syndromes, Gold *et al.* (1983b) reported improvement of hematologic parameters in five patients, including normalization of bone marrow blast count, platelet count, and hemoglobin concentration. A randomized trial of isotretinoin has been initiated by this group to confirm these findings.

Warrell *et al.* (1982) are studying isotretinoin in patients with lymphoid malignancies. Currently, no patient with nodular or diffuse non-Hodgkin's lymphoma has shown a therapeutic response. Two patients with "prolymphocytic leukemia" (an aggressive variant of chronic lymphocytic leukemia) demonstrated > 25% reduction in abnormal peripheral leukocytes. No changes were observed in lymphocyte morphology nor were there any changes in cell-surface markers in the one patient tested. Two of six patients with mycosis fungoides refractory to electron beam radiation have shown significant improvement in cutaneous disease.

Several clinical trials of isotretinoin have been initiated. Ongoing studies in advanced cancers include chronic myelogenous leukemia, carcinoma of the head and neck, and myelodysplastic syndromes. Adjuvant therapy trials are ongoing in non-small cell lung cancer, head and neck carcinoma, and malignant melanoma. Most of these trials utilize a randomized double-blind design. Unfortunately, for most of these studies (especially the adjuvant trials) definitive results will not be available for several years.

c. SUMMARY. Most of the early clinical trials are largely uninterpretable for efficacy because of small patient numbers, lack of control groups, variability in response definitions, and imprecise reporting of data. Nevertheless, evidence of activity is apparent in several malignant conditions such as mycosis fungoides and basal cell carcinomas. Other areas of reported activity, such as myelodysplastic syndromes and carcinomas of the head and neck, must await verification by randomized, controlled clinical trials.

C. Prevention

One of the most exciting potential uses for the retinoids lies in the area of cancer prevention. The rationale for the use of retinoids in cancer prevention derives from their ability to control normal differentiation in many epithelial tissues. Animal studies that document the efficacy of retinoids in cancer prevention have been reviewed in detail earlier in this chapter. To date, clinical trials have concentrated primarily on the reversal of premalignant conditions or the prevention of recurrent lesions in patients with certain malignant diagnoses.

The earliest trials were conducted in the 1960s using all-*trans*-retinoic acid. Activity was reported in the topical treatment of actinic keratoses (Stuttgen, 1962; Bollag and Ott, 1971), in the systemic therapy of oral leukoplakia (Ryssel *et al.,* 1971; Koch and Schettler, 1973), and in the prevention of recurrent bladder papillomas (Evard and Bollag, 1972). However, marked hypervitaminosis A toxicity generally prohibited the use of effective doses.

Subsequently, most clinical trials used either etretinate or isotretinoin because of a more favorable therapeutic index. Oral treatment with etretinate has produced regression in several malignant conditions including actinic keratoses (Berretti *et al.,* 1981; Schnitzler and Verret, 1981; Moriarity *et al.,* 1982), arsenic keratoses (Berretti *et al.,* 1981), epidermodysplasia verrucaformis (Jablonska *et al.,* 1981; Lutzner *et al.,* 1981), keratoacanthomas (Berretti *et al.,* 1981), and leukoplakia of the oral cavity (Koch, 1978; Ehrl, 1980) and larynx (Bichler and Spoendlin, 1979).

By means of serial bronchoscopy and bronchial biopsies, one study (Gouveia *et al.,* 1982) indicated the ability of etretinate to reverse metaplastic lesions in the lungs of heavy smokers at low and nontoxic dose levels. However, these results are preliminary and longer follow-up will be required before a definitive statement can be made regarding effectiveness in the prevention of lung cancer. Two European studies of etretinate in patients with recurrent superficial bladder tumors have begun (Alfthan *et al.,* 1983; Studer *et al.,* 1982). Preliminary analysis of these double-blind trials suggests a decrease in the rate of recurrences for patients receiving etretinate relative to the group receiving placebo.

Isotretinoin has also been reported to cause remissions in oral leukoplakia (Koch, 1978; Shah *et al.,* 1983), epidermodysplasia verrucaformis (Meyskens *et al.,* 1982), and keratoacanthoma (Haydey *et al.,* 1980; Meyskens *et al.,* 1982). A prevention trial of isotretinoin in patients with recurrent bladder cancer was initiated by the National Bladder Cancer Collaborative Group (1981). However, because of unacceptable toxicity, this study was terminated after accession of only a small number of patients and response could not be accurately assessed. In a preliminary trial of patients with multiple basal cell carcinomas, isotretinoin was effective in preventing the recurrence of new lesions (Peck *et al.,* 1982).

The United States National Cancer Institute (NCI) has initiated a comprehensive program to evaluate the potential of certain natural and synthetic agents (including the retinoids and β-carotene) in reducing the incidence of cancer. This program began in response to increasing evidence that indicated that retinoids and β-carotene could arrest or reverse the natural progression of precancerous conditions and reduce the incidence of cancer in human populations (Committee on Diet, Nutrition, and Cancer, National Academy of Sciences, 1982).

The epidemiologic studies conducted under the auspices of the NCI include identification of banks of previously collected human blood samples that can be correlated with present or future cancer incidence. Approximately 15,000 adults who participated in the National Health and Nutrition Examination Survey from 1971 to 1975 will be evaluated in an extensive epidemiologic survey. This study will examine the relation between diet and incidence of site-specific cancer. Additionally, studies which specifically evaluate the relationship between vitamin A ingestion and lung cancer risk are ongoing in Hawaii and New Orleans.

Although many high-risk populations are being evaluated, several intervention trials have already begun. Two trials of retinyl acetate and all-*trans*-retinoic acid are being conducted in women with cervical dysplasia (Romney *et al.*, 1981; Surwit *et al.*, 1982). In these studies, the retinoid is delivered directly to the cervix by means of a vaginal applicator or a collagen sponge within a cervical cap. Once toxicity and dose schedules are established, comparative double-blind studies are planned to assess the cancer-preventive effect.

A randomized double-blind trial of isotretinoin versus placebo in a population of approximately 1800 patients with multiple basal cell carcinomas is also planned. This trial will use low doses of isotretinoin (10 mg/day). Patients will be treated for 3 years and will be followed closely for the development of new lesions.

Since it is unclear whether the protective effects of β-carotene are due to its partial conversion to retinol or to some other factor, additional studies are planned to evaluate the ability of β-carotene to reduce cancer risk. In a large cooperative study sponsored jointly by NCI and the National Heart, Lung and Blood Institute, approximately 20,000 male physicians over 40 years of age will be followed to evaluate the potential of β-carotene in lowering cancer incidence.

Another NCI-sponsored trial will assess the protective effects of orally administered β-carotene in combination with canthaxanthin in a population of albino Africans in Tanzania. This trial will study whether skin cancers induced by ultraviolet light can be prevented in this high-risk population. Other proposals to study high-risk populations using retinoids are currently under review. Since prevention studies generally require a large population and long follow-up, results from these studies cannot be anticipated for several years.

The present recommendations of the National Academy of Sciences, regarding vitamin A intake for cancer prevention, is summarized as follows:

> The epidemiological evidence is sufficient to suggest that foods rich in carotenes or vitamin A are associated with reduced risk of cancer. The toxicity of vitamin A in doses exceeding those required for optimum nutrition, and the difficulty of epidemiological studies to distinguish the effects of carotenes from those of vitamin A, argue against increasing vitamin A intake by the use of supplements.

D. Directions for Future Clinical Investigations

There are several areas where retinoids may prove useful in clinical oncology: (1) for the treatment of advanced disease; (2) for treatment which is adjuvant to other anticancer therapy; (3) for the reversal or differentiation of preneoplastic lesions; (4) for the prevention of cancer in high-risk populations and; (5) ultimately, for the prevention of cancer in the general population.

In the management of advanced disease, retinoids are most likely to find use in combination with other anticancer therapies. The relative lack of toxic reactions that are commonly associated with standard anticancer therapies should allow for easier use in combination regimens. As yet, the specific roles are undefined. However, certain possibilities would include use as a maintenance therapy in the acute leukemias following induction of remission with conventional drugs, or the addition of retinoids to standard combination regimens for the treatment of certain epithelial cancers. They may also be useful as single agents for the management of certain hematologic disorders.

Retinoids represent a unique approach to the problem of adjuvant therapy in cancer. Since the retinoids have a mechanism of action different from conventional cytotoxic therapies, they may be effective in situations where tumor burden is reduced. Additionally, these compounds are likely to be well received by both patient and physician since they cause little toxicity at low doses. Successful adjuvant treatment for minimal residual cancer represents one of the greatest challenges in cancer management. Admittedly, the retinoids currently in clinical testing do not possess the ultimately desirable attributes of total efficacy with little toxicity. However, the need for newer retinoid analogs with improved therapeutic index will be clearly verified if studies with currently available drugs indicate beneficial effect.

The available data, although preliminary, are highly suggestive that retinoids can reverse several preneoplastic lesions. While this information is of great interest, especially when these lesions are clinically significant, the true effect of such intervention on the ultimate development of cancer will require long follow-up, similar to that needed for prevention studies in other high-risk populations. It is unknown whether the reversal of what is commonly considered a "pre-malignant" lesion will affect the ultimate occurrence of malignancy.

It is clear that the retinoid derivatives chosen for use as cancer preventives must cause minimal toxicity. It is possible that synergistic combination of retinoids with other chemopreventive agents will diminish toxicity and increase therapeutic activity. Since it is likely that treatment will be of long (possibly life-long) duration, the potential long-term biological effects of retinoids on lipid metabolism and reproductive function must be carefully considered. Additionally, constant vigilance must be exercised in the identification of as yet unknown effects that may be associated with long-term use.

Careful analysis of structure–function relationships may lead to the synthesis of newer retinoids with a superior therapeutic index that may be more suitable for cancer prophylaxis in healthy patients. Nevertheless, ongoing trials in cancer treatment and prevention with currently available retinoids will provide valuable information. The prevention of cancer in high-risk populations and the reversal of premalignant disease would clearly represent major advances in public health. While specific indications are not yet established, cautious optimism is warranted.

REFERENCES

Abels, J. C., Gorham, A. T., Pack, G. T., and Rhoads, C. P. (1941). *J. Clin. Invest.* **20,** 749–764.

Alfthan, O., Tarkkanen, J., Pyrhonen, S., Grohn, P., Heinonen, E., and Saila, K. (1983). *Eur. Urol.* **9,** 6–9.

Atukorala, S., Basu, T. K., Dickerson, J. W. T., Donaldson, D., and Sakula, A. (1979). *Br. J. Cancer* **40,** 927–931.

Basu, T. K., Donaldson, D., Jenner, M., Williams, D. C., and Sakula, A. (1976). *Br. J. Cancer* **33,** 119–121.

Becci, P. J., Thompson, H. J., Grubbs, C. J., Squire, R. A., Brown, C. C., Sporn, M. B., and Moon, R. C. (1978). *Cancer Res.* **38,** 4463–4466.

Becci, P. J., Thompson, H. J., Grubbs, C. J., Brown, C. C., and Moon, R. C. (1979). *Cancer Res.* **39,** 3141–3144.

Becci, P. J., Thompson, H. J., Strum, J. M., Brown, C. C., Sporn, M. B., and Moon, R. C. (1981). *Cancer Res.* **41,** 927–932.

Berenblum, I., and Shubik, P. (1947). *Br. J. Cancer* **1,** 379–382.

Berretti, B., Grupper, Ch., Edelson, Y., and Bermejo, D. (1981). *In* "Retinoids, Advances in Basic Research and Therapy" (C. E. Orfanos, O. Braun-Falco, E. M. Farber, C. Grupper, M. K. Polano, and R. Schuppli, eds.), pp. 397–399. Springer-Verlag, Berlin and New York.

Bichler, E., and Spoendlin, H. (1979). *Arch. Oto-Rhino-Laryngol.* **225,** 9–14.

Birt, D. F., Sayed, S., Davies, M. H., and Pour, P. (1981). *Cancer Lett. (Shannon, Irel.)* **14,** 13–21.

Bjelke, E. (1975). *Int. J. Cancer* **15,** 561–565.

Bjelke, E. (1978). *Klin. Prax. Suppl.* **2,** 10–17.

Bollag, W. (1970). *Int. J. Vitam. Res.* **40,** 299–314.

Bollag, W. (1971a). *Cancer Chemother. Rep.* **55,** 53–58.

Bollag, W. (1971b). *Experientia* **27,** 90–92.

Bollag, W. (1972a). *Eur. J. Cancer* **8,** 689–693.

Bollag, W. (1972b). *Experientia* **28,** 1219–1220.

Bollag, W. (1974). *Eur. J. Cancer* **10,** 731–737.

Bollag, W. (1975a). *Chemotherapy (Basel)* **21,** 236–247.

Bollag, W. (1975b). *Eur. J. Cancer* **11,** 721–724.

Bollag, W., and Ott, F. (1971). *Cancer Chemother. Rep.* **55,** 59–60.

Boutwell, R. K. (1964). *Prog. Exp. Tumor Res.* **4,** 207–250.

Boutwell, R. K. (1974). *CRC Crit. Rev. Toxicol.* **2,** 419–443.

Boutwell, R. K. (1977). *In* "Origins of Human Cancer, Book B" (H. H. Hiatt, J. D. Watson, and J. A. Winsten, eds.), pp. 773–783. Cold Spring Harbor Lab., Cold Spring Harbor, New York.

Breitman, T. R., Selonick, S. E., and Collins, S. J. (1980). *Proc. Natl. Acad. Sci. U.S.A.* **77,** 2936–2940.

Breitman, T. R., Collins, S. J., and Keene, B. R. (1981). *Blood* **57,** 1000–1004.

Breitman, T. R., Keene, B. R., and Hemmi, H. (1983). *Cancer Surv.* **2,** 263–291.

Burns, F. J., Sargent, E. V., McCormick, D. L., and Albert, R. E. (1982). *Proc. Am. Assoc. Cancer Res.* **23,** 89.

Capel, I. D., and Williams, D. C. (1979). *IRCS Med. Sci. Libr. Compend.* **7,** 361.
Cassidy, J., Lippman, M., Lacroix, A., and Peck, G. (1982). *Eur. J. Cancer Clin. Oncol.* **18,** 925–928.
Chaudhy, N. A., Jafarey, N. A., and Ibrahim, K. (1980). *J. Pakistan Med. Assn.* pp. 221–223.
Chu, E. W., and Malmgren, R. A. (1965). *Cancer Res.* **25,** 883–895.
Clifford, P. (1972). *Proc. R. Soc. Med.* **65,** 682–686.
Cohen, M. H., Primack, A., Broder, L. E., and Williams, L. R. (1977). *Cancer Lett. (Shannon Irel.)* **4,** 51–54.
Cohen, S. M., Wittenberg, J. F., and Bryan, G. T. (1976). *Cancer Res.* **36,** 2334–2339.
Committee on Diet, Nutrition, and Cancer; Assembly of Life Sciences; National Research Council (1982). "Diet, Nutrition, and Cancer." Natl. Acad. Sci., Washington, D. C.
Cone, M. V., and Nettesheim, P. (1973). *JNCI, J. Natl. Cancer Inst.* **50,** 1599–1606.
Cook-Mozaffari, P. (1979). *Nutr. Cancer* **1,** 51–60.
Cook-Mazaffari, P. J., Azordegan, F., Day, N. E., Ressicand, A., Sabai, C., and Aramesh, B. (1979)., *Br. J. Cancer* **39,** 293–309.
Daoud, A. H., and Griffin, A. C. (1980). *Cancer Lett. (Shannon, Irel.)* **9,** 299–304.
Davies, R. E. (1967). *Cancer Res.* **27,** 237–241.
DeHoog, J. V., and Welsch, C. W. (1982). *Proc. Am. Assoc. Cancer Res.* **23,** 87.
Ehrl, P. A. (1980). *Dtsch. Zahnarztl. Z.* **35,** 554–558.
Epstein, J. H. (1977a). *Aust. J. Dermatol.* **18,** 57–61.
Epstein, J. H. (1977b). *Photochem. Photobiol.* **25,** 211–213.
Epstein, J. H. (1981). *J. Invest. Dermatol.* **77,** 144–146.
Epstein, J. H., and Grekin, D. A. (1981). *J. Invest. Dermatol.* **76,** 178–180.
Evard, J. P., and Bollag, W. (1972). *Schweiz. Med. Wochenschr.* **102,** 1880–1833.
Forbes, P. D. (1981). *J. Invest. Dermatol.* **77,** 139–143.
Forbes, P. D., Urbach, F., and Davies, R. E. (1979). *Cancer Lett. (Shannon, Irel.)* **7,** 85–90.
Gilmore, W., and Giunta, J. L. (1981). *Oral Surg., Oral Med. Oral Pathol.* **51,** 256–265.
Gold, E., Bosl, G. J., Whitmore, W., Myers, W. P. L., and Golbey, R. (1983a). *Proc. Am. Assoc. Cancer Res.* **24,** 150.
Gold, E. J., Mertelsmann, R. H., Itri, L. M., Gee, T., Arlin, Z., Kempin, S., Clarkson, B., and Moore, M. A. S. (1983b). *Cancer Treat. Rep.* **67,** 981–986.
Goodman, D. S. (1979). *Fed. Proc., Fed. Am. Soc. Exp. Biol.* **38,** 2501–2503.
Gouveia, J., Mathe, G., Hercend, T., Gros, F., Lemaigre, G., Santelli, G., Homasson, J. P., Gaillard, J. P., Angebault, M., Bonniot, J. P., Lededente, A., Marsac, J., Parrot, R., and Pretet, S. (1982). *Lancet* **1,** 710–712.
Graham, S., Schotz, W., and Martino, P. (1972). *Cancer (Philadelphia)* **30,** 927–938.
Graham, S., Mettlin, C., Marshall, J., Priore, R., Rzepka, T., and Shedd, D. (1981). *Am. J. Epidemiol.* **113,** 675–680.
Gregor, A., Lee, P. N., Roe, F. J. C., Wilson, M. J., and Melton, A. (1980). *Nutr. Cancer* **2,** 93–97.
Grubbs, C. J., Moon, R. C., Sporn, M. B., and Newton, D. L. (1977a). *Cancer Res.* **37,** 599–602.
Grubbs, C. J., Moon, R. C., Squire, R. A., Farrow, G. M., Stinson, S. F., Goodman, D. G., Brown, C. C., and Sporn, M. B. (1977b). *Science* **198,** 743–744.
Grubbs, C. J., Becci, P. J., and Moon, R. C. (1980). *Proc. Am. Assoc. Cancer Res.* **21,** 102.
Haenszel, W., Kurihara, M., Segi, M., and Lee, R. K. C. (1972). *JNCI, J. Natl. Cancer Inst.* **49,** 969–988.
Haydey, R. P., Reed, M. L., Dzubow, L. M., and Shupack, J. L. (1980). *N. Engl. J. Med.* **303,** 560–562.
Hedenberg, I. (1954). *Acta Chir. Scand., Suppl.* No. 192, pp. 1–87.
Heilman, C., and Swarm, R. L. (1975). *Fed. Proc., Fed. Am. Soc. Exp. Biol.* **34,** 882.
Hicks, R. M. (1968). *J. Ultrastruct. Res.* **22,** 206–230.

Hicks, R. M. (1980). *Br. Med. Bull.* **36,** 39–46.

Hicks, R. M.,Chowaniec, J., Turton, J. A., Massey, E. D., and Harvey, A. (1982). *In* "Molecular Interrelations of Nutrition and Cancer" (M. S. Arnott *et al.*, eds.), pp. 419–447. Raven, New York.

Hill, D. L., and Grubbs, C. J. (1982). *Anticancer Res.* **2,** 111–124.

Hirayama, T. (1967). *In* "Proceedings of the 9th International Cancer Congress" (R. J. C. Harris, ed.), UICC Monography Series, Vol. 10, pp. 37–48. Springer-Verlag, Berlin and New York.

Hirayama, T. (1977). *In* "Origins of Human Cancer, Book A: Incidence of Cancer in Humans" (H. H. Hiatt, J. D. Watson, and J. A. Winsten, eds.), pp. 55–75. Cold Spring Harbor Lab., Cold Spring Harbor, New York.

Hirayama, T. (1979). *Natl. Cancer Inst. Mongr.* No. 53, pp. 149–155.

Hirayama, T. (1982). *Nutr. Cancer* **1,** 67–81.

Hixson, E. J., and Denine, E. P. (1979). *JNCI, J. Natl. Cancer Inst.* **63,** 1359–1364.

Hormozdiari, H., Day, N. E., Aramesh, B., and Mahboubi, E. (1975). *Cancer Res.* **35,** 3493–3498.

Ibrahim, K., Jafarey, N. A., and Zuberi, S. J. (1977). *Clin. Oncol.* **3,** 203–207.

Iizuka, T., Ichimura, S., and Kawachi, T. (1979). *Proc. Jpn. Cancer Assoc. 38th Annu. Meet.*, Tokyo, p. 321.

Ip, C., and Ip, M. M. (1981). *Carcinogenesis* **2,** 915–918.

Israel, L., and Aguilera, J. (1980). *Pathol. Biol.* **28,** 253–259.

Jablonska, S., Obalek, S., Wolska, H., and Harzabek-Chorzelska, M. (1981). *In* "Retinoids, Advances in Basic Research and Therapy" (C. E. Orfanos, O. Braun-Falco, E. M. Farber, C. Grupper, M. K. Polano, and R. Schuppli, eds.), pp. 401–405. Springer-Verlag, Berlin and New York.

Janss, D. H., Moon, R. C., and Irving, C. C. (1972). *Cancer Res.* **32,** 254–258.

Joint Iran-International Agency for Research on Cancer Study Group (1977). *JNCI, J. Natl. Cancer Inst.* **59,** 1127–1138.

Kark, J. D., Smith, A. H., Switzer, B. R., and Hames, C. G. (1981). *JNCI, J. Natl. Cancer Inst.* **66,** 7–16.

Koch, F. H. (1978). *J. Maxillary-Facial Surg.* **6,** 59–63.

Koch, H., and Schettler, D. (1973). *Dtsch. Zahnarztl. Z.* **28,** 623–627.

Komiyama, S., Hiroto, I., Ruy, S., Nakashima, T., Kuwano, M., and Endo, H. (1978). *Oncology* **35,** 253–257.

Kucera, H. (1980). *Wien. Klin. Wochenschr.* **92,** (Suppl. 118), 20 pp.

Lasnitzki, I. (1955). *Br. J. Cancer* **9,** 434–441.

Levij, I. S., and Polliack, A. (1968). *Cancer (Philadelphia)* **22,** 300–306.

Levij, I. S., and Polliack, A. (1969). *Pathol. Microbiol.* **34,** 282–288.

Lichti, U., Patterson, E., Hennings, H., and Yuspa, S. H. (1981). *Cancer Res.* **41,** 49–54.

Longnecker, D. S., Roebuck, B. D., Yager, J. D., Jr., Lilja, S. D., and Siegmund, B. (1981). *Cancer (Philadelphia)* **47,** 1562–1572.

Longnecker, D. S., Curphey, T. J., Kuhlmann, E. T., and Roebuck, B. D. (1982). *Cancer Res.* **42,** 19–24.

Lutzner, M. A., Blanchet-Bardon, C., and Puissant, A. (1981). *In* "Retinoids, Advances in Basic Research and Therapy" (C. E. Orfanos *et al.*, eds.), pp. 407–410. Springer-Verlag, Berlin and New York.

MacLennan, R., Da Costa, J., Day, N. E., Law, C. H., Ng, Y. K., and Shanmugaratnam, K. (1977). *Int. J. Cancer* **20,** 854–860.

McCormick, D. L., and Moon, R. C. (1982a). *Cancer Res.* **42,** 2639–2643.

McCormick, D. L., and Moon, R. C. (1982b). *Proc. Int. Cancer Congr. 13th,* p. 682.

McCormick, D. L., Burns, F. J., and Albert, R. E. (1980). *Cancer Res.* **40,** 1140–1143.

McCormick, D. L., Burns, F. J., and Albert, R. E. (1981). *JNCI, J. Natl. Cancer Inst.* **66,** 559–564.

McCormick, D. L., Becci, P. J., and Moon, R. C. (1982a). *Carcinogenesis* **3,** 1473–1477.
McCormick, D. L., Mehta, R. G., Thompson, C. A., Dinger, N., Caldwell, J. A., and Moon, R. C. (1982b). *Cancer Res.* **42,** 508–512.
Mahmoud, L. A. N., and Robinson, W. A. (1982). *Int. J. Cancer* **30,** 143–145.
Maiorana, A., and Gullino, P. M. (1980). *JNCI, J. Natl. Cancer Inst.* **64,** 655–663.
Mathews-Roth, M. M. (1982). *Oncology* **39,** 33–37.
Mayer, H., Bollag, W., Hänni, R., and Rüegg, R. (1978). *Experientia* **34,** 1105–1119.
Mehta, R. G., and Moon, R. C. (1980). *Cancer Res.* **40,** 1109–1111.
Mehta, R. G., Cerny, W. L., and Moon, R. C. (1983). *Carcinogenesis* **4,** 23–26.
Merriman, R. L., and Bertram, J. S. (1979). *Cancer Res.* **39,** 1661–1666.
Mettlin, C., and Graham, S. (1979). *Am. J. Epidemiol.* **110,** 255–263.
Mettlin, C., Graham, S., and Swanson, M. (1979). *JNCI, J. Natl. Cancer Inst.* **62,** 1435–1438.
Mettlin, C., Graham, S., Priore, R., Marshall, J., and Swanson, M. (1981). *Nutr. Cancer* **2,** 143–147.
Meyskens, F. L., Aapro, M. S., Voakes, J. B., Moon, T. E., and Gilmartin, E. (1981). *In* "Adjuvant Therapy of Cancer III" (S. E. Salmon and S. E. Jones, eds.), pp. 217–224. Grune & Stratton, New York.
Meyskens, F. L., Gilmartin, E., Alberts, D. S., Levine, N. S., Brooks, R., Salmon, S. E., and Surwit, E. A. (1982). *Cancer Treat. Rep.* **66,** 1315–1319.
Micksche, M., Cerni, C., Kokron, O., Titscher, R., and Wrba, H. (1977). *Oncology* **34,** 234–239.
Miyata, Y., Tsuda, H., Matayoshi-Miyasato, K., Fukushima, S., Murasaki, G., Ogiso, T., and Ito, N. (1978). *Gann* **69,** 845–848.
Miyata, Y., Nakatsuka, T., Arai, M., Murasaki, G., Nakanishi, K., and Ito, N. (1980). *Gann* **71,** 341–348.
Modan, B., Barell, V., Lubin, R., Modan, M., Greenberg, R. A., and Graham, S. (1975). *JNCI, J. Natl. Cancer Inst.* **55,** 15–18.
Moon, R. C. (1981). *In* "Banbury Report 8: Hormones and Breast Cancer" (M. C. Pike, P. K. Siiteri, and C. W. Welsch, eds.), pp. 353–364, Cold Spring Harbor Lab., Cold Spring Harbor, New York.
Moon, R. C., and McCormick, D. L. (1982). *J. Am. Acad. Dermatol.* **6,** 809–814.
Moon, R. C., and Mehta, R. G. (1982). *In* "Hormones and Cancer" (W. W. Leavitt, ed.), pp. 231–249, Plenum, New York.
Moon, R. C., Grubbs, C. J., and Sporn, M. B. (1976). *Cancer Res.* **36,** 2626–2630.
Moon, R. C., Grubbs, C. J., Sporn, M. B., and Goodman, D. G. (1977). *Nature (London)* **267,** 620–621.
Moon, R. C., Thompson, H. J., Becci, P. J., Grubbs, C. J., Gander, R. J., Newton, D. L., Smith, J. M., Phillips, S. L., Henderson, W. R., Mullen, L. T., Brown, C. C., and Sporn, M. B. (1979). *Cancer Res.* **39,** 1339–1346.
Moon, R. C., McCormick, D. L., Becci, P. J., Shealy, Y. F., Frickel, F., Paust, J., and Sporn, M. B. (1982). *Carcinogenesis* **3,** 1469–1472.
Moon, R. C., McCormick, D. L., and Mehta, R. G. (1983a). *Cancer Res.* **43,** 2469s–2475s.
Moon, R. C., Mehta, R. G., and McCormick, D. L. (1983b). *In* "Cancer: Etiology and Prevention" (R. G. Crispen, ed.), pp. 275–288. Elsevier, Amsterdam.
Moriarty, M., Dunn, J., Darragh, A., Lambe, R., and Brick, I. (1982). *Lancet* **1,** 364–365.
Murasaki, G., Miyata, Y., Babaya, K., Arai, M., Fukushima, S., and Ito, N. (1980). *Gann* **71,** 333–340.
Narisawa, T., Reddy, B. S., Wong, C. Q., and Weisburger, J. H. (1976). *Cancer Res.* **36,** 1379–1383.
National Bladder Cancer Collaborative Group A: Koontz, W. W., Jr., Cutler, S., Penick, G., Prout, G. R., Jr., Shipley, W. U., and Smith, W. K. (1981). *Program 76th Annu. Meet., Am. Urol. Assoc., May 10–14, 1981, Boston, Massachusetts,* Abstr. No. 96, p. 115.

Neering, H., and Zonnenberg, B. (1981). *Ned. Tijdschr. Geneeskd.* **125,** 124.
Nettesheim, P., and Williams, M. L. (1976). *Int. J. Cancer* **17,** 351–357.
Nettesheim, P., Cone, M. W., and Snyder, C. (1976). *Cancer Res.* **36,** 966–1002.
Newberne, P. M., and Rogers, A. E. (1973). *JNCI, J. Natl. Cancer Inst.* **50,** 439–448.
Newberne, P. M., and Suphakarn, V. (1977). *Cancer (Philadelphia)* **40,** 2553–2556.
Newton, D. L., Henderson, W. R., and Sporn, M. B. (1980). *Cancer Res.* **40,** 3413–3425
O'Brien, T. G. (1976). *Cancer Res.* **36,** 2644–2653.
O'Brien, T. G., Simsiman, R. C., and Boutwell, R. K. (1975a). *Cancer Res.* **35,** 1662–1670.
O'Brien, T. G., Simsiman, R. C., and Boutwell, R. K. (1975b). *Cancer Res.* **35,** 2426–2433.
Pawson, B. A., Chan, K. K., DeNoble, J., Han, R. L., Piermattie, V., Specian, A. C., Srisethnil, S., Trown, P. W., Bohoslawec, O., Machlin, L. J., and Gabriel, E. (1979). *J. Med. Chem.* **22,** 1059.
Pawson, B. A., Ehmann, C. W., Itri, L. M., and Sherman, M. I. (1982). *J. Med. Chem.* **25,** 1269–1277.
Paymaster, J. C., Sanghvi, L. D., and Gangadharan, P. (1968). *Cancer (Philadelphia)* **21,** 279–288.
Peck, G. L., Yoder, F. W., Olsen, T. G., Pandya, M., and Butkus, D. (1978). *Dermatologica, Suppl. 1,* No. 157, pp. 11–12.
Peck, G. L., Olsen, T. G., Butkus, D., Pandya, M., Arnaud-Battandier, J., Yoder, F., and Levis, W. R. (1979). *Proc. Am. Assoc. Cancer Res.* **20,** 56.
Peck, G. L., Gross, E. G., Butkus, D., and DiGiovanna, J. J. (1982). *J. Am. Acad. Derm.* **6,** 817–823.
Peto, R., Doll, R., Buckley, J. D., and Sporn, M. B. (1981). *Nature (London)* **290,** 201–208.
Polliack, A., and Levij, I. S. (1968). *Oncology* **22,** 129–136.
Polliack, A., and Levij, I. S. (1969). *Cancer Res.* **29,** 327–332.
Port, C. D., Sporn, M. B., and Kaufman, D. G. (1975). *Proc. Am. Assoc. Cancer Res.* **16,** 21.
Presant, C. A., and Bearman, R. (1981). *Proc. Am. Soc. Clin. Oncol.* **22,** 488.
Rogers, A. E., Herndon, B. J., and Newberne, P. M. (1973). *Cancer Res.* **33,** 1003–1009.
Romney, S. L., Palan, P. R., Duttagupta, C., Wassertheil-Smoller, S., Wylie, J., Miller, G., Slagle, N. S., and Lucido, D. (1981). *Am. J. Obstet. Gynecol.* **141,** 890–894.
Ryssel, H. J., Brunner, K. W., and Bollag, W. (1971). *Schweiz. Med. Wochenschr.* **101,** 1027–1030.
Saccomanno, G., Moran, P. G., Schmidt, R. D., Hartshorn, D. F., Brian, D. A., Dreher, W. H., and Sowada, B. J. (1982). *Acta Cytol.* **26,** 78–85.
Saffiotti, U., Montesano, R., Sellakumar, A. R., and Bork, S. A. (1967). *Cancer (Philadelphia)* **20,** 857–864.
Schill, W. B., Wagner, A., Nikolowski, J., and Plewig, G. (1981). *In* "Retinoids, Advances in Basic Research and Therapy" (C. E. Orfanos, O. Braun-Falco, E. M. Farber, C. Grupper, M. K. Polano, and R. Schuppli, eds.), pp. 389–395. Springer-Verlag, Berlin and New York.
Schmähl, D., Kruger, C., and Preissler, P. (1972). *Arzneim. Forsch.* **22,** 946–949.
Schmähl, D., and Habs, M. (1978). *Arzneim. Forsch.* **28,** 49–51.
Schmähl, D., and Habs, M. (1981). *Arzneim. Forsch.* **31,** 677–679.
Schmähl, D., Danisman, A., Habs, M., and Diehl, B. (1976). *Arzneim. Forsch.* **86,** 89–94.
Schnitzler, L., and Verret, J. L. (1981). *In* "Retinoids, Advances in Basic Research and Therapy" (C. E. Orfanos, O. Braun-Falco, E. M. Farber, C. Grupper, M. K. Polano, and R. Schuppli, eds.), pp. 385–388. Springer-Verlag, Berlin and New York.
Schreiber, H., Nettesheim, P., Lijinsky, W., Richter, C. B., and Walburg, H. E., Jr. (1972). *JNCI, J. Natl. Cancer Inst.* **49,** 1107–1114.
Schuman, L. M., Mandell, J. S., Radke, A., Seal, U., and Halberg, F. (1982). *In* "Trends in Cancer Incidence: Causes and Practical Implications" (K. Magnus, ed.), pp. 345–354. Hemisphere Publ. Corp., Washington, New York, and London.

Shah, J. P., Strong, E. W., DeCosse, J. J., Itri, L., and Sellers, P. (1983). *Am. J. Surg.* **146,** 466–470.

Shamberger, R. J. (1971). *JNCI, J. Natl. Cancer Inst.* **47,** 667–673.

Shekelle, R. B., Liu, S., Raynor, W. J., Jr., Lepper, M., Maliza, C., and Rossof, A. H. (1981). *Lancet* **2,** 1185–1189.

Shklar, G., Flynn, E., Szabo, G., and Marefat, P. (1980a). *JNCI, J. Natl. Cancer Inst.* **65,** 1307–1316.

Shklar, G., Schwartz, J., Grau, D., Trickler, D. P., and Wallace, K. D. (1980b). *Oral Surg.* **50,** 45–52.

Silverman, J., Katayama, S., Zelenakas, K., Lauber, J., Musser, T. K., Reddy, M., Levenstein, M. J., and Weisburger, J. H. (1981). *Carcinogenesis* **2,** 1167–1172.

Sluyser, M., and Van Nie, R. (1974). *Cancer Res.* **34,** 3253–3257.

Smith, D. M., Rogers, A. E., Herndon, B. J., and Newberne, P. M. (1975a). *Cancer Res.* **35,** 11–16.

Smith, D. M., Rogers, A. E., and Newberne, P. M. (1975b). *Cancer Res.* **35,** 1485–1488.

Smith, P., Jungi, W. F., Mayr, Ca. A., Nagel, G. A., and Senn, H. (1974). *Eur. J. Cancer* **10,** 57–61.

Smith, P. G., and Jick, H. (1978). *Cancer (Philadelphia)* **42,** 808–811.

Souteyrand, P., Thivolet, J., and Fulton, R. (1981). *In* "Retinoids, Advances in Basic Research and Therapy" (C. E. Orfanos, O. Braun-Falco, E. M. Farber, CH. Grupper, M. K. Polano, and R. Schuppli, eds.), pp. 313–316. Springer-Verlag, Berlin and New York.

Sporn, M. B., Squire, R. A., Brown, C. C., Smith, J. M., Wenk, M. L., and Springer, S. (1977). *Science* **195,** 87–489.

Squire, R. A., Sporn, M. B., Brown, C. C., Smith, J. M., Wenk, M. L., and Springer, S. (1977). *Cancer Res.* **37,** 2930–2936.

Stinson, S. F., Reznik, G., and Donahoe, R. (1981). *JNCI, J. Natl. Cancer Inst.* **66,** 947–951.

Strickland, S. (1981). *Cell* **24,** 277–278.

Strickland, S., and Mahdavi, V. (1978). *Cell* **15,** 393–403.

Studer, U. E., Biedermann, C., Chollet, D., Karrer, P., Kraft, R., Toggenburg, H., and Vonbank, R. (1982). *In* "5th Congress of the European Association of Urology" Egermann, Vienna.

Stuttgen, G. (1962). *Dermatologica* **124,** 65–80.

Surwit, E. A., Graham, V., Droegemueller, W., Alberts, D., Chvapil, M., Dorr, R. T., Davis, J. R., and Meyskens, F. L., Jr. (1982). *Am. J. Obstet. Gynecol.* **143,** 821–823.

Thatcher, N., Blackledge, G., and Crowther, D. (1980). *Cancer (Philadelphia)* **46,** 1324–1328.

Thompson, H. J., Becci, P. J., Brown, C. C., and Moon, R. C. (1979). *Cancer Res.* **39,** 3977–3980.

Thompson, H. J., Becci, P. J., Grubbs, C. J., Shealy, Y. F., Stanek, E. J., Brown, C. C., Sporn, M. B., and Moon, R. C. (1981a). *Cancer Res.* **41,** 933–936.

Thompson, H. J., Meeker, L. D., and Becci, P. J. (1981b). *Cancer Res.* **41,** 1413–1416.

Trown, P. W., Buck, M. J., and Hansen, R. (1976). *Cancer Treat. Rep.* **60,** 1647–1653.

Trown, P. W., Palleroni, A. V., Bohoslawec, O., Richelo, B. N., Halpern, J. M., Gizzi, N., Geiger, R., Lewinski, C., Machlin, L. J., Jetten, A., and Jettern, M. E. R. (1980). *Cancer Res.* **40,** 212–220.

Verma, A. K., and Boutwell, R. K. (1977). *Cancer Res.* **37,** 2196–2201.

Verma, A. K., and Boutwell, R. K. (1980). *In* "Polyamines in Biomedical Research" (J. M. Gaugas, ed.), p. 474. Wiley (Interscience), New York.

Verma, A. K., Shapas, B. G., Rice, H. M., and Boutwell, R. K. (1979a). *Cancer Res.* **39,** 419–425.

Verma, A. K., Lowe, N. J., and Boutwell, R. K. (1979b). *Cancer Res.* **39,** 1035–1040.

Verma, A. K., Conrad, E. A., and Boutwell, R. K. (1980). *Carcinogenesis* **1,** 607–611.

Wahi, P. N., Bodkke, R. R., Arora, S., and Srivastava, M. C. (1962). *Indian J. Pathol. Bacteriol.* **5,** 10–16.

Wald, N., Idle, M., Boreham, J., and Bailey, A. (1980). *Lancet* **2,** 813–815.

Ward, J. M., Sporn, M. B., Wenk, M. L., Smith, J. M., Feser, D., and Dean, R. J. (1978). *JNCI, J. Natl. Cancer Inst.* **60,** 1489–1493.

Warrell, R. P., Coonley, C. J., and Itri, L. M. (1982). *Clin. Res.* **30,** Abstr. No. 692A, p. 701.

Weber, W., Arnold, H., Drings, P., Germano, G. H., Muller, K. M., Queisser, W., Westerhausen, M., and Obrecht, J. P. (1983) *Onkologie* **6,** 62–65.

Welsch, C. W., Goodrich-Smith, M., Brown, C. K., and Wilson, M. (1979). *Int. J. Cancer* **24,** 92–96.

Welsch, C. W., Brown, C. K., Goodrich-Smith, M., Chuisano, J., and Moon, R. C., (1980). *Cancer Res.* **40,** 3095–3098.

Welsch, C. W., Goodrich-Smith, M., Brown, C. K., and Crowe, N. (1981). *JNCI, J. Natl. Cancer Inst.* **67,** 935–938.

Wenk, M. L., Ward, J. M., Reznik, G., and Dean, J. (1981). *Carcinogenesis* **2,** 1161–1166.

White, F. H., and Al-Azzawi, B. (1981). *12th Int. Congr. Chemother., Florence,* p. 301.

Windhorst, D. B., and Peck, G. L. (1983). *In* "Update: Dermatology and General Medicine" (T. B. Fitzpatrick, A. Z. Eisen, K. Wolff, I. M. Freedberg, and K. F. Austen, eds.), pp. 226–237. McGraw-Hill New York.

Wolbach, S. B., and Howe, P. R. (1925). *J. Exp. Med.* **42,** 753–778.

Wynder, E. L., and Bross, I. J. (1961). *Cancer* **14,** 389–413.

Yarita, T., Nettesheim, P., and Mitchell, T. J. (1980). *Carcinogenesis* **1,** 255–262.

Zachariae, H., Grunnet, E., Thestrup-Pedersen, K., Molin, L., Schmidt, H., Starfelt, F., and Thomsen, K. (1982). *Acta Derm. Venereol.* **62,** 162–164.

Zonnenberg, B. A., Neering, H., and Van der Esch, E. P. (1980). *Ned. Tijdschr. Geneeskd.* **124,** 1546–1550.

15

Retinoids and the Immune System: Immunostimulation by Vitamin A

GUNTHER DENNERT

The Salk Institute for Biological Studies
San Diego, California

I. INTRODUCTION

This chapter attempts to summarize current knowledge about the effects of retinoids on the immune system. Although more than 100 publications spanning a time period of 10 years can be cited on this topic, the information available is still quite fragmentary and rather incomplete. Therefore, it seems appropriate at this point in time to review the data available, evaluate their significance, ask why research is not progressing as quickly as it might, and indicate those areas of research that require more attention in the future.

THE RETINOIDS, VOL. 2

ISBN 0-12-658102-9

II. EVIDENCE FOR IMMUNOSTIMULATORY ACTION OF RETINOIDS

A. Prevention of Infection by Retinoids

Historically there are several areas of investigation that point to potential effects of retinoids on the immune system. One is the vitamin A nutritional status, both in humans and experimental animals. Bloch (1928), for instance, reported that humans with vitamin A deficiency may have an increase in severity and frequency of various infections. Scrimshaw *et al.* (1968) reviewed 50 individual studies and concluded that the severity and incidence of viral, bacterial, and parasitic infections are higher in vitamin A-deficient animals. Retinoid supplementation, on the other hand, could be shown to afford protection from infection. For example, Hof and Emerling (1979) reported an increase in resistance to *Listeria* infections as a result of retinoid supplementation. Cohen and Elin (1974) demonstrated protection against *Pseudomonas* and *Candida* infections in mice by retinyl palmitate (A4). While in all these studies the causal relationship between retinoid deficiency and infection remains obscure, it was an *a priori* reasonable assumption that a malfunctioning immune system resulting from retinoid deficiency might be the reason for the preponderance of infections. Hence, what is missing in these studies is the direct proof that humoral or cell-mediated immunity to the infectious agents is in fact deficient, and that this is the result of vitamin A deficiency.

B. Inhibition of Tumors by Retinoids

A second area relevant to this discussion is that of prevention of tumor induction or growth by retinoids. As documented in this book (see Chapter 14), research in this area has expanded quite rapidly, making use of a variety of different model systems, not only naturally occurring but also synthetic retinoids. The topics studied can be classified as follows:

1. Inhibitory effects of retinoids on chemical carcinogenesis *in vivo*
2. The effects of retinoids on viral tumorigenesis
3. The possible influence of retinoids on the rejection of transplantable tumors

1. Effects of Retinoids on Chemical Carcinogenesis

The effects of retinoids on development of carcinogen-induced tumors *in vivo* have been studied with the aim of demonstrating inhibitory effects on tumor growth. Davies (1967), for instance, reported that retinol (A1) given orally to mice may significantly reduce the incidence of skin papillomas after 9,10-dimethyl-1,2-benzanthracene (DMBA) induction. This observation was reproduced and extended by Bollag (1972) who showed that Swiss albino mice treated with DMBA and retinoic acid (D1) express a delayed appearance, retarded growth, and reduced incidence of skin papillomas. Interestingly, the incidence of subsequent skin carcinomas was also reduced and some carcinomas even regressed. Sonis and Shklar (1981) demonstrated a delay of development of DMBA-induced tumors in the hamster cheek pouch as a result of 13-*cis*-retinoic acid (E6) treatment.

Several very systematic studies on the effects of retinoids on chemical carcinogenesis were published by Sporn and colleagues (1976, 1977). Sprague–Dawley rats previously injected intravenously with *N*-methylnitrosourea (NMU) and then treated with retinyl acetate (A3), not only showed a delay of mammary carcinoma appearance but also a reduced incidence of tumors (Moon *et al.*, 1977). Instillation of NMU into the bladder of rats induced transitional and squamous cell carcinomas, which were inhibited by feeding the animals with 13-*cis*-retinoic acid (Sporn *et al.*, 1977). In a related study, oral doses of *N*-butyl-*N*-(4-hydroxybutyl)nitrosamine induced transitional cell carcinoma in the bladder of rats, which again could be significantly reduced by feeding 13-*cis*-retinoic acid (Grubbs *et al.*, 1977).

Whereas all these experiments are highly encouraging from a clinical point of view, the mechanism by which retinoids bring about these effects is not known. The hypothesis of immunosurveillance of neoplasia states that arising neoplastic cells are eliminated by the immune system. One could therefore hypothesize that neoplastic cells are eliminated by an immune system stimulated by retinoids. This reasoning, however, would require several assumptions yet to be proved. First, the tumors induced must be immunogenic, i.e., effective in sensitizing lymphocytes, and second, they must be susceptible to immune-mediated effector mechanisms. Since neither of the two parameters has been investigated or established in most of the model systems employed, alternative explanations for the inhibitory effects of retinoids on carcinogen-induced neoplasia have to be considered as well. It is well known that retinoids are required for normal cell differentiation, particularly in epithelial tissue. It is therefore possible that the effects seen with retinoids subsequent to chemical carcinogenesis are due to direct effects of these compounds on preneoplastic cells. In fact, it has been previously shown by Wolbach and Howe (1925), Lasnitzki (1955), and Chopra and Wilkoff (1976) with organ cultures of prostate glands, that formation of hyperplastic and anaplastic epithelial lesions induced with carcinogens can be reversed by retinoids. Similarly, Merriman and Bertram (1979) demonstrated that chemical transformation of mouse embryo fibroblasts by methylcholanthrene (MCA) *in vitro* is inhibitable by retinoids. Because effects of the immune system are excluded in cell culture, these particular effects of retinoids cannot be immune-mediated. Inhibition of carcinogenesis by retinoids *in vivo* may therefore be due to mechanisms similar to the ones operative *in vitro*. Hence, the inhibitory effects on carcinogenesis of these compounds *in vivo* should not be used as an argument for their immunostimulatory action.

2. *Effects of Retinoids on Viral Tumorigenesis*

A somewhat different situation may exist in the other tumor model systems employed, viral tumorigenesis and rejection of transplantable tumors, because in these systems the role of the immune system in tumor control has been well documented. McMichael (1965) and later Frankel *et al.* (1980) showed that systemic administration of retinoids [retinoic acid and etretinate (F1), respectively] inhibited the growth of papillomas that were induced in rabbits by Shope rabbit papilloma virus. Seifter *et al.* (1973a) induced tumors in mice with Moloney sarcoma virus (MSV) and reported that retinyl palmitate prolonged the tumor latency period as well as reduced the tumor incidence and tumor growth (Seifter *et al.*, 1973; Levine *et al.*, 1975). Frankel *et al.* (1980) induced tumors in hamsters by Rous sarcoma virus (RSV) and showed that weekly treatment with the synthetic retinoid etretinate dramatically inhibited tumor incidence. The same retinoid caused tumor regression in chickens with RSV-induced tumors. Although the effects of retinoids on immune functions have not been investigated in any of the viral systems used, it is well-known that MSV-induced tumors are recognized and destroyed by cytotoxic T cells *in vivo* and *in vitro* (LeClerc *et al.*, 1973; Lamon *et al.*, 1972). It is therefore possible that the tumor-suppressive effects of retinoids are, at least in some of these models, due to immunostimulation.

3. *Effects of Retinoids on Transplantable Tumors*

Transplantable tumors have been used frequently to examine the protective effects of retinoids. Reported results have been quite variable, however, and many were negative. Bollag (1972) tested six transplantable tumors in mice treated with retinoic acid and observed no inhibitory effects on tumor growth. Similarly, Felix *et al.* (1976) could not demonstrate growth inhibition with retinyl palmitate in BALB/c mice transplanted with a syngeneic MCA-induced tumor. Patek *et al.* (1979) were unable to show inhibitory or curative effects by retinoic acid in BALB/c mice transplanted with four different syngeneic tumors. Baron *et al.* (1981) reported that injection of retinoic acid into P388 tumors growing in DBA/2 mice did not slow tumor growth. Clearly, negative results of this sort are difficult to interpret because the dosing of the drug with respect to its potential toxicity, as well as the timing of injection, are important parameters. The dose of the tumor inoculum is equally important so that the results of experiments depend on the manipulation of several variables. Furthermore, in order to be able to demonstrate that immune-mediated effectors stimulated by retinoids inhibit tumor growth, the tumor must express antigenic determinants that are recognizable by and immunogenic for lymphocytes. In the cases cited above, immunogenicity of the tumors has not yet been demonstrated, and the failure of these tumors to be inhibited by retinoids may be due to their lack of immunogenic cell-surface determinants (Patek *et al.*, 1979).

More interesting for this discussion may be the cases in which retinoids were used in conjunction with other drugs or treatments to cause tumor inhibition. For instance, Nathanson *et al.* (1969) reported that retinyl palmitate inhibits tumor growth when given with cytoxan but not by itself. Similarly, Meltzer and Cohen (1974) using an MCA-induced sarcoma obtained effects with retinyl palmitate in conjunction with BCG (Bacille Calmette Guérin) but not with the retinoid alone. Kurata and Mickshe (1977) also reported that Lewis lung tumors in syngeneic C57BL/6 mice are inhibited by a combination of BCG and retinyl palmitate but not by the retinoid only. Similarly, Pavelic *et al.* (1980) found that retinyl palmitate alone had no effect on Lewis lung tumor growth, but in conjunction with *Corynebacterium* showed a significant lag in tumor appearance, a doubling of survival time, and some permanent cures.

In the examples discussed so far it was shown that a tumor was inhibitable *in vivo* by some modalities and that retinoids, in conjunction with these modalities, i.e., cytoxan, BCG, or *Corynebacterium,* increase the inhibitory effect. It is therefore pertinent to ask how these various modalities might work in conjunction with retinoids. Do they act on immune effectors, directly on the tumor, or both? Cytoxan has toxic effects on both tumor cells and lymphocytes and is able to shift the percentage of the two major classes of immunocompetent lymphocytes from B cells to T cells (Poulter and Turk, 1972), thereby causing an enhanced cell-mediated immune response under certain experimental conditions. Retinoids exert inhibitory effects on the growth of a number of tumor cell lines *in vitro,* as reviewed by Lotan (1980), and as discussed in more detail in Chapter 12. Retinoids may also influence the cellularity of lymphoid organs by either decreasing cell numbers at toxic doses (Dennert and Lotan, 1978) or increasing them at subtoxic doses (Seifter *et al.*, 1981; Taub *et al.*, 1970). Whether these changes affect certain classes or subclasses of lymphocytes and whether this may influence a subsequent immune response has not yet been determined. The immunoadjuvants BCG and *Corynebacterium,* which have been used in conjunction with retionoids, may further influence possible retinoid-induced immune effects. Thus, tumor inhibition in these experimental systems is a very complex event. The combined inhibitory effect of retinoids in conjunction with immunopotentiators on tumor growth may support the hypothesis that retionids activate immunomediated mechanisms but by no means proves it.

Therefore, one must consider the less-complex situations in which retinoids per se have effects on transplantable tumors. Apparently these tumors are in the minority, and only a few papers on such models have been published. Deneufbourg (1979) reported that etretinate inhibited the growth of a sarcoma in syngeneic C57BL/6 mice as well as increased animal survival. Interestingly, upon challenge of cured mice, a high percentage of the mice that previously received retinoid were resistant to a secondary tumor challenge. This was in contrast to the mice that did not receive retinoid and suggests that the retinoid might have an effect on the priming of immunological memory rather than a direct toxic effect on the tumor. In a prior study, Tannock *et al.* (1972) observed that mice that received oral doses of retinyl palmitate required a reduced dose of local tumor-directed irradiation to reject a syngeneic fibrosarcoma than mice that did not receive retinoid. While this result could be caused by direct toxic effects of the retinoid on the tumor, in control experiments it was seen that the effects could not be obtained in mice that were immunosuppressed by a prior sublethal dose of whole-body irradiation. This finding therefore provides additional support for the immunoenhancing effects of retinoids.

Patek *et al.* (1979) studied the effects of retinoic acid in seven tumor models and found that inhibitory effects with the drug could be accomplished in the following three models: (1) the appearance of the EL4 tumor transplanted into syngeneic C57BL/6 mice was delayed, but the tumor growth rate was not decreased nor was animal survival increased; (2) the DBA/1 melanoma S91 transplanted into allogeneic BALB/c mice showed inhibited growth; and (3) the L33 tumor transplanted into syngeneic BALB/c mice showed a markedly reduced tumor incidence. The S91 model had been studied previously by Felix *et al.* (1975, 1976), and two interesting observations had been made. First, inhibition of S91 growth by retinoid in BALB/c mice could be abrogated by prior injection of mice with antilymphocyte serum (ALS), an antiserum raised in rabbits against mouse T cells and able to suppress T cell-mediated immune responses. Second, the inhibition by retinoid did not occur in (BALB/c $\times$ DBA/1)F_1 mice, which, of course, share the major histocompatibility antigens (H-2) with the DBA/1 tumor S91. This indicated that the retinoid-induced tumor suppression in BALB/c mice was due to retinoid-stimulated allo-H-2-specific T killer cells. Therefore, these two observations very strongly suggested immunomediated effects by the retinoid and do not support direct toxic effects responsible for the tumor growth inhibition *in vivo*.

Further direct proof that retinoid-effected inhibition of tumor growth and lower incidence of tumors is due to stimulation of the immune system rather than by toxic effects of retinoid directly on the tumor was provided by the experiments of Patek *et al.* (1979). Animals were thymectomized, lethally irradiated, and reconstituted with fetal liver (ATxFL) in order to compromise their humoral and cell-mediated immune system by the elimination of T cells. Under these conditions it was demonstrated that ATxFL mice are not able to suppress the growth of S91 or L33 tumor growth when treated with retinoid. Therefore, for these two tumor models, it seems highly likely that the retinoid effect requires the participation of immunocompetent cells, in particular, T cells.

III. ATTEMPTS TO DELINEATE THE IMMUNOSTIMULATORY EFFECTS OF RETINOIDS

Because it appears that retinoids may act in an immunostimulatory capacity, at least in some tumor systems, the next question that can be asked is whether humoral, i.e., antibody-mediated, or cellular immunity is stimulated. In particular, one can ask specifically which immunological cell types are influenced by retinoids. Experiments with this objective require *in vivo* and *in vitro* immune response assays, cell culture techniques, and cell separation procedures that are quite laborious. Consequently, there are only a small number of reports that adequately deal with this question, albeit, only in a rather preliminary way. In the following sections, we attempt to evaluate the effects of retinoids on the two major arms of the immune system—the humoral and the cell-mediated. This division may seem somewhat artificial because of the overlap between the two systems, but it does provide some classification, enabling us to evaluate more easily the work in this field.

A. Effects of Retinoids on Humoral Immunity

The humoral immune response culminates in the formation of antibody and is the consequence of as yet insufficiently understood interactions between various cell types. The antigen is thought to be processed and presented to B cells by so-called antigen-presenting cells, often referred to as adhering cells or macrophages. The antigen is recognized by B cells, which subsequently produce antibody, as well as by helper T cells, which provide stimulatory signals to the B cells. The humoral antibody response can be measured in several ways; for instance, by direct assay of antibody in the serum, i.e., agglutination or radioactive binding assays, or by determining the number of lymphocytes in lymphoid organs capable of producing antibody using the well-known Jerne plaque technique. Because these assays for antibody production are relatively easy to perform, initial work on the immunostimulatory effects of retinoids was directed primarily toward demonstrating augmentation by retinoids on the humoral arm of the immune system. Publications discussed in the next section deal with the effects of retinoids on cell-mediated immunity.

One of the first studies of the effects of retinoids on humoral immunity was that of Dresser (1968) who found that retinol stimulates the humoral immune response in mice to nonaggregated bovine gamma globulin (BGG), an antigen of relatively low immunogenicity. A very similar finding was reported by Spitznagel and Allison (1970) who observed that the primary and secondary serum antibody response to bovine serum albumin (BSA) can be stimulated by retinol. Jurin and Tannock (1972) reported an increase in serum antibody of mice that received retinyl palmitate and sheep erythrocytes as antigen. Similarly, Cohen and Cohen (1973a,b), using a plaque assay for sheep erythrocytes reported that prior doses of retinyl palmitate induced a fivefold increase of plaque-forming cells to sheep erythrocytes in the spleen. Using a radioactive binding assay, they also reported that the serum antibody response to dinitrophenyl-coupled ovalbumin (DNP–OVA) is enhanced owing to prior treatment of mice with retinyl palmitate. Falchuck *et al.* (1977) confirmed that retinyl palmitate stimulates the serum antibody response to BSA and showed, by assaying mucosal extracts from the intestine, that the local immune response (i.e., IgA) was also enhanced. In a study pertinent to this observation, Sirisinha *et al.* (1980) reported that retinoid-deficient rats have a low level of IgA antibody in their intestinal fluid, and that the humoral response to DNP–BGG is depressed in the intestine but apparently not in the serum. Also using rats, Lapin *et al.* (1974) showed that retinyl palmitate stimulates the plaque-forming response to sheep eryhrocytes. Barnett and Bryant (1980) emulsified retinol in Tween 80 as adjuvant and found that the secondary but not the primary antibody response to OVA was stimulated. In a subsequent study, it was shown that oral administration of 13-*cis*-retinoic acid resulted in stimulation of the IgE response after secondary OVA injection (Barnett, 1982). From these results it appears that retinol, retinyl palmitate, and 13-*cis*-retinoic acid may stimulate the humoral immune response to soluble or particulate antigens.

In all these studies, either serum antibody or the plaque-forming response was tested with the exception of IgE, which was tested by a passive cutaneous anaphylaxis test. In none of the experiments was an attempt made to delineate the cell types that may be stimulated by the retinoid. Therefore, we do not know presently whether the drug acts on B cells, T cells, antigen-presenting cells or other cell types. Three different retinoids were employed and it is not known whether any one in particular may be superior to the others for immunostimulation because no comparative studies have been done.

In the various experiments reported, the drugs were administered, in most cases by the oral route and in some cases by injection, with no apparent differences. Oral administration, of course, has the inherent problem of not knowing with certainty how much retinoid each animal receives and therefore, accurate dosing studies require injection. Dose–response studies to explore the relationship between the optimal dose and the humoral antibody response have also not been performed yet, but it seems clear from the previously cited reports that subtoxic doses of retinoids are required in order to obtain stimulatory effects. Another problem often not addressed is the retinoid content of the food given to the animals *ad libitum*. Clearly, a relatively low amount of retinoid in the food may bring about a higher response in animals given additional retinoid.

In our own experiments, varying doses of retinoic acid injected into mice had no effect on the primary plaque-forming response to sheep erythrocytes nor was there a stimulation of carrier priming (Dennert and Lotan, 1978). Carrier priming is a measure for the induction of T helper cells. In this assay, the degree of priming is assessed by a subsequent challenge with hapten-coupled carrier and measurement of the antibody response to the hapten. It is difficult to explain why in our experiments no stimulation of the humoral antibody response was seen, particularly since under the same experimental conditions a very significant stimulation of the cell-mediated response was observed (see below). It appears that much work has yet to be done, determining both the best type of retinoid to use and optimal conditions for its use, before experiments aimed at analyzing the cell type stimulated by and mode of action of retinoids can be addressed. At present it appears well founded, however, to conclude that retinoids may stimulate the humoral antibody response to various antigens in experimental animals.

B. Cell-Mediated Immunity Is Stimulated by Retinoids

The term cell-mediated immunity encompasses a number of different immunological phenomena in which the response is caused by sensitized cells rather than by humoral antibody. Common examples for cell-mediated immunity are delayed-type hypersensitivity, transplant rejection, and *in vitro* cell-mediated cytotoxicity, all of which can be caused by cell suspension but not by antibody-containing sera.

1. Delayed-Type Hypersensitivity

Induction of the delayed-type hypersensitivity reaction (DTH) requires prior sensitization by antigen. DTH is measured as a skin-swelling reaction 24 h after antigen challenge. Although histologically the reaction is characterized by heavy neutrophil infiltration, the cell type that appears crucial in the reaction and possesses immunological memory is a T cell the precise function of which is not known. The intensity of the DTH reaction is usually considered to be a reflection of the degree of T cell priming. Jayalakshmi and Gopalan (1958) reported that children with vitamin A deficiency have a depressed DTH reaction to BCG. We tried to verify this finding in a mouse model in which animals were sensitized to sheep erythrocytes after intraperitoneal injection of various doses of retinoic acid. Using a footpad-swelling assay, no stimulatory effects of retinoic acid in the DTH reaction to sheep erythrocytes were observed (Dennert and Lotan, 1978).

Athanassiades (1981) published a more comprehensive study also using sheep erythrocytes as antigen. Four different retinoids were tested, all in subtoxic doses: retinyl palmitate, motretinid (F3), etretinate, and 13-*cis*-retinoic acid. All retinoids, with the exception of etretinate, stimulated DTH when injected together with the sensitizing antigen. This protocol had not been used in our experiments, which may explain our negative results. To date, this is the only report of the direct effects of retinoids on the DTH reaction, with the exception of that by Micksche (1978) who observed that cancer patients treated with retinoids develop an increase in skin reactions to various antigens. Because the retinoids act in the sensitization phase of DTH induction, it is likely that T cells are directly or indirectly stimulated by the retinoid. Evidence for this conclusion, however, requires further experimentation.

2. *Skin Graft Rejection*

Another parameter of cell-mediated immunity is the rejection of either skin or tumor grafts. Since both responses appear to be effected primarily by cytotoxic T cells in allogeneic systems, they provide good models for assessing the effects of retinoids on the cellular immune response. Jurin and Tannock (1972) observed that skin grafts from male C57BL/6 mice transplanted onto syngeneic female mice, were rejected significantly faster when the transplanted mice were given retinyl palmitate. In a very similar system, Floersheim and Bollag (1972) reported that CBA skin transplanted onto C3H mice was rejected earlier (i.e., at 20 rather than 40 days) if mice were given retinoic acid. While skin graft rejection in the absence of retinoids is most likely due to T killer cells, the effects caused by the retinoids could be due to either stimulation of the cell-mediated immune system or the direct toxic effects on the skin graft resulting in a more efficient rejection by the T killer cells. This latter possibility can not be excluded since Floersheim and Bollag (1972) observed that doubling the dose of retinoid may cause the rejection of autografts. Also, Boss *et al.* (1966) demonstrated that skin autografts may be rejected by rabbits given high doses of retinoid. Theoretically, the immune system should not recognize or respond against autologous tissue.

In another study, Medawar and Hunt (1981) studied the effects of dietary retinyl acetate on the rejection of male skin transplanted onto syngeneic female C57BL/6 mice and confirmed that this drug stimulates skin transplant rejection. It was also observed that there is a difference in second-set transplant rejection between animals that received or did not receive retinoid during the first rejection. Animals that did not receive retinoid during the first rejection showed a fast second-set rejection, whereas animals previously given the drug showed a slow first-set rejection on regrafting with male skin. This result appears to be at variance with that of Deneufbourg (1979) who showed an enhanced second-set tumor rejection in retinoid-treated animals rechallenged with tumor grafts. Medawar and Hunt's observation may suggest that the enhancement of primary skin graft rejection is caused by direct toxicity rather than by enhanced cell-mediated immunity. This reasoning is derived from the observation that there is apparently no immunological memory in the retinoid-treated mice. Hence, the primary rejection may be caused mainly by retinoid toxicity and only partially by T killer cells. Owing to this effect, little priming of memory T killer cells might have occurred. Another explanation for this result might be that the retinoid, besides stimulating cytotoxic immune effectors in the primary reaction, also induces suppresser cells for the secondary rejection. It is therefore apparent that stimulation of skin graft rejection by retinoids cannot be used as an argument for the immunostimulatory action of these compounds, even when it was shown that retinoid-stimulated skin graft rejection is inhibitable by antilymphocyte serum (Medewar and Hunt, 1981). The elimination of one immune effector, presumably the T killer, by antilymphocyte serum would obviously prolong the time required for skin graft rejection, regardless of whether retinoids act by immunostimulation or by direct toxicity on the graft.

3. Tumor Models

One of the first pieces of evidence suggesting that immunocompetent cells play a role in retinoid-enhanced rejection of tumor transplants was the observation that immunocompromised ATxFL mice fail to reject tumors when treated with retinoid (Section I,B,3). Glaser and Lotan (1979) used a somewhat different approach to demonstrate that cell-mediated immunity to tumors can be stimulated by retinoic acid. BALB/c mice were treated with retinoic acid and subsequently immunized with the syngeneic SV40 transformed tumor KSA-TU5. Spleen cells of immunized mice were harvested, mixed at various ratios with tumor cells, and injected into normal mice. In this assay (Winn assay) the incidence of tumors growing is a function of the ratio of spleen cells to tumor cells and degree of sensitization of spleen cells injected together with the tumor. Results showed that the tumor-neutralizing efficacy of sensitized spleen cells was strongly stimulated as a result of retinoic acid injection during the sensitization step. It was also observed that retinoic acid-stimulated immunity was markedly enhanced, initiated earlier, and of an increased duration. Furthermore, it was shown that the neutralizing activity was specific for the tumor and sensitive to anti-Thy-1 plus complement treatment. These results demonstrated several important points. Immunity stimulated by retinoic acid is cell-mediated and involves Thy-1^+ cells. Also, since immunity is specific, it is likely that specific T killer cells are the effectors that are subject to retinoid stimulation and active in the neutralization assay.

To further explore which cell types are involved in the retinoid-stimulated tumor rejection, a number of different *in vivo* and *in vitro* systems were employed. First, an allogeneic tumor model was chosen in which it was known that T killer cells are the effectors: C57BL/6 (H-2^b) mice transplanted with the BALB/c (H-2^d) tumor S194. Injection of mice with 25–100 μg/mouse/day of retinoic acid for 6 days prior to tumor challenge resulted in a strongly stimulated induction of cytotoxic cells as assayed by an *in vitro* cell-mediated cytotoxicity assay using S194 targets. This observation showed that killer-cell activity could be stimulated *in vivo* by retinoic acid and provided the basis for further experiments in which retinoic acid was injected into mice, the spleen cells harvested and challenged *in vitro* with S194 cells. Again, in this situation, there was a strong stimulation of killer activity as a result of prior retinoid injection. This result suggested that retinoic acid treatment of mice may make the cell-mediated immune system more alert to a subsequent challenge. In order to verify that the retinoid-stimulated killer response is T cell-mediated, C57BL/6 mice were treated with retinoid, challenged with S194, and the cytolytic activity tested on S194 targets (Dennert *et al.*, 1979). Treatment of sensitized spleen cells with anti-Thy-1 plus complement completely abrogated the cytolytic activity. This proved that Thy-1^{+} cells are the effectors for lysis and that the retinoid stimulation, whatever its mechanism, results in either higher activity per cell or a greater number of cytolytic Thy-1^{+} cells.

This result envisages two possibilities: either the effector cells are specific T killer cells or they are nonspecific natural killer (NK) cells because both cell types can express the Thy-1 antigen. In order to distinguish between NK cells and T killer cells, specificity experiments were done. Results showed that C57BL/6 (H-2^{b}) effector cells sensitized to S194(H-2^{d}) lysed S194(H-2^{d}), P815(H-2^{d}), and BALB/c(H-2^{d}) spleen blast cells. NK cells, in contrast to T killer cells, are not able to lyse spleen blast cells or the tumor P815 (Warner and Dennert, 1982). Therefore, the lysis of BALB/c spleen cells and P815 cells shows that the retinoid-stimulated cells are specific T killer cells and not NK cells. These results provided the first direct proof that retinoic acid stimulates T killer cells specific for major histocompatibility (H-2) antigens in the sensitization step.

The next important step was to ascertain that the same effects of retinoids, i.e., stimulation of T killer cells, can be observed in syngeneic tumor models. This was accomplished in two systems: C57BL/6 spleen cells sensitized to the syngeneic tumor EL4 and BALB/c spleen cells sensitized to the syngeneic tumor S194. The two models are interesting to compare because we know that both EL4 and S194 induce T killer cells in syngeneic situations. But S194, in contrast to EL4, can be lysed by NK cells, which provided another possibility to determine if NK cells are stimulated by retinoic acid. C57BL/6 and BALB/c mice were injected with retinoid as before, the spleen cells harvested and cultured with EL4 and S194 tumor cells, respectively. Results showed that in both cases cytolysis was strongly stimulated as a result of prior retinoid treatment of the mice. Since EL4 targets are not lysable by NK cells, by inference, the lysis is due to T killer cells. In the case of S194, the situation is not clear because this tumor is lysable by NK cells. Further experiments, however, showed that lysis was specific for S194 in that other BALB/c tumor targets, also lysable by NK cells, were not lysed by S194-induced killer cells. These experiments conclusively showed that retinoic acid stimulates the induction of T killer cells specific for syngeneic tumors (Dennert *et al.*, 1979).

C. Mechanism of Retinoid Action

At present, we can only speculate how retinoids stimulate the immune system because so little is known about the cell types that interact with retinoids. Moreover, more information is available for the cell-mediated than the humoral immune system. We know that pretreatment of mice with retinoid continuously for several months (Dennert *et al.*, 1979) may give rise to higher cell-mediated immune responses on antigenic challenge. We have not observed any stimulation of lymphocyte proliferation induced by retinoids *in vitro* (Dennert and Lotan, 1978) nor at the effector level (i.e., postsensitization) of cell-mediated lysis (Dennert *et al.*, 1979). The latter result had been reported earlier by Stavy *et al.* (1973) using retinol. From these results, it seems clear that retinoids act during or before the sensitization step of cell-mediated immunity. It was therefore of interest to examine whether retinoids may be able to stimulate other primary T cell responses.

As a first step in this direction, we tested whether retinoic acid *in vitro* would influence the *in vitro* proliferative response of spleen cells to T cell mitogens like concanavalin A (Con A) and phytohemagglutinin (PHA). Our results were negative (Dennert and Lotan, 1978). This, however, does not appear to be a universal finding. Sidell (1981), for instance, using human thymocytes and tonsil lymphocytes, found that retinoic acid stimulates both the proliferative response to allogeneic lymphocytes and to PHA. Interestingly, when spleen cells or peripheral blood lymphocytes were used, these effects could not be reproduced. This is in contrast to the work by Abb and Deinhardt (1980), who observed that retinoic acid stimulates the PHA response of human peripheral blood lymphocytes. It is not clear at present what is the reason for this discrepancy.

Another approach to studying the effects of retinoids on primary T cell responses is to examine cells from animals, either deficient in or supplemented with retinoid, for their ability to respond by cell proliferation to allogeneic cells or mitogens. Nauss *et al.* (1979) found that retinoid-deficient rats have marked lymphopenia and a low proliferative response to PHA and Con A. All responses increased when animals were fed retinyl acetate. In a similar study, Sonis and Shklar (1981) observed that splenocytes from hamsters treated with 13-*cis*-retinoic acid show a higher response to PHA in culture. Cohen *et al.* (1979) reported that human patients, after surgery, may have a normal PHA response but a depressed mixed-lymphocyte reaction for several weeks. Retinoid therapy increased reactivity in the mixed-lymphocyte reaction even beyond preoperative levels. Micksche *et al.* (1977; see also Micksche, 1978) used both retinyl palmitate and 13-*cis*-retinoic acid in patients with squamous cell carcinoma. While there was little effect of these drugs on the rate of survival of these patients, they showed a significantly higher responsiveness to PHA. Identical results were reported by Kucera (1980) in a study of patients with cervical carcinoma. It therefore appears to be a well-established observation that retinoid treatment of normal or retinoid-deficient humans or animals may result in a heightened reactivity to either T cell mitogens or to allogeneic lymphocytes. Regarding the speculations as to how retinoids may influence cell-mediated immunity, one could hypothesize that the T cell subclass responsible for cell proliferation in mitogen and mixed-lymphocyte reaction responses also plays a role in the induction of T killer cells, for instance, by performing helper functions.

Of an equally speculative nature are the possible effects of retinoids on other cell types in the cell-mediated immune system. In preliminary experiments, we examined the effects of retinoic acid on NK activity. While in some experiments *in vivo* treatment of mice with retinoids tended to increase *in vitro* NK cytotoxicity, in other experiments it did not (G. Dennert, unpublished results). Goldfarb and Herberman (1981) reported weak stimulation of *in vitro* activity of NK cells from some but not all human donors by *in vitro* incubation with retinoic acid. Further experiments are needed to clarify what the significance of this result may be since it could be due to an artifact of *in vitro* testing. Rhodes and Oliver (1980) studied effects of retinoic acid on human peripheral blood monocytes and observed inhibition of Fc receptor expression, another observation with a significance that cannot be assessed at present. One could envisage that retinoids act on cells by influencing lymphokine production and thereby regulating humoral or cell-mediated immune responsiveness. Only a few publications have dealt with this aspect of retinoid action. Blalock and Glifford (1976) found that retinoic acid inhibits interferon production of Newcastle disease virus-infected L cells, and Abb and Deinhardt (1980) observed that interferon secretion of human peripheral blood lymphocytes stimulated by PHA is inhibited by retinoic acid. Therefore, at present, no clear hypothesis of how retinoids may regulate immune responsiveness has been formulated.

IV. CONCLUSIONS

From the literature discussed above, it is apparent that our knowledge on the effects of retinoids on the immune system is still rather incomplete. Nevertheless, there are a number of conclusions we can draw at this point.

1. Retinoids have effects on the cellularity of lymphoid organs. It is well-known that retinoids in high doses are very toxic (Mallia *et al.*, 1975; Bollag, 1974) and may cause cell loss from lymphoid organs, in particular from the thymus (Dennert and Lotan, 1978). On the other hand, moderate subtoxic doses of retinoids have been reported to result in a significant increase in thymus

weight, to stimulate thymic involution (Seifter *et al.*, 1981), and to increase cellularity of lymph nodes (Taub *et al.*, 1970).

2. Retinoids inhibit or stimulate the immune system. The histological changes in lymphoid organs caused by retinoids are reflected in their inhibitory or stimulatory effects on various immune responses. Thus, high doses of retinoids quite effectively may inhibit both humoral and cell-mediated immunity (Uhr *et al.*, 1963; Dennert and Lotan, 1978), whereas subtoxic doses may stimulate them. From the very diverse results reported in the literature on stimulatory or inhibitory effects of retinoids, it is clear that timing, dose, and mode of administration of the drugs are crucial in causing either stimulatory or inhibitory effects on immune responses. It is also important to note that stimulatory or inhibitory effects observed in an *in vitro* assay may not necessarily be found *in vivo* (Dennert and Lotan, 1978) and that even effects with one lymphocyte population may not necessarily be reproducible with one from another organ or another individual. Great care, therefore, must be exercised in interpreting experimental results.

3. Synthetic retinoids have effects similar to those of natural retinoids. A number of different natural or synthetic retinoids have been used in the experiments discussed. Judging from the limited number of comparative studies, no dramatic differences between different retinoids are found. In the induction of cell-mediated cytotoxicity, retinoic acid, 13-*cis*-retinoic acid, retinyl palmitate, and etretinate all had quite comparable stimulatory effects (Lotan and Dennert, 1979). In the DTH reaction, retinyl palmitate, motretinid (F3), and 13-*cis*-retinoic acid had stimulatory effects, but etretinate (F1) had no effect (Athanassiades, 1981). At present, it is therefore not clear what the molecular constraints on the retinoid molecule may be for stimulatory effects on the immune system.

4. Retinoids appear to act in the induction phase of immunity. There is conclusive evidence that retinoids stimulate the induction of T killer activity *in vivo* and *in vitro* both to allogeneic and syngeneic (tumor) cells. Since retinoids also stimulate the proliferative responsiveness of T cells to PHA, Con A, or allogeneic lymphocytes, it is possible that retinoids act on proliferating helper-type T cells that participate in the inductive phase of T killer sensitization. Under certain experimental conditions, retinoids also have stimulatory effects on the sensitization step of DTH. Although at present we do not know which cell type is affected, chances are that it is the DTH T cell. If this were correct, one could correlate stimulation of DTH and stimulation of humoral immunity, i.e., T cells causing DTH and T cells stimulating B cells for antibody synthesis may be identical (Bianchi *et al.*, 1981; Dennert *et al.*, 1981) and hence may cause retinoid-stimulated enhancement of humoral and cell-mediated immunity.

5. There are many unresolved questions that could be subject to further experimentation. There are areas of retinoid research that have not been explored,

for instance, the potential effects of retinoids on suppressor cells. Patients with Dariers disease, for example, have lymphocytes that express unusual hyperreactivity to Con A. If these patients are treated with etretinate, the response levels normalize, which may be due to induction of suppressor cells (Soppi *et al.*, 1981). Another area of potential experimentation is that of the effects of retinoids on interleukins, lymphokines which regulate lymphocyte function. With the recent advances in the lymphokine field, it is possible that this approach may lead to some insights into the action of retinoids.

6. Retinoids may provide important tools to influence the immune system in clinical situations. Concerning the potential application of retinoids in infectious disease, many results available were obtained under situations of vitamin A deficiency. In situations where no vitamin A deficiency is detectable, protection from bacterial and fungal infections in animals has been reported. Since it seems difficult to extrapolate these findings to humans, epidemiological studies in this respect may be informative. At present the few data available in animals do not permit conclusions either way. A very similar situation exists for the question of the effectiveness of retinoids in tumor growth or immune surveillance. On the one hand, it is clear that a variety of transplantable tumors exhibit inhibited growth in experimental animals owing to retinoid-enhanced immune mechanisms. Also, carcinogen-induced tumor formation in experimental animals can be inhibited by retinoids, although the mechanisms by which this occurs are not known and may not be immunologically mediated. On the other hand, attempts aimed at inhibiting established tumors in humans have shown few promising results (Micksche *et al.*, 1977; Kucera, 1980). It may turn out that retinoids are much more effective in preventing the appearance of neoplasia. It seems that studies in both areas, therapy, and prevention, are desperately needed. The finding that retinoids have stimulatory effects on the immune system add an important aspect to future studies.

ACKNOWLEDGMENTS

This work was supported by United States Public Health Service Grants CA 15581 and CA 19334 and American Cancer Society Grant IM-284. I thank Marian Marra for help in writing this chapter and Dr. John Warner for useful discussions.

REFERENCES

Abb, J., and Deinhardt, F. (1980). *Exp. Cell Biol.* **48,** 169–179.

Athanassiades, T. J. (1981). *JNCI, J. Natl. Cancer Inst.* **67,** 1153–1156.

Barnett, J. B. (1982). *Int. Arch. Allergy Appl. Immun.* **67,** 287–290.

Barnett, J. B., and Bryant, R. L. (1980). *Int. Arch. Allergy Appl. Immun.* **63,** 145–152.

Baron, S., Kleyn, K. M., Russel, J. K., and Blalock, J. E. (1981). *JNCI, J. Natl. Cancer Inst.* **67,** 95–97.

Bianchi, A. T. J., Hooijkaas, H., Benner, R., Tees, R., Nordin, A. S., and Schreier, M. H. (1981). *Nature (London)* **290,** 62–63.

Blalock, J. E., and Gifford, G. (1976). *J. Gen. Virol.* **32,** 143–147.

Bloch, C. E. (1928). *Acta Paediatr. (Stockholm)* **7,** 61–64.

Bollag, W. (1972). *Eur. J. Cancer* **8,** 689–693.

Bollag, W. (1974). *Eur. J. Cancer* **10,** 731–737.

Boss, J. H., Bitterman, W., and Gabriel, M. (1966). *Transplantation* **4,** 293–299.

Chopra, D. P., and Wilkoff, L. J. (1976). *JNCI, J. Natl. Cancer Inst.* **56,** 583–589.

Cohen, B. E., and Cohen, I. K. (1973a). *Surg. Forum* **24,** 276–278.

Cohen, B. E., and Cohen, I. K. (1973b). *J. Immunol.* **111,** 1376–1380.

Cohen, B. E., and Elin, R. J. (1974). *J. Infect. Dis.* **129,** 597–600.

Cohen, B. E., Gill, G., Cullen, P. R., and Morris, P. J. (1979). *Surg. Gynecol. Obstet.* **149,** 658–662.

Davies, R. E. (1967). *Cancer Res.* **27,** 237–241.

Deneufbourg, J. M. (1979). *Biomed. Express* **31,** 122–123.

Dennert, G., and Lotan, R. (1978). *Eur. J. Immunol.* **8,** 23–29.

Dennert, G., Crowley, C., Kouba, J., and Lotan, R. (1979). *JNCI, J. Natl. Cancer Inst.* **62,** 89–94.

Dennert, G., Weiss, S., and Warner, J. (1981). *Proc. Natl. Acad. Sci. U.S.A.* **78,** 4540–4543.

Dresser, D. W. (1968). *Nature (London)* **217,** 527–529.

Falchuck, K. R., Walker, W. A., Perotto, J. L., and Isselbucher, K. J. (1977). *Infect. Immun.* **17,** 361–365.

Felix, E., Loyd, B., and Cohen, M. H. (1975). *Science* **189,** 836–837.

Felix, E. L., Cohen, M. H., and Loyd, B. C. (1976). *J. Surg. Res.* **21,** 307–312.

Floersheim, G. L., and Bollag, W. (1972). *Transplantation* **15,** 564–567.

Frankel, J. W., Horton, E. J., Winters, A., Samis, H. V., and Ito, Y. (1980). *Intervirology* **14,** 321–325.

Glaser, M., and Lotan, R. (1979). *Cell. Immunol.* **45,** 175–181.

Goldfarb, R., and Herberman, R. B. (1981). *J. Immunol.* **126,** 2129–2135.

Grubbs, C., Moon, R. C., Squire, R. A., Farrow, G. M., Stinton, S. F., Goodman, D. G., Braun, C. C., and Sporn, M. B. (1977). *Science* **198,** 743–744.

Hof, H., and Emerling, P. (1979). *Ann. Immunol. (Paris)* **130,** 587–594.

Jayalakshmi, V. T., and Gopalan, C. (1958). *Indian J. Med. Res.* **46,** 87–92.

Jurdin, M., and Tannock, I. F. (1972). *Immunology* **23,** 283–287.

Kucera, H. (1980). *Wien. Klin. Wochenschr.* **92,** Suppl. 118, 3–20.

Kurata, T., and Micksche, M. (1977). *Oncology* **32,** 212–215.

Lamon, E. W., Skurzak, H. M., Klein, E., and Wigzell, H. (1972). *J. Exp. Med.* **136,** 1072–1079.

Lapin, V., Rella, W., and Wrba, H. (1974). *Öster. Zeitschr. f. Onkol.* **3,** 96–99.

Lasnitzki, I. (1955). *Br. J. Cancer* **9,** 434–441.

LeClerc, J. C., Gomard, E., Plata, F., and Levy, J. P. (1973). *Int. J. Cancer* **11,** 426–432.

Levine, N. S., Salisbury, R. E., Seifter, E., Walker, H. L., Madison, A. D., Jr., and Pruitt, B. A., Jr. (1975). *Experentia* **31,** 1309–1312.

Lotan, R. (1980). *Biochim. Biophys. Acta* **605,** 33–91.

Lotan, R., and Dennert, G. (1979). *Cancer Res.* **39,** 55–58.

McMichael, H. (1965). *Cancer Res.* **25,** 947–952.

Mallia, A. K., Smith, J. E., and Goodman, D. S. (1975). *J. Lipid Res.* **16,** 180–188.

Medawar, P. B., and Hunt, R. (1981). *Immunology* **42,** 349–353.

Meltzer, M. S., and Cohen, B. E. (1974). *JNCI, J. Natl. Cancer Inst.* **53,** 585–587.

Merriman, R. L., and Bertram, J. S. (1979). *Cancer Res.* **39,** 1661–1666.

Micksche, M., Cerni, C., Kokron, O., Titscher, P., and Wrba, H. (1977). *Oncology* **34,** 234.
Micksche, V. M. (1978). *Wien Klin. Wochenschr.* **90,** 3–28.
Moon, R. C., and Grubbs, C. J., Sporn, M. B., and Goodman, D. G. (1977). *Nature (London)* **267,** 620–621.
Nathanson, L., Maddock, C. L., and Hall, T. C. (1969). *J. Clin. Pharm.* **9,** 359.
Nauss, K. M., Mark, D. A., and Suskind, R. M. (1979). *J. Nutr.* **109,** 1815–1823.
Patek, P. Q., Collins, J. L., Yogeeswaran, G., and Dennert, G. (1979). *Int. J. Cancer* **24,** 624–628.
Pavelic, Z. P., Sulochana, D., Bialkowski, S., Priore, R. L., and Greco, W. R. (1980). *Cancer Res.* **40,** 4617–4621.
Poulter, L. W., and Turk, J. L. (1972). *Nature (London), New Biol.* **238,** 17–18.
Rhodes, J., and Oliver, S. (1980). *Immunology* **40,** 467–472.
Sirisinha, S., Darip, M. D., Moongkarndi, P., Ongsakul, M., and Lamb, A. J. (1980). *Clin. Exp. Immunol.* **40,** 127.
Scrimshaw, N. S., Taylor, C. E., and Gordon, J. E. (1968). *WHO Monogr. Ser. No. 57* pp. 40–41, 87–96, 143–180.
Seifter, E., Rettura, G., Seifter, J., Davidson, H., and Levenson, S. M. (1973a). *Fed. Proc., Fed. Am. Soc. Exp. Biol.* **32,** 947.
Seifter, E., Zisblatt, M., Levine, N., and Returra, G. (1973b). *Life Sci.* **13,** 945–952.
Seifter, E., Rettura, G., and Levenson, S. M. (1981). *JNCI, J. Natl. Cancer Inst.* **67,** 467–472.
Sidell, N., Famatiga, E., and Golub, S. H. (1981). *Exp. Cell Biol.* **49,** 239–245.
Sonis, S., and Shklar, G. (1981). *J. Oral Med.* **36,** 117–120.
Soppi, A. M., Soppi, E., Escola, J., and Jansen, C. T. (1982). *Br. J. Dermatol.* **106,** 141–152.
Spitznagel, J. K., and Allison, A. C. (1970). *J. Immunol.* **101,** 119–127.
Sporn, M. B., Dunlop, N. M., Newton, D. L., and Smith, J. L. (1976). *Fed. Proc., Fed. Am. Soc. Exp. Biol.* **35,** 1332–1338.
Sporn, M. B., Squire, R. A., Brown, C. C., Smith, J. M., Wenk, M. L., and Springer, S. (1977). *Science* **195,** 487–489.
Stavy, L., Cohen, I. R., and Feldman, M. (1973). *Cell. Immunol.* **7,** 302–312.
Tannock, I. F., Suit, H. O., and Marshall, N. (1972). *JNCI, J. Natl. Cancer Inst.* **48,** 731–741.
Taub, R. N., Krantz, A. R., and Dresser, D. W. (1970). *Immunology* **18,** 171–185.
Uhr, J. W., Weissmann, G., and Thomas, W. (1963). *Proc. Soc. Exp. Biol. Med.* **112,** 287–291.
Warner, J. F., and Dennert, G. (1982). *In* ''Lymphokines in Antibody and Cytotoxic Responses'' (S. B. Mizell, ed.), pp. 165–184. Academic Press, New York.
Wolbach, S. B., and Howe, P. R. (1925). *J. Exp. Med.* **42,** 753–777.

16

Synthetic Retinoids in Dermatology

GARY L. PECK
Dermatology Branch
National Cancer Institute
National Institutes of Health
Bethesda, Maryland

I. INTRODUCTION

During the past year (1982–1983), two synthetic retinoids have been introduced into the clinical practice of dermatology. In the United States, isotretinoin (E6) (13-*cis*-retinoic acid, Accutane) was approved for the treatment of severe

THE RETINOIDS, VOL. 2

ISBN 0-12-658102-9

cystic acne, and in Europe, etretinate (F1) (Ro 10-9359, Tigason) is in use, either alone or in combination with other agents, for the treatment of psoriasis and related cutaneous disorders of keratinization. In addition, there is preliminary evidence that synthetic retinoids may be of value in the treatment and prevention of skin cancer.

These accomplishments represent the culmination of a decade of clinical and laboratory investigation into the therapeutic spectrum and the mechanisms of action of retinoids in the skin. This interest arose from the use of oral vitamin A as therapy for a variety of dermatoses for the past 40 years. This usage was based initially on the similarity between the follicular keratoses seen in vitamin A deficiency and those in previously treatment-resistant skin disorders, such as Darier's disease (keratosis follicularis) and pityriasis rubra pilaris. Vitamin A deficiency thus provides a conceptual link in understanding how retinoids may be effective as therapy for a wide range of dermatologic disorders.

Vitamin A deficiency is characterized by squamous metaplasia of a variety of epithelia with increased cell proliferation and hyperkeratosis. These changes are also features of some benign dermatoses, for example, psoriasis. Once beneficial effects of oral vitamin A were observed, its use spread to the treatment of other diseases of the epidermis and epidermal appendages, including acne, psoriasis, and basal cell carcinoma. Since the hypervitaminosis A syndrome (see Chapter 13) interfered with long-term therapy with vitamin A, the need arose for synthetic derivatives that could be at least as efficacious as vitamin A and yet be less toxic. The use of isotretinoin and etretinate, as described above, represents the first development of this concept in clinical practice.

II. BIOLOGICAL EFFECTS OF RETINOIDS ON THE SKIN

A. Morphologic Effects

Since Fell and Mellanby (Fell and Mellanby, 1953) initially observed the inhibition of keratinization in chick embryo skin and its subsequent transformation into a mucous-secreting structure by vitamin A, many studies have been concerned with the mode of action of vitamin A on a variety of epithelia. The initial effects of vitamin A were focal and reversible, selectively altering the differentiation of germinative layers of epithelia.

The morphologic changes observed in cultures of embryonic chick skin exposed to all-*trans*-retinoic acid (tretinoin, D1) were reversible and both dose- and time-dependent. Keratinization was inhibited and fewer desmosomes and tonofilaments were seen. Golgi elements, rough endoplasmic reticulum, and polyribosomes were unusually prominent. Mucin granules formed and gland-like structures developed with intercellular canaliculi characterized by tight junctions, brush borders, and dense secretory contents (Peck *et al.*, 1977).

Several possible mechanisms were considered by which retinoic acid could alter epidermal differentiation in this system. Retinoic acid-induced gaps in the basal lamina allowed direct contact between epidermal basal cells and fibroblasts and collagen fibers, which could result in inappropriate dermal signals reaching the epidermis. In younger embryos the entire epidermis, including the mitotically inactive surface cells, appeared to respond to retinoic acid, and this could imply an epigenetic modulation of cell phenotype. Finally, after the formation of a stratum corneum in older embryos, only the relatively undifferentiated basal layer showed a metaplastic response, indicating that retinoic acid could be acting directly on the genome (Peck *et al.*, 1977).

In contrast to the observed effects of retinoids in embryonic chick skin, retinoids do not produce mucous metaplasia in postembryonic mammalian epidermis. However, both oral and topical retinoids profoundly affect epidermal morphology. Topical retinoic acid enhances cell proliferation and hyperplasia leading to epidermal acanthosis. Initially, after both topical and oral retinoids, there is diminution of the granular layer, but with long-term administration pronounced hypergranulosis occurs (Elias and Williams, 1981). Retinoid treatment decreases stratum corneum cohesiveness, with resultant impaired function of the permeability barrier with increased transepidermal water loss and enhanced fragility of the upper epidermis. (Elias *et al.*, 1981a). This interference with permeability barrier function enhances percutaneous absorption of topical agents, which may either result in therapeutic benefit or potentially in toxicity from unsuspected drug absorption (Fritsch *et al.*, 1981).

Ultrastructurally, there are decreased numbers of tonofilaments and desmosomal attachments, emergence of tight junctions, an increased number of keratinosomes, mitochondria, ribosomes, and endoplasmic reticulum, and an increased amount of amorphous material that is seen within widened intercellular spaces (Fritsch, 1981; Williams and Elias, 1981).

Another important effect of retinoid treatment is the stimulation of gap junction hyperplasia and hypertrophy in the epidermis (Elias and Williams, 1981). This stimulation occurs before other effects are observed, such as tight junction proliferation. Gap junction proliferation has been observed *in vitro* in embryonic chick skin exposed to retinoic acid, and *in vivo* in the rabbit keratoacanthoma and in human basal cell carcinomas after topical application of retinoic acid (Elias *et al.*, 1981b). When observed with the scanning electron microscope, stratum corneum cells were irregular and possessed abnormal and decreased intercellular attachments after treatment with retinoic acid (Papa, 1976).

The histological effects of etretinate in psoriatic skin included regression of the inflammatory cell infiltrate, the extracellular accumulation of an amorphous material considered by some to be mucous-like, an increase in size and number of keratohyalin granules, a widened intercellular space, the reappearance of the stratum corneum layer where diminished, and normalization of other pathologic changes, not only in the epidermis but also in the dermis, particularly the appearance of the capillaries of the dermal papillae (Ward *et al.*, 1983).

The histologic changes in hairless mice receiving etretinate and in patients with disorders of keratinization treated with isotretinoin were similar. The stratum corneum appeared loose, disorganized, and fragmented and demonstrated patchy areas of reduced thickness. Loss of desmosomes occurred owing to shedding of these structures at the level of the stratum spinosum (Williams and Elias, 1981). Consequently, there were numerous regions in the stratum granulosum where desmosomes were absent. Perinuclear and intercellular deposits of amorphous material that did not stain with mucin stains were present.

A dramatic (~ 75%) reduction in size and number of desmosomes was observed in a freeze-fracture analysis of keratinocyte membranes in patients with psoriasis treated with etretinate (Kitajima and Mori, 1983). This reduction in desmosomal size and number (Williams and Elias, 1981), which was particularly evident in the stratum corneum but also observed in the stratum spinosum of both lesional and nonlesional skin, appeared to be of sufficient magnitude to contribute significantly to the keratolytic effect of retinoids in hyperkeratotic disorders. Other factors, for example, decreased tonofilaments and decreased glycocalyx cohesion, may also be contributory. This effect also has been observed *in vitro* as retinoids cause increased shedding of squames from stratified cultures of human foreskin epidermal cells (McGuire *et al.*, 1982; Kubilus, 1980).

Etretinate, however, may not completely return the keratinocyte membrane from a psoriatic to a normal state despite clinical improvement. This is based on the failure of etretinate to reduce the greater than normal intramembranous particle density of the fracture face of the protoplasmic half-layer (P face) of the spinous cell plasma membrane in lesional skin after the lesions were clinically resolved (Kitajima and Mori, 1983).

Cross-fractures of keratinocytes in the posttreatment lesional skin revealed an increase in vesicular components, a considerable development of Golgi complex, and a decrease in tonofilaments. Cross-fractures of the cell boundaries showed an irregular appearance with microvilli and cell processes and widened intercellular spaces after therapy. Increased thickness of the granular layer occurred. The granular cells were rich in vesicles and poor in tonofilaments. The frequency of gap junctions is higher in psoriatic skin than in normal skin, particularly in the upper spinous layer. However, no significant changes in size or frequency of gap junctions were found in lesional posttreatment skin. Gap junctions were frequently noticed on the basal cell plasma membranes in nonlesional posttreatment skin (Kitajima and Mori, 1983).

The fine granular, amorphous, "mucous-like" deposits induced by treatment with oral retinoids began to appear both within and between keratinocytes in the upper stratum spinosum and persisted into the stratum corneum. By transmission electron microscopy, no evidence of active secretion or endocytosis was noted.

By freeze-fracture analysis, this amorphous material was found to be associated with the openings of plasma membrane vesicles. The amorphous material did not stain with PAS or Alcian blue or fluorescent lectins and was thereby thought unlikely to be mucin or other glycoprotein (Williams and Elias, 1981).

In this regard, even when hairless mice were treated with a high dose of etretinate, such as 50 mg/kg/day, no mucous metaplasia could be detected by lectin staining. Lectin staining of mammalian epidermis reveals a pattern of increased sugar complexity during normal keratinocyte maturation. Retinoids disrupt this pattern but only at high doses and late in the course of treatment, suggesting that retinoid-induced alterations may be secondary. For example, in the etretinate-treated hairless mice, abnormalities in lectin staining developed only after 15 days of therapy. By that time dramatic alterations in epidermal structure and function had already occurred, including abnormal transepidermal water loss, mild acanthosis, and focal loss of stratification (Nemanic *et al.*, 1982; Elias *et al.*, 1983).

Both the accumulation of amorphous material and the diminution of desmosomes induced by retinoids seemed to be responsible for the enhanced fragility of the upper epidermis to frictional trauma. This enhanced fragility led to an intraepidermal cleavage plane that traversed intercellular spaces filled with the amorphous material and desmosome-depleted surfaces. Since keratinocytes containing these amorphous deposits seemed to be fragile, it is possible that the intercellular deposits accumulated as a result of rupture and leakage of cells containing the amorphous material. Additionally, it has been suggested that this amorphous material could be secondary to serum infiltration into the epidermis, with macropinocytosis accounting for the intracellular localization (Ellis *et al.*, 1982).

The diminution of desmosomes induced by oral retinoids with subsequent loss of horny cell layers may result in shedding of chemotactic factors contained in these horny cells and represents one mechanism by which etretinate is effective in pustular psoriasis, a condition in which neutrophils infiltrate the epidermis. In addition to possibly decreasing the expression of chemotactic factors in the skin, it is possible that retinoids may directly modify neutrophil function.

B. Neutrophil Migration

One of the first cellular events observed after oral etretinate therapy of psoriasis was the loss of neutrophil migration from dermal capillaries to the epidermis. Oral etretinate also inhibited the migration of neutrophils out of suction blister bases into overlying skin chambers in normal patients (Dubertret *et al.*, 1982). Topically applied etretinate at 0.1 mg/ml also inhibited neutrophil migration in this system. However, inhibition of neutrophil migration occurs in psoriasis after clearing of skin lesions, whatever the treatment used. Similarly, isotretinoin led to almost complete inhibition of neutrophil and monocyte migration in acne patients (Norris *et al.*, 1983). This reversible anti-inflammatory effect suggests that retinoids may be of value in the treatment of neutrophil-dependent diseases, in addition to acne and psoriasis. Retinoids also inhibit other discrete polymorphonuclear leukocyte functions *in vitro*. Retinoids may exert their anti-inflammatory effects by interacting with neutrophil membranes to inhibit a variety of responses, such as lysosomal enzyme release and superoxide generation (Camisa *et al.*, 1982).

C. Lysosomes

Much of the early research on vitamin A emphasized the detergent-like effect of vitamin A on cell membranes that can, in turn, lead to destabilization or labilization of lysosomal membranes, with release of lysosomal enzymes and subsequent cytotoxicity. There are arguments, however, against the statement that lysosomal labilization and subsequent cytotoxicity underly the antitumor and other effects of the retinoids. For instance, in pityriasis rubra pilaris and in Darier's disease treated with oral isotretinoin, there was a highly significant decrease in the specific activity of the lysosomal hydrolases cathepsin D and β-glucuronidase. This indicates that clinical remissions are not due to increasing the intracellular concentration of lysosomal enzymes (Farb *et al.*, 1980). However, clinical findings in hypervitaminosis A are consistent with the results of a detergent-like effect on cell membranes.

D. Langerhans Cells

Langerhans cells are known to play a major role in skin immune reactions and may also be involved in the regulation of epidermal differentiation. The distribution and number of Langerhans cells, as detected by monoclonal antibodies against human HLA-DR antigens and OKT6 and by ATPase cytochemical staining, are altered in involved skin of psoriasis and are corrected after successful therapy with oral etretinate (Haftek *et al.*, 1983). It is not known whether this normalization of Langerhans cell distribution in psoriasis is due to a direct effect of etretinate or represents a secondary effect of correcting the altered epidermal differentiation. In contrast, other successful therapies in psoriasis, such as UVB, PUVA, and topical corticosteroids, deplete Langerhans cells from the skin.

E. Epidermal Cell Proliferation, Polyamines, Ornithine Decarboxylase

The changes in cell kinetic parameters induced in hairless mice by etretinate (1–10 mg/kg/day) were dose-dependent and consisted of transitory cellular hypertrophy, persistent epidermal hyperplasia, and increased labeling indices. The mean basal cell generation time was greatly accelerated owing to a shortening of all cell-cycle phases tested (Fritsch *et al.*, 1981). The fraction of noncycling basal cells was reduced during therapy. The nuclear enlargement and increased labeling indices occurred prior to the onset of retinoid-induced desquamation of the horny layer, indicating that these events were not simply secondary to the mitogenic stimulus that occurs after the loss of the stratum corneum, an effect observed after adhesive tape stripping of the skin.

It may at first appear paradoxical that diseases such as psoriasis and lamellar ichthyosis, which are characterized by a hyperproliferative epidermis, can benefit from drugs such as the retinoids, which can stimulate epidermal proliferation under certain experimental conditions. However, when tested in patients with psoriasis, etretinate led to decreased ornithine decarboxylase activity, decreased levels of urinary and cutaneous polyamines, and decreased epidermal DNA synthesis (Kaplan *et al.*, 1983).

Polyamines and their rate-limiting enzyme ornithine decarboxylase (ODC) are elevated in untreated psoriatic skin. Since retinoids were known to interfere with ODC activity and polyamine biosynthesis in experimental cutaneous carcinogenesis (see Chapters 5, 12, and 14), it was anticipated that retinoids would lead to similar effects in psoriasis. Within 4 weeks of etretinate therapy, cutaneous levels of ODC activity and polyamines were significantly inhibited in both the diseased and uninvolved skin of psoriatics (Lowe *et al.*, 1982; Lauharanta *et al.*, 1981b; Kaplan *et al.*, 1983). The accelerated polyamine biosynthesis in psoriasis was normalized prior to any significant inhibition of epidermal DNA synthesis. Epidermal DNA synthesis did not decrease in response to etretinate until week 8 of therapy (Krueger *et al.*, 1980). Oral etretinate led to a dose-dependent decrease in urinary excretion of polyamines in psoriasis patients. Mean putrescine levels fell by 27%, spermidine by 34%, and spermine by 37% by the end of a 16-week treatment period, indicating an inhibition of polyamine biosynthesis (Grekin *et al.*, 1983). It is known that clinical improvement in psoriasis, by whatever therapy, lowers cutaneous and urinary levels of polyamines, suggesting that these changes are secondary to disease improvement. However, one exception was observed when a patient, who failed to improve clinically in response to etretinate, developed reduced levels of urinary polyamines in a pattern similar to responding patients.

F. Epidermal Transglutaminase Activity and Cornified Envelope Formation

Retinoids are well-known inhibitors of phorbol ester-mediated promotion of skin tumor formation (see Chapters 5 and 14). In one suggested explanation of this effect, the induction of terminal differentiation of normal epidermal cells by phorbol esters would allow compensatory expansion of a population of initiated cells (Yuspa, 1983). Epidermal cell cultures exposed to phorbol esters show a significant increase in transglutaminase activity and a marked increase in cornified cells that are lost from the culture dish. As these cornified cells are lost, transglutaminase activity returns to basal levels. Tranglutaminase is a calcium-dependent, cytosolic enzyme that catalyzes the formation of the (–glutamyl–lysine) dipeptide bond in the cross-linked cornified envelope in epidermal cells, and thus is a critical regulator of epidermal differentiation.

Retinoic acid, surprisingly, is also a potent inducer of epidermal transglutaminase. But in contrast to the transient increase with phorbol esters, the elevated epidermal transglutaminase activity remains high throughout the course of exposure to retinoic acid. Retinoids were also observed to inhibit terminal cornification of epidermal cells induced to differentiate by calcium. Retinoic acid, furthermore, prevented phorbol ester-induced cell loss from culture dishes. Thus, it appears that retinoids, by inhibiting the cornification process can interfere with an essential biological function of phorbol ester tumor promoters, the induction of terminal differentiation (Yuspa, 1983).

Similarly, retinyl acetate suppressed the ability of cultured keratinocytes, derived from a human squamous cell carcinoma, to form cross-linked envelopes at the cell periphery (Rice *et al.*, 1983). Cell envelope protein production was inhibited also by the retinoidal benzoic acid derivative TTNPB ethyl ester (G6) (Stadler *et al.*, 1983).

G. Connective Tissue Effects

Both isotretinoin and tretinoin inhibited collagenase and gelatinase production in skin fibroblast cultures derived from normal skin and from patients with recessive dystrophic epidermolysis bullosa (Bauer *et al.*, 1982). The inhibition of collagenase activity was paralleled by a reduction in immunoreactive enzyme, suggesting that these retinoids act by inhibiting synthesis and/or secretion of the enzyme. Retinoids, therefore, might be useful therapeutic agents in recessive dystrophic epidermolysis bullosa, a disease in which the pathogenesis of blistering is in part related to connective tissue destruction. In another report, increasing concentrations of all-*trans*-retinoic acid had an inhibitory effect on collagen sysnthesis but enhanced the activity of procollagen amino protease (Nelson and Balian, 1983).

III. CYSTIC ACNE

Cystic acne is a severe scarring form of acne characterized by the presence of large, deep-dermal, and subcutaneous nodules and cysts. In the initial report demonstrating the effectiveness of isotretinoin in the treatment of cystic acne, a disease which is unlikely to undergo spontaneous remission, 13 of 14 patients had complete clearance of their acne (Figs. 1 and 2) with an average maximum dosage of 2 mg/kg/day for 4 months (Peck *et al.*, 1979a). The 14th patient had an 89% improvement when seen 3 years posttreatment. In those patients whose acne had not cleared completely during the 4-month course of therapy, continued healing leading to complete clearing was noted after therapy had been withdrawn. Prolonged remissions, currently lasting over 6 years in duration, were observed in all patients, and only one patient required a second successful course of therapy for a slight relapse 12 months after discontinuing therapy.

In a parallel, double-blind protocol, it was clearly demonstrated that the therapeutic effects of isotretinoin were not a placebo response (Peck *et al.*, 1982a). The 17 placebo-treated patients showed a 57% worsening of their condition in 2 months, while during the same period, the 16 isotretinoin-treated patients improved by 32%. Of the 32 patients who received isotretinoin in this study, including 16 of the placebo patients who crossed over to active drug, 27 showed complete clearance, leading to an overall improvement of 97%. Doses of isotretinoin as low as 0.5 mg/kg/day were found to be effective. Facial acne responded more rapidly and at a lower dosage level than truncal acne. Five patients in this study developed mild relapses requiring additional therapy. All five were less than 20 years of age at time of entry into the study and 0.5 mg/kg/day had been used as their predominant dosage schedule. Lesions of hidradenitis suppuritiva in the axillae and groin, and keloidal acne on the nape of the neck also responded to the treatment, but more slowly and only at higher dosages (2 mg/kg/day).

Data from unpublished studies (G. L. Peck *et al.*) indicate that high initial doses (1–2 mg/kg/day) need to be given for only 2 weeks to achieve comparable therapeutic results if they are followed by a low maintenance dose (0.25–0.50 mg/kg/day) for the remainder of a 16-week treatment period. The higher doses were used for patients with predominantly truncal acne and the lower doses for facial acne patients. The high-initial, low-maintenance dose schedule was superior to both a 2-week high dose followed by placebo and to a constant low-dosage schedule. Specifically, the constant low-dosage schedule (0.5 mg/kg/day) led to an initial 20% increase in the lesion count at 2 weeks and, at the end of the 16-week treatment period, only a 50% reduction in lesions.

Seventy-six patients with acne of a lesser severity than those included in the above studies responded well to constant low doses (0.1, 0.5, and 1.0 mg/kg/day) of isotretinoin (Jones *et al.*, 1983). An 80% decrease in the number of facial acne lesions was noted with 0.1 and 0.5 mg/kg/day after 16 weeks of therapy. The use of 1.0 mg/kg/day led to an 89% decrease in facial lesions but only a 55% reduction of back lesions. Equal reductions in lesion counts were seen for non-inflamed (comedones), superficial inflamed (papules, pustules), and deep inflamed (nodules, cysts) lesions. Since patients treated with 0.1 mg/kg/day had a greater incidence of relapse, a dose of 0.5 mg/kg/day was recommended as initial therapy. The therapeutic effectiveness of low dose (0.05–0.02 mg/kg/day) isotretinoin along with the greater tendency for acne to recur after the use of these low doses has been noted in other studies (Farrell *et al.*, 1980; Plewig *et al.*, 1982).

No initial flare of nodulocystic acne was observed at any dose level in a multicenter trial of isotretinoin at the dosages of 0.1, 0.5, and 1.0 mg/kg/day for 16–20 weeks (Rapini *et al.*, 1983). The maximum improvement reported was 88% reduction in number of lesions at 1.0 mg/kg/day, 89% at 0.5, and 82% at 0.1, indicating that the lowest dosage was virtually as effective as the higher dosages. In this report, no significant differences were observed in the average reduction in the numbers of facial versus truncal lesions at the various dose levels. However, more relapses were observed at 0.5 and 0.1 mg/kg/day than at 1.0, and since there was no significant reduction in incidence of toxicity observed with the lower dose levels, 1.0 mg/kg/day was the recommended dosage.

In another multicenter study of 198 acne patients, only a group of patients treated with isotretinoin at 1.0 mg/kg/day for 12 weeks and 0.2 mg/kg/day for another 12 weeks had more than 75% improvement. Patients treated with initial doses of 0.2 and 0.5 mg/kg/day had less favorable responses. In addition, the number of patients in remission appeared to be dose-related (Gollnick *et al.*, 1983).

In addition to cystic acne, isotretinoin therapy is effective in gram-negative folliculitis, acne fulminans, acne conglobata, hidradenitis suppuritiva, dissecting cellulitis of the scalp, and acne rosacea (Plewig *et al.*, 1982). However, the treatment of hidradenitis may require prolonged therapy with 2 mg/kg/day and the response may be partial. Isotretinoin is not only effective in treating the papules, pustules, and nodules of acne rosacea but also has been reported to improve the telangiectasia and shrink rhinophyma (Plewig *et al.*, 1982). Concerning gram-negative folliculitis, the disappearance of bacteria (*Klebsiella, Proteus, Escherichia, Enterobacteria*) from the skin and nasal mucosa is not considered to be a direct antibacterial effect of isotretinoin but rather a secondary effect of alterations in the microenvironment (Plewig *et al.*, 1982; King *et al.*, 1982).

Topical isotretinoin, 0.2% in a vanishing cream base and applied twice daily for 20 weeks, led to decreased lesion counts, no cutaneous irritation, and a 30% decrease in skin surface lipids in 20 patients with acne (Plewig *et al.*, 1983). Topical administration of the retinoidal benzoic acid derivative TTNPB ethyl ester (G6) was toxic to hamsters in concentrations of 0.001–0.3% causing dermatitis, hair loss, bone deformation, and fracture. Concentrations of 0.0001% and lower were nontoxic. This suggests that topical applications of this agent in man may not be of clinical use because of the observed toxicity.

Inhibition of sebum production with alterations in skin-surface lipid film chemistry may represent a key mechanism of action of isotretinoin leading to clinical improvement in acne. Certainly, isotretinoin is the most effective inhibitor of sebum production that has been found, being superior to estrogen and X irradiation. Inhibition of quantitative sebum production (or sebum excretion rate) is almost maximal by the fourth week of treatment with isotretinoin. Inhibition is dose-dependent in that 0.05 mg/kg/day reduces sebum production by about 50%. Treatment with 0.1 mg/kg/day leads to 67–75% inhibition, 0.5 mg/kg/day leads to 82–89% inhibition, and 1.0 mg/kg/day leads to 88–91% inhibition after 12–16 weeks of therapy (Farrell *et al.*, 1980; Jones *et al.*, 1983). However, 16 weeks after treatment has stopped, the inhibition observed with 0.1 mg/kg/day has almost totally been reversed, whereas the higher doses have led to a persistent 34–40% inhibition (Jones *et al.*, 1983). Longer-term evaluations (20–99 weeks) after stopping therapy show an overall persistent 38% (range: 0–80%) inhibition (Strauss and Stranieri, 1982). In another report, 12 patients who received two courses of therapy with isotretinoin for a total of 10 months showed a persistent 60% decrease in quantitative sebum production, and 28 patients who received only one 4-month course of therapy had a prolonged 39% decrease in quantitative sebum production when measured 1 year or more after treatment. The differences between these two groups is significant, indicating dose-dependency (Gross *et al.*, 1983). These data suggest that prolonged remission in some patients may be related at least in part to continued sebaceous gland inhibition. The histological changes parallel and reflect the inhibition of sebum production and reveal that the sebaceous glands virtually disappear during treatment with isotretinoin and gradually recover after discontinuation of therapy.

Reduction in sebum production occurs prior to clinical improvement. At peak levels of sebum suppression, the relative percentage of the skin-surface lipid film comprising wax esters and squalene is reduced, and the percentage of cholesterol and cholesterol esters is increased. Etretinate is less effective than isotretinoin in the treatment of acne and is capable of suppressing sebum production by approximately 30% (Goldstein *et al.*, 1982), which is similar to the effect achieved with 300,000 units of vitamin A given orally.

Using the bentonite clay technique, the wax ester secretion rate was studied in response to 0.1, 0.5, and 1.0 mg/kg/day of isotretinoin (Stewart *et al.*, 1983). Wax esters are a component of the skin-surface lipid film and are thought to be of purely sebaceous origin. The mean rates of wax ester secretion, greatly elevated in untreated acne patients, were reduced to normal values or below by the fourth week of therapy. Isotretinoin at dose levels of 0.5 and 1.0 mg/kg/day suppressed wax ester secretion below the non-acne range, and the inhibition seen with 1.0 mg/kg/day was significantly greater than that observed with 0.1 mg/kg/day. However, since the posttreatment secretion rates rose above the non-acne range at all dose levels, other factors must contribute to the continued healing of acne and to the prolonged remissions observed after stopping therapy.

In addition to inhibition of sebum production, anti-inflammatory, antibacterial, inhibitory effects on microbial enzyme activity and desquamative effects on poral occlusion should be considered as possible mechanisms by which isotretinoin is effective in the treatment of acne. Reduced erythema and pustule formation in the potassium iodide patch test during isotretinoin therapy indicates an anti-inflammatory action of the drug (Plewig *et al.*, 1982). Isotretinoin reduces the number of *Propionibacterium acnes* on the skin surface that probably reflects decreased follicular colonization of *P. acnes* secondary to the decrease in sebaceous secretion (Plewig *et al.*, 1982; King *et al.*, 1982). The therapeutic effect of isotretinoin on comedones provides clinical evidence for an effect on follicular keratinization. There is no increased affinity of binding proteins for this retinoid in the sebaceous glands when compared with the epidermis. Thus the inhibitory effect of isotretinoin on sebaceous glands cannot be explained on this basis (Puhvel and Sakamoto, 1983).

Isotretinoin does not appear to be acting as an antiandrogen since no change has been noted in serum testosterone levels or gonadotropins, nor have there been any signs of feminization in males during therapy. Furthermore, the androgen-sensitive parts of the hamster flank organ, aside from the sebaceous component, do not involute during treatment with isotretinoin (Gomez, 1982).

The aromatic retinoid etretinate is less effective than isotretinoin in the treatment of acne. In a double-blind study, 56 men received either isotretinoin or etretinate at a dosage of 1 mg/kg/day for 8 weeks. At 8 weeks posttreatment, there was a 69% reduction in the number of acne cysts and nodules in the isotretinoin group but only a 33% decrease in the etretinate group (Goldstein *et al.*, 1982).

Isotretinoin, therefore, has proved to be the most effective therapy for cystic acne. Most cases of even the greatest severity, being previously treatment-

Fig. 1. Cystic and conglobate lesions of the face prior to treatment with isotretinoin. [Reprinted with permission from Peck *et al.* (1982a).]

Fig. 2. Complete clearing and prolonged remission are seen 3 years after therapy. [Reprinted with permission from Peck *et al.* (1982a).]

Fig. 3. Widespread plaques of psoriasis vulgaris prior to treatment with oral etretinate, 1 mg/kg/day. [Reprinted with permission from Peck (1982).]

Fig. 4. Complete healing of all lesions is observed at the end of a 4-month course of therapy with etretinate as the sole therapeutic agent. [Reprinted with permission from Peck (1982).]

Fig. 5. Generalized lamellar ichthyosis prior to treatment with isotretinoin. [Reprinted with permission from Peck and Yoder (1976).]

Fig. 6. Almost complete clearing is noted after 4 weeks of therapy at 120 mg/day. Reprinted with permission from Peck and Yoder (1976).]

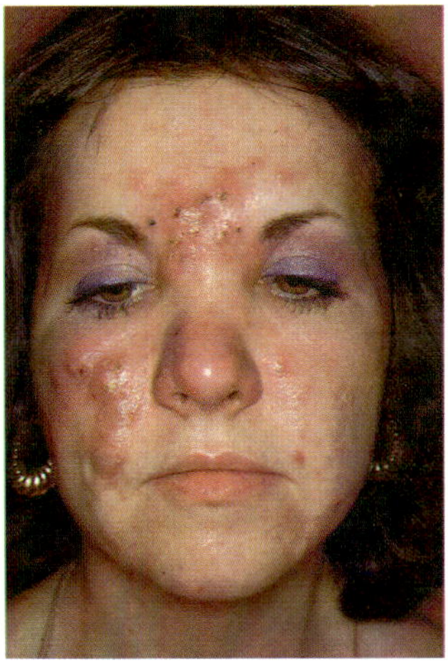

1

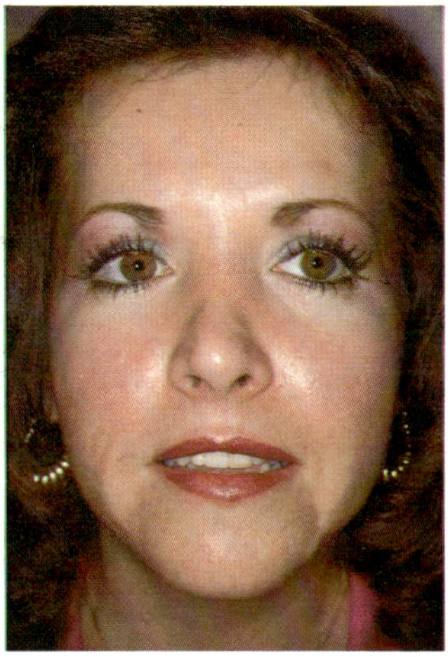

2

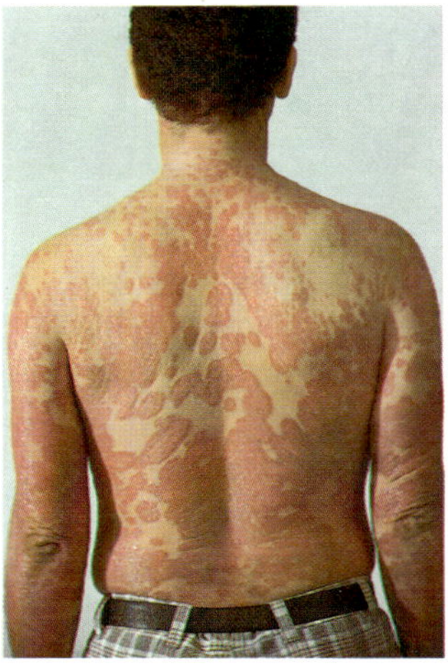

3

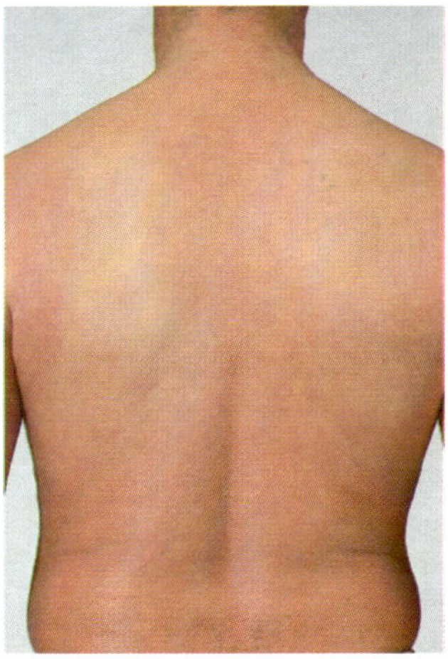

4

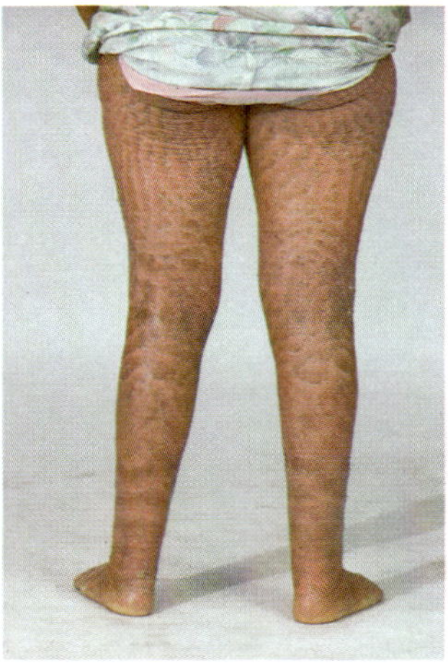

5

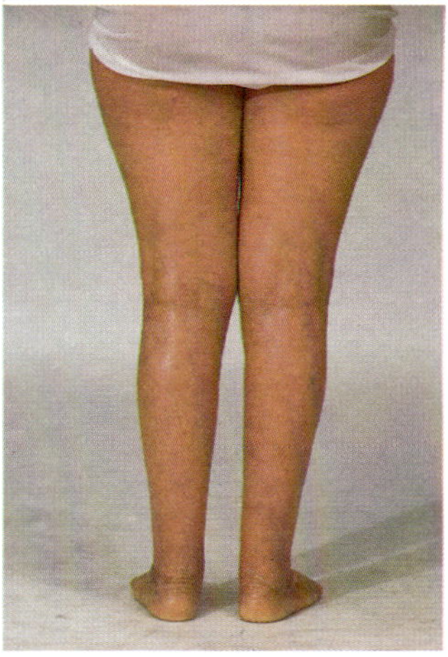

6

resistant, can be treated successfully with only one 4- or 5-month course at doses of 0.5–2.0 mg/kg/day. The higher doses may be required for patients with cystic acne located predominantly on the trunk. Only about one-third of patients need a second course and only a handful require a third to completely clear. Because of the continued healing seen after discontinuing therapy, 2-month treatment-free evaluation periods are useful in determining which patients require additional therapy.

IV. PSORIASIS

Psoriasis is an inherited chronic dermatosis characterized by epidermal hyperplasia and an accelerated rate of epidermal turnover. Lesions are typically sharply outlined red papules and plaques with firmly adherent white scales. As mentioned above, a single 4- or 5-month course of therapy with retinoids for acne can lead to prolonged remissions in most cases. The treatment of psoriasis with retinoids, however, usually requires long-term administration because of the relapses that regularly occur if therapy is stopped. The prolonged administration of retinoids thus places psoriasis patients at greater potential risk of developing chronic toxicity than acne patients. The major emphasis of the treatment of psoriasis with synthetic retinoids has focused on etretinate (Ehmann and Voorhees, 1982). Psoriasis patients who had been treated initially with isotretinoin and subsequently with etretinate responded more completely to etretinate (Peck *et al.*, 1981). Etretinate has been used both as monotherapy (Figs. 3 and 4) and in combination with other active agents such as anthralin, ultraviolet light (UVB, 280–320 nm), photochemotherapy (PUVA), and topical and systemic corticosteroids. The initial dosage of etretinate in a multicenter study involving 291 patients was 70–100 mg/day (1 mg/kg/day) and was reduced to 50 mg/day when significant improvement occurred, usually at 3 weeks (Goerz and Orfanos, 1978). In addition to psoriasis vulgaris characterized by plaques, erythrodermic and pustular psoriasis also responded dramatically to etretinate. A maintenance dose of 25–50 mg/day was administered to patients with good or excellent responses (61% of the total). However, relapses occurred in approximately 75% of patients within 6 months, despite maintenance therapy with etretinate with or without topical anthralin.

Forty-two patients with palmoplantar pustulosis, including 21 with psoriasis elsewhere on the body, responded well to etretinate at an initial dosage of 75 mg/day. Pustules healed within 2 weeks but recurred in some patients within 2 weeks of stopping therapy. Other patients with complete remissions had prolonged remissions lasting as long as 6 months (Thune, 1982).

Etretinate markedly augmented the response of psoriatic patients to photochemotherapy, oral methoxsalen, and long-wave ultraviolet light (UVA,

320–400 nm) or PUVA. This regimen combining a retinoid with photochemotherapy has been termed "RePUVA." In the initial report describing RePUVA, etretinate was given for 7–10 days, followed by the addition of PUVA. Complete clearing of the psoriasis was observed after an average of six PUVA treatments over 11 days (Fritsch *et al.*, 1978). The combined treatment considerably decreased the total amount of UVA required for clearing and accelerated the response of psoriasis to PUVA. Moreover, it was effective in patients who had been PUVA failures previously. Two additional reports confirmed the efficacy of this combination therapy and found that pretreatment with etretinate for a 2- to 4-week period prior to beginning RePUVA gave optimal results (Grupper and Berretti, 1981; Lauharanta *et al.*, 1981a). RePUVA was noted to produce far longer remissions than PUVA. Fewer side effects from etretinate were seen during RePUVA than with etretinate used alone because of the lower doses employed.

The combination therapy of isotretinoin at a dose of 1 mg/kg/day and PUVA was compared prospectively with RePUVA (Honigsmann and Wolff, 1983). The retinoids were given alone for 5 days prior to adding PUVA and were discontinued once psoriasis had cleared completely, at which time the patients were placed on PUVA maintenance. No significant difference between the two treatment regimens was observed in regard to duration of treatment required for clearing, number of UVA exposures required, and cumulative UVA dose, even though etretinate is superior to isotretinoin when used alone in the treatment of psoriasis (Peck *et al.*, 1981). Although birth defects have been observed with both isotretinoin and etretinate, the beneficial response of psoriasis to isotretinoin–PUVA may be of particular importance to fertile women since etretinate is a more potent teratogen that is much more slowly eliminated from the body (see Chapter 13). In another study using combination therapy, isotretinoin, at a dose of 0.75 mg/kg/day, has been reported to enhance the effect of short-contact anthralin in the treatment of psoriasis (Lowe and Ashton, 1983).

Etretinate at dosages of 0.5–1 mg/kg/day enhanced ultraviolet light (UVB) phototherapy of psoriasis. In one study, 66% of patients who had previously failed to respond to UVB phototherapy could be maintained successfully on etretinate and UVB treatments given once a week or less (Orfanos *et al.*, 1979). In a multicenter double-blind trial, etretinate in initial doses of 0.5–0.66 mg/kg/day was found to produce still better results if used in conjunction with a topical corticosteroid (0.1% triamcinolone acetonide). Thirty-one patients were maintained in complete remission for over 2 years by use of a low-maintenance dose of etretinate (25 mg/day) in addition to the topical corticosteroid (van der Rhee and Polano, 1981).

The maintenance of good therapeutic responses and the absence of chronic toxicity were noted after 6 years of continuous treatment of psoriasis with etretinate (10–50 mg/day) (Ott, 1981). Temporary elevations of liver function tests were more frequent early in therapy. Recent dosage recommendations for the treatment of psoriasis with etretinate were that initial therapy of erythrodermic psoriasis consist of 25–35 mg/day increasing to 50–60 mg/day within 2–4 weeks. Pustular psoriasis may require initial doses of 75 mg/day, while chronic psoriasis vulgaris should be treated with 50 mg/day in combination with other active agents. Women, in particular, should not be given more than 50 mg/day, if possible, in order to minimize hair loss from etretinate (Orfanos *et al.*, 1981).

Several small uncontrolled pilot studies have indicated that most patients with psoriatic arthritis improve when treated with etretinate at 0.5–1 mg/kg/day (Brackertz and Muller, 1979; Rosenthal 1979; Stollenwerk *et al.*, 1981; Kaplan *et al.*, 1983). In one report the etretinate-induced improvement allowed patients to decrease or discontinue their use of nonsteroidal anti-inflammatory agents. In another report patients with chronic polyarthritis similarly improved. RePUVA also led to improvement in five of six patients with psoriatic arthritis. However, because healing of cutaneous lesions of psoriasis with PUVA is associated with an improvement in joint disease, controlled studies are needed to confirm that etretinate is effective in psoriatic arthritis and that the effect is not secondary to cutaneous improvement.

V. CUTANEOUS DISORDERS OF KERATINIZATION

As in the case of etretinate in psoriasis, it required the demonstration of isotretinoin's effectiveness in previously recalcitrant cases of disorders of keratinization, such as Darier's disease, lamellar ichthyosis (Figs. 5 and 6), nonbullous congenital ichthyosiform erythroderma, and pityriasis rubra pilaris to stimulate interest in the use of synthetic retinoids in these diseases (Peck and Yoder, 1976, 1978). Since 1976, numerous reports have indicated that these and other disorders of keratinization respond both to isotretinoin and to etretinate. In contrast to results in acne, for which isotretinoin is more effective than etretinate, however, etretinate and isotretinoin gave comparable responses in Darier's disease, lamellar ichthyosis, nonbullous congenital ichthyosiform erythroderma, and pityrisis rubra pilaris. Etretinate was superior to isotretinoin in the treatment of psoriasis, epidermolytic hyperkeratosis, keratoderma palmaris et plantaris, x-linked ichthyosis, ichthyosis vulgaris, erythrokeratodermia variabilis, and lichen planus (Peck *et al.*, 1981).

Patients with the dry, brown, hyperkeratotic type of Darier's disease respond better and may have more prolonged remissions than those with the red, inflamed, infected variety of Darier's disease who also have marked intertrigenous involvement. These latter patients are much more difficult to treat and relapse very quickly after therapy is stopped. Although there was no initial worsening of disease, as occasionally occurs with psoriasis, isomorphic reactions did occur in one-third of patients with Darier's disease treated with etretinate in one report. (Lowhagen *et al.*, 1982). Since etretinate, unlike isotretinoin, is stored in the body, intermittent courses of etretinate at 0.5 mg/kg/day or less were recommended as therapy for Darier's disease in order to minimize the likelihood of chronic toxicity.

In a report of 45 patients with pityriasis rubra pilaris of varying duration prior to therapy with intermittent courses of isotretinoin, long-term remissions were noted after discontinuation of treatment. Although most cases of adult-onset pityriasis rubra pilaris spontaneously clear within 3 years, this finding could indicate either that isotretinoin induced or accelerated a spontaneous remission or was merely coincidental with it. In patients who did not have a complete remission after a course of therapy, new areas of involvement did not occur and the return of disease did not reach the pretreatment degree of severity, as had been observed in Darier's disease after stopping treatment with isotretinoin (Goldsmith *et al.*, 1982). However, two patients with chronic pityriasis rubra pilaris, characterized by childhood onset, myriads of follicular papules, and a duration of longer than 10 years, responded very dramatically to treatment initially with isotretinoin and subsequently with etretinate and relapsed dramatically and completely after each 4- to 6-month course of therapy over a more than 6-year period of retinoid therapy (Peck and Yoder, 1978; Peck *et al.*, 1981).

In addition to reduction in scale, patients with lamellar ichthyosis treated with retinoids also notice increased heat tolerance and ability to sweat and improved ectropion (Baden *et al.*, 1982). Clearing in these patients is usually not complete and may be greater in the summer than in the winter.

Since, unlike acne, disorders of keratinization may require long-term therapy with retinoids, it must be emphasized that the safety of chronic administration of retinoids has not been determined (Windhorst and Nigra 1982). Of particular concern is bone toxicity in children (DiGiovanna and Peck, 1983) with regard to premature closure of epiphyses (Milstone *et al.*, 1982) and fractures (Tamayo and Ruiz-Maldonado, 1981).

Unusual responses to therapy have been that palmoplantar blistering was enhanced by etretinate during therapy of patients with keratoderma palmaris et plantaris (epidermolytic type), epidermolytic hyperkeratosis, and pachyonychia congenita. Patients with Hailey–Hailey disease and atopic dermatitis have worsened with both retinoids.

VI. SKIN CANCER—THERAPY AND PREVENTION

Oral isotretinoin led to complete clinical and histological regression of approximately 10% of 248 tumors in 11 patients with multiple basal cell carcinomas (Peck *et al.*, 1979b). In 5 of 11 patients, inflammation of the tumors sparing the surrounding, apparently uninvolved, skin developed during therapy. After discontinuation of therapy, the inflammation subsided and most tumors were noted to have undergone either complete or partial regression. However, in the other six patients, tumor regression occurred during therapy without the clinical appearance of inflammation. Three of these patients who received continuous courses of therapy for 4–5 years developed no new lesions (Peck *et al.*, 1982b). On the other hand, patients who discontinued therapy did develop new tumors, usually after an 8- to 18-month interval off therapy.

The therapeutic effectiveness of etretinate in the treatment of actinic keratoses was demonstrated in a placebo-controlled, double-blind cross-over study. Thirty-seven of 44 patients with multiple actinic keratoses had a complete or partial response when treated with etretinate at 75 mg/day for 2 months. Most responses occurred within the first month of therapy (Moriarty *et al.*, 1982).

An 11-year-old girl with xeroderma pigmentosum was treated with etretinate at a dosage of 1 mg/kg/day with resultant disappearance of actinic keratoses and basal cell carcinomas. No new tumors were noted over a 7-month treatment period. When the dose was decreased to 0.4 mg/kg given on alternate days, four new basal cell carcinomas and one keratoacanthoma appeared within 6 weeks, indicating the need for a higher maintenance dose (Braun-Falco *et al.*, 1982).

A probable prophylactic effect of etretinate was described in one patient with porokeratosis of Mibelli with malignant degeneration. No new tumors developed during 2 years of retinoid therapy, whereas three squamous cell carcinomas had occurred in the preceding 2 years. However, new keratoses and Bowen's disease were seen 3 months after stopping therapy, indicating that chronic therapy may be needed for maintenance of a prophylactic effect (Schnitzler and Verret, 1981).

Etretinate in doses of 0.5–1 mg/kg/day led to complete regression of three of three keratoacanthomas, 20 of 29 actinic keratoses, and two of 18 basal cell carcinomas (Berretti *et al.*, 1981). Eight of 18 basal cell carcinomas underwent partial regression. Relapse of basal cell carcinomas was noted when therapy was either reduced or discontinued; a large keratoacanthoma of the nose relapsed 27 months after discontinuation, again indicating that maintenance therapy is mandatory.

Etretinate has also been found to be effective in the treatment of epidermodysplasia verruciformis induced by the oncogenic human papilloma virus (HPV) type 5 and the nononcogenic type 3. However, in one patient, the wart virus could still be detected during therapy. Relapse followed a decrease in dosage (Lutzner *et al.*, 1981).

Isotretinoin arrested new tumor formation in a patient with multiple keratoacanthomas of the Ferguson–Smith type. Therapeutic effects on existing lesions were also noted as lesions of recent onset did undergo complete regression, but older lesions, although becoming less tender and less scaly, did not regress. New lesions began to appear within 8 weeks of discontinuing treatment with isotretinoin (Haydey *et al.*, 1980).

Complete and partial regressions of cutaneous metastases of malignant melanoma occurred when two patients were treated daily with a 0.05% solution of retinoic acid under occlusion for 12 weeks (Levine and Meyskens, 1980). In another study, oral isotretinoin treatment resulted in partial regressions of cutaneous metastases of malignant melanoma in some patients (Cassidy *et al.*, 1981).

Etretinate when added to a combination chemotherapy program of bleomycin, cyclophosphamide, prednisone, and transfer factor enhanced therapeutic efficacy and survival in mycosis fungoides. There were eight of ten complete regressions in the etretinate group and none in the six-patient control group receiving only the chemotherapeutic agents and transfer factor. All in the control group died within 3–12 months, whereas nine of ten in the etretinate group are still alive (Zachariae *et al.*, 1982). Etretinate, when used alone in one case of mycosis fungoides led to improvement of nodular lesions with relapse observed 4 months after withdrawal of therapy (Claudy *et al.*, 1982).

Similarly, four patients with previously treatment-resistant mycosis fungoides (cutaneous T cell lymphoma) responded well to oral isotretinoin, 2–3 mg/kg/day, with production of prolonged partial remission (Kessler *et al.*, 1983). Tumors and plaques underwent near complete clearing. The skin lesions of cutaneous T cell lymphoma are characterized by epidermal infiltration with atypical mononuclear cells, mainly helper T cells but lower than normal numbers of killer T cells. In this regard, retinoic acid has been shown to enhance antigen-specific cytotoxic T cell activity (Dennert and Lotan, 1978), suggesting one immune mechanism by which retinoids could affect mycosis fungoides.

Combination therapy with etretinate (50 mg/day) and 0.1% etretinate paste led to complete or partial remission in 58% of patients with oral leukoplakia. This combination therapy was as effective and less toxic than oral etretinate at 75 mg/day (Koch, 1981).

VII. CONCLUSIONS

Synthetic retinoids are beneficial in a variety of cutaneous disorders. Isotretinoin is the drug of choice for severe cystic acne. The development of optimum dosage schedules for acne of varying severity and for facial and truncal locations is continuing.

The use of etretinate either alone or in combination with presently available therapies for psoriasis has been very effective, especially for the typically treatment-resistant pustular and erythrodermic varieties. Unlike isotretinoin in acne, maintenance therapy with etretinate is necessary for most psoriatic patients. This raises questions about the long-term safety and the potential for teratogenicity of this retinoid. So far, no serious toxicity has been observed in patients who have received etretinate for 6 years.

Synthetic retinoids must now also be considered the most effective treatment for Darier's disease and certain other disorders of keratinization.

The use of synthetic retinoids in cancer prevention and therapy for both cutaneous and internal tumors is potentially the most significant clinical use of these drugs, requiring further investigation and clarification. Based on the results of preliminary studies, it appears that chronic maintenance therapy is needed for successful chemoprevention of cancer with retinoids. One must now conclude that the future of the retinoids is most promising, particularly with the continuing development of new synthetic compounds that may improve still further their efficacy or tolerability.

REFERENCES

Baden, H. P., Buxman, M. M., Weinstein, G. D., and Yoder, F. W. (1982). *J. Am. Acad. Dermatol.* **6,** 716–720.

Bauer, E. A., Seltzer, J. L., and Eisen, A. Z. (1982). *J. Am. Acad. Dermatol.* **6,** 603–607.

Berretti, B., Grupper, C., Edelson, Y., and Bermejo, D. (1981). *In* "Retinoids: Advances in Basic Research and Therapy" (C. E. Orfanos *et al.*, eds.), pp. 397–399. Springer-Verlag, Berlin and New York.

Brackertz, D., and Muller, W. (1979). *Verhandl. Deuts. Gesel. Inn. Med.* **85,** 1343–1346.

Braun-Falco, O., Galosi, A., Dorn, M., and Plewig, G. (1982). *Hautarzt* **33,** 445–448.

Camisa, C., Eisenstat, B., Ragaz, A., and Weissman, G. (1982). *J. Am. Acad. Dermatol.* **6,** 620–629.

Cassidy, J., Lippman, M., Lacroix, A., and Peck, G. L. (1981). *Proc. Am. Soc. Clin. Oncol.* **22,** 441.

Claudy, A., Delomier, Y., and Hermier, C. (1982). *Arch. Dermatol. Res.* **273,** 37–42.

Dennert, G., and Lotan, R. (1978). *Europ. J. Immunol.* **8,** 23–29.

DiGiovanna, J. J., and Peck, G. L. (1983). *Pediatr. Dermatol.* **1,** 77–88.

Dubertret, L., Lebreton, C., and Touraine, R. (1982). *Brit. J. Dermatol.* **107,** 681–685.

Ehmann, C. W., and Voorhees, J. J. (1982). *J. Am. Acad. Dermatol.* **6,** 692–696.

Elias, P. M., and Williams, M. L. (1981). *Arch. Dermatol.* **117,** 160–180.

Elias, P. M., Fritsch, P., Lampe, M., Williams, M., Brown, B., Nemanic, M., and Grayson, S. (1981a). *Lab. Invest.* **44,** 531–540.

Elias, P. M., Grayson, S., Gross, E. G., Peck, G. L., and McNutt, N. S. (1981b). *Cancer* **48,** 932–938.

Elias, P. M., Chung, J.-C., Orozco-Topete, R., and Nemanic, M. K. (1983). *J. Invest. Dermatol.* **81,** 81s–85s.

Ellis, C. N., Gold, R. C., Grekin, R. C., Anderson, T. F., Swanson, N. A., and Voorhees, J. J. (1982). *J. Am. Acad. Dermatol.* **6,** 699–704.

Farb, R. M., Lazarus, G. S., Chiaramonti, A., Goldsmith, L. A, Gilgor, R. S., and Balakrishnan, C. V. (1980). *J. Invest. Dermatol.* **75,** 133–135.

Farrell, L. N., Strauss, J. S., and Stranieri, A. M. (1980). *J. Am. Acad. Dermatol.* **3,** 602–611.

Fell, H. B., and Mellanby, E. (1953). *J. Physiol.* **119,** 470–488.

Fritsch, P. (1981). *Internat. J. Dermatol.* **20,** 314–329.

Fritsch, P. O., Honigsmann, H., Jaschke, E., and Wolff, K. (1978). *J. Invest. Dermatol.* **70,** 178–182.

Fritsch, P. O., Pohlin, G., Langle, U., and Elias, P. M. (1981). *J. Invest. Dermatol.* **77,** 287–291.

Goerz, G., and Orfanos, C. E. (1978). *Dermatologica (Suppl. 1)* **157,** 38–44.

Goldsmith, L. A., Weinrich, A. E., and Shupack, J. (1982). *J. Am. Acad. Dermatol.* **6,** 710–715.

Goldstein, J. A., Socha-Szott, A., Thomsen, R. J., Pochi, P. E., Shalita, A. R., and Strauss, J. S. (1982). *J. Am. Acad. Dermatol.* **6,** 760–765.

Gollnick, H., Meigel, W., Plewig, G., and Wokalek, H. (1983). *J. Invest. Dermatol.* **80,** 376.

Gomez, E. C. (1982). *J. Am. Acad. Dermatol.* **6,** 746–750.

Grekin, R. C., Ellis, C. N., Goldstein, N. G., Swanson, N. A., Anderson, T. F., Duell, E. A., and Voorhees, J. J. (1983). *J. Invest. Dermatol.* **80,** 181–184.

Gross, E. G., Peck, G. L., Gantt, P. G., and Wesley, M. N. (1983). *J. Invest. Dermatol.* **80,** 357–358.

Grupper, C., and Berretti, B. (1981). *Dermatologica* **162,** 404–413.

Haftek, M., Faure, M., Schmitt, D., and Thivolet, J. (1983). *J. Invest. Dermatol.* **81,** 10–14.

Haydey, R. P., Reed, M. L., Dzubow, L. M., and Shupack, J. L. (1980). *N. Engl. J. Med.* **303,** 560–562.

Honigsmann, H., and Wolff, K. (1983). *Lancet* **i** (8318), 236.

Jones, D. H., King, K., Miller, A. J., and Cunliffe, W. J. (1983). *Brit. J. Dermatol.* **108,** 333–343.

Kaplan, R. P., Russell, D. H., and Lowe, N. J. (1983). *J. Am. Acad. Dermatol.* **8,** 95–102.

Kessler, J. F., Meyskens, F. L., Jr., Levine, N., Lynch, P. J., and Jones, S. E. (1983). *Lancet* **i** (8338), 1345–1347.

King, K., Jones, D. H., Daltrey, D. C., and Cunliffe, W. J. (1982). *Brit. J. Dermatol.* **107,** 583–590.

Kitajima, Y., and Mori, S. (1983). *J. Invest. Dermatol.* **80,** 174–180.

Koch, H. F. (1981). *In* "Retinoids: Advances in Basic Research and Therapy" (C. E. Orfanos *et al.*, eds.), pp. 201–203. Springer-Verlag, Berlin and New York.

Krueger, G. G., Shelby, N. J., Hansen, C. D., and Taylor, M. D. (1980). *Clin. Res.* **28,** 21A.

Kubilus, J. (1980). *J. Invest. Dermatol.* **81,** 55s–58s.

Lauharanta, J., Juvakoski, T., and Lassus, A. (1981a). *Brit. J. Dermatol.* **104,** 325–332.

Lauharanta, J., Kousa, M., Kapyaho, K., Linnamaa, K., and Mustakallio, K. (1981b). *Brit. J. Dermatol.* **105,** 267–272.

Levine, N., and Meyskens, F. L. (1980). *Lancet* **ii** (8188), 224–226.

Lowe, N. J., and Ashton, R. E. (1983). *Clin. Res.* **31,** 265A.

Lowe, N. J,, Kaplan, R., and Breeding, J. (1982). *J. Am. Acad. Dermatol.* **6,** 697–698.

Lowhagen, G. B., Michaelsson, G., Mobacken, H., Pettersson, U., and Vahlquist, A. (1982). *Dermatologica* **165,** 123–130.

Lutzner, M. A., Blanchet-Bardon, C., and Puissant, A. (1981). *In* "Retinoids: Advances in Basic Research and Therapy" (C. E. Orfanos *et al.*, eds.), pp. 407–410. Springer-Verlag, Berlin and New York.

McGuire, J., Fedarko, N., Johanssen, E., La Vigne, J., Lyons, G., Milstone, L., and Osber, M. (1982). *J. Am. Acad. Dermatol.* **6,** 630–639.

Milstone, L. M., McGuire, J., and Ablow, R. C. (1982). *J. Am. Acad. Dermatol.* **7,** 663–666.

Moriarty, M., Dunn, J., Darragh, A., Lambe, R., and Brick, I. (1982). *Lancet* **i** (8268), 364–365.

Nelson, D. L., and Balian, G. (1983). *J. Invest. Dermatol.* **80,** 357.

Nemanic, M. K., Fritsch, P. O., and Elias, P. M. (1982). *J. Am. Acad. Dermatol.* **6,** 801–808.

Norris, D. A., Tonnesen, M. G., Lee, L. A., Robinson, W. R., and Johnston, R. J. (1983). *Clin. Res.* **31,** 593A.

Orfanos, C. E., Steigleder, G. K., Pullman, H., and Bloch, P. H. (1979). *Acta Derm. Venereol.* **59,** 241–244.

Orfanos, C. E., Gollnick, H., and Tsambaos, D. (1981). *Hautarzt* **32,** 275–280.

Ott, F. (1981). *In* "Retinoids: Advances in Basic Research and Therapy" (C. E. Orfanos *et al.*, eds.) pp. 355–357. Springer-Verlag, Berlin and New York.

Papa, C. M. (1976). *Cutis* **17,** 575–580.

Peck, G. L. (1982). *Drugs* **24,** 341–351.

Peck, G. L., and Yoder, F. W. (1976). *Lancet* **ii,** (7996), 1172–1174.

Peck, G. L., and Yoder, F. W. (1978). *In* "The Ichthyoses" (R. Marks and P. J. Dykes, eds.), pp. 193–201. MTP Press, Lancaster.

Peck, G. L., Elias, P. M., and Wetzel, B. (1977). *J. Invest. Dermatol.* **69,** 463–476.

Peck, G. L., Olsen, T. G., Yoder, F. W., Strauss, J. S., Downing, D. T., Pandya, M., Butkus, D., and Arnaud-Battandier, J. (1979a). *N. Engl. J. Med.* **300,** 329–333.

Peck, G. L., Olsen, T. G., Butkus, D., Pandya, M., Arnaud-Battandier, J., Yoder, F., and Levis, W. R. (1979b). *Proc. Am. Assoc. Cancer Res.* **20,** 56.

Peck, G. L., Gross, E. G., and Butkus, D. (1981). *In* "Retinoids: Advances in Basic Research and Therapy" (C. E. Orfanos *et al.*, eds.), pp. 279–286. Springer-Verlag, Berlin and New York.

Peck, G. L., Olsen, T. G., Butkus, D., Pandya, M., Arnaud-Battandier, J., Gross, E. G., Windhorst, D. B., and Cheripko, J. (1982a). *J. Am. Acad. Dermatol.* **6,** 735–745.

Peck, G. L., Gross, E. G., Butkus, D., and DiGiovanna, J. J. (1982b). *J. Am. Acad. Dermatol.* **6,** 815–823.

Plewig, G., Nikolowski, J., and Wolff, H. H. (1982). *J. Am. Acad. Dermatol.* **6,** 766–785.

Plewig, G., Ruhfus, A., and Klovekorn, W. J. (1983). *J. Invest. Dermatol.* **80,** 357.

Puhvel, S. M., and Sakamoto, M. (1983). *J. Invest. Dermatol.* **80,** 355.

Rapini, R. R., Konecky, E. A., Schillinger, B., Comite, H., Exner, J. H., Strauss, J. S., Shalita, A. R., and Pochi, P. E. (1983). *J. Invest. Dermatol.* **80,** 358.

Rice, R. H., Cline, P. R., and Coe, E. L. (1983). *J. Invest. Dermatol.* **81,** 176s–178s.

Rosenthal, M. (1979). *Schweiz. Med. Wochenschr.* **109,** 1912–1914.

Schnitzler, L., and Verret, J. L. (1981). *In* "Retinoids: Advances in Basic Research and Therapy" (C. E. Orfanos *et al.*, eds.), pp. 385–388. Springer-Verlag, Berlin and New York.

Stadler, R., Marcelo, C., Voorhees, J., and Orfanos, C. (1983). *J. Invest. Dermatol.* **80,** 356.

Stewart, M. E., Benoit, A. M., Stranieri, A. M., Rapini, R. P., Strauss, J. S., and Downing, D. T. (1983). *J. Am. Acad. Dermatol.* **8,** 532–538.

Stollenwerk, R., Fischer-Hoinkes, H., Komenda, K., and Schilling, F. (1981). *In* "Retinoids: Advances in Basic Research and Therapy" (C. E. Orfanos *et al.*, eds.), pp. 205–209. Springer-Verlag, Berlin New York.

Strauss, J. S., and Stranieri, A. M. (1982). *J. Am. Acad. Dermatol.* **6,** 751–756.

Tamayo, L., and Ruiz-Maldonado, R. (1981). *In* "Retinoids: Advances in Basic Research and Therapy" (C. E. Orfanos *et al.*, eds.) pp. 287–294. Springer-Verlag, Berlin and New York.

Thune, P. (1982). *Dermatologica* **164,** 67–72.

van der Rhee, H. J., and Polano, M. K. (1981). In "Retinoids: Advances in Basic Research and Therapy" (C. E. Orfanos *et al.*, eds.), pp. 193–199. Springer-Verlag, Berlin and New York.

Ward, A., Brogden, R. N., Heel, R. C., Speight, T. M., and Avery, G. S. (1983). *Drugs* **26,** 9–43.

Williams, M. L., and Elias, P. M. (1981). *Arch. Dermatol.* **117,** 611–619.

Windhorst, D. B., and Nigra, T. (1982). *J. Am. Acad. Dermatol.* **6,** 675–682.

Yuspa, S. H. (1983). *In* "Diet, Nutrition and Cancer: From Basic Research to Policy Implications" (D. Roe, ed.), pp. 95–109. Alan R. Liss, Inc., New York.

Zachariae, H., Grunnet, E., Thestrup-Pedersen, K., Molin, L., and Thomsen, K. (1982). *Acta Derm. Venereol.* **62,** 162–164.

Appendix

This appendix provides structural formulas for retinoids that are referred to frequently within the book. Trivial names, together with formal nomenclature, are provided; in addition, an indexing number used in *Chemical Abstracts* is given for each retinoid. The commonly used system for numbering the carbon atoms of the naturally occurring retinoids, retinol, retinaldehyde, and retinoic acid, as well as those derivatives whose carbon skeletons can be derived from these three retinoids, is shown immediately below. For a detailed discussion of *Chemical Abstracts* and systematic nomenclature, see Volume 1, page 12 and following. It is recommended that the stereochemistry of all double bonds should be cited when the *E,Z* system is used. However, for simplicity sake, only the double bonds with *Z* configuration are identified here.

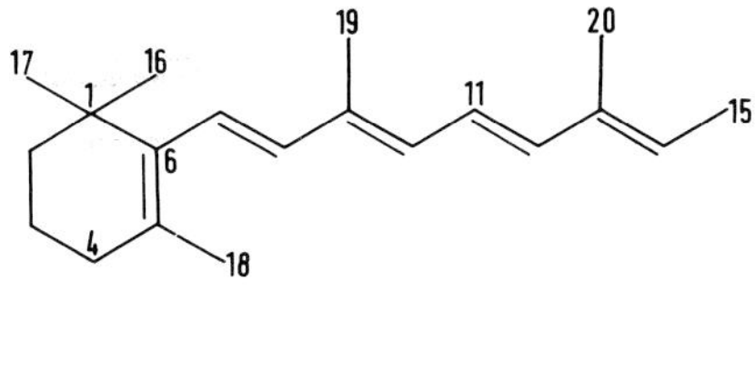

A. Retinol and Derivatives

A1 all-*trans*-Retinol [68-26-8]; (all-*E*)-3,7-Dimethyl-9-(2,6,6-trimethyl-1-cyclohexen-1-yl)-2,4,6,8-nonatetraen-1-ol; Retinol; all-*trans*-Vitamin A_1; Vitamin A; Vitamin A alcohol; Axerophthol or Axerol; β-Retinol

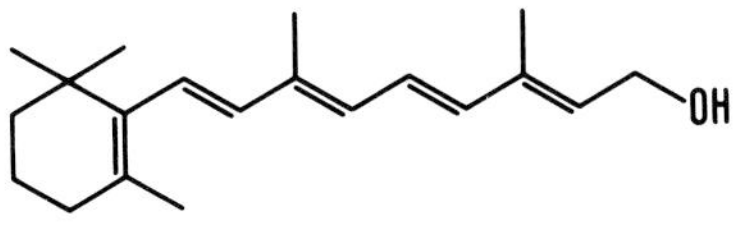

$C_{20}H_{30}O$

A2 11-*cis*-Retinol [22737-96-8]; (11*Z*)-Retinol

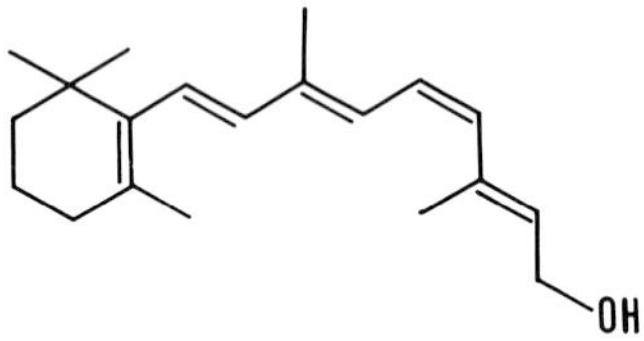

$C_{20}H_{30}O$

A3 all-*trans*-Retinyl acetate [127-47-9]; (all-*E*)-Retinyl acetate

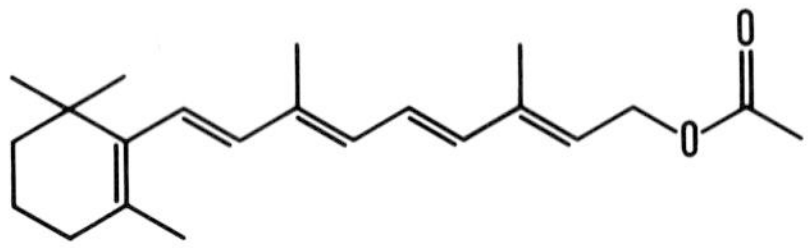

$C_{22}H_{32}O_2$

A4 Retinyl hexadecanoate [79-81-2]; (all-*E*)-Retinyl hexadecanoate; Retinyl palmitate

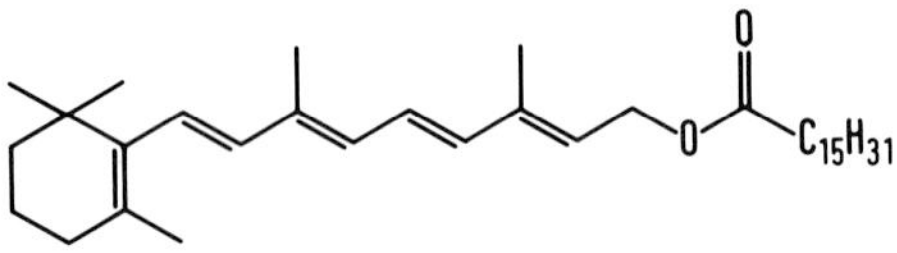

$C_{36}H_{60}O_2$

A5 α-Retinol [54375-76-7; 67772-69-4]; 4,6-Retroretinol; 4,5-Didehydro-5,6-dihydroretinol

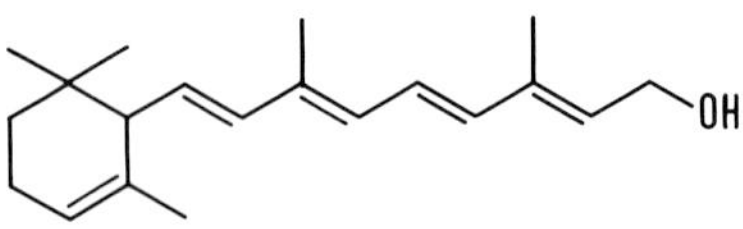

$C_{20}H_{30}O$

A6 4,14-Retroretinol [16729-22-9]

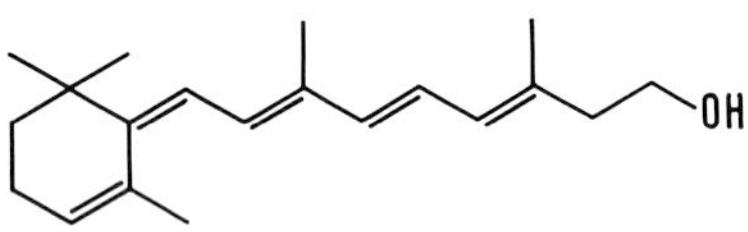

$C_{20}H_{30}O$

A7 Axerophthene [6895-29-0]; (all-*E*)-2-(3,7-Dimethyl-1,3,5,7-nonatetraenyl)-1,3,3-trimethylcyclohexene; Deoxyretinol; (all-*E*)-3,7-Dimethyl-1-(2,6,6-trimethyl-2-cyclohexen-1-ylidene)-2,4,6-nonatriene

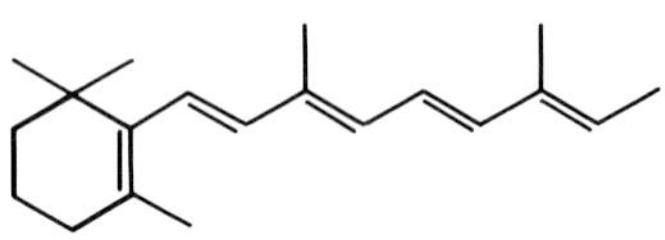

$C_{20}H_{30}$

A8 Anhydrovitamin A[1224-78-8]; (all-*E*)-6-(3,7-dimethyl-2,4,6,8-nonatetraenylidene)-1,5,5,-trimethylcyclohexene; 4,5-Didehydro-15,5-retrodeoxyretinol; Anhydroretinol

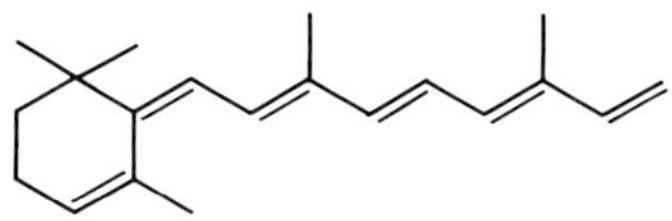

$C_{20}H_{28}$

A9 3,4-Didehydroretinol [79-80-1]; (all-*E*)-3,7-dimethyl-9-(2,6,6-trimethyl-1,3-cyclohexadien-1-yl)-2,4,6,8-nonatetraen-1-ol; 3-Dehydroretinol; Vitamin A_2

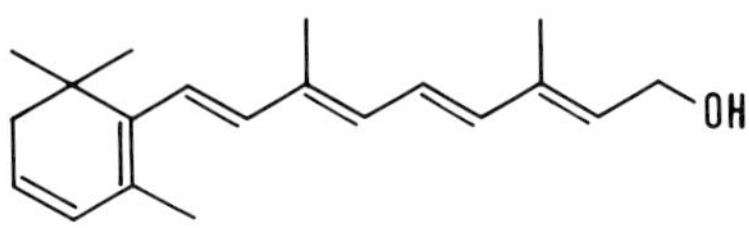

$C_{20}H_{28}O$

A10 15,15-Dimethyl retinol [67517-39-9]

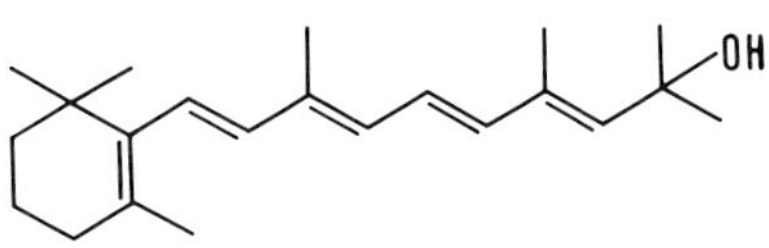

$C_{22}H_{34}O$

A11 Retinyl methyl ether [32450-56-9]; (all-*E*)-2-(9-Methoxy-3,7-dimethyl-1,3,5,7-nonatetraenyl)-1,3,3-trimethylcyclohexene

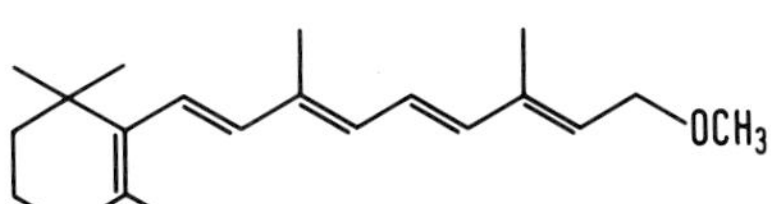

$C_{21}H_{32}O$

A12 Retinyl dihydrogen phosphate [53859-19-1]; Retinyl phosphate

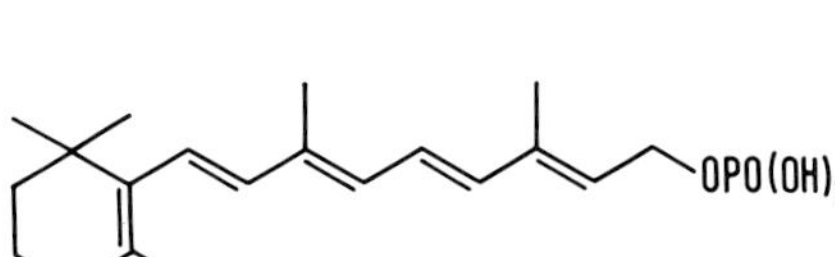

$C_{20}H_{31}O_4P$

A13 Retinyl mannosyl hydrogen phosphate [61152-15-6]; Mannosyl retinyl phosphate; Retinyl β-D-mannopyranosyl hydrogen phosphate

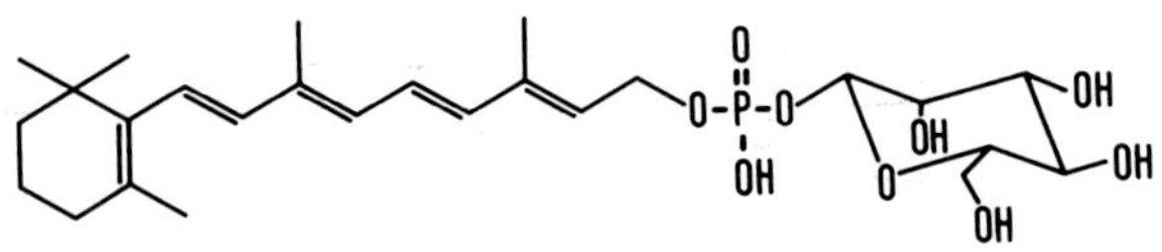

$C_{26}H_{41}O_9P$

A14 Retinyl thioacetate [72053-19-1]

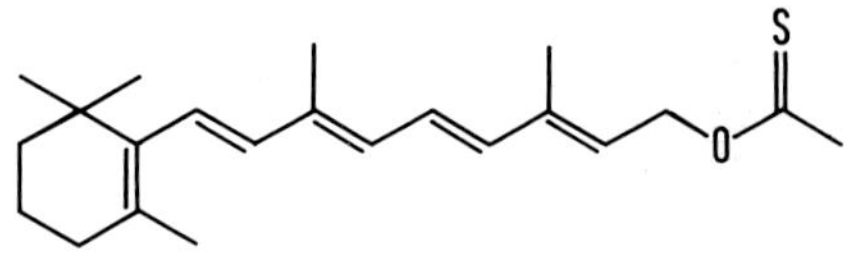

$C_{22}H_{32}OS$

A15 β-Ionylideneethanol bromoacetate [81112-41-6]; IEBA

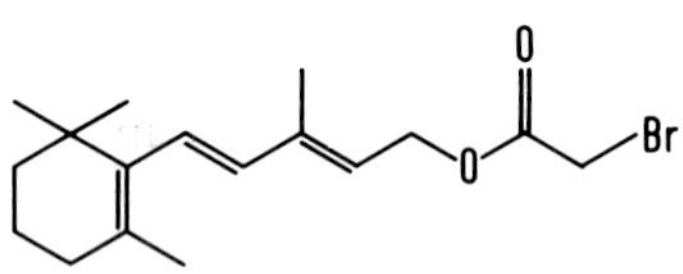

$C_{17}H_{25}BrO_2$

B. Retinylamine Derivatives

B1 *N*-Acetylretinylamine [43219-29-0]; (all-*E*)-*N*-[3,7-Dimethyl-9-(2,6,6-trimethyl-1-cyclohexen-1-yl)-2,4,6,8-nonatetraenyl]acetamide

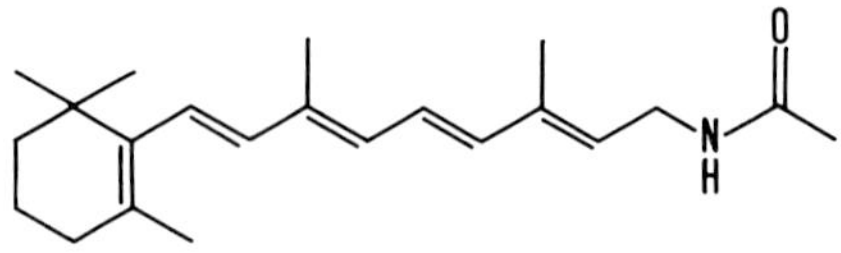

$C_{22}H_{33}NO$

B2 *N*-Benzoylretinylamine [74193-18-3]; (all-*E*)-*N*-[3,7-Dimethyl-9-(2,6,6-trimethyl-1-cyclohexen-1-yl)-2,4,6,8-nonatetraenyl]benzamide

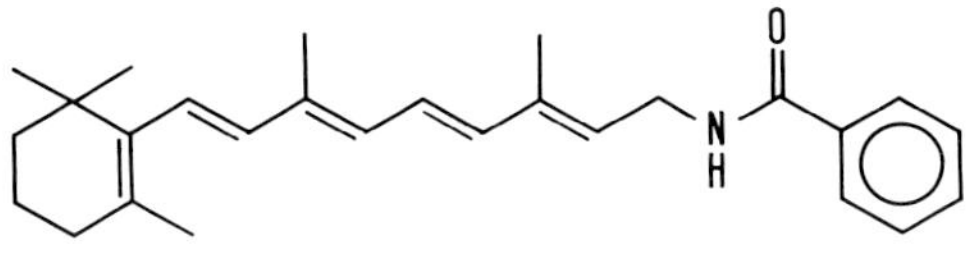

$C_{27}H_{35}NO$

C. Retinaldehyde and Derivatives

C1 all-*trans*-Retinaldehyde [116-31-4]; (all-*E*)-3,7-Dimethyl-9-(2,6,6-trimethyl-1-cyclohexen-1-yl)-2,4,6,8-nonatetraenal; all-*trans*-Retinal; (all-*E*)-Retinal; (all-*E*)-Retinaldehyde; Retinene; Vitamin A aldehyde; Vitamin A_1 aldehyde

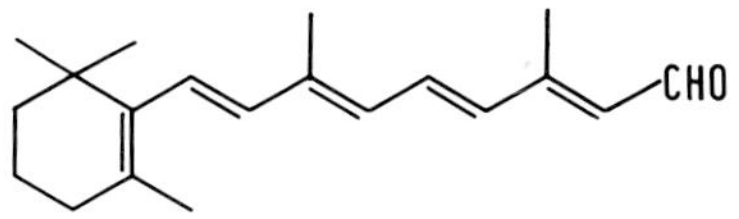

$C_{20}H_{28}O$

C2 11-*cis*-Retinaldehyde [564-87-4]; 11-*cis*-Retinal; (11*Z*)-Retinal; (11*Z*)-Retinaldehyde

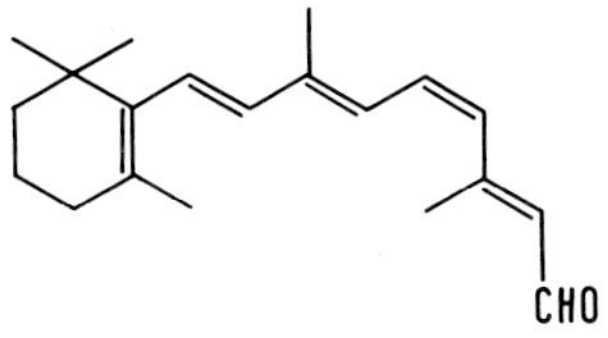

$C_{20}H_{28}O$

C3 11-*cis*-3,4-Didehydroretinaldehyde [41470-05-7]; (11*Z*)-3,4-Didehydroretinaldehyde; (11*Z*)-3,4-Didehydroretinal

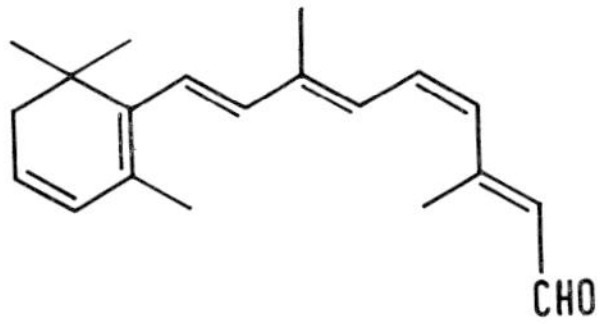

$C_{20}H_{26}O$

C4 all-*trans*-Retinylidene-2,4-pentanedione [6991-16-8]; (all-*E*)-3-[3,7-Dimethyl-9-(2,6,6-trimethyl-1-cyclohexen-1-yl)-2,4,6,8-nonatetraenylidene]-2,4-pentanedione; all-*trans*-Retinylidene acetylacetone

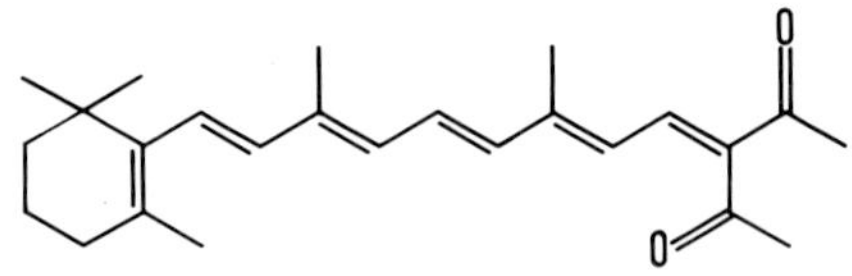

$C_{25}H_{34}O_2$

C5 all-*trans*-Retinylidene-1,3-cyclopentanedione [70359-69-2]; (all-*E*)-2-[3,7-dimethyl-9-(2,6,6-trimethyl-1-cyclohexen-1-yl)-2,4,6,8-nonatetraenylidene]-1,3-cyclopentanedione

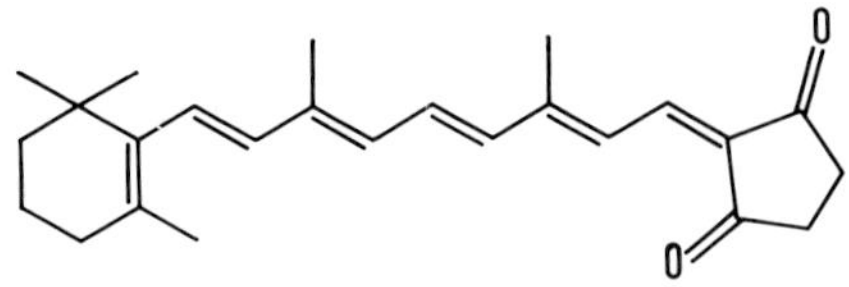

$C_{25}H_{32}O_2$

C6 all-*trans*-Retinylidene dimedone [70424-15-6]; (all-*E*)-2-[3,7-Dimethyl-9-(2,6,6-trimethyl-1-cyclohexen-1-yl)-2,4,6,8-nonatetraenylidene]-5,5-dimethyl-1,3-cyclohexanedione

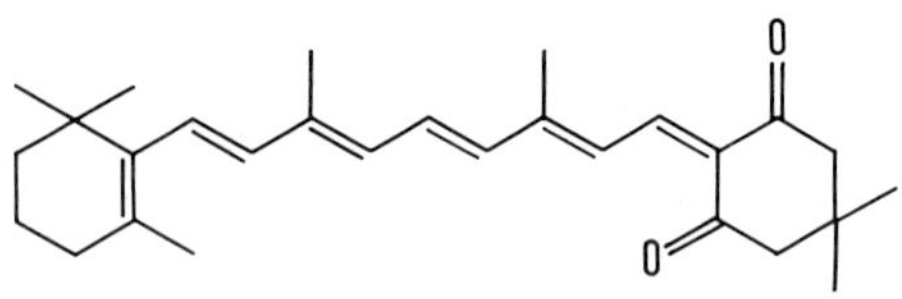

$C_{28}H_{38}O_2$

C7 all-*trans*-Retinaldehyde oxime [17672-05-8]; Retinal oxime

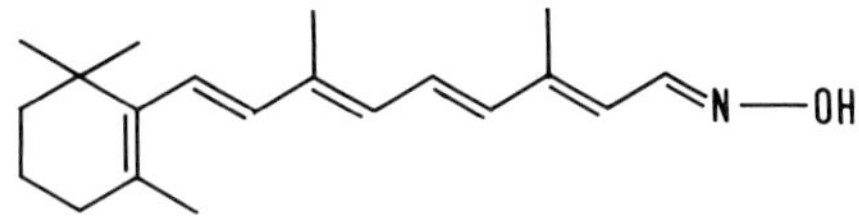

$C_{20}H_{29}NO$

C8 all-*trans*-Retinaldehyde acetylhydrazone [69460-76-0; 61630-48-6]; Acetic acid (all-*E*)-[3,7-dimethyl-9-(2,6,6-trimethyl-1-cyclohexen-1-yl)-2,4,6,8-nonatetraenylidene] hydrazide

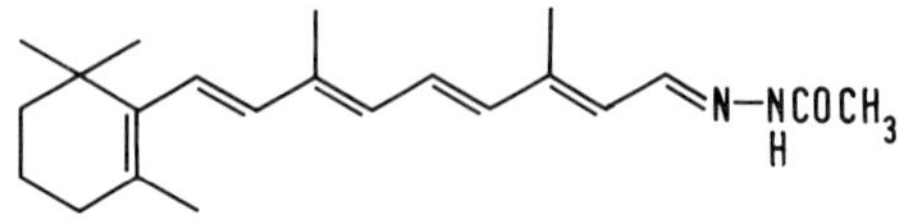

$C_{22}H_{32}N_2O$

D. *Retinoic Acid and Derivatives with Modified Ring*

D1 all-*trans*-Retinoic acid [302-79-4]; Retinoic acid; (all-*E*)-3,7-Dimethyl-9-(2,6,6-trimethyl-1-cyclohexen-1-yl)-2,4,6,8-nonatetraenoic acid; (all-*E*)-Retinoic acid; all-*trans*-Vitamin A acid; Vitamin A acid; Tretinoin; β-Retinoic acid

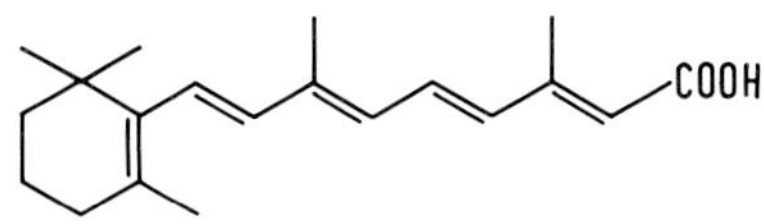

$C_{20}H_{28}O_2$

D2 α-Retinoic acid [52978-64-0]; 4,6-Retroretinoic acid; α-Vitamin A acid; 4,5-Didehydro-5,6-dihydro-retinoic acid

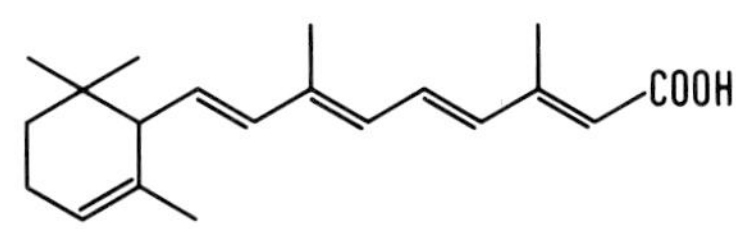

$C_{20}H_{28}O_2$

D3 4-Hydroxyretinoic acid [66592-72-1]

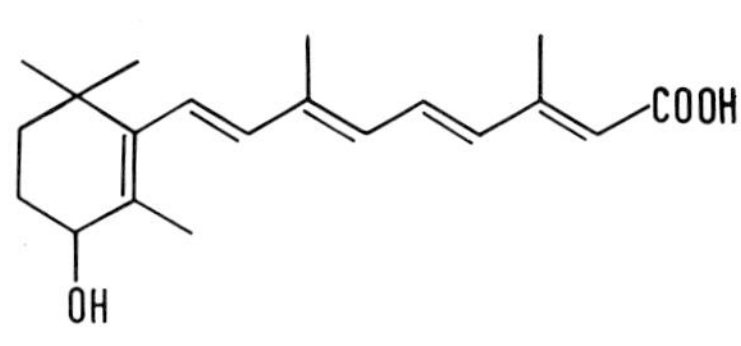

$C_{20}H_{28}O_3$

D4 4-Oxoretinoic acid [38030-57-8]; 4-Ketoretinoic acid

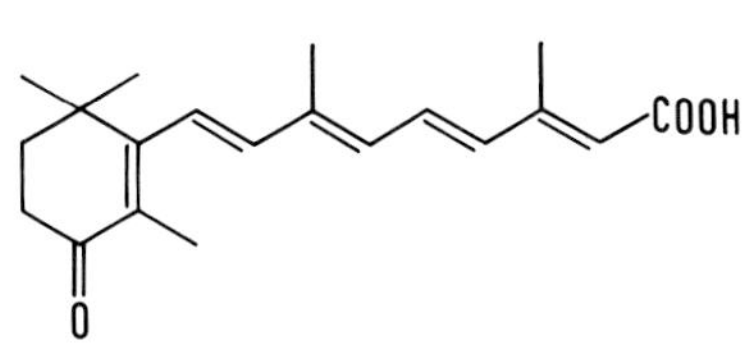

$C_{20}H_{26}O_3$

D5 5,6-Dihydroretinoic acid [19907-33-6]

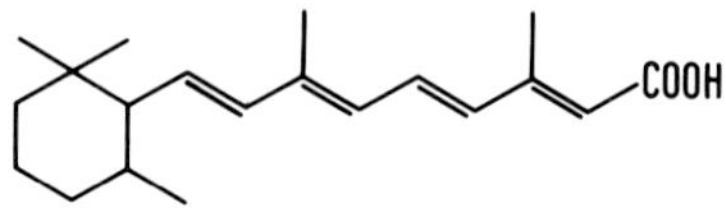

$C_{20}H_{30}O_2$

D6 5,6-Epoxyretinoic acid [13100-69-1]; 5,6-Epoxy-5,6-dihydroretinoic acid

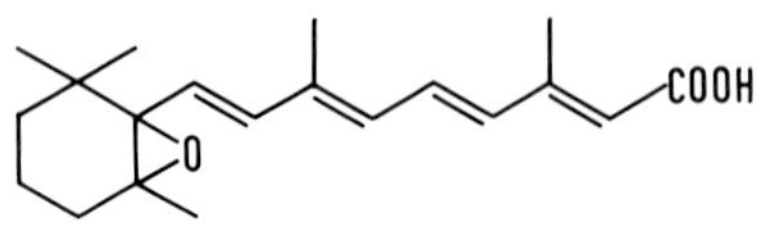

$C_{20}H_{28}O_3$

D7 5,8-Epoxyretinoic acid [3012-76-8]; 5,8-Epoxy-5,8-dihydroretinoic acid

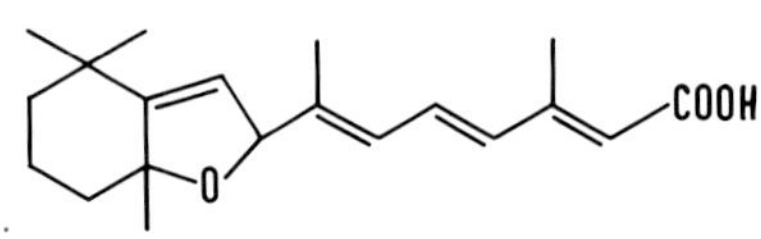

$C_{20}H_{28}O_3$

D8 DACP analog of retinoic acid [50890-42-1]; (all-*E*)-9-(2-Acetyl-5,5-dimethyl-1-cyclopenten-1-yl)-3,7-dimethyl-2,4,6,8-nonatetraenoic acid; 5-Acetyl-4,18-dinorretinoic acid

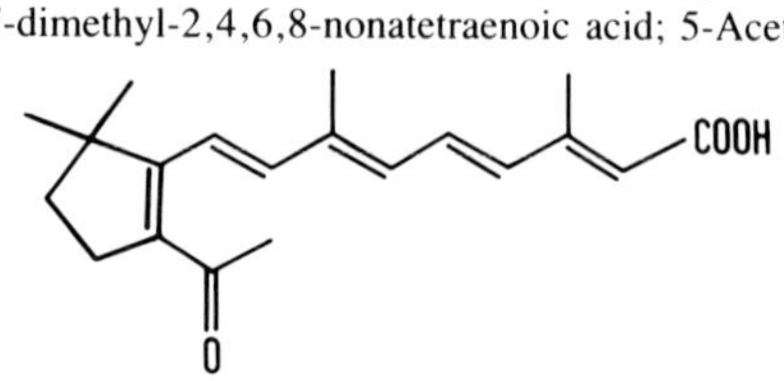

$C_{20}H_{26}O_3$

D9 DMECP analog of retinoic acid [50890-38-5]; (all-*E*)-9-[2-(1-methoxyethyl)-5,5-dimethyl-1-cyclopenten-1-yl]-3,7-dimethyl-2,4,6,8-nonatetraenoic acid; 5-(1-Methoxyethyl)-4,18-dinorretinoic acid

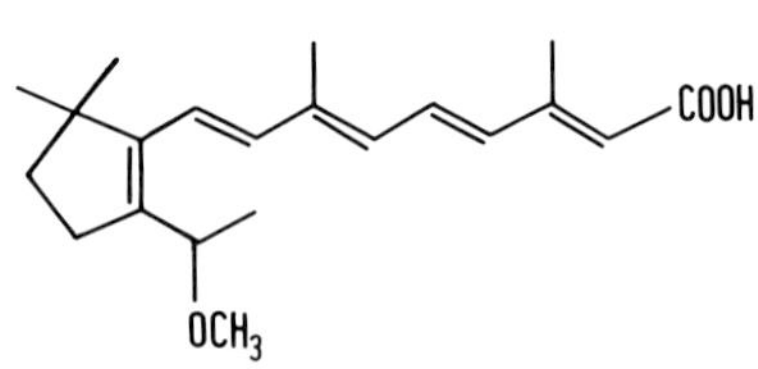

$C_{21}H_{30}O_3$

D10 2-Furyl analog of retinoic acid [55079-85-1]; (all-*E*)-9-(2-Furanyl)-3,7-dimethyl-2,4,6,8-nonatetraenoic acid

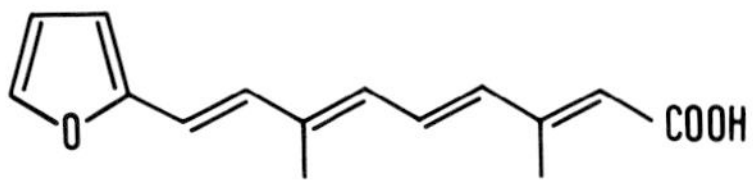

$C_{15}H_{16}O_3$

D11 3-Thienyl analog of retinoic acid [75664-65-2]; (all-*E*)-3,7-Dimethyl-9-(3-thienyl)-2,4,6,8-nonatetraenoic acid

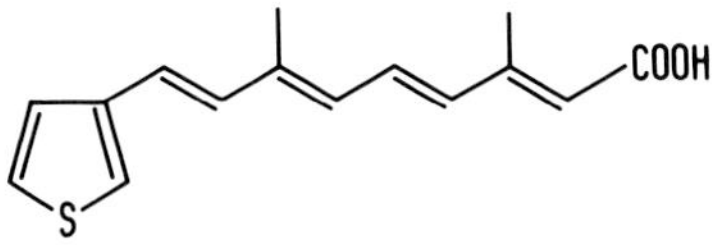

$C_{15}H_{16}O_2S$

D12 3-Pyridyl analog of retinoic acid [55079-86-2]; (all-*E*)-3,7-Dimethyl-9-(3-pyridinyl)-2,4,6,8-nonatetraenoic acid

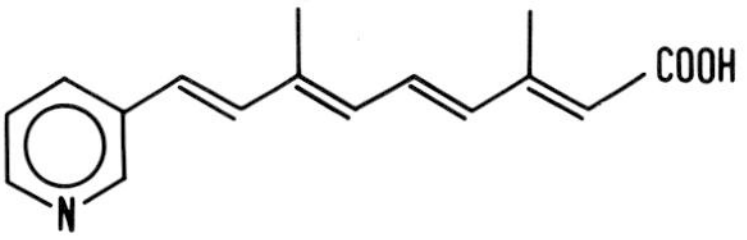

$C_{16}H_{17}NO_2$

D13 Phenyl analog of retinoic acid [55079-84-0]; (all-*E*)-3,7-Dimethyl-9-phenyl-2,4,6,8-nonatetraenoic acid

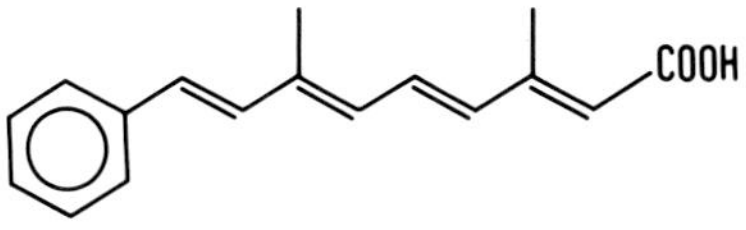

$C_{17}H_{18}O_2$

D14 Open-chain C_{20} analog of retinoic acid; (all-*E*)-3,7,11,15-Tetramethyl-2,4,6,8,10,12,14-hexadecaheptaenoic acid

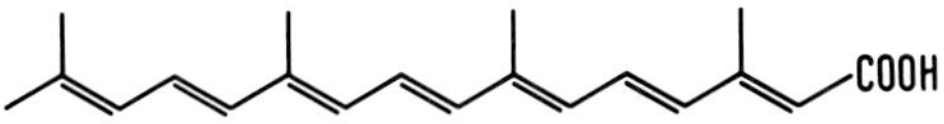

$C_{20}H_{26}O_2$

E. Retinoic Acid Derivatives with Modified Side Chain and/or Polar Terminus

E1 7,8-Didehydroretinoic acid [74193-14-9]

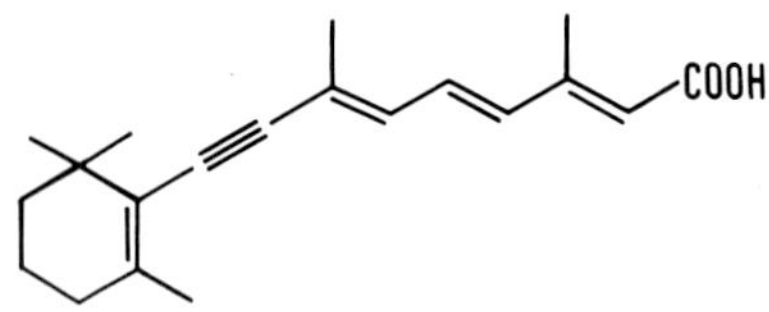

$C_{20}H_{26}O_2$

E2 7,8-Dihydroretinoic acid [51077-50-0]

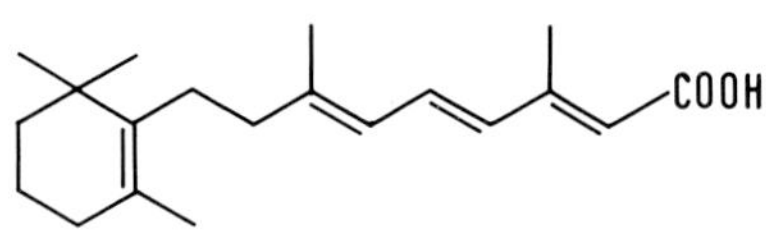

$C_{20}H_{30}O_2$

E3 "C_{15} Acid" [14398-42-6]; (*E,E*)-3-Methyl-5-(2,6,6-trimethyl-2-cyclohexen-1-yl)-2,4-pentadienoic acid

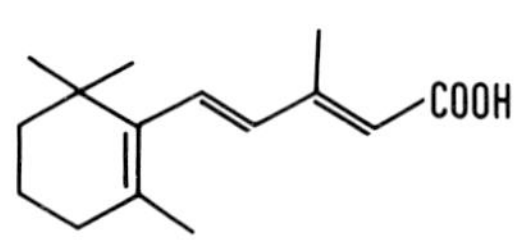

$C_{15}H_{22}O_2$

E4 "C_{17} Acid" [37836-28-5]; (*E,E,E*)-5-Methyl-7-(2,6,6-trimethyl-1-cyclohexen-1-yl)-2,4,6-heptatrienoic acid

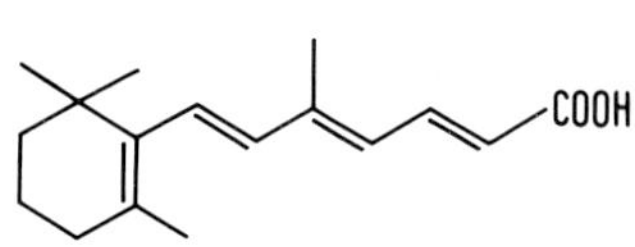

$C_{17}H_{24}O_2$

E5 "C_{22} Acid" [6722-00-5]; 14′-Apo-β,ψ-carotenoic acid

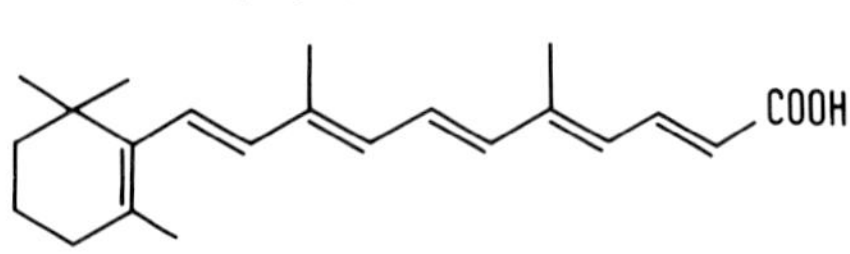

$C_{22}H_{30}O_2$

E6 13-*cis*-Retinoic acid [4759-48-2]; (13*Z*)-Retinoic acid; Isotretinoin

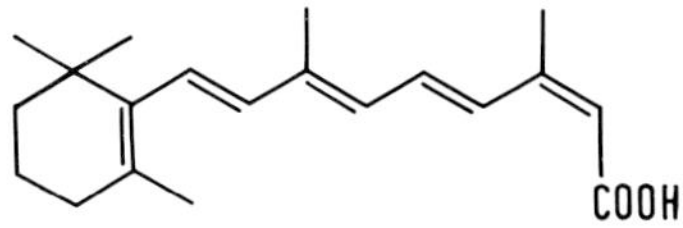

$C_{20}H_{28}O_2$

E7 Retinoic acid methyl ester [339-16-2]; Methyl retinoate

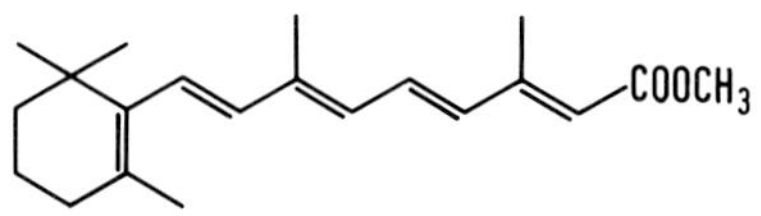

$C_{21}H_{30}O_2$

E8 Retinoic acid ethyl ester [3899-20-5]; Ethyl retinoate

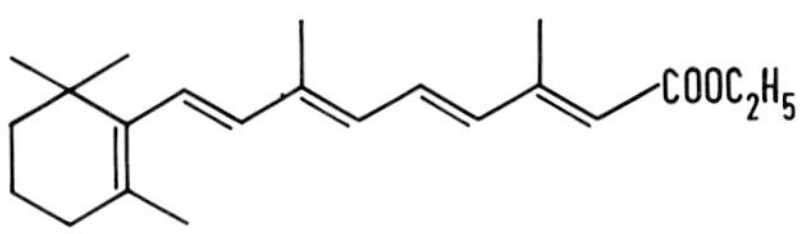

$C_{22}H_{32}O_2$

E9 12-Fluororetinoic acid ethyl ester [3887-00-1]; (11*Z*)-12-Fluororetinoic acid ethyl ester; Ethyl 12-fluororetinoate

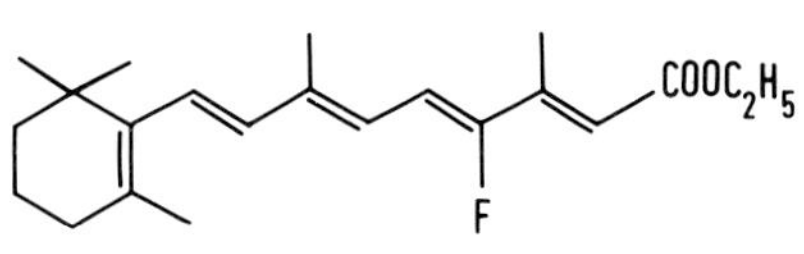

$C_{22}H_{31}FO_2$

E10 14-Fluororetinoic acid ethyl ester [2609-26-9]; (13*Z*)-14-Fluororetinoic acid ethyl ester; Ethyl 14-fluororetinoate

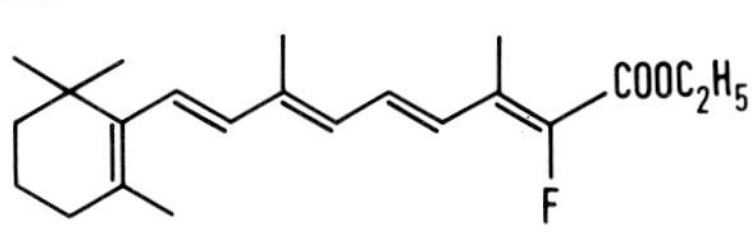

$C_{22}H_{31}FO_2$

E11 *N*-Ethylretinamide [33631-41-3]; Retinoic acid ethylamide

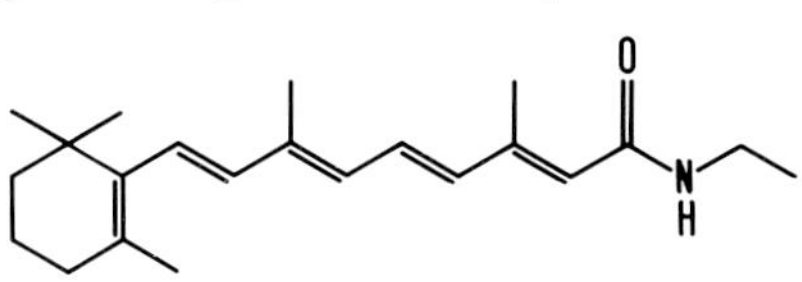

$C_{22}H_{33}NO$

E12 *N*-(2-Hydroxyethyl)retinamide [33631-47-9]; Retinoic acid 2-hydroxyethylamide

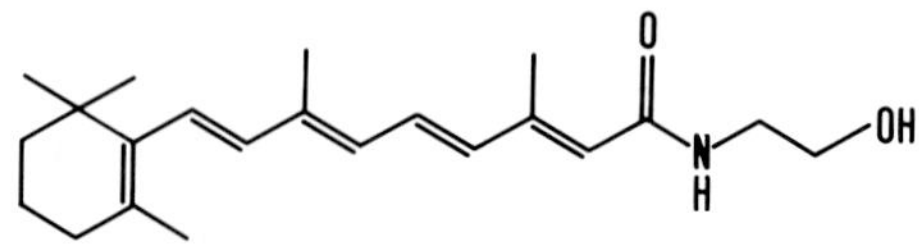

$C_{22}H_{33}NO_2$

E13 *N*-Butylretinamide [33631-44-6]

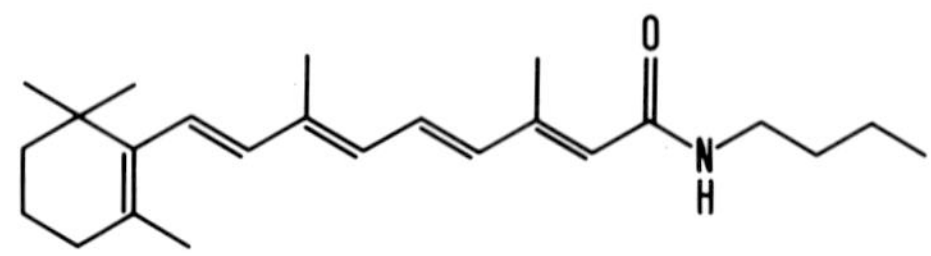

$C_{24}H_{37}NO$

E14 *N-tert*-Butylretinamide [65646-91-5]; *N*-(1,1-Dimethylethyl)retinamide

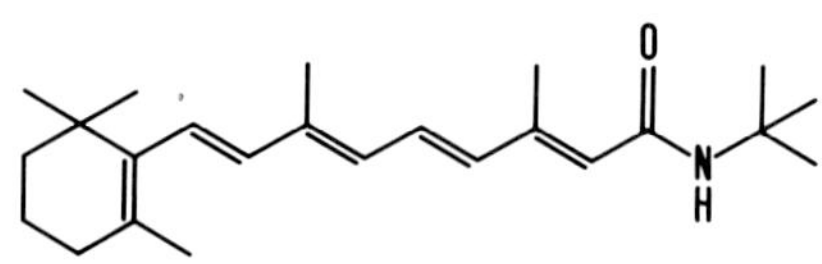

$C_{24}H_{37}NO$

E15 *N*-Phenylretinamide [33631-48-0]; Retinanilide

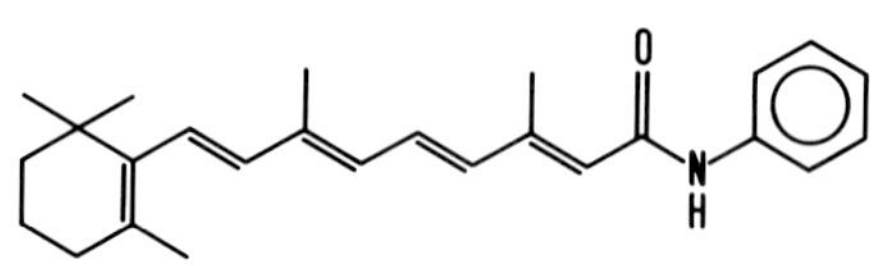

$C_{26}H_{33}NO$

E16 *N*-(4-Hydroxyphenyl)retinamide [65646-68-6]; 4-HPR

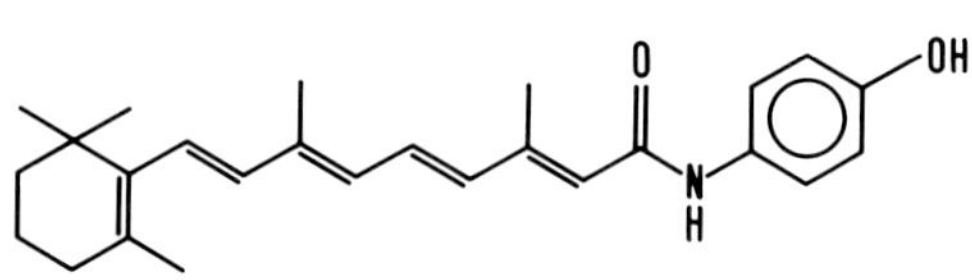

$C_{26}H_{33}NO_2$

E17 *N*-(1*H*-Tetrazol-5-yl)retinamide [75664-79-8]

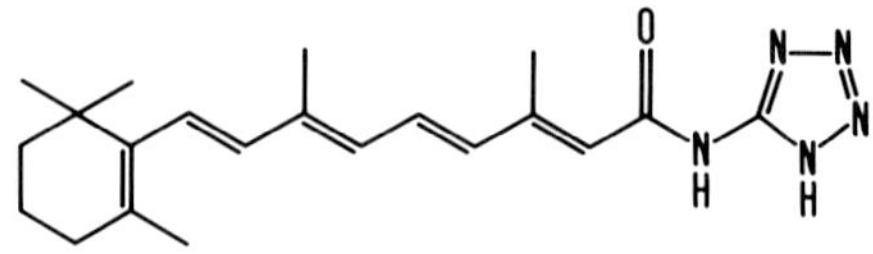

$C_{21}H_{29}N_5O$

E18 *N*-2-Retinamidoethyl sodium sulfate [75664-80-1]; *N*-[2-(Sulfoxy)ethyl]retinamide monosodium salt

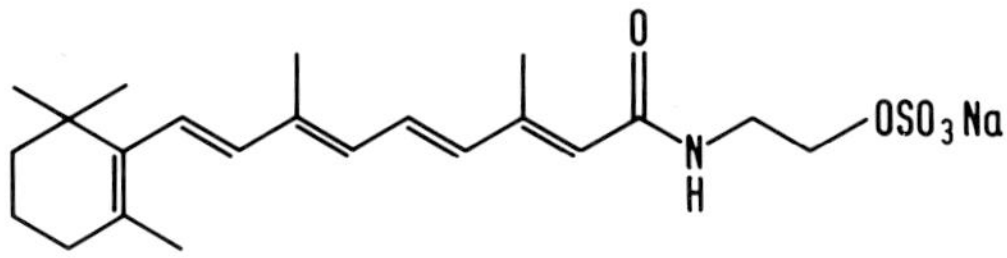

$C_{22}H_{32}NO_5SNa$

E19 13-*cis*-*N*-Ethylretinamide [75686-04-3]; (13*Z*)-*N*-Ethylretinamide

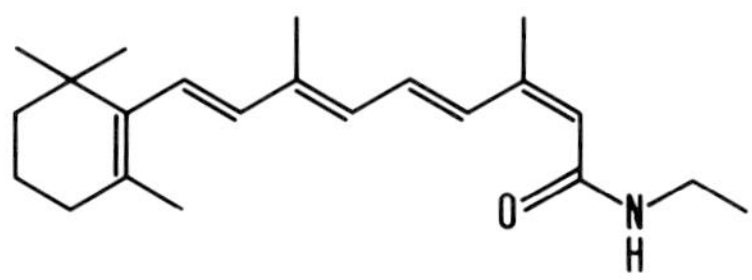

$C_{22}H_{33}NO$

E20 13-*cis*-*N*-(2-Hydroxyethyl)retinamide [75686-05-4]; (13*Z*)-*N*-(2-Hydroxyethyl)retinamide

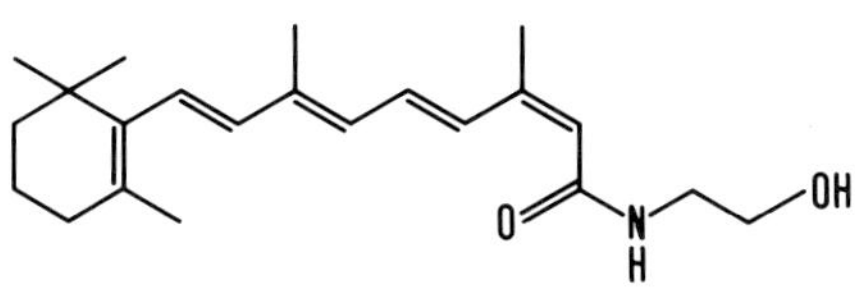

$C_{22}H_{33}NO_2$

E21 13-*cis*-*N*-(4-Hydroxyphenyl)retinamide [75686-07-6]; (13*Z*)-*N*-(4-Hydroxyphenyl)retinamide

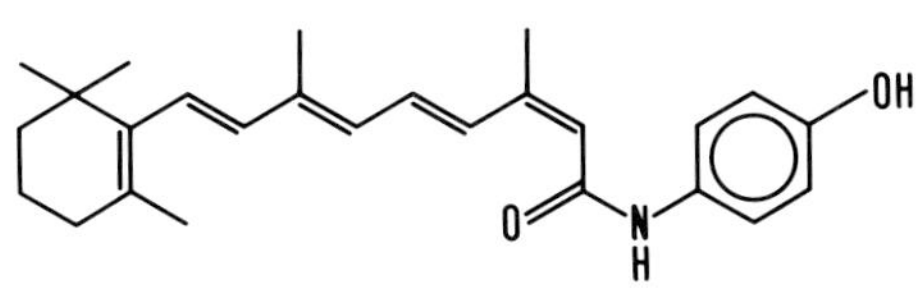

$C_{26}H_{33}NO_2$

E22 Methyl retinone [67517-37-7]; (all-*E*)-4,8-Dimethyl-10-(2,6,6-trimethyl-1-cyclohexen-1-yl)-3,5,7,9-decatetraen-2-one

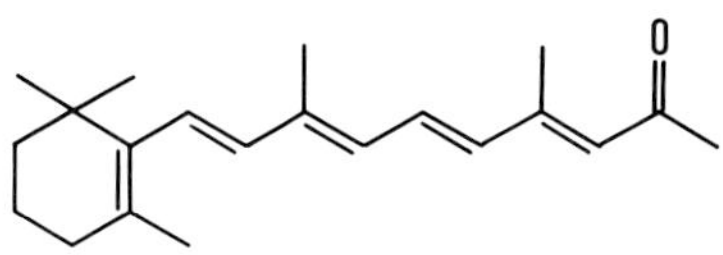

$C_{21}H_{30}O$

E23 "C_{18} Ketone"]17974-57-1;1220-77-5]; (*E,E,E*)-6-Methyl-8-(2,6,6-trimethyl-1-cyclohexen-1-yl)-3,5,7-octatrien-2-one

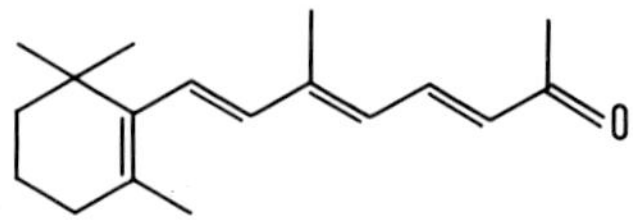

$C_{18}H_{26}O$

E24 Retinoyl-β-glucuronide [401-10-5]; 1-*O*-Retinoyl-β-D-glucopyranuronic acid

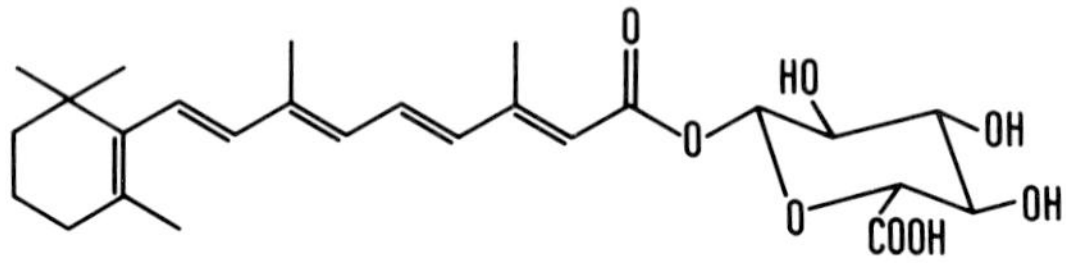

$C_{26}H_{36}O_8$

F. Etretinate and Derivatives

F1 Etretinate [54350-48-0]; (all-*E*)-9-(4-Methoxy-2,3,6-trimethylphenyl)-3,7-dimethyl-2,4,6,8-nonatetraenoic acid ethyl ester; Ethyl 3-methoxy-2-methyl-17-nor-1,2,3,4-tetradehydroretinoate; TMMP analog of retinoic acid ethyl ester

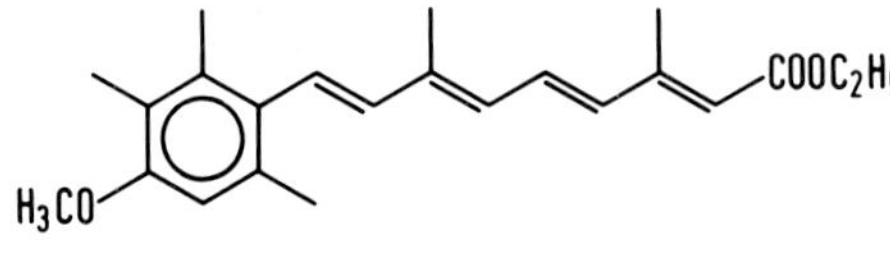

$C_{23}H_{30}O_3$

F2 TMMP analog of retinoic acid [55079-83-9]; (all-*E*)-9-(4-Methoxy-2,3,6-trimethylphenyl)-3,7-dimethyl-2,4,6,8-nonatetraenoic acid; 3-Methoxy-2-methyl-17-nor-1,2,3,4-tetradehydroretinoic acid

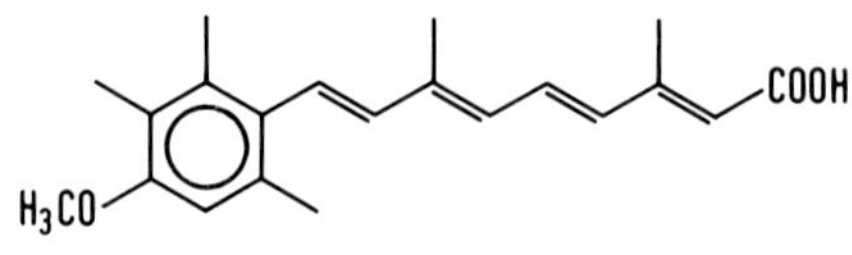

$C_{21}H_{26}O_3$

F3 Motretinid [54757-59-4]; *N*-Ethyl-9-(4-methoxy-2,3,6-trimethylphenyl)-3,7-dimethyl-2,4,6,8-nonatetraenamide; TMMP analog of ethylretinamide

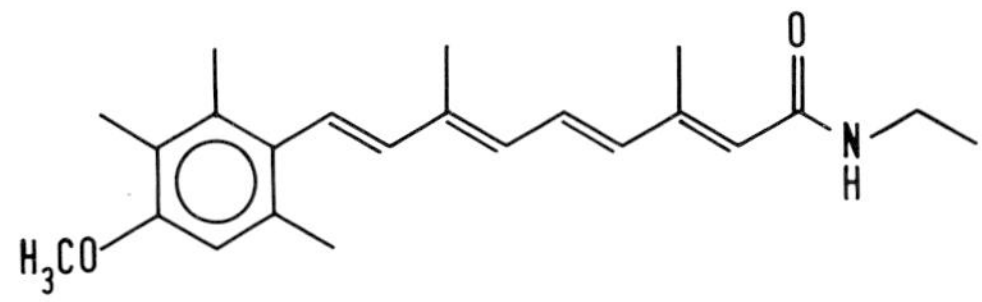

$C_{23}H_{31}NO_2$

F4 4-Fluoro analog of etretinate [63700-90-3]; (*E,Z,E,E*)-4-Fluoro-9-(4-methoxy-2,3,6-trimethylphenyl)-3,7-dimethyl-2,4,6,8-nonatetraenoic acid ethyl ester

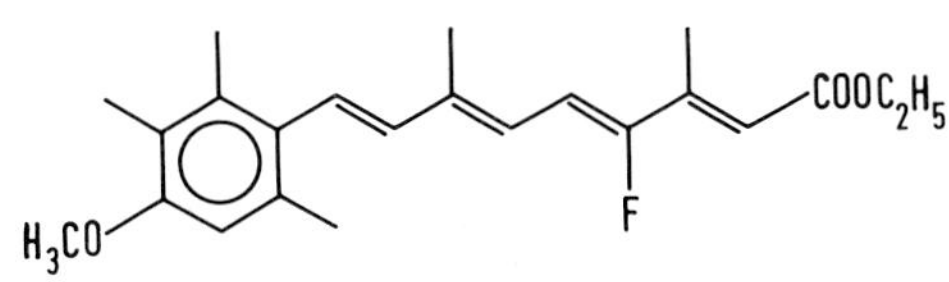

$C_{23}H_{29}FO_3$

F5 6-Fluoro analog of etretinate; (*E,E,Z,E*)-6-Fluoro-9-(4-methoxy-2,3,6-trimethylphenyl)-3,7-dimethyl-2,4,6,8-nonatetraenoic acid ethyl ester

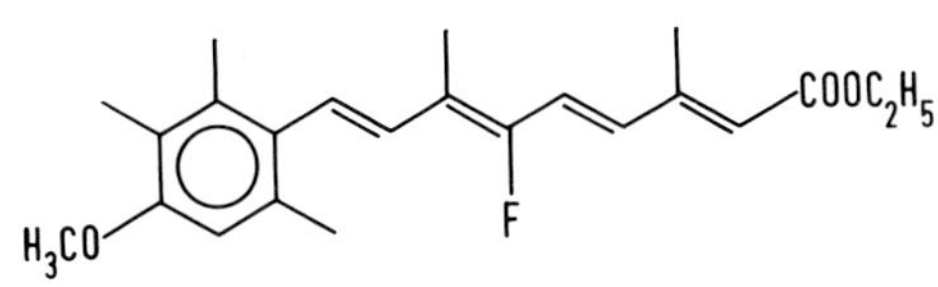

$C_{23}H_{29}FO_3$

G. *Retinoidal Benzoic Acid Derivatives*

G1 "Aryltriene analog of retinoic acid (*trans*)" [75664-66-3]; (*E,E*)-4-[2-Methyl-4-(2,6,6-trimethyl-1-cyclohexen-1-yl)-1,3-butadienyl]benzoic acid

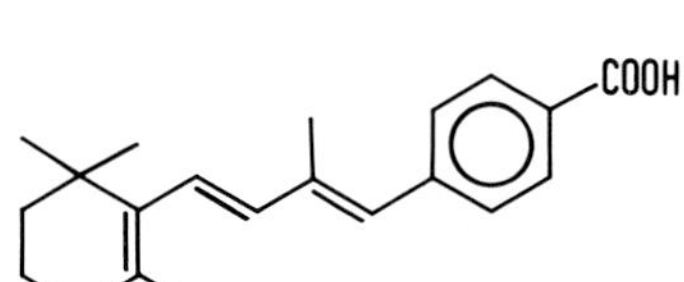

$C_{21}H_{26}O_2$

G2 "Aryltriene analog of retinoic acid (*cis*)" [75664-67-4]; (*Z,E*)-4-[2-Methyl-4-(2,6,6-trimethyl-1-cyclohexen-1-yl)-1,3-butadienyl]benzoic acid

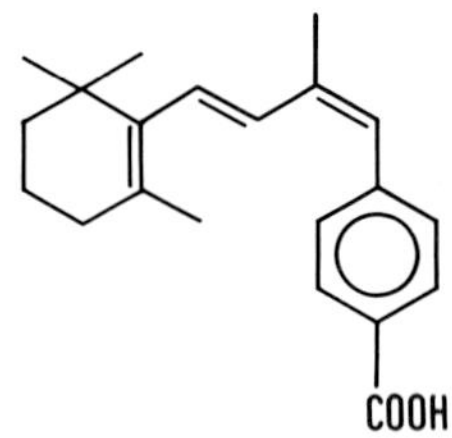

$C_{21}H_{26}O_2$

G3 TTNPB (Tetrahydrotetramethylnaphthalenylpropenylbenzoic acid) [71441-28-6]; (*E*)-4-[2-(5,6,7,8-Tetrahydro-5,5,8,8-tetramethyl-2-naphthalenyl)-1-propenyl]benzoic acid

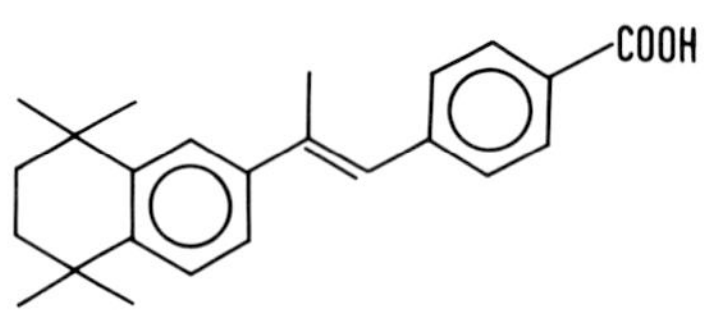

$C_{24}H_{28}O_2$

G4 *m*-Carboxy analog of TTNPB; (*E*)-3-[2-(5,6,7,8-Tetrahydro-5,5,8,8-tetramethyl-2-naphthalenyl)-1-propenyl]benzoic acid

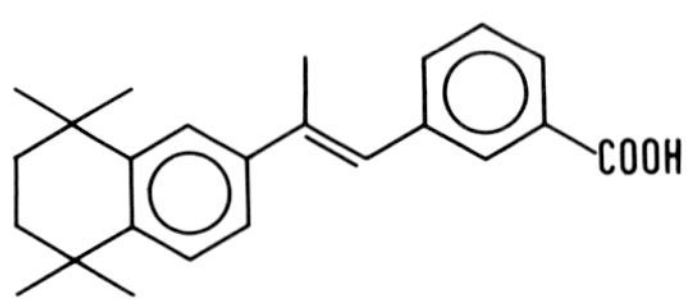

$C_{24}H_{28}O_2$

G5 Decarboxylated analog of TTNPB [75078-91-0]; (*E*)-1,2,3,4-Tetrahydro-1,1,4,4-tetramethyl-6-(1-methyl-2-phenylethenyl)naphthalene

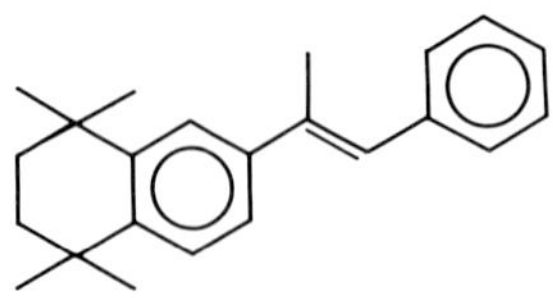

$C_{23}H_{28}$

G6 TTNPB ethyl ester [71441-09-3]; (*E*)-4-[2-(5,6,7,8-Tetrahydro-5,5,8,8-tetramethyl-2-naphthalenyl)-1-propenyl] benzoic acid ethyl ester

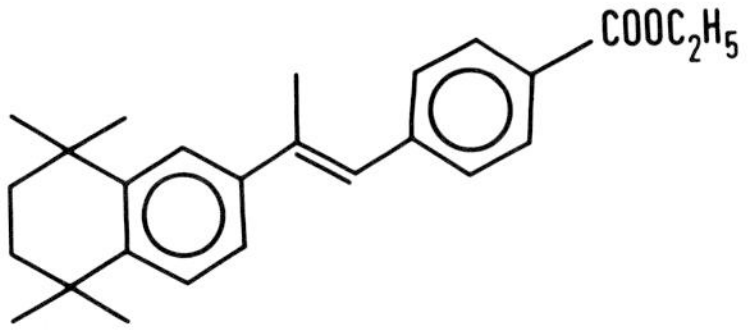

$C_{26}H_{32}O_2$

G7 TTNPB ethylamide [71441-29-7]; (*E*)-N-Ethyl-4-[2-(5,6,7,8-tetrahydro-5,5,8,8-tetramethyl-2-naphthalenyl)-1-propenyl]benzamide

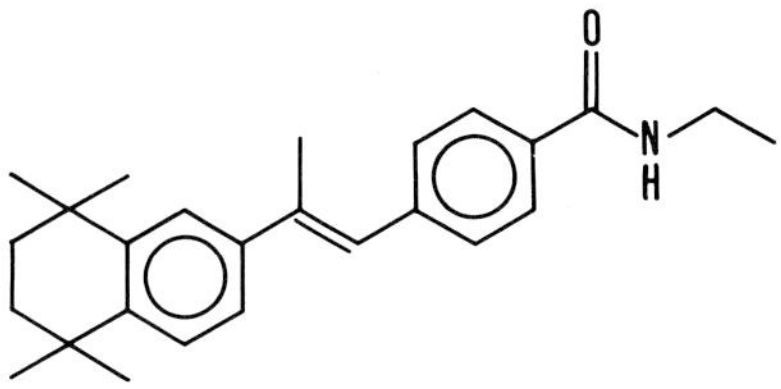

$C_{26}H_{33}NO$

G8 Toluene analog of TTNPB [72441-45-7]; (*E*)-1,2,3,4-Tetrahydro-1,1,4,4-tetramethyl-6-[1-methyl-2-(4-methylphenyl)ethenyl]naphthalene

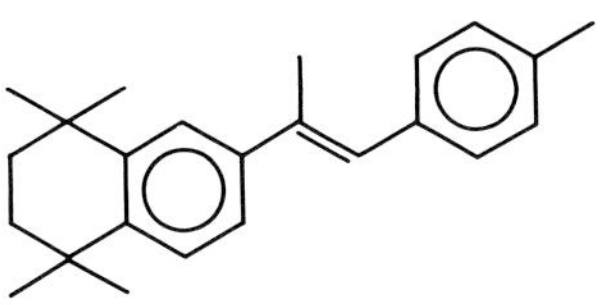

$C_{24}H_{30}$

G9 Primary alcohol analog of TTNPB [71441-30-0]; (*E*)-4-[2-(5,6,7,8-tetrahydro-5,5,8,8-tetramethyl-2-naphthalenyl)-1-propenyl]benzenemethanol

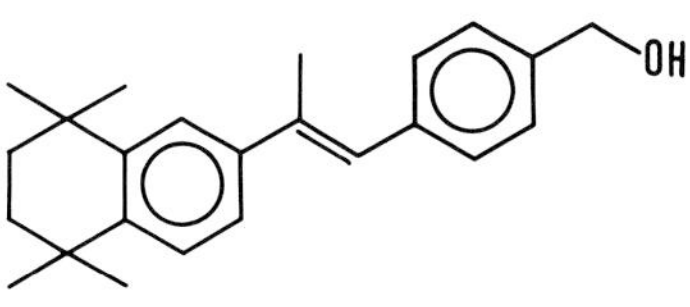

$C_{24}H_{30}O$

G10 Methyl ether analog of TTNPB [71441-32-2]; (*E*)-1,2,3,4-Tetrahydro-6-[2-[4-(methoxymethyl)phenyl]-1-methylethenyl]-1,1,4,4-tetramethylnaphthalene

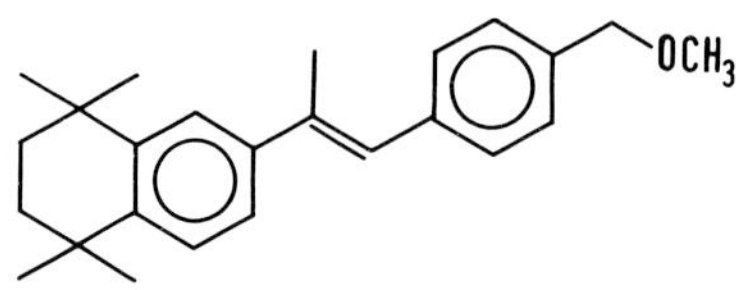

$C_{25}H_{32}O$

G11 "TDNPB" (Tetrahydrodimethylnaphthalenylpropenylbenzoic acid); (*E*)-4-[2-(5,6,7,8-Tetrahydro-8,8-dimethyl-2-naphthalenyl)-1-propenyl] benzoic acid

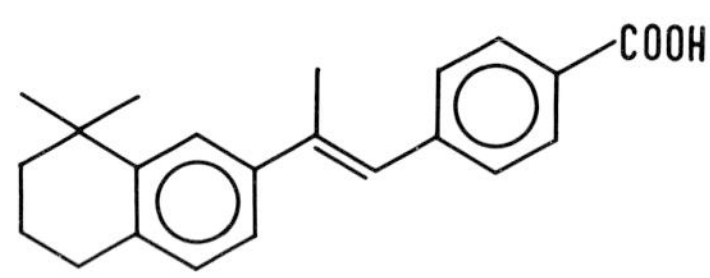

$C_{22}H_{24}O_2$

G12 3-Methyl TTNPB; (*E*)-4-[2-(5,6,7,8-Tetrahydro-3,5,5,8,8-pentamethyl-2-naphthalenyl)-1-propenyl]benzoic acid

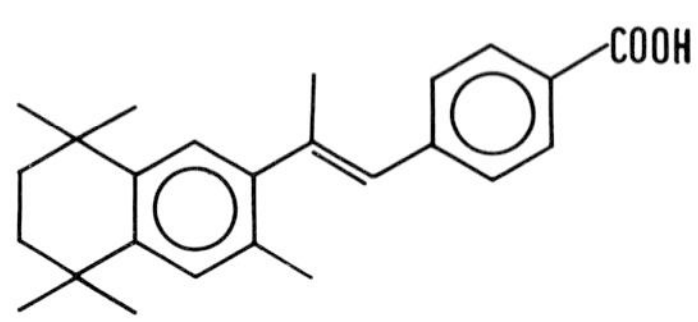

$C_{25}H_{30}O_2$

G13 3-Ethyl TTNPB; (*E*)-4-[2-(5,6,7,8-Tetrahydro-3-ethyl-5,5,8,8-tetramethyl-2-naphthalenyl)-1-propenyl]benzoic acid

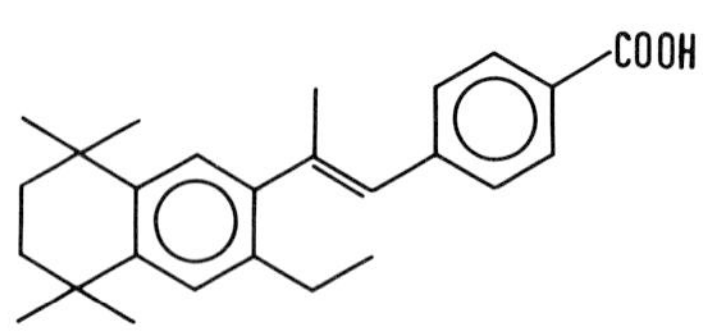

$C_{26}H_{32}O_2$

G14 3-Methoxy TTNPB; (*E*)-4-[2-(5,6,7,8-Tetrahydro-3-methoxy-5,5,8,8-tetramethyl-2-naphthalenyl)-1-propenyl]benzoic acid

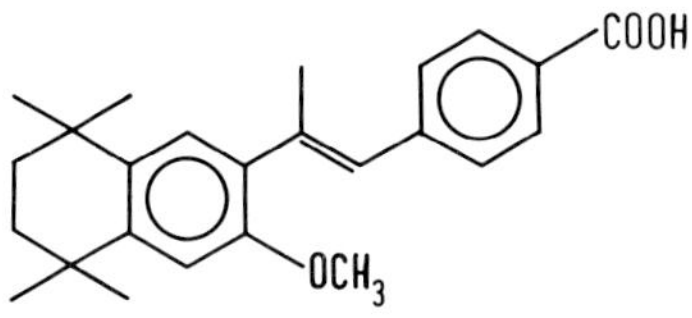

$C_{25}H_{30}O_3$

G15 3-Isopropyl TTNPB; (*E*)-4-[2-(5,6,7,8-Tetrahydro-3-(1-methylethyl)-5,5,8,8-tetramethyl-2-naphthalenyl)-1-propenyl]benzoic acid

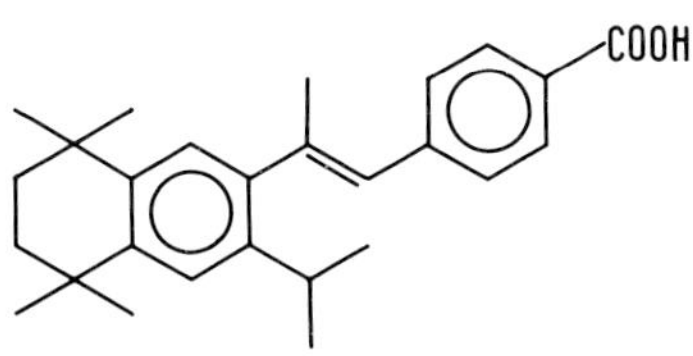

$C_{27}H_{34}O_2$

G16 Pentamethyl indanoic acid; (*E*)-4-[2-(2,3-Dihydro-1,1,2,3,3-pentamethyl-1*H*-inden-5-yl)-1-propenyl]benzoic acid

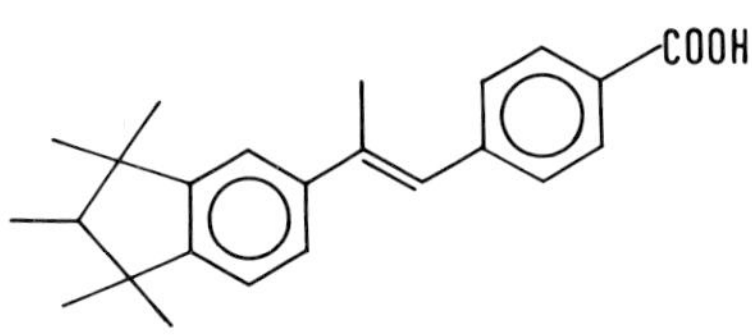

$C_{24}H_{28}O_2$

G17 4-Hydroxyphenyl analog of pentamethylindanoic acid; (*E*)-*N*-(4-Hydroxyphenyl)-4-[2-(2,3-dihydro-1,1,2,3,3-pentamethyl-1*H*-inden-5-yl)-1-propenyl]benzamide

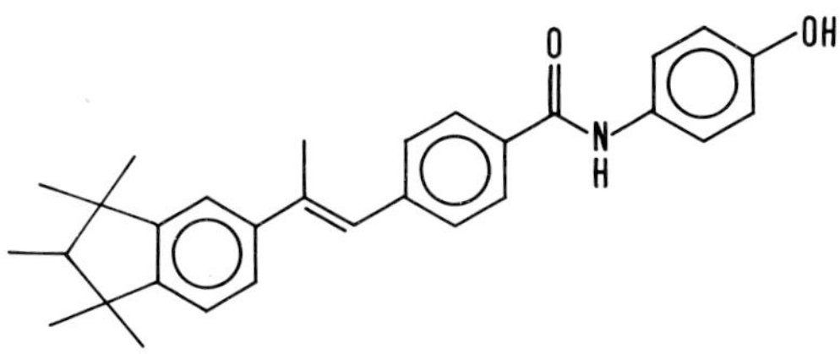

$C_{30}H_{33}NO_2$

G18 TTNN (Tetrahydrotetramethylnaphthalenylnaphthoic acid); 6-(1,2,3,4-Tetrahydro-1,1,4,4-tetramethyl-6-naphthyl)-2-naphthalenecarboxylic acid; 5′,6′,7′,8′-Tetrahydro-5′,5′,8′,8′-tetramethyl-[2,2′-binaphthalene]-6-carboxylic acid

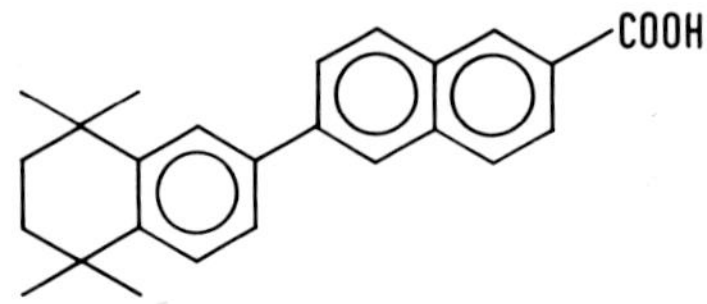

$C_{25}H_{26}O_2$

Index

D

N

O

P

S

T